183
Advances in Biochemical Engineering/Biotechnology

Aims and Scope

This book series reviews current trends in modern biotechnology and biochemical engineering. Its aim is to cover all aspects of these interdisciplinary disciplines, where knowledge, methods and expertise are required from chemistry, biochemistry, microbiology, molecular biology, chemical engineering and computer science.

Volumes are organized topically and provide a comprehensive discussion of developments in the field over the past 3–5 years. The series also discusses new discoveries and applications. Special volumes are dedicated to selected topics which focus on new biotechnological products and new processes for their synthesis and purification.

In general, volumes are edited by well-known guest editors. The series editor and publisher will, however, always be pleased to receive suggestions and supplementary information. Manuscripts are accepted in English.

In references, Advances in Biochemical Engineering/Biotechnology is abbreviated as *Adv. Biochem. Engin./Biotechnol.* and cited as a journal.

Katja Bühler • Pia Lindberg
Editors

Cyanobacteria in Biotechnology

Applications and Quantitative Perspectives

With contributions by

I. M. Axmann · K. Bühler · R. L. Burnap · A. Y. Chen · C. Deepika · B. Hankamer · B. C. Hung · J. J. Hung · P. R. Jones · A. Kenkel · W. Khetkorn · S. Klähn · J. Kollmen · J. Krömer · J. T. Ku · B. Lai · E. I. Lan · P. Lindberg · P. Lindblad · C. Maneeruttanarungroj · K. Muffler · F. Opel · W. Raksajit · J. Roles · I. Ross · P. Sattayawat · H. Schneider · A. Schwarz · D. Strieth · T. P. Tsai · R. Ulber · M. Witthohn · J. Wolf · I. S. Yunus

Springer

Editors
Katja Bühler
Helmholtz Center for Environmental Research
Leipzig, Germany

Pia Lindberg
Department of Chemistry-Ångström
Uppsala University
Uppsala, Sweden

ISSN 0724-6145 ISSN 1616-8542 (electronic)
Advances in Biochemical Engineering/Biotechnology
ISBN 978-3-031-33273-9 ISBN 978-3-031-33274-6 (eBook)
https://doi.org/10.1007/978-3-031-33274-6

This Springer imprint is published by the registered company Springer Nature Switzerland AG
The registered company address is: Gewerbestrasse 11, 6330 Cham, Switzerland

Preface

Cyanobacteria are fascinating, highly diverse organisms, which fulfill essential roles on our planet. As inventors of oxygenic photosynthesis, they have been responsible for the great oxygen event, which changed our atmosphere from an anaerobic oxidizing to an aerobic reducing one with dramatic consequences for the evolution of life. Nowadays, they are notorious for the role they play in harmful cyanobacterial blooms. The occurrence and disastrous impact of cyanobacterial blooms is expanding and becoming more intense, promoted by increasing global temperatures, extreme rainfall events, and protracted droughts in the wake of climate change and by the continuing anthropogenic eutrophication of limnetic systems. On the other hand, cyanobacteria are recognized as primary producers playing a crucial role in the maintenance of the global food webs. Most importantly, they supply huge microbial consortia with organic carbon compounds by simply utilizing light energy, water, and carbon dioxide. The ability to upgrade abundant carbon dioxide in this way also makes them highly interesting as host organisms for biotechnology. Especially in the time of climate change and global resource shortage, this ability holds the key for a truly sustainable approach of producing basic and fine chemicals as well as energy carriers in a CO_2-neutral way.

Cyanobacterial photosynthetic production of various chemicals, originating from many different pathways in cyanobacterial metabolism, has been demonstrated in the last two decades. However, despite increasing efforts in photo-biotechnological research, the big breakthrough in terms of productivity and efficiency is still to come, with the majority of studies performed so far on a proof-of-concept level. Processes based on cyanobacteria are still hampered by low space-time yields as a consequence of slow, light-dependent growth. This results in low cell densities and low reaction rates in often non-optimal photo-bioreactor setups, and further technological development of both per-cell productivities and cultivation systems to improve efficiency is needed. Furthermore, one must consider that compared to widely used heterotrophic microbes, the basic knowledge about these organisms is still largely in its infancy. Especially on the regulatory level, cyanobacteria are a black box in many aspects. Concepts that we have developed over decades for heterotrophic chassis strains cannot simply be transferred one-to-one to cyanobacteria, as they largely differ in their metabolism and its regulation. However, it is inspiring to see how many new tools and concepts are being published to close this gap.

This book provides a comprehensive overview on cyanobacteria and their utilization as solar cell factories. This involves two major aspects: the biological whole-cell catalyst and the technical environment in which the catalyst is applied. In the biocatalyst we can differentiate modules, which are important for utilizing these organisms as production hosts and need to be tackled in an integrative approach as they are closely intertwined. It is essential to consider both electron and carbon flow, for successful biocatalyst development. In the first chapters of this volume, cyanobacterial biotechnology and the fundamentals of cyanobacterial bioenergetics are introduced in Chaps. 1 and 2, followed by chapters on tools and strategies for engineering cyanobacteria, Chaps. 3 and 4. Further on, examples of applications, engineering and production of different industrially relevant compounds in these organisms are provided in Chaps. 5 through 8, and finally process technology specific for cyanobacteria is covered in Chap. 9. Thus, this book will provide interested students and researchers in the area of photo-biotechnology with a deeper understanding of the cyanobacterial cell-factory, the latest achievements and persisting challenges.

Leipzig, Germany Katja Bühler
Uppsala, Sweden Pia Lindberg

Editorial Letter

Professor Scheper
(by Eberhard Franke)

Professor Ulber
© University of Kaiserslautern

The Hour of Parting

In 1970, the first volume of the *Advances in Biochemical Engineering/Biotechnology* was edited by Tarun Ghose and Armin Fiechter. It was the declared aim of the editors to bring together all the different players in the emerging new field of "biotechnology" which had just begun to reveal its vast potential in applications for the large-scale production of antibiotics and amino acids. Within the next 25 years, under the editorship of Armin Fiechter, more than 50 volumes of the Advances series provided a prominent showcase for this fascinating field and highlighted its growing importance in areas such as medicine, the food and fodder industry, renewable energy, and personal care. The series was a reflection of the rapid development of biotechnology, especially in the interaction between engineering and modern biology, especially genetic engineering, and showed how these areas mutually inspired each other.

In 1995, it was a great honor for me to take over the role of Managing Editor of this series from Armin Fiechter. Because biotechnology had become so complex in its different areas of application, we decided to focus on special topic volumes of ABE to really delve into all facets of biotechnological research and development. These special volumes presented collaborative work from researchers from various fields, ranging from fundamental molecular studies to industrial engineering, and shone a light on the pathways from basic research to industrial application and realization. The main focal topics were novel fields in medical research, such as tissue engineering, stem cell research or antibody production, bioanalytics, bioprocess monitoring or usage of biotechnological products in daily life, the development and production of new products in the modern food industry, such as aromas or nutraceuticals, or the usage of agricultural by-products for the production of biofuels or in biorefineries. More than 130 special volumes were published within the last 27 years, documenting the tremendous significance and impact of modern biotechnology not only for the transition to a sustainable biobased industry, but also for personalized medicine.

We are now entering the next period of ABE and heralding in a new era in biotechnology where sustainability issues and digitalization are set to be game changers for the field. Under sustainability aspects, cells and biomolecules will be designed from scratch in silico and modern genetic engineering tools, such as CRISPR/CAS9, will bring these digital solutions to reality. Artificial intelligence and machine learning tools will help to design stable, state-of-the-art production processes within a fraction of the time currently needed. Digital technologies have the potential to disrupt and reshape biotechnology with exciting perspectives for the future of the field: this future has only just begun.

As fascinating new opportunities emerge in biotechnology, it's time to hand over the role of managing editor to younger scientists. I am extremely happy that Prof. Roland Ulber from the University of Kaiserslautern will take over the role of Managing Editor. A chemist by education and a professor of bioprocess engineering, he represents the generation of scientists bridging the gap between molecular life sciences and engineering. I am firmly convinced that Prof. Roland Ulber will do an excellent job as Managing Editor and will continue to successfully lead the ABE together with the Editorial Board. My warmest wishes go to him for this wonderful and challenging task.

In addition, I would like to sincerely thank all members of the editorial board and the editorial team at Springer for the wonderful cooperation over the past 27 years. Without them, the series would not have been such a huge success. Thank you to all of you!

Hannover, Germany Thomas Scheper
November 2022

Changes

A Note of Thanks and an Announcement of Future Prospects

Professor Thomas Scheper joined the series *Advances in Biochemical Engineering/Biotechnology* as managing editor in 1995 and for 25+ years helped us establish the series as an important outlet in the scientific community.

He has already pointed out the impact of the volumes published in *Advances in Biochemical Engineering/Biotechnology* during this period, documenting the development of modern biotechnology not only for the transition to a sustainable biobased industry, but also for personalized medicine.

We are very pleased to have Prof. Roland Ulber as the successor to Prof. Thomas Scheper. He is a scientist who represents the new generation in Biotechnology. As the publisher, we are looking forward to cooperating with him and we have no doubt that this cooperation will bring stimulating new ideas to our series.

Heidelberg
December 2022

The Publisher

Contents

Adv Biochem Eng Biotechnol (2023) 183: 1–24
https://doi.org/10.1007/10_2023_217

Published online: 3 April 2023

Introduction to Cyanobacteria

Pia Lindberg, Amelie Kenkel, and Katja Bühler

Contents

Abstract Cyanobacteria are highly interesting microbes with the capacity for oxygenic photosynthesis. They fulfill an important purpose in nature but are also potent biocatalysts. This chapter gives a brief overview of this diverse phylum and shortly addresses the functions these organisms have in the natural ecosystems. Further, it introduces the main topics covered in this volume, which is dealing with the development and application of cyanobacteria as solar cell factories for the production of chemicals including potential fuels. We discuss cyanobacteria as industrial workhorses, present established chassis strains, and give an overview of the current target products. Genetic engineering strategies aiming at the

P. Lindberg
Department of Chemistry-Ångström, Uppsala University, Uppsala, Sweden
e-mail: pia.lindberg@kemi.uu.se

A. Kenkel and K. Bühler (✉)
Helmholtzcenter for Environmental Research, Leipzig, Germany
e-mail: amelie.kenkel@ufz.de; katja.buehler@ufz.de

photosynthetic efficiency as well as approaches to optimize carbon fluxes are summarized. Finally, main cultivation strategies are sketched.

Graphical Abstract

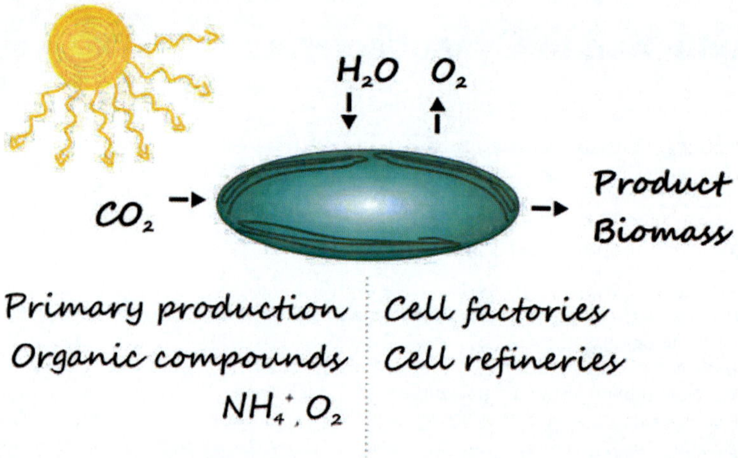

Keywords Cyanobacteria, Metabolic engineering, Photobioreactors, Solar cell factories

1 Introduction

Global warming, leading to expected upheavals on the environmental as well as on the social level, represent one of the major challenges humankind is facing today. Accordingly, measures to mitigate global warming have found their way into several sustainable development goals (SDGs) of the EU, such as SDG 7 "Affordable and Clean Energy" or SDG 13 "Climate Action." Already today, the consequences of climate change are noticeable, exemplified by the shrinkage of the Artic Sea ice by 1.07 million km^2 every decade, which in turn is accompanied by rising sea levels [1]. It is generally accepted that anthropogenic CO_2 emissions due to the combustion of fossil, carbon-based fuels mostly used for energy supply are the key-driver of this process [2]. Nevertheless, CO_2 emissions have increased by tremendous 50% since 1990 (https://www.globalcarbonproject.org/). Accordingly, a successful transition towards a sustainable, CO_2-neutral economy based on alternative, renewable resources is essential. Energy carriers are in the focus of such a future scenario, as their combustion constitutes the major share of the global CO_2-budget. In addition, it also needs alternative sources for organic carbon compounds required for chemical syntheses.

In nature, almost 385 billion tons of CO_2 are annually converted to organic materials using a light-driven reduction process termed oxygenic photosynthesis [3]. In particular, light energy is used to oxidize water, and the obtained electrons drive an autotrophic metabolism based on CO_2 fixation. As a byproduct of water oxidation, molecular oxygen accumulates. This fundamental metabolic process has its origin in the prokaryotic phylum of Cyanobacteria. Via endosymbiosis, it was also conveyed to the eukaryotic domain and established in microalgae and plants. The production capacity of the latter is the fundament of human agriculture and nutrition.

Cyanobacteria are increasingly recognized as potential biocatalysts independent of organic carbon. In the context of carbon neutral production processes and land use, this is a key feature, as all production processes based on chemo-heterotrophic organisms need a high-energy organic carbon source, mostly glucose, which in turn is commonly produced by sugar cane, or sugar beet. Although based on renewables, the area efficiency and the ecological impact of the respective farming procedure need to be considered, when evaluating the ecological footprint of such approaches. In contrast, oxygenic photosynthesis of cyanobacteria and microalgae may represent a key-technology to make inorganic carbon available for the production of value-added chemicals and fuels without competing with agricultural resources and food production [4]. During oxygenic photosynthesis, the high-energy molecules ATP and NADPH are generated. They are mainly utilized for the assimilation of CO_2 via the Calvin–Benson–Bassham (CBB) cycle, where, the enzyme ribulose-1,5-bisphosphate carboxylase/oxygenase (RuBisCO) catalyzes CO_2 fixation. C3- and finally C6-sugars, which constitute precursors for the synthesis of all carbon-based cell bricks and carbon storage compounds like glycogen [5] or polyhydroxybutyrate [6] are synthesized. Thus, energy and carbon are invested to produce biomass. This biomass may be exploited for its natural products like cyanophycin in a kind of biorefinery approach [7, 8], in which case maximizing biomass formation is the ultimate goal for process optimization. On the other hand, cyanobacteria can be engineered to function as microbial cell factories (Fig. 1). In such an approach, production pathways are coupled to junction points in the central carbon metabolism with the goal of redirecting the majority of the carbon and energy flow towards synthesis of a target product while minimizing biomass production. For an excellent overview on photosynthetic electron fluxes see Chap. 2; Water oxidation bioenergetics in cyanobacteria.

This chapter will introduce the phylum of cyanobacteria from an evolutionary and microbiology perspective, give a short overview on the major cyanobacterial work-horses and their products and will briefly address the still prevailing challenges of CyanoBioTechnology.

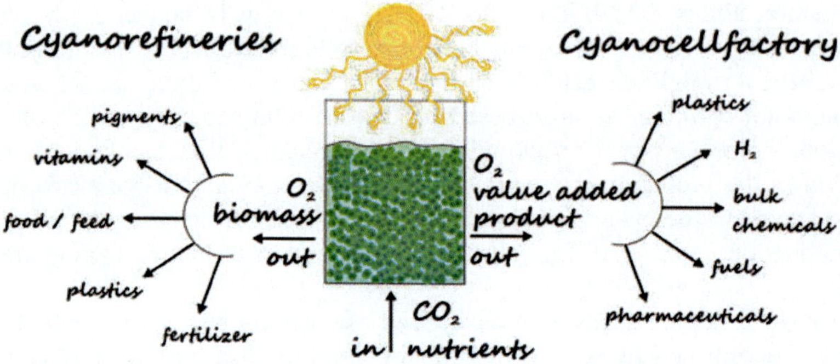

Fig. 1 Scheme comparing the two major approaches of CyanoBioTechnology. (**a**) In cyanorefineries the main product is biomass, which is then processed to be used either directly or to extract certain compounds. Mainly wildtype strains are utilized. Key objective is biomass maximization. (**b**) The cyanobacterial cell is operated as cellular factory, shuttling the major part of the electrons and the carbon derived from the water splitting reaction and the CBB cycle, respectively, into product synthesis. Engineered strains are applied. Key objective is maximization of the space-time yield

2 Microbiological Perspective on Cyanobacteria and Their Role in Nature and Technology

Cyanobacteria are an ancient, highly diverse group of photoautotrophic organisms, evolving a wide variety of morphologies reaching from unicellular to filamentous organization and thereby represents one of the most diverse prokaryotic phyla. Based on their morphology, cyanobacteria have been divided into five different morphological sections. While sections i and ii comprise unicellular species showing binary or multiple fission, respectively, sections iii–iv refer to multicellular species including ones with cellular differentiation and / or branching morphologies [9].

Cyanobacteria are considered the inventors of oxygenic photosynthesis. This ability is unique in the prokaryotic kingdom, and makes this phylum responsible for the so called "Great Oxygen Event" around 2.4 billions of years ago. On our planet, around 20–30% of the primary photosynthetic activity converting solar energy into biomass-based chemical energy is accomplished by cyanobacteria [10]. Furthermore, the ability of performing oxygenic photosynthesis was transferred to the eukaryotic clades via endosymbiosis of a cyanobacterium within a eukaryotic unicellular organism, giving rise to today's biodiversity [11]. Apart from performing oxygenic photosynthesis, many cyanobacteria are able to fix atmospheric nitrogen, adding to their importance for natural ecosystems [12].

As diverse as their morphologies are the habitats and ecological niches colonized by cyanobacteria. They adapted to a wide range of ecosystems, ranging from various aquatic systems like fresh, brackish, and marine waters, as well as hot springs to cold Arctic environments, but also to terrestrial ecosystems. Thereby cyanobacteria are exposed to multiple stresses such as solar ultraviolet radiations and variations in light intensity and quality, inorganic nutrients availabilities, temperatures, salinity, pH, drought, and pollutants. Consequently, they developed numerous traits to survive and succeed in such diverse environments. Due to their ability to perform oxygenic photosynthesis and fix carbon dioxide they fulfill a most important role as primary producers in the global food web in planktonic as well as in an immobilized lifestyle. Especially their role in the formation of microbial mats has been well investigated due to the importance these structures have for the benthic ecosystem, for terraforming by stabilizing the sediment surface and increasing the sediment erosion threshold [13]. Stromatolites are considered to be the fossil analogs of microbial mats, dating back to about 3.5 billion years representing one of the oldest ecosystems known [14]. Microbial mats comprise communities of multiple functional groups of microorganisms embedded in a self-produced, extracellular polymeric matrix. Due to their versatile composition, they represent a self-sustaining, nearly closed ecosystem, which includes the major element cycles and different trophies, and features various models of microbial cooperation. In extreme cases, microbial mats can be up to several centimeters thick, and the activities of the inhabiting microbes generate and maintain dynamic physicochemical gradients (Fig. 2).

The organic matter formed through primary production by cyanobacteria in the top layer is the basis of the microbial food web. Via dark respiration, active secretion of metabolites or cell lysis of the photoautotrophic microbes, organic matter like carbohydrates or organic acids becomes available to the other inhabitants of the mat. The cycling of carbon and nutrients through microbial components of pelagic aquatic communities is also termed the microbial loop [16]. Due to respiration activities, especially in the lower regions of the mat, the oxygen partial pressure significantly decreases until anoxic zones develop in the deeper parts of the mat, leading to a switch from respiration to fermentation metabolism. Fermentation activities then add to the amount of reduced organic carbon compounds like organic

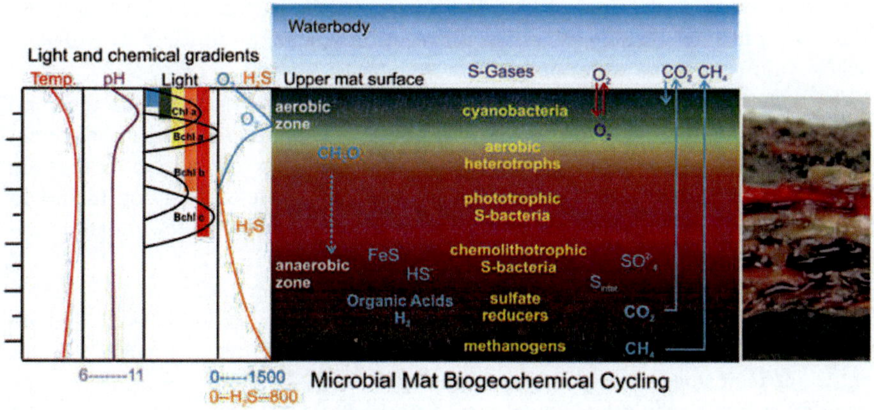

Fig. 2 Typical organization of a microbial mat due to the physical gradients developing in an illuminated environment. Figure modified from [15]

acids supplying methanogenic and sulfate-reducing bacteria, often in a syntrophic mode with other microorganisms, with carbon substrates [17].

Apart from being essential organisms in microbial mat formation, cyanobacteria are also important in the free water body as phytoplankton. One of the most abundant representatives of the cyanobacteria phylum, and in fact of all photosynthetic microorganisms, are the picocyanobacteria, which are composed of only two genera, namely *Prochlorococcus* and *Synechococcus* [18]. These organisms are among the smallest microbes known, with a minimum cell diameter around 0.6 μm [19]. The small cell volume results in an enhanced surface area-to-volume ratio, improving nutrient uptake. Moreover, *Prochlorococcus* possesses one of the smallest genomes sequenced so far with approximately 1.65 Mbp comprising ~1,700 genes [20]. Minimized genomes as in *Prochlorococcus*, are otherwise only known from organelles and host dependent bacteria. Two counteracting processes mainly drive genome size development: gene acquisition, e.g., by horizontal gene transfer, and the deletion of non-essential genes. Genomic flux by these gains and losses alters gene content and drives divergence of bacterial species and eventually adaptation to new ecological niches. It is hypothesized, that the small genomes may be important for the abundance of *Prochlorococcus* in oligotrophic open ocean environments, compared to other cyanobacterial species [21, 22]. *Prochlorococcus* is thriving in the euphotic zone of the tropical and subtropical oligotrophic ocean, which represents a highly dilute habitat. It is estimated that 4 gigatons carbon are fixed each year by these organisms [18] which equals the net primary productivity of the global croplands [23]. A unique feature of *Prochlorococcus* is its pigmentation and the organization of its photosynthetic apparatus. Instead of phycobilisomes, it harbors the prochlorophyte chlorophyll-binding protein (Pcb). Pcb binds divinyl chlorophyll a and divinyl chlorophyll b as an accessory pigment, forming the main light-harvesting antenna complex. Thereby *Prochlorococcus* is able to absorb blue light, the dominant wavelength in deep waters, and thus prosper in such typical low-light zones deep in the water column [24]. Nevertheless, *Prochlorococcus* is frequently found also in typical high-light zones on the surface of the water body, which demonstrates its huge diversity allowing for adaptation to various ecological niches within the entire water body.

Synechococcus, like *Prochlorococcus*, also belongs to the picocyanobacteria and both share a common ancestor. However, they developed into genetically different groups with own strategies to survive in the multifaceted marine ecosystem [18, 19]. *Synechococcus* grows in complementary though overlapping niches to *Prochlorococcus* [25] and can be found basically everywhere in the marine environment from the tropical regions up to the polar circle. Several strains from this genus are by now established in both basic as well as applied research, namely *S. elongatus* strain UTEX2973, *Syn.* sp. PCC 7002, and *Syn.* sp. PCC 7942. They do not compete for freshwater resources due to their ability to grow in seawater, survive high light intensities, and temperatures up to 40°C [26]. Furthermore, they grow rather fast compared to other cyanobacteria [27] with UTEX 2973 [28], PCC 11901 [29], and PCC 11801 [30] being amongst the fastest-growing cyanobacterial strain described up to now. Well-annotated genome sequences [31] and various available molecular tools allow for genetic engineering and systemic understanding of these strains [32]. For more details on molecular tools see Chap. 3; The molecular toolset required to build cyanobacterial cell factories.

Besides *Synechococcus*, *Synechocystis* sp. PCC 6803 is one of the most extensively studied species, meeting many constrains necessary for becoming a chassis strain in biotechnology [33]. *Synechocystis* strains exhibit a flexible carbon metabolism, growing under various conditions like photoautotrophic, mixotrophic, and heterotrophic [34]. It also established itself as an easy-to-handle model organism for stress responses in higher plants as it is biochemically highly similar to plant chloroplasts [35]. It was the first photoautotrophic organism to be sequenced in the late nineties [36]. Surprisingly, despite being recognized widely as a potent photobiotech chassis strain, only around 1,200 coding sequences (~30%) have an assigned function up to date, which is less than half compared with *Escherichia coli* [33]. Only a small proportion of these coding sequences have been characterized. Despite the fact, that there might be huge differences between phototrophic and heterotrophic bacteria most of the functions assigned are based on homologous sequences in other bacteria and there are several examples of *Synechocystis* genes already experimentally validated as having functions different to the original assigned ones [33].

More and more cyanobacterial genomes are becoming available and currently there are 290 draft genomes and 84 full genomes available online in the CyanoBase database (http://genome.microbedb.jp/cyanobase [37]). This promotes the development of genome-scale models (GSMs) and allows to adopt a systems biology approach to propose and predict the outcomes of engineering strategies [38].

Besides being potentially interesting organisms for photo-biotech applications, cyanobacteria are also notorious for the role they play in harmful cyanobacterial blooms, so called CyanoHABs. The global proliferation of CyanoHABs continues to increase in prevalence, intensity, and toxicity. Key drivers for this development are increasing global temperatures due to climate change combined with extreme rainfall events and protracted droughts and the continuing anthropogenic eutrophication of limnetic systems specifically with phosphorus (P) and nitrogen (N) [39]. These scenarios have led to perfect conditions promoting CyanoHABs; increases in pulsed nutrient loading events, followed by persistent low-flow, long water residence times, favoring bloom formation and proliferation. CyanoHABs are dominated by toxigenic cyanobacterial genera, e.g., Cylindrospermopsis, Dolichospermum (formerly Anabaena), Microcystis, and Planktothrix [40]. They produce a high number of bioactive molecules, among which some are cyanotoxins. These include anatoxin (ATX), cylindrospermopsin (CYN), microcystin (MC), nodularin (NOD), and saxitoxin (STX). The types and concentrations are largely determined by interactions between environmental factors that promote toxigenic genotypes and toxin gene expression. The effects of cyanotoxins are manifold [41]. Numerous cases of lethal poisonings have been associated with cyanotoxins ingestion in wild animal and livestock. In humans few episodes of lethal or severe human poisonings have been recorded after acute or short-term exposure, but the repeated/chronic exposure to low cyanotoxin levels remains a critical issue [41]. Most cyanotoxins are endotoxins, and their release into the water body is dependent on environmental parameters and bloom development stage [42]. CyanoHABs are typically not susceptible to physical forms of cell lysis from breaking wave action or shear stress, but only release toxins into the water column during cell senescence, lysis through viral activity, or remediation processes such as algaecide treatments or exposure to heightened salinity along estuarine gradients [41].

The problem of CyanoHABs is severe, posing a serious threat to aquatic marine and freshwater ecosystems. However, in this book, we will focus on potential industrial workhorses, and chances and pitfalls connected to their establishment in photo-biotechnology.

3 Cyanobacteria as Industrial Workhorses

In recent decades, the idea of using cyanobacteria as microbial cell factories has gained increasing attention. The central process is photosynthesis, which allows the organism to grow in minimal media without addition of a carbon feedstock. A phototrophic microorganism can therefore act as a biocatalyst, converting CO_2 and water into industrially useful compounds in a direct process driven by the energy in sunlight. This has long been regarded as a potential route for production of biofuels, since it would bypass the need for energy and arable land for production of crops to be used as feedstock for fermentative fuel production processes. However, in the last 15–20 years, focus has shifted to include also products other than fuels, with cyanobacteria now seen as potential hosts for production of sustainable chemicals of many different kinds.

The target products for cyanobacterial biotechnology can be compounds produced naturally by cyanobacteria, including biomass, pigments, storage molecules such as cyanophycin or PHB, and also hydrogen gas. Bioactive natural products from cyanobacteria are also of biotechnological interest [43, 44]. However, in recent years a large focus in cyanobacterial research has been on genetic engineering of cyanobacteria to introduce new metabolic capabilities, enabling the production of compounds which are not native to the host cells. These efforts have been developing greatly during the last decades, and now production of a wide range of different products have been demonstrated in cyanobacteria [45–48].

Most native products, such as pigments, vitamins, or biomass find their use in nutrition, as food supplements or coloring agents [49] (more details about pigments in Chap. 6; Production and isolation of pigments from Cyanobacteria). While the use of cyanobacteria for food and animal feed dates back thousands of years, large-scale commercial cultivation of cyanobacteria for these purposes has been in operation for the last 50 years [50]. Other products have been targeted for their potential use as fuel, such as hydrogen gas, which was identified as a potential fuel in the early stages of algal and cyanobacterial biotechnology [51] (more details about hydrogen production in Chap. 7; Hydrogenases in cyanobacteria), as well as lipids, hydrocarbons, and alcohols. For a product that will be used as fuel, large production facilities and a very high conversion efficiency from solar influx to fuel are required to make the production energetically and commercially viable [52, 53]. Cyanobacteria have also been engineered for photosynthetic production of a range of other chemicals of industrial or pharmaceutical interest, including small alkanes and alkenes or alcohols suitable as raw material for polymer synthesis, and organic acids with use as platform chemicals with many applications [47, 54]. For such products, the scale required for their production would still need to be large, but smaller than that required for fuels.

4 Model Strains and Genetic Engineering

Most studies on engineering cyanobacteria have been performed in a handful of model strains. These include the filamentous *Nostoc* PCC 7120, the unicellular freshwater strains *Synechocystis* PCC 6803, *Synechococcus elongatus* PCC 7942, and the marine strain *Synechococcus* PCC 7002. For these strains, numerous genetic engineering tools have been developed. *Nostoc* PCC 7120, as well as other strains, can be transformed using conjugation with *E. coli.* The unicellular model strains mentioned are also capable of natural transformation, where the cells spontaneously take up exogenously added DNA [55]. Synthetic biology tools such as vectors and promoters for control of gene expression are available for these strains. Vectors for transformation can be replicative vectors, which are maintained in the cyanobacterial cell after introduction, or integrative vectors carrying a gene construct which is incorporated in the cyanobacterial genome via homologous recombination. However, neither vectors nor promoters are compatible with all used strains, decreasing portability of developed tools [56]. Furthermore, tools developed for other organisms, such as the multitude of promoters and regulatory elements available for use in *E. coli*, often do not work, or do not work as efficiently, in cyanobacteria due to differences in initiation and regulation of transcription. Nevertheless, many different tools have been developed for the different model strains, including tools for regulation of gene expression such as anti-sense RNA or riboswitch-based methods, as well as CRISPRi for multiplexed downregulation of several genes simultaneously [57–60]. See Chap. 3; The molecular toolset required to build cyanobacterial cell factories, for more details on tools for engineering cyanobacteria.

The four strains mentioned above are the most commonly used in the literature of cyanobacterial biotechnology. *Synechocystis* PCC 6803 was the first photosynthetic organism to have its genome sequenced [36]. *Nostoc* PCC 7120 is the most used model organism for studying nitrogen fixation and cell differentiation in heterocystous cyanobacteria, while *S. elongatus* PCC 7942 is the prominent model for studies of the cyanobacterial circadian clock, and *Synechococcus* 7002 is a well-characterized salt and high light tolerant strain [26, 61, 62]. As these strains have been the focus of many fundamental studies of cyanobacteria, they are now also the best characterized strains, and therefore they have been the first option for most studies aiming to develop cyanobacteria as host organisms for biotechnological production of various compounds. However, in recent years a number of other unicellular strains with traits desirable for large-scale cultivation have been described, and methods for engineering them established. These include the fast-growing sibling of *S. elongatus* PCC 7942, *Synechococcus elongatus* UTEX 2973, as well as *Synechococcus* PCC 11801, PCC 11802, and PCC 11901. These strains have in common a faster growth rate as well as a tolerance to and ability to benefit from higher temperatures and light intensities compared to the more used *Synechocystis* and *Synechococcus* strains [28–30, 63].

5 Target Products

For products which are native for cyanobacteria, the most used organism in industry today is *Arthrospira platensis*, a nitrogen fixing filamentous strain which has been used traditionally for food in several parts of the world [64]. This strain is cultivated in open ponds with high alkalinity which keeps the cultivation relatively free from contaminants. Biomass is harvested and can be used fresh or dried as food supplement. Phycobiliproteins isolated from *A. platensis* are used as dyes, and phycocyanin is approved in both Europe and the US as food coloring agent (more details in Chap. 6; Production and isolation of pigments from Cyanobacteria) [65].

Biomass and pigments from *A. platensis* are the primary commercial application of cyanobacteria today. However, many different products, native and non-native, are being developed for production. These can be grouped according to the metabolic pathways, which lead to their synthesis (Fig. 3). Products derived from pyruvate include ethanol [66, 67], lactate [68], 2,3-butanediol [69], isobutanol [70, 71], 2-methyl-1-butanol [72], 3-methyl-1-butanol [73]. Products derived from acetyl-CoA include the storage polymer polyhydroxybuturate (PHB), which has

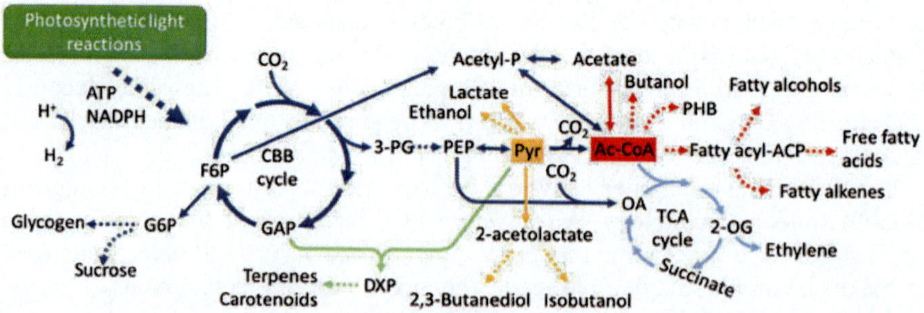

Fig. 3 Pathways relevant for selected products generated in cyanobacteria. Yellow – pathways for products derived from pyruvate; Red – pathways for products derived from acetyl-CoA; Green – terpenoid biosynthesis pathway; Light blue – TCA- cycle. *F6P* fructose-6-phosphate, *G6P* glucose-6-phosphate, *GAP* glyceraldehyde-3-phosphate, *3-PG* 3-phosphoglycerate, *PEP* phosphoenol pyruvate, *Pyr* pyruvate, *Ac-CoA* acetyl-CoA, *PHB* polyhydroxybuturate, *OA* oxaloacetate, *2-OG* 2-oxo-glutarate, *DXP* 1-deoxy-D-xylulose 5-phosphate

potential use for production of plastics [6, 74, 75], butanol [76–78], acetate, fatty acids, fatty alcohols, and fatty acid methyl esters (for more details on fatty acid production see Chap. 5; Production of fatty acids and derivatives using cyanobacteria) [79–81], alkenes and alkanes [82]. TCA-cycle derived compounds include organic acids succinate and fumarate, as well as ethylene, formed from 2-oxoglutarate. Terpenes and terpenoids are derived in cyanobacteria via the MEP pathway, and include many different compounds ranging from small volatile hydrocarbons to carotenoids [83]. Sugars and sugar alcohols can be produced from native sugars in the cyanobacterial metabolism [84]. Furthermore, aromatic amino acids and their derivatives may be produced from cyanobacteria and find potential use in nutrition or as feedstock for the chemical industry [85, 86].

6 Engineering Strategies

Common for all cyanobacterial production so far, with exception of pigments and biomass from *A. platensis*, is that yields are too low for commercial application. In some cases, such as lactic acid, ethanol, and butanol, the yields are approaching those that would be required for upscaling, but so far the cost of production remains too high for commercial success [87]. This is especially true for products with relatively low value, which has caused a shift in interest from commercial actors, from applications such as fuels or bulk chemicals to other applications such as pigments and vitamins, which may be sold for a higher price. In order to reach the needed productivities also for lower value products, there are several areas to which engineering strategies may be applied (Fig. 4).

In order to reach higher productivities, strategies may aim to improve the last steps of a production pathway, creating a sink for cellular resources and thereby provoking an adaptive response where the cell compensates by upregulating the upstream reactions. This may be referred to as a "pull" strategy (Fig. 4, 1). In addition to this, strategies may aim to enhance the flux of precursors to the desired product. This may be done in the immediate upstream metabolism, by enhancing expression of enzymes in related pathways, or by down-regulating or knocking out enzymes in competing pathways (Fig. 4, 2). In the case of cyanobacteria this may include enhancing carbon fixation to increase the availability of carbon precursors in the cell (Fig. 4, 3). These strategies may be considered as "push" strategies. Furthermore, engineering of cofactor supply may be a possibility to steer flux towards product formation rather than growth (Fig. 4, 4) [88]. Strategies to enhance overall cell productivity, for growth as well as the desired product formation, may include improvements in photosynthesis performance. Pigment type and content could potentially be adjusted to improve light utilization in individual cells as well as in mass culture (Fig. 4, 5).

If the above strategies to enhance formation of the product are successful, the product may be accumulating in the cell, limiting the effectiveness of host cell engineering and product generation. Thus, a crucial cell engineering strategy

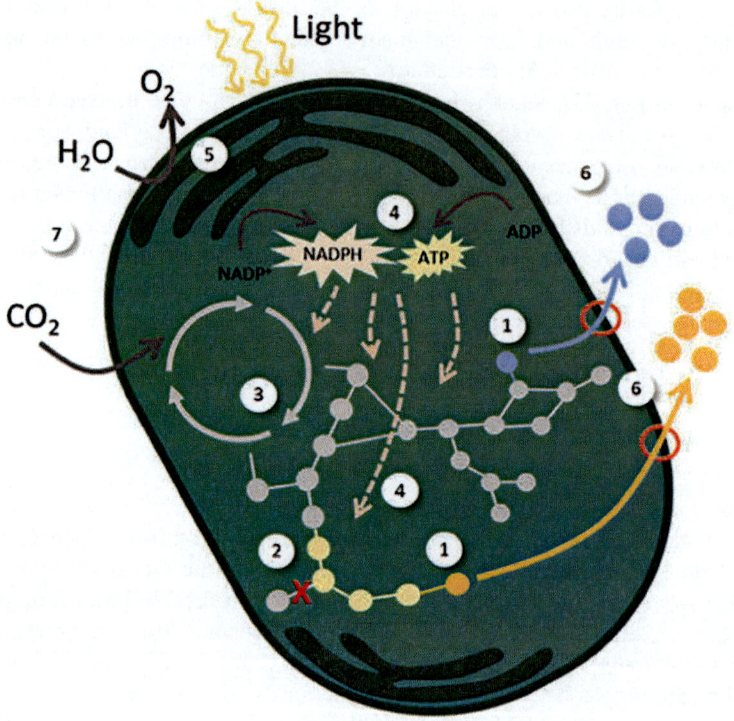

Fig. 4 Targets for engineering and optimization of enhanced production include biosynthetic enzyme activity and expression levels (1). Increasing precursor supply via upstream pathway engineering, deletion, or downregulation of competing pathways (2). Carbon fixation and central carbon metabolism (3). Adjusting cofactor supply (4). Light harvesting (5). Export and harvest of product (6). External conditions: light, nutrient supply including CO_2, temperature (7). (Adapted from Rodrigues and Lindberg 2021) [83]

would be to enable and enhance product export and removal from the culture, also a "pull" strategy, for example by expression of transporters (Fig. 4, 6) [89]. Finally, external conditions, such as cultivation medium, gas supply, light availability and temperature should be optimized to reach the highest possible productivities (Fig. 4, 7).

The strategies outlined above have been employed individually or in combination in many studies during the last decades of cyanobacterial research. In recent years, genome scale metabolic models of cyanobacterial cells have come into use for guiding engineering efforts as well as for interpreting experimental results and models in combination with metabolomics and metabolic flux analysis are powerful tools for optimizing and evaluating engineered cells [90]. Below, we will highlight a few of the engineering strategies, regarding engineering photosynthetic efficiency, carbon fixation, and growth environment. Engineering approaches focusing on carbon fluxes are discussed in detail in Chap. 4; Metabolic engineering design strategies for increasing carbon fluxes relevant for biosynthesis in cyanobacteria.

7 Engineering Photosynthetic Efficiency

Cyanobacteria are phototrophic organisms utilizing light as energy source. Light is a most complex substrate consisting of multiple wavelengths of which only a defined fraction, the photosynthetically active radiation (PAR) can be partly exploited by the microbes. Wavelength in the range of 400–700 nm is generally defined as PAR, which represents about 50% of the incoming solar energy [91]. The rest of the radiation is mostly absorbed as heat. The intensities of the wavelength differ, as well as the pigmentation of the bacteria enabling them to make use of the various wavelength. This significantly influences the conversion yield of light energy to chemical energy, which is in the range of 1–3%. Several discoveries open possibilities to broaden the exploitable light spectrum, for instance, chlorophyll *d* absorbs light with a wavelength of 700–750 nm and chlorophyll *f* has an absorption spectrum around 706 nm. Implementing these pigments in cyanobacteria used for biotechnological applications may enhance their light utilization [92, 93].

Another approach focused on the use of the phenomena, that cyanobacterial growth can be tuned to a certain degree by applying light of a specific wavelength instead of the whole spectrum. Red light in the range of 620–645 nm corresponding to the absorption peak of Chl *a* and phycocyanobilin, increases growth rates in *Spirulina platensis* by 37.5% compared to white light [94]. On the contrary, growth and oxygen evolution rate of *Synechocystis* sp. PCC 6803 seem to be impaired when supplying blue light. It was suggested that non absorbed blue light causes an imbalance between Photosystem I and Photosystem II [95]. When only using orange-red light (625–660 nm) up to 500 μmol m^{-2} s^{-1}, the opposite effect was reported. Growth rate and oxygen concentrations increased, correlating with elevated light intensity up to 500 μE m^{-2} s^{-1}, although the efficiency of photosynthesis declined concomitantly to approximately 1% [96]. Source-sink imbalances can be relieved by introducing heterologous electron sinks into the cell, as has been shown for *S. elongatus* PCC 7942. By implementation of a sucrose production pathway or a cytochrome P450 oxygenase consuming NADH, the electron flux through the PET enhanced and thus the quantum yield of PSII increased [97].

For green algae it was demonstrated that the truncation of the light harvesting complex resulted in higher biomass yields. Authors hypothesized that this effect was fostered by the diminished shading effect of the cells and less heat dissipation at the layer closest to the light source [98, 99]. In cyanobacteria however, this approach gave the opposite effect; lower biomass and lower PSII activity [100].

8 Engineering Carbon Fixation

Cyanobacteria fix carbon via the CBB cycle. Rubisco (ribulose-biphosphate-carboxylase-oxygenase) catalyzes the carboxylation of ribulose-1,5-bisphosphate, yielding an instable C6-compound which immediately splits into two molecules of 3-phosphoglycerate. This is the rate-limiting step of carbon fixation, and Rubisco with an average turnover rate of 1–10 s^{-1} a rather slow enzyme.

As Rubiscos of different origin differ in their activities, expressing more efficient Rubiscos in typical chassis strains like *Synechocystis* sp. may be an approach to enhance carbon fixation rates. However, since Rubisco synthesis is complex and relying on chaperons varying between species, this approach is challenging [101]. Overexpression of Rubisco from *S. elongatus* PCC 6301 in *S. elongatus* PCC 7942 was successfully achieved, resulting in an increased activity of 21.3 ± 2.9 nmol min^{-1} mg^{-1} $_{Rubisco}$ combined with the production of isobutyraldehyde [70].

Other approaches consider improvement of the carbon concentrating mechanisms or enhancing the regeneration of Ribulose-1,5-biphosphate. In the first case, overexpressing an inorganic carbon transporter doubled the growth rate and biomass yield of *Synechocystis* sp. PCC 6803 compared to the wild type [102]. Secondly, by overexpression of CBB cycle enzymes it was possible to increase growth and oxygen evolution rates in *Synechocystis* sp. PCC 6803. Cultivating the mutant *Synechocystis* sp. PCC 6803 EtOH-fbaA under light intensities of 65 μmol m^{-2} s^{-1} boosted ethanol production from CO_2 [101]. Same could be observed for the overexpression of other CBB enzymes like fructose-1,6/sedoheptulose-1,7-biphosphate, aldolase, and transketolase [103].

9 Comparison to Heterotrophic Production Hosts

Photosynthetic efficiency and carbon uptake are thus two major targets to improve the performance of solar cell factories, in combination with other strategies as outlined above (Fig. 4). Although there has been significant development of the field in the last decade the difference to established microbial workhorses is still large, especially for bulk products. Table 1 compares ethanol production in *Synechocystis* sp. PCC 6803 to an established production route via fermentation of glucose utilizing *Saccharomyces diastaticus*.

The product titer achieved in yeast fermentation is 16 times higher than via the cyanobacterial route. For lactate, a product like ethanol derived from pyruvate and thus metabolically close to the central carbon metabolism, the most productive cyanobacterial strain reported so far reached a titer of 1.45 g L^{-1} [68] while current heterotrophic producers can reach above 200 g L^{-1} of lactate [108, 109]. Similarly for butanol, another compound where cyanobacterial production has reached relatively high levels, the best performing strain reached 4.8 g L^{-1} of n-butanol, while in

Table 1 Ethanol production in *Synechocystis* sp. PCC 6803 and *Saccharomyces diastaticus*

Production strain	Ethanol titer [g/L]	Added precursor	Engineering performed	Reference
Synechocystis PCC 6803	0.46	CO_2	Ethanol biosynthetic enzymes pyruvate decarboxylase and alcohol dehydrogenase expressed	[104]
Synechocystis PCC 6803	0.2	50 mM $NaHCO_3$	Ethanol biosynthetic enzymes expressed, enhanced expression of CBB cycle enzymes	[101]
Synechocystis PCC 6803	1.2	50 mM $NaHCO_3$	Ethanol biosynthetic enzymes expressed, further enhanced expression of CBB cycle enzymes	[105]
Synechocystis PCC 6803	5.5	5% CO_2	Ethanol biosynthetic enzymes expressed, overexpressing several copies of alcohol dehydrogenase, inactivating PHB synthesis	[106]
Saccharomyces diastaticus	90	100 g/L glucose	Selection and optimization of strains for fermentations above 40°C	[107]

E. coli titers of up to 30 g L^{-1} have been reported [110]. While these differences may seem too large to make cyanobacterial production of such compounds attractive, the benefit of the photosynthetic approach is the independence of organic carbon. The knowledge on cyanobacterial metabolism and its regulation is also not yet at the same level as for established industrial workhorses like *E. coli* or *Saccharomyces*. Taking into account the progress made in recent years, we are convinced that the field is only at the beginning and that the potential of these organisms is far from being realized.

10 Optimizing the Growth Environment

Optimizing cultivation conditions will be an important step for enhancing production. As mentioned above the field of cyano-photobiotechnology is still in its infancy. This is also reflected in the variety of applied cultivation systems used in different studies. Standardization of highly important parameters like light source and cultivation device is not yet implemented in the field, which makes it difficult to compare the studies and draw solid conclusions. The range is large, starting from classical shake flask cultures in simple benchtop systems, via climate controlled light incubators with a defined CO_2 atmosphere to highly regulated conditions in automated photobioreactor systems. The choice of cultivation system and method of illumination significantly influences light quality and intensity, availability of carbon and other nutrients, pH, and oxygen removal, and thereby affects the physiology and productivity of the organisms.

A major bottleneck for larger scale applications are the low cell densities reached in established larger volume photobioreactors due to severe light limitation by self-shading effects and long light paths. On average, planktonic growing cyanobacterial cultures reach about 3–6 g_{CDW} L^{-1}, which is about factor 10–20 lower than for standard heterotrophic cultures using established industrial strains, and part of the reason for the low volumetric productivities in cyanobacteria compared to such systems (Table 1). The possible penetration depth of the light into the bulk phase of a photobioreactor defines the scaling limit and reactor geometry [91]. This issue is included in all PBR designs from flat plate to tubular reactors and can only partly be relieved by mixing the culture [111]. How fluctuations in light intensity during mixing, when cells are moved from well illuminated to darker zones in the reactor, influence photosynthetic growth, is still a matter of debate. Most authors claim that a higher frequency of light fluctuations is beneficial for biomass growth [112–114]. Mixing also prevents the formation of gradients and dead zones, however, many cyanobacteria, especially those growing filamentously, may be sensitive to the accompanying shear forces.

Apart from light, also carbon availability may be a challenge. It is not sufficient to supply CO_2 via ambient air, as the CO_2 content of 0.04% is not enough for optimal biomass growth. When using CO_2 enriched air or flue gas, the pH of the medium has to be closely monitored. CO_2 is in equilibrium with dissolved carbonic acid, bicarbonate, and carbonate. This equilibrium is highly pH depended with a shift to bicarbonate at high pH, which is beneficial for carbon uptake by the microbes [115]. When using sodium bicarbonate as carbon source, it will fulfill a dual function as substrate and buffer for the medium pH [116].

Recently, a small-scale reactor system based on transport of CO_2 into the culture via a membrane has been developed, where cyanobacterial cultures can grow to high cell densities in a short time due to a non-limiting supply of inorganic carbon [8]. Utilizing this setup, which also requires an increased supply of other nutrients as well as high light intensities to support growth, volumetric productivities can be strongly enhanced while per cell productivities remain constant or are lower than in more dilute, standard conditions [86, 117]. While it is not clear how this system could be scaled up, the results demonstrate that the production capacity for cyanobacteria may actually be higher than what is mostly observed, and given optimal conditions production may eventually come closer to that of heterotrophic hosts.

In addition to substrate supply, also the removal of the key product, oxygen, is an important issue to consider when designing photobioreactors. During photosynthetic growth, cyanobacteria release significant amounts of oxygen, which can accumulate in the broth depending on the design of the cultivation device. High oxygen concentration will on the one hand trigger the photorespiration activity of Rubisco and on the other hand toxify the organisms due to the formation of radical oxygen species [118]. Thus, the accumulation of oxygen in the reactor needs to be prevented. Sufficient degassing methods include tilted reactors, so gas bubbles can be collected and removed at the upper part [119], the use of oxygen permeable materials to prevent oxygen accumulation [87], or co-cultivation of oxygen-respiring bacteria [120].

In order to profit from the ability of cyanobacteria to utilize light energy, reactors should in the ideal case be operated outside in natural light. However, climate dynamics pose a great challenge to the cultivation, especially regarding temperature and light regime. Energy necessary for cooling the reactors adds significantly to the operating costs. Several ideas have been tested, for instance the use of shading devices [121], submerging the reactor in water basins [122] or regulating the temperature of the medium via a heat exchanger [123]. A totally different concept was introduced by Kim and colleagues [124], who designed a PBR floating in the sea. Using the mixing energy by the waves and cooling by the surrounding water, the energy input into the reactors could be drastically reduced.

Commonly used photobioreactor designs include flat plate, column and tubular reactors (Fig. 5). Flat panel photo-bioreactors combine a high surface to volume ratio for good illumination with the possibility of CO_2-supply via spargers and mixing via the airlift principle [125]. Tubular photobioreactors with up to 1,000 km length (BGG World, China) are in operation. The scale is limited by light penetration over

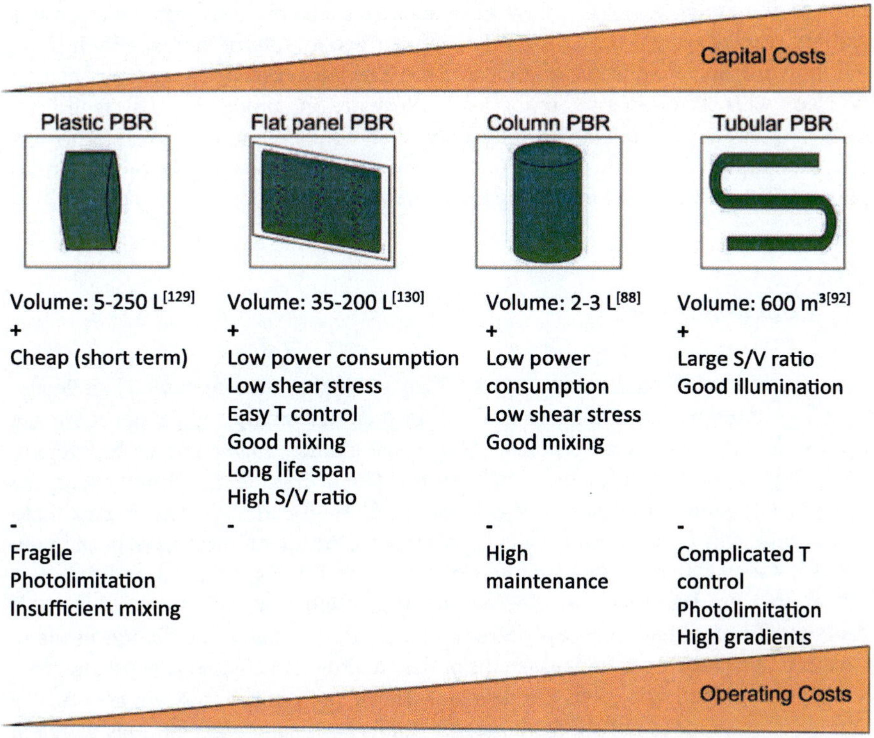

Fig. 5 Different reactor geometries and their advantages and disadvantages. Given volume is referring to published dimensions of one unit. Scaling via numbering up is not considered. Adapted from [91]

the tube radius, as well as the substrate supply and oxygen removal [87]. Column photo-bioreactors use the airlift principle for CO_2 supply, but as with tubular systems the diameter is limited due the possible light penetration depth [126]. Newer concepts include plastic bag photo-bioreactors, which offer cheap production costs on one hand, but the cells suffer from photo-limitation and inadequate mixing. Also, it was reported, that the bags are relatively fragile [127]. The larger the scale, the more difficult it becomes to ensure sufficient mixing of the system. The energy needed for pumping / mixing the bulk fluid adds a major share to the overall operating costs. This includes supply of gas and removal of oxygen. The addition of static mixers like baffles or by changing the shape of the bioreactor (Subitech GmbH, https://www.subitec.com/en/), will reduce costs. However, in tubular PBRs pumping is essential to move the bulk fluid through the tubes and adds to the operating costs. Finally, the maintenance and cleaning of the reactors has to be considered, which is more or less complicated depending on the reactor geometry [91].

A completely novel cultivation format based on biofilm formation was recently introduced by Hoschek et al. [120]. In this concept the ability of many organisms to attach to surface structures was utilized and the phototrophic workhorses were cultured as biofilms instead of planktonic cultures. Thereby exceptionally high cell densities of up to 60 $g_{CDW}\,L^{-1}$ have been achieved. The biofilms were cultured in a capillary reactor system providing very high surface to volume ratios. Biofilms are well known from wastewater treatment. They are very robust, the reactors can be operated with low energy input and downstream processing is facilitated [130, 131]. However, this is currently only a lab concept and the question how to scale such a system remains open. For more details on process development strategies see Chap. 9; Process technologies of cyanobacteria.

11 Conclusions

Cyanobacteria are fascinating organisms. They are highly diverse, which is reflected in their phenotypic expression as well as in their physiology. Cyanobacteria are responsible for toxic algal blooms. At the same time, many strains are highly interesting for biotechnological applications. The ability to use light energy to carry out oxygenic photosynthesis and reduce CO_2 to carbohydrates makes them potent candidates for biotechnological applications. Although there have been many studies in the past to open up cyanobacteria for biotechnology, the major difficulties such as low product titers, insufficient stabilities and low cell densities are still unsolved. On the other hand, cyanobacteria are still not really understood in many respects. Especially regarding a systems understanding, gene and protein regulation, electron and carbon flow, and the development of molecular tools, we are still far from the level as for established production strains such as *E. coli*, and thus there are abundant opportunities for improvement in cyanobacterial production systems. When current limitations can be overcome, cyanobacterial biotechnology has the potential to provide a truly sustainable source of renewable chemicals made directly from CO_2 powered by solar light.

In this book we cover basic research investigating water oxidation bioenergetics in cyanobacteria (Chap. 2), development of molecular tools (Chap. 3) and metabolic engineering strategies focusing on the carbon fluxes (Chap. 4). In the second part we look at application examples, covering established studies like fatty acids (Chap. 5) and pigments (Chap. 6), but also emerging fields like hydrogen production (Chap. 7) and biophotovoltaics (Chap. 8). Finally, we discuss bioprocess development strategies (Chap. 9).

References

1. Lindsey R, Scott M (2020) Climate change: Arctic Sea ice summer minimum. https://www.climate.gov/news-features/understanding-climate/climate-change-arctic-sea-ice-summer-minimum
2. Stips A, Macias D et al (2016) On the causal structure between CO_2 and global temperature. Sci Rep 6:21691
3. Appel AM, Bercaw JE et al (2013) Frontiers, opportunities, and challenges in biochemical and chemical catalysis of CO_2 fixation. Chem Rev 113:6621–6658
4. Zhang W, Fernández-Fueyo E et al (2018) Selective aerobic oxidation reactions using a combination of photocatalytic water oxidation and enzymatic oxyfunctionalizations. Nat Catal 1:55–62
5. Zilliges Y (2014) Glycogen: a dynamic cellular sink and reservoir for carbon. In: Herrero EFaA (ed) The cell biology of cyanobacteria. Caister Academic Press
6. Koch M, Doello S et al (2019) PHB is produced from glycogen turn-over during nitrogen starvation in *Synechocystis* sp. PCC 6803. Int J Mol Sci 20:1942
7. Prabha S, Vijay AK et al (2022) Cyanobacterial biorefinery: towards economic feasibility through the maximum valorization of biomass. Sci Total Environ 814:152795
8. Lippi L, Bähr L et al (2018) Exploring the potential of high-density cultivation of cyanobacteria for the production of cyanophycin. Algal Res 31:363–366
9. Schirrmeister BE, Anisimova M et al (2011) Evolution of cyanobacterial morphotypes: taxa required for improved phylogenomic approaches. Commun Integr Biol 4:424–427
10. Pisciotta JM, Zou Y, Baskakov IV (2010) Light-dependent electrogenic activity of cyanobacteria. PLoS One 5:e10821
11. Sagan L (1967) On the origin of mitosing cells. J Theor Biol 14:225-IN6
12. Welsh EA, Liberton M et al (2008) The genome of *Cyanothece* 51142, a unicellular diazotrophic cyanobacterium important in the marine nitrogen cycle. Proc Natl Acad Sci 105:15094–15099
13. Hoehler TM, Bebout BM, Des Marais DJ (2001) The role of microbial mats in the production of reduced gases on the early earth. Nature 412:324–327
14. Bolhuis H, Cretoiu MS, Stal LJ (2014) Molecular ecology of microbial mats. FEMS Microbiol Ecol 90:335–350
15. David C, Karande R, Bühler K (2021) Cyanobacterial biofilms in natural and synthetic environments. In: Hudson P (ed) Cyanobacteria biotechnology. Wiley-VCH, pp 477–504
16. Wetzel RG (2001) Planktonic communities: zooplankton and their interaction with fish. In: Wetzel RG (ed) Limnology. Academic Press, San Diego, pp 395–488
17. Stal LJ, Moezelaar R (1997) Fermentation in cyanobacteria. FEMS Microbiol Rev 21:179–211

18. Flombaum P, Gallegos JL et al (2013) Present and future global distributions of the marine cyanobacteria *Prochlorococcus* and *Synechococcus*. Proc Natl Acad Sci 110:9824–9829
19. Morel A, Ahn Y-H et al (1993) *Prochlorococcus* and *Synechococcus*: a comparative study of their optical properties in relation to their size and pigmentation. J Mar Res 51:617–649
20. Dufresne A, Salanoubat M et al (2003) Genome sequence of the cyanobacterium *Prochlorococcus marinus* SS120, a nearly minimal oxyphototrophic genome. Proc Natl Acad Sci 100:10020–10025
21. Dufresne A, Garczarek L, Partensky F (2005) Accelerated evolution associated with genome reduction in a free-living prokaryote. Genome Biol 6:R14
22. Strehl B, Holtzendorff J et al (1999) A small and compact genome in the marine cyanobacterium *Prochlorococcus marinus* CCMP 1375: lack of an intron in the gene for tRNA(Leu) (UAA) and a single copy of the rRNA operon. FEMS Microbiol Lett 181:261–266
23. Observatory NE. https://neo.gsfc.nasa.gov/view.php?datasetId=MOD17A2_M_PSN
24. Ralf G, Repeta DJ (1992) The pigments of *Prochlorococcus marinus*: the presence of divinylchlorophyll a and b in a marine procaryote. Limnol Oceanogr 37:425–433
25. Scanlan DJ, Ostrowski M et al (2009) Ecological genomics of marine Picocyanobacteria. Microbiol Mol Biol Rev 73:249–299
26. Ludwig M, Bryant DA (2012) *Synechococcus* sp. strain PCC 7002 transcriptome: acclimation to temperature, salinity, oxidative stress, and mixotrophic growth conditions. Front Microbiol 3:354
27. Clark RL, McGinley LL et al (2018) Light-optimized growth of cyanobacterial cultures: growth phases and productivity of biomass and secreted molecules in light-limited batch growth. Metab Eng 47:230–242
28. Yu J, Liberton M et al (2015) *Synechococcus elongatus* UTEX 2973, a fast growing cyanobacterial chassis for biosynthesis using light and CO_2. Sci Rep 5:8132
29. Włodarczyk A, Selão TT et al (2020) Newly discovered *Synechococcus* sp. PCC 11901 is a robust cyanobacterial strain for high biomass production. Commun Biol 3:215
30. Jaiswal D, Sengupta A et al (2018) Genome features and biochemical characteristics of a robust, fast growing and naturally transformable cyanobacterium *Synechococcus elongatus* PCC 11801 isolated from India. Sci Rep 8:16632
31. Ludwig M, Bryant DA (2011) Transcription profiling of the model cyanobacterium *Synechococcus* sp. strain PCC 7002 by next-gen (SOLiD™) sequencing of cDNA. Front Microbiol 2:41
32. Markley AL, Begemann MB et al (2015) Synthetic biology toolbox for controlling gene expression in the cyanobacterium *Synechococcus* sp. strain PCC 7002. ACS Synth Biol 4: 595–603
33. Mills LA, McCormick AJ, Lea-Smith DJ (2020) Current knowledge and recent advances in understanding metabolism of the model cyanobacterium *Synechocystis* sp. PCC 6803. Biosci Rep 40:BSR20193325
34. Vermaas W (1996) Molecular genetics of the cyanobacterium *Synechocystis* sp. PCC 6803: principles and possible biotechnology applications. J Appl Phycol 8:263–273
35. Los DA, Zorina A et al (2010) Stress sensors and signal transducers in cyanobacteria. Sensors (Basel) 10:2386–2415
36. Kaneko T, Sato S et al (1996) Sequence analysis of the genome of the unicellular cyanobacterium *Synechocystis* sp. strain PCC6803. II. Sequence determination of the entire genome and assignment of potential protein-coding regions. DNA Res 3:109–136
37. Fujisawa T, Narikawa R et al (2016) CyanoBase: a large-scale update on its 20th anniversary. Nucleic Acids Res 45:D551–D5D4
38. Gale GAR, Schiavon Osorio AA et al (2019) Emerging species and genome editing tools: future prospects in cyanobacterial synthetic biology. Microorganisms:7
39. Burford MA, Carey CC et al (2020) Perspective: advancing the research agenda for improving understanding of cyanobacteria in a future of global change. Harmful Algae 91:101601

40. Plaas HE, Paerl HW (2021) Toxic cyanobacteria: a growing threat to water and air quality. Environ Sci Technol 55:44–64
41. Buratti FM, Manganelli M et al (2017) Cyanotoxins: producing organisms, occurrence, toxicity, mechanism of action and human health toxicological risk evaluation. Arch Toxicol 91:1049–1130
42. Pearson L, Mihali T et al (2010) On the chemistry, toxicology and genetics of the cyanobacterial toxins, microcystin, nodularin, saxitoxin and cylindrospermopsin. Mar Drugs 8:1650–1680
43. Tan LT (2007) Bioactive natural products from marine cyanobacteria for drug discovery. Phytochemistry 68:954–979
44. Demay J, Bernard C et al (2019) Natural products from cyanobacteria: focus on beneficial activities. Mar Drugs 17
45. Savakis P, Hellingwerf KJ (2015) Engineering cyanobacteria for direct biofuel production from CO_2. Curr Opin Biotechnol 33:8–14
46. Treece TR, Gonzales JN et al (2022) Synthetic biology approaches for improving chemical production in cyanobacteria. Front Bioeng Biotechnol 10:869195
47. Angermayr SA, Gorchs Rovira A, Hellingwerf KJ (2015) Metabolic engineering of cyanobacteria for the synthesis of commodity products. Trends Biotechnol 33:352–361
48. Jaiswal D, Sahasrabuddhe D, Wangikar PP (2022) Cyanobacteria as cell factories: the roles of host and pathway engineering and translational research. Curr Opin Biotechnol 73:314–322
49. Liu D, Liberton M et al (2021) Engineering biology approaches for food and nutrient production by cyanobacteria. Curr Opin Biotechnol 67:1–6
50. Borowitzka MA (1999) Commercial production of microalgae: ponds, tanks, tubes and fermenters. J Biotechnol 70:313–321
51. Weaver PF, Lien S, Seibert M (1980) Photobiological production of hydrogen. Sol Energy 24: 3–45
52. Wijffels RH, Barbosa MJ (2010) An outlook on microalgal biofuels. Science 329:796–799
53. Nilsson A, Shabestary K et al (2020) Environmental impacts and limitations of third-generation biobutanol: life cycle assessment of n-butanol produced by genetically engineered cyanobacteria. J Ind Ecol 24:205–216
54. Knoot CJ, Ungerer J et al (2018) Cyanobacteria: promising biocatalysts for sustainable chemical production. J Biol Chem 293:5044–5052
55. Heidorn T, Camsund D et al (2011) Synthetic biology in cyanobacteria engineering and analyzing novel functions. Methods Enzymol 497:539–579
56. Vavitsas K, Kugler A et al (2021) Doing synthetic biology with photosynthetic microorganisms. Physiol Plant 173:624–638
57. Yao L, Cengic I et al (2016) Multiple gene repression in cyanobacteria using CRISPRi. ACS Synth Biol 5:207–212
58. Yao L, Shabestary K et al (2020) Pooled CRISPRi screening of the cyanobacterium *Synechocystis* sp PCC 6803 for enhanced industrial phenotypes. Nat Commun 11:1666
59. Taton A, Ma AT et al (2017) NOT gate genetic circuits to control gene expression in cyanobacteria. ACS Synth Biol 6:2175–2182
60. Sun T, Li S et al (2018) Re-direction of carbon flux to key precursor malonyl-CoA via artificial small RNAs in photosynthetic *Synechocystis* sp. PCC 6803. Biotechnol Biofuels 11:26
61. Herrero A, Stavans J, Flores E (2016) The multicellular nature of filamentous heterocyst-forming cyanobacteria. FEMS Microbiol Rev 40:831–854
62. Cohen SE, Golden SS (2015) Circadian rhythms in cyanobacteria. Microbiol Mol Biol Rev 79: 373–385
63. Jaiswal D, Sengupta A et al (2020) A novel cyanobacterium *Synechococcus elongatus* PCC 11802 has distinct genomic and metabolomic characteristics compared to its neighbor PCC 11801. Sci Rep 10:191
64. Abdulqader G, Barsanti L, Tredici MR (2000) Harvest of *Arthrospira platensis* from Lake Kossorom (Chad) and its household usage among the Kanembu. J Appl Phycol 12:493–498

65. Lauceri R, Chini Zittelli G, Torzillo G (2019) A simple method for rapid purification of phycobiliproteins from *Arthrospira platensis* and *Porphyridium cruentum* biomass. Algal Res 44:101685
66. Deng M-D, Coleman JR (1999) Ethanol synthesis by genetic engineering in cyanobacteria. Appl Environ Microbiol 65:523–528
67. Kopka J, Schmidt S et al (2017) Systems analysis of ethanol production in the genetically engineered cyanobacterium *Synechococcus* sp. PCC 7002. Biotechnol Biofuels 10:56
68. Selão TT, Jebarani J et al (2020) Enhanced production of D-lactate in cyanobacteria by re-routing photosynthetic cyclic and pseudo-cyclic electron flow. Front Plant Sci:10
69. Kanno M, Carroll AL, Atsumi S (2017) Global metabolic rewiring for improved CO_2 fixation and chemical production in cyanobacteria. Nat Commun 8:14724
70. Atsumi S, Higashide W, Liao JC (2009) Direct photosynthetic recycling of carbon dioxide to isobutyraldehyde. Nat Biotechnol 27:1177–1180
71. Miao R, Liu X et al (2017) Isobutanol production in *Synechocystis* PCC 6803 using heterologous and endogenous alcohol dehydrogenases. Metab Eng Commun 5:45–53
72. Shen CR, Liao JC (2012) Photosynthetic production of 2-methyl-1-butanol from CO_2 in cyanobacterium *Synechococcus elongatus* PCC7942 and characterization of the native acetohydroxyacid synthase. Energy Environ Sci 5:9574–9583
73. Kobayashi S, Atsumi S et al (2022) Light-induced production of isobutanol and 3-methyl-1-butanol by metabolically engineered cyanobacteria. Microb Cell Factories 21:7
74. Koch M, Bruckmoser J et al (2020) Maximizing PHB content in *Synechocystis* sp. PCC 6803: a new metabolic engineering strategy based on the regulator PirC. Microb Cell Factories 19: 231
75. Hein S, Tran H, Steinbüchel A, Synechocystis sp. (1998) PCC6803 possesses a two-component polyhydroxyalkanoic acid synthase similar to that of anoxygenic purple sulfur bacteria. Arch Microbiol 170:162–170
76. Lan EI, Liao JC (2011) Metabolic engineering of cyanobacteria for 1-butanol production from carbon dioxide. Metab Eng 13:353–363
77. Lan EI, Liao JC (2013) Microbial synthesis of n-butanol, isobutanol, and other higher alcohols from diverse resources. Bioresour Technol 135:339–349
78. Liu X, Miao R et al (2019) Modular engineering for efficient photosynthetic biosynthesis of 1-butanol from CO_2 in cyanobacteria. Energy Environ Sci 12:2765–2777
79. Yunus IS, Jones PR (2018) Photosynthesis-dependent biosynthesis of medium chain-length fatty acids and alcohols. Metab Eng 49:59–68
80. Ruffing AM (2014) Improved free fatty acid production in cyanobacteria with *Synechococcus* sp. PCC 7002 as host. Front Bioeng Biotechnol 2:17
81. Tan X, Yao L et al (2011) Photosynthesis driven conversion of carbon dioxide to fatty alcohols and hydrocarbons in cyanobacteria. Metab Eng 13:169–176
82. Wang W, Liu X, Lu X (2013) Engineering cyanobacteria to improve photosynthetic production of alka(e)nes. Biotechnol Biofuels 6:69
83. Rodrigues JS, Lindberg P (2021) Engineering cyanobacteria as host organisms for production of terpenes and terpenoids. Cyanobacteria Biotechnol:267–300
84. Frigaard N-U (2018) Chapter 2 – Sugar and sugar alcohol production in genetically modified cyanobacteria. In: Holban AM, Grumezescu AM (eds) Genetically engineered foods. Academic Press, pp 31–47
85. Brey LF, Włodarczyk AJ et al (2020) Metabolic engineering of *Synechocystis* sp. PCC 6803 for the production of aromatic amino acids and derived phenylpropanoids. Metab Eng 57:129–139
86. Kukil K, Lindberg P (2022) Expression of phenylalanine ammonia lyases in *Synechocystis* sp. PCC 6803 and subsequent improvements of sustainable production of phenylpropanoids. Microb Cell Factories 21:8
87. Posten C (2009) Design principles of photo-bioreactors for cultivation of microalgae. Eng Life Sci 9:165–177

88. Shabestary K, Hudson EP (2016) Computational metabolic engineering strategies for growth-coupled biofuel production by *Synechocystis*. Metab Eng Commun 3:216–226
89. Abramson BW, Kachel B et al (2016) Increased photochemical efficiency in cyanobacteria via an engineered sucrose sink. Plant Cell Physiol 57:2451–2460
90. Hendry JI, Bandyopadhyay A et al (2020) Metabolic model guided strain design of cyanobacteria. Curr Opin Biotechnol 64:17–23
91. Huang Q, Jiang F et al (2017) Design of photobioreactors for mass cultivation of photosynthetic organisms. Engineering 3:318–329
92. Chen M, Blankenship RE (2011) Expanding the solar spectrum used by photosynthesis. Trends Plant Sci 16:427–431
93. Chen M, Schliep M et al (2010) A red-shifted chlorophyll. Science 329:1318–1319
94. Wang C-Y, Fu C-C, Liu Y-C (2007) Effects of using light-emitting diodes on the cultivation of *Spirulina platensis*. Biochem Eng J 37:21–25
95. Luimstra VM, Schuurmans JM et al (2018) Blue light reduces photosynthetic efficiency of cyanobacteria through an imbalance between photosystems I and II. Photosynth Res 138:177–189
96. Cordara A, Re A et al (2018) Analysis of the light intensity dependence of the growth of *Synechocystis* and of the light distribution in a photobioreactor energized by 635 nm light. PeerJ 6:e5256
97. Santos-Merino M, Torrado A et al (2021) Improved photosynthetic capacity and photosystem I oxidation via heterologous metabolism engineering in cyanobacteria. Proc Natl Acad Sci 118:e2021523118
98. de Mooij T, Janssen M et al (2015) Antenna size reduction as a strategy to increase biomass productivity: a great potential not yet realized. J Appl Phycol 27:1063–1077
99. Shin W-S, Lee B et al (2016) Truncated light-harvesting chlorophyll antenna size in *Chlorella vulgaris* improves biomass productivity. J Appl Phycol 28:3193–3202
100. Luimstra VM, Schuurmans JM et al (2019) Exploring the low photosynthetic efficiency of cyanobacteria in blue light using a mutant lacking phycobilisomes. Photosynth Res 141:291–301
101. Liang F, Englund E et al (2018) Engineered cyanobacteria with enhanced growth show increased ethanol production and higher biofuel to biomass ratio. Metab Eng 46:51–59
102. Kamennaya NA, Ahn S et al (2015) Installing extra bicarbonate transporters in the cyanobacterium *Synechocystis* sp. PCC 6803 enhances biomass production. Metab Eng 29:76–85
103. Raines CA (2003) The Calvin cycle revisited. Photosynth Res 75:1–10
104. Dexter J, Fu P (2009) Metabolic engineering of cyanobacteria for ethanol production. Energy Environ Sci 2:857–864
105. Roussou S, Albergati A et al (2021) Engineered cyanobacteria with additional overexpression of selected Calvin-Benson-Bassham enzymes show further increased ethanol production. Metab Eng Commun 12:e00161
106. Gao Z, Zhao H et al (2012) Photosynthetic production of ethanol from carbon dioxide in genetically engineered cyanobacteria. Energy Environ Sci 5:9857–9865
107. D'Amore T, Celotto G et al (1989) Selection and optimization of yeast suitable for ethanol production at 40°C. Enzym Microb Technol 11:411–416
108. Huang S, Xue Y et al (2021) A review of the recent developments in the bioproduction of polylactic acid and its precursors optically pure lactic acids. Molecules 26:6446
109. Tsuge Y, Kato N et al (2019) Metabolic engineering of *Corynebacterium glutamicum* for hyperproduction of polymer-grade l- and d-lactic acid. Appl Microbiol Biotechnol 103:3381–3391
110. Ferreira S, Pereira R et al (2020) Metabolic engineering strategies for butanol production in *Escherichia coli*. Biotechnol Bioeng 117:2571–2587
111. Krujatz F, Illing R et al (2015) Light-field-characterization in a continuous hydrogen-producing photobioreactor by optical simulation and computational fluid dynamics. Biotechnol Bioeng 112:2439–2449

112. Abu-Ghosh S, Fixler D et al (2015) Continuous background light significantly increases flashing-light enhancement of photosynthesis and growth of microalgae. Bioresour Technol 187:144–148
113. Huang J, Li Y et al (2014) Novel flat-plate photobioreactors for microalgae cultivation with special mixers to promote mixing along the light gradient. Bioresour Technol 159:8–16
114. Iluz D, Abu-Ghosh S (2016) A novel photobioreactor creating fluctuating light from solar energy for a higher light-to-biomass conversion efficiency. Energy Convers Manag 126:767–773
115. Summerfield TC, Sherman LA (2008) Global transcriptional response of the alkali-tolerant cyanobacterium *Synechocystis* sp. strain PCC 6803 to a pH 10 environment. Appl Environ Microbiol 74:5276–5284
116. Vasumathi KK, Premalatha M, Subramanian P (2012) Parameters influencing the design of photobioreactor for the growth of microalgae. Renew Sust Energ Rev 16:5443–5450
117. Dienst D, Wichmann J et al (2020) High density cultivation for efficient sesquiterpenoid biosynthesis in *Synechocystis* sp. PCC 6803. Sci Rep 10:5932
118. Latifi A, Ruiz M, Zhang C-C (2009) Oxidative stress in cyanobacteria. FEMS Microbiol Rev 33:258–278
119. Tredici MR (2003) Mass production of microalgae: photobioreactors. In: Handbook of microalgal culture, pp 178–214
120. Hoschek A, Heuschkel I et al (2019) Mixed-species biofilms for high-cell-density application of *Synechocystis* sp. PCC 6803 in capillary reactors for continuous cyclohexane oxidation to cyclohexanol. Bioresour Technol 282:171–178
121. Ugwu CU, Aoyagi H (2008) Influence of shading inclined tubular photobioreactor surfaces on biomass productivity of *C. sorokiniana*. Photosynthetica 46:283
122. Carlozzi P, Pushparaj B et al (2006) Growth characteristics of *Rhodopseudomonas palustris* cultured outdoors, in an underwater tubular photobioreactor, and investigation on photosynthetic efficiency. Appl Microbiol Biotechnol 73:789–795
123. Watanabe Y, de la Noüe J, Hall DO (1995) Photosynthetic performance of a helical tubular photobioreactor incorporating the cyanobacterium *Spirulina platensis*. Biotechnol Bioeng 47: 261–269
124. Kim ZH, Park H et al (2016) Development of a floating photobioreactor with internal partitions for efficient utilization of ocean wave into improved mass transfer and algal culture mixing. Bioprocess Biosyst Eng 39:713–723
125. Sierra E, Acién FG et al (2008) Characterization of a flat plate photobioreactor for the production of microalgae. Chem Eng J 138:136–147
126. Mirón AS, Gómez AC et al (1999) Comparative evaluation of compact photobioreactors for large-scale monoculture of microalgae. In: Osinga R et al (eds) Progress in industrial microbiology. Elsevier, pp 249–270
127. Wang B, Lan CQ, Horsman M (2012) Closed photobioreactors for production of microalgal biomasses. Biotechnol Adv 30:904–912
128. Jacob-Lopes E, Cacia Ferreira Lacerda LM, Franco TT (2008) Biomass production and carbon dioxide fixation by *Aphanothece microscopica* Nägeli in a bubble column photobioreactor. Biochem Eng J 40:27–34
129. Rasoul-Amini S, Montazeri-Najafabady N et al (2011) *Chlorella* sp.: a new strain with highly saturated fatty acids for biodiesel production in bubble-column photobioreactor. Appl Energy 88:3354–3356
130. Fux CA, Costerton JW et al (2005) Survival strategies of infectious biofilms. Trends Microbiol 13:34–40
131. Pesce S, Lissalde S et al (2010) Evaluation of single and joint toxic effects of diuron and its main metabolites on natural phototrophic biofilms using a pollution-induced community tolerance (PICT) approach. Aquat Toxicol 99:492–499

Adv Biochem Eng Biotechnol (2023) 183: 25–64
https://doi.org/10.1007/10_2022_215

Published online: 11 February 2023

Cyanobacterial Bioenergetics in Relation to Cellular Growth and Productivity

Robert L. Burnap

Contents

Abstract Cyanobacteria, the evolutionary originators of oxygenic photosynthesis, have the capability to convert CO_2, water, and minerals into biomass using solar energy. This process is driven by intricate bioenergetic mechanisms that consist of interconnected photosynthetic and respiratory electron transport chains coupled. Over the last few decades, advances in physiochemical analysis, molecular genetics, and structural analysis have enabled us to gain a more comprehensive understanding of cyanobacterial bioenergetics. This includes the molecular understanding of the primary energy conversion mechanisms as well as photoprotective and other dissipative mechanisms that prevent photodamage when the rates of photosynthetic output, primarily in the form of ATP and NADPH, exceed the rates that cellular

This work was supported by the U.S. Department of Energy Basic Energy Sciences; grant no. DE-FG02-08ER15968.

R. L. Burnap (✉)
Department of Microbiology and Molecular Genetics, Oklahoma State University, Stillwater, OK, USA
e-mail: robert.burnap@okstate.edu

assimilatory processes consume these photosynthetic outputs. Despite this progress, there is still much to learn about the systems integration and the regulatory circuits that control expression levels for optimal cellular abundance and activity of the photosynthetic complexes and the cellular components that convert their products into biomass. With an improved understanding of these regulatory principles and mechanisms, it should be possible to optimally modify cyanobacteria for enhanced biotechnological purposes.

Keywords ATP, Anabolic, Catabolic, Chlorophyll fluorescence, Cyanobacteria, Cyclic electron flow, Electron transfer, Light harvesting, Membrane bioenergetics, Metabolic flux analysis, NADPH, Optimal growth theory, Oxygenic photosynthesis, Phosphorylation, Photosystem, Phycobilisomes, Reduction, Sink capacity

Abbreviations

2PG	Initial product of photorespiration due to the oxygenation reaction of RuBP by Rubisco
CBB	Calvin–Bassham–Benson cycle of photosynthetic carbon fixation
C_i	Inorganic carbon, primarily [HCO_3^- + CO_2]
Rubisco	Ribulose bisphosphate carboxylase/oxygenase
RuBP	Ribulose bisphosphate

1 Introduction

The ability to use solar energy to convert CO_2, water, and minerals into biomass is the signature characteristic of oxygenic photosynthesis. Cyanobacteria (or their ancestors), the evolutionary originators of this process, are now the focus of efforts to engineer their metabolism to produce valuable products. The engine driving this metabolism is the cyanobacterial bioenergetic mechanism comprised of the interlaced photosynthetic and respiratory electron transport chains, which cooperate during the diurnal cycle to capture and store solar energy to maintain a "dynamic energy bank" of reducing and phosphorylating power (Fig. 1). Progress toward understanding cyanobacterial bioenergetics has been truly remarkable over the last few decades. This is the consequence of physiochemical analyses, molecular genetics, and structural analyses, which, together, have produced an increasingly detailed picture of the main components of the photosynthetic and respiratory mechanisms in cyanobacteria. Moreover, the photoprotective mechanisms that have evolved to cope with the unique challenges of oxygenic photosynthesis are also increasingly well understood. That is, the components of the photosynthetic electron transport chains are increasingly well understood in terms of their molecular structure and function. Nevertheless, our understanding of the principals and mechanisms that determine the optimum cellular abundance and activity is still rather unknown. What are the features of the regulatory circuits ensuring the expression of optimal cellular levels of the photosynthetic and respiratory complexes, for example? This type of regulatory information along with knowledge of what constitutes an optimal balance of the

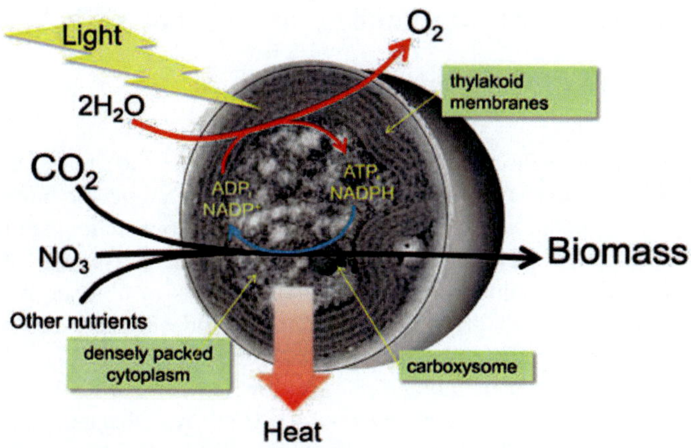

Fig. 1 Cyanobacteria are Gram-negative bacteria that are capable of performing oxygenic photosynthesis. The densely packed cyanobacterial cell contains extensive photosynthetic membranes to produce the reducing and phosphorylating power necessary to drive the assimilation of mineral nutrients to form the complex biomolecules necessary for autotrophic growth and cellular reproduction. As discussed here, the dense packing of the cell interior in terms of both the cytosol and the membranes contributes to the "proteomic" constraint that is argued to be important for evolutionary and engineering optimizations of productivity [1–3]

various bioenergetic complexes will be critical in developing a deeper insight into cyanobacterial bioenergetics. This will allow for modification and therefore harnessing of the biotechnological potential of cyanobacteria.

This chapter will focus on recent developments in cyanobacterial bioenergetics, especially in terms of the structure, function, and regulation of selected photosynthetic and respiratory complexes that form the energy generation systems for driving metabolism. An understanding of these systems is important from the biotechnological context given the development of an increasing array of molecular genetic manipulations to divert photosynthate into valuable products. Overall, these recent studies are showing us that diversion of the products of photosynthesis to desired products typically corresponds to an increase in the sink capacity of the cyanobacterial cell and, remarkably, the natural regulatory mechanisms that govern the expression of the bioenergetic mechanisms results in characteristic responses as the cells adjust to the increased sink capacity. From these energetic-metabolic source-sink relationships, a simplified view of cyanobacterial bioenergetics is described. These intrinsic, yet still poorly understood homeostatic responses will eventually provide insight into the function and regulation of the bioenergetic mechanisms and give clues as to how to better optimize biotechnological engineering designs which aim to maximize the production of desired products. The question of how cyanobacterial bioenergetics relates to overall cellular productivity is handled in two parts: first, we will consider an outline of the basic bioenergetic macromolecular components and mechanisms that have been steadily revealed with increasing detail. Secondly, we will consider how these complexes work as a unit to drive cellular metabolism and achieve persistent, successful growth through changing environments.

2 Energy Production Mechanisms of Cyanobacteria

2.1 Membrane Bioenergetics

At the foundation of cyanobacterial energy production are the processes of photosynthetic and respiratory electron transfer and the associated proton pumping. Like other bacteria, cyanobacteria are surrounded by a cytoplasmic membrane that mediates the uptake of nutrients and the transmembrane movement of ions. To accomplish this, the cytoplasmic membrane (CM) needs to be energized with an electrochemical gradient to drive these exchanges. Indeed, there are specialized respiratory chain complexes embedded in the cytoplasmic membrane contributing to this essential cytoplasmic membrane energization. However, with the exception of the primitive cyanobacterium *Gloeobacter* [4, 5], all cyanobacteria are endowed with an extensive internal membrane system, the thylakoid membranes (TM), where the vast majority of energy production occurs both for photosynthesis and respiration.

2.1.1 Cytoplasmic Membrane (CM) Bioenergetics

Nutritional substrate and ion uptake across the cytoplasmic membrane requires ion gradients of both sodium and proton electrochemical potentials, the orientations of which adhere to the classic "negative inside" rule where the periplasm represents the electrochemically positive (P-side) of the membrane, whereas the cytoplasmic side is electrochemically negative (N-side) (Fig. 2). Sodium and proton gradients dominate the CM energetics and transport situation, although oppositely directed potassium gradients are important especially in regard to osmotic homeostasis. Few studies have attempted to measure the magnitude of the proton gradients [4, 6–9] across the CM and fewer still measured the sodium gradients (ΔNa^+) [7–9]. These studies confirm the mutual exchangeability of ΔpH and $\Delta\Psi$ components of the trans-CM proton energy potential ($\Delta\mu^{H+}$) in conformance with basic chemiosmotic theory. The net proton electrochemical potential across the CM is experimentally estimated using inhibitors and ionophores to correspond to approximately 100–200 mV, depending on internal and external conditions. However, the physiological ΔpH may be considerably smaller due to the efficient conversion of the proton gradient into a sodium gradient through the activity of Na^+/H^+ antiporters (Fig. 2). The Na^+/H^+ antiporters can use the proton gradient to drive sodium extrusion, establishing an inwardly directed Na^+ electrochemical gradient. The Na^+ gradient can, in turn, be exploited by Na^+-coupled symporter proteins also in the CM, such as bicarbonate (HCO_3^-) transporters (e.g., SbtA, BicA). However, the action of other energized

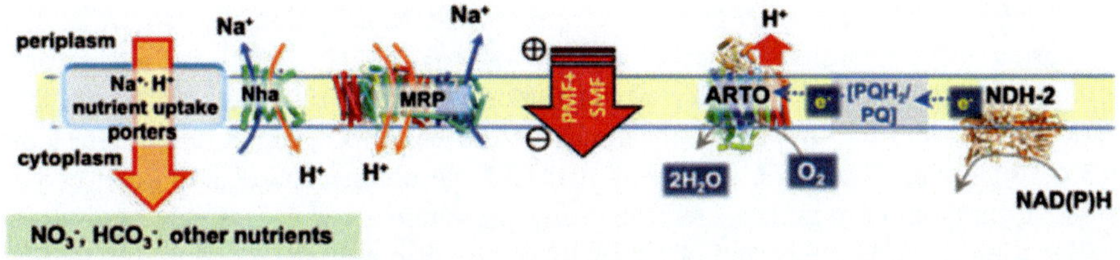

Fig. 2 Overview of the major bioenergetic membrane complexes of the cytoplasmic membrane (CM) The Type-2 NAD(P)H dehydrogenase, NDH-2, and possibly other reductases provide plastoquinol to proton-pumping respiratory terminal oxidases (ARTO), although detailed information on these complexes are lacking and their activities are largely inferred based upon homology with more well-defined complexes. The resultant proton motive force (PMF) can be energetically interchanged through the action of Na^+/H^+ antiporters (Nha) in the Nha and MRP (multi-resistance and pH) families to create a parallel sodium motive force (SMF). Both the PMF and the SMF across the CM can be used for the import of nutrients into the cytoplasm through an array of Na^+- or H^+-linked transporters. For example, inorganic carbon uptake in the form of bicarbonate relies upon the activity of Na^+/HCO_3^- symporters, SbtA and BicA

Na^+ export mechanisms besides antiporters are not excluded and, overall, the exact mechanisms by which proton and sodium gradients are generated remain poorly understood.

How much of the cyanobacterial cell's energy resources are dedicated to ion fluxes and the dependent nutrient transport processes? The few biophysical studies of cyanobacterial ion transport suggest that massive cellular energy expenditures are required to power the uptake and ion homeostasis systems, with the overall estimated cost of ion transport in the light being about 20–30% of the total power available from photosynthesis depending upon the external pH with Na^+ and HCO_3^- fluxes being the dominant factors [7–9]. Again, more biophysical work in these areas is needed, since such information augments and may help with quantitative checks of emerging metabolic flux analyses that model overall metabolism. However, the estimate of 20–30% of the photosynthetic total power generation is not out of line with estimates obtained by other analyses considering bacterial bioenergetics. Those analyses estimate that the main consumer of energy is protein synthesis, which is estimated to consume ~50% of the total energy budget [10] as discussed in detail in Sect. 3.

Somewhat more is known about the molecular biology of the CM based upon studies combining mutational, physiological, and -omics methods. The cyanobacterial CM appears to have a relatively simple electron transport chain contributing to the generation of transmembrane proton gradients. Several complexes with PQ reductase activities, including succinate dehydrogenase (SDH) and Type 2 NAD(P)H dehydrogenase (NDH-2), transfer electrons to the PQH_2/PQ pool using reduced soluble substrates from the cytoplasm. To harvest the redox energy of the PQH_2/PQ pool a proton-pumping alternative terminal respiratory oxidase (ARTO) complex, which is a cytochrome *bo*-type quinol oxidase, directly oxidizes the PQ pool localized only to the cytoplasmic membrane in *Synechocystis* sp. PCC6803 [11, 12]. Importantly, the uptake of external substrates is adequately supported by the proton-pumping activity of ARTO. Nevertheless, respiratory proton-pumping by the TM-located respiratory terminal oxidases (RTO) in the dark also powers the uptake of substrates across the CM, even in the absence of ARTO function. Therefore, it appears that CM membrane energization can occur without participation in the electron transport chain of the CM. One possible explanation for CM energization under these conditions is the operation of F_oF_1-type ATP synthase situated in the CM since recent proteomic analyses suggest significant levels of ATP synthase are found in the CM in *Synechocystis*, although the majority partitions to the TM fractions [12]. In this scenario, it is conceivable that ATP synthases could operate in reverse, consuming ATP produced in the cytoplasm and pumping protons across the CM into the periplasm. This could be one explanation for the apparently energetic coupling between TM energization and CM energization. However, other studies, also with *Synechocystis,* conclude that ATP synthase is located exclusively in the TM [11, 13], thus it remains difficult to make firm conclusions. It is also important to keep in mind potential species-specific differences. In a convincing biochemical study, the alkalophilic cyanobacterium *Aphanothece halophytica* was found to have a Na^+-dependent F_oF_1 ATP synthase which, working as an ATPase in the CM, has an active Na^+ efflux pump driven by the hydrolysis of ATP. This was found to be encoded by a second operon, with high homology to, but distinct from the operon encoding the canonical H^+-dependent F_oF_1-type ATP synthase. Interestingly, such a second operon was bioinformatically identified in Na^+-ATPases from *Synechococcus* sp. PCC 7002, *A. marina* MBIC11017, and *Cyanothece* sp. ATCC 51142. Moreover, it is also possible that certain F_oF_1-type ATP synthase orthologs are promiscuous regarding their Na^+ versus H^+ dependence and, additionally, that its distribution on the CM may be dependent upon nutritional conditions. Nitrogen-starved, chlorotic cells were shown to utilize an RTO-generated trans-CM ΔNa^+ to drive ATP synthesis in *Synechocystis,* but this was not observed in nutrient-replete vegetative cells [14]. In that same study, the requirement for Na^+ in the growth medium was only necessary for HCO_3^- uptake in vegetative cells, in contrast to the absolute requirement in chlorotic cells. The ARTO system can, in principal, drive the formation of a sodium gradient, ΔNa^+, via the action of Na^+/H^+ antiporters; there are at least three NhaS paralogs located in the CM in *Synechocystis* (NhaS2, NhaS4, and NhaS5) [12, 15]. In addition to these comparatively simple antiporters, most cyanobacteria appear to have the more complex, MRP-type (Multi-resistance and pH, reviewed in [16]) antiporter situated in the CM. Besides being important for salt tolerance [17], this antiporter is implicated in driving the formation of a trans-CM Na^+ gradient, because it is part of a regulon controlled by NdhR (CcmR). The regulon is upregulated by inorganic carbon starvation and includes the gene for the Na^+-HCO_3^- symporter SbtA [18]. Key subunits of the MRP are located in the CM and are annotated as NdhD5/6 based upon homology to the antiporter-like ion proton pumping subunits of the NDH-1 complex, but are more likely to form in a different type of complex that either has no redox component or one very different from the NDH-1 complex [19]. Thus, while much has been learned about the bioenergetics of the CM, much remains. Notably, the composition of the CM electron transport pathway leading to ARTO, probable coupling between energy outputs of the TM and the CM that supports transport in the absence of ARTO, and the overall bioenergetics of ion fluxes that will be important to integrate into flux-balance and computational modeling of cyanobacterial metabolism.

2.1.2 Thylakoid Membrane Bioenergetics

The extensive internal TM membranes are the bioenergetic core of the cyanobacterial cell, housing the photosynthetic and respiratory electron transport chains along with ATP synthases and a large variety of less abundant complexes including transporters and selective ion channels (Fig. 3). Through the action of the

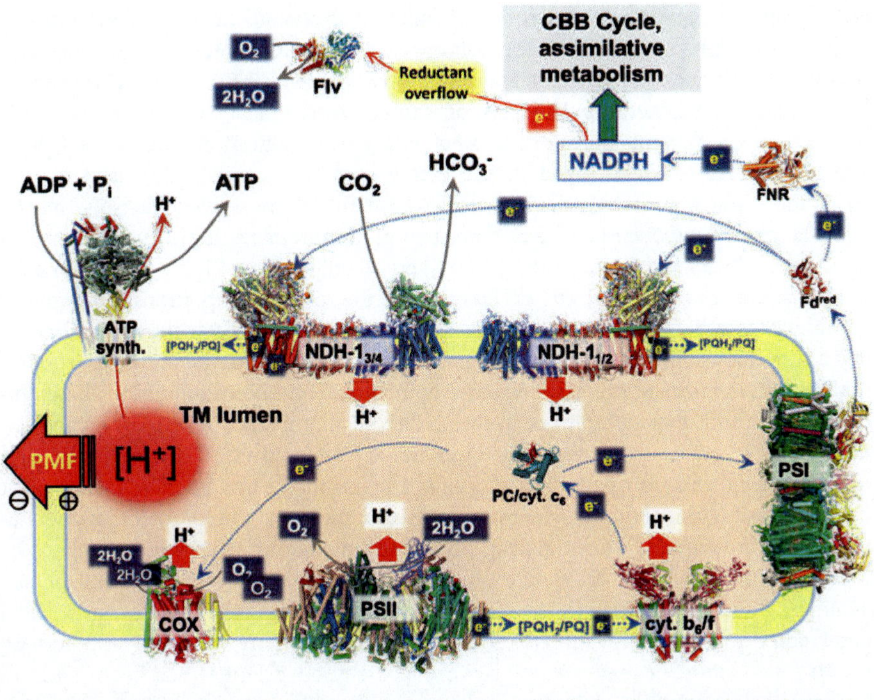

Fig. 3 Overview of the major bioenergetic membrane complexes of the thylakoid membranes (TM). The TMs contain both photosynthetic and respiratory electron transport chains and their proton pumping activities generates a PMF that can be used to synthesize ATP via a shared ATP synthase. Respiratory electron transport and the majority of the photosynthetic cyclic electron flow (CEF) involve the NDH-$1_{1/2}$ isoforms of the NDH-1 complexes containing the NdhF1 and NdhD1/D2 and are supplied high energy electrons from reduced ferredoxin (Fd^{red}). The CO_2-hydration activity involving the NDH-$1_{3/4}$ isoforms of the NDH-1 complexes is also driven by Fd^{red} mediated CEF. These complexes contain the unique CupA/B proteins and the NdhF3/4 and NdhD3/D4 subunits

photosynthetic electron transport chain, the major production of electrochemical gradients for ATP production and the generation of reducing power in the form of reduced ferredoxin (Fd^{red}) and NADPH to drive nutrient assimilation and biosynthetic metabolism is accomplished. The respiratory chain is necessary for the exploitation of energy reserves for metabolism at night for both maintenance metabolism, but also for the biosynthetic processes that occur as the darkened cells prepare for dawn and the next solar period [20, 21]. Moreover, the respiratory complexes of the TM have an essential role in modulating the relative outputs of reducing and phosphorylating power to match metabolic demands as well as to provide reductant overflow "escape valves" to safely dissipate excess reductant, for example, during high and/or fluctuating light conditions [22].

Although subject to considerable variation depending upon the physiological conditions, the area of the CM of the model organism *Synechocystis* is roughly 10 μm^2. In contrast, the area of its TM, again subject to considerable physiological variation is in the range of 50–70 μm^2 [23], and thus is more than fivefold greater. An exception is *Gloeobacter*, which has no internal thylakoid membrane system and maintains the photosynthetic and respiratory electron transport chains in the cytoplasmic membrane [24]. The photosynthetic complexes contribute to ATP production through the generation of a proton motive force and contribute to NADPH production via electron transport starting with the oxidation of water — substrate water by PSII — which provides the electrons that are eventually used for the reduction of ferredoxin by PSI. The thylakoid membranes are very densely packed with the photosynthetic complexes dominating, though intermingled with the respiratory complexes and sundry transporters and channels. This high density of membrane proteins in cyanobacterial thylakoids and the relative proportions of the major complexes is schematically represented in Fig. 4. These would include ion channel and antiporter proteins, not to mention protein complexes involved in the insertion of the integral membrane polypeptides and cofactors. The intermingling of photosynthetic and respiratory complexes in the thylakoid membranes results in a complex network of electron fluxes and functional interactions. This network of interconnected electron pathways accounts for the many unique characteristics of cyanobacterial bioenergetics, ranging from the multiple photoprotective mechanisms that are capable of dissipating potentially damaging excess reductant to the regulatory processes that modulate the relative activities of respiratory and photosynthetic fluxes. Compared to the metabolically streamlined thylakoids of plant and algal chloroplasts, the cyanobacterial thylakoids are more complex and remain poorly understood in many ways. Nevertheless, good progress has been made in defining the foundational features of electron transport, proton-pumping, and membrane energization in cyanobacterial membranes. The high density of membrane complexes has at least two important functional consequences: firstly, it places constraints on the lateral mobility of complexes, although a high degree of mobility is nevertheless observed according to fluorescence recovery after bleaching experiments [26]. The second consequence has function implications relating to optimal growth theory [1–3, 27, 28]. Namely, it is an example where space and proteomic constraints limit expression levels of functional complexes and consequently,

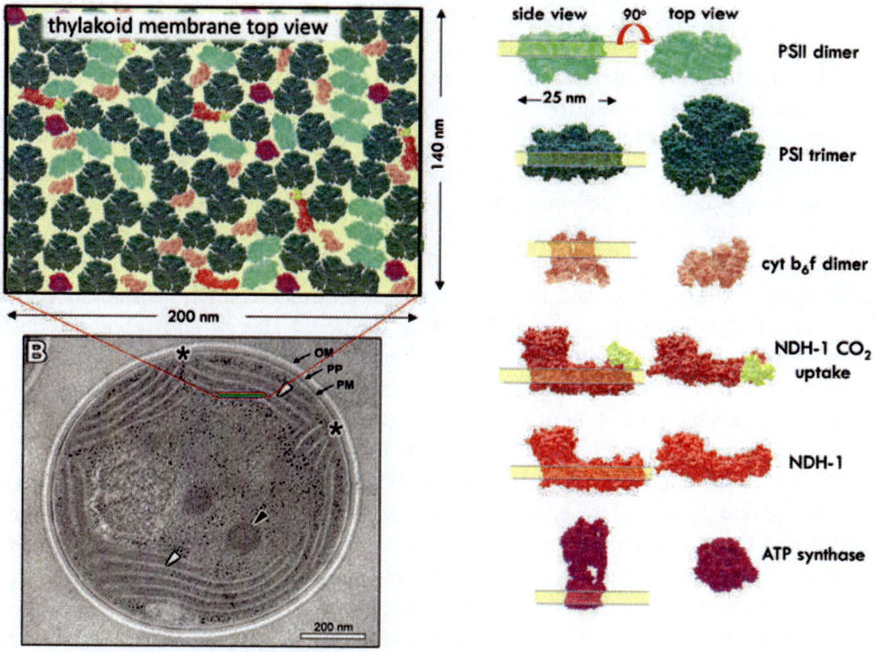

Fig. 4 The main photosynthetic complexes densely pack the thylakoid membrane. Based upon the calculations and figures of [23]. The electron micrograph adapted from [25]

optimal cell growth depends upon the optimal expression levels of the various complexes in the finite membrane space to maximize the outputs such as linear electron flow and minimizing bottlenecks in flow.

2.2 *Photochemical Charge Separation*

The capture and conversion of solar energy into chemical energy is the primary driver of cyanobacterial metabolism. Although some cyanobacterial species have the capability to use organic substrates as energy and carbon source inputs, photosynthesis remains the bioenergetic foundation for all cyanobacteria. In general terms, solar energy capture is mediated by antenna pigments that absorb photons and transfer the excitation energy to the photochemical reaction centers (RCs). Excitation energy is transferred to a "special pair" of chlorophyll (Chl) in the RCs of photosystem II (PSII) and photosystem I (PSI), where the excitation energy induces charge separation. Primary charge separation involves the loss of an electron from the excited electron donor Chl (primary donor, P) with the transfer of the energized electron to an acceptor cofactor within the reaction center. Upon transfer, the energized electron loses some of the original excitation energy, which is the price of stabilizing forward electron flow and decreasing the probability of wasteful back-reactions. In general terms, the oxidized primary donor represents an electron hole that is filled by a secondary electron donor to restore the photochemical reaction center to its original condition ready for the next light absorption and charge separation. Excitation of the primary donor Chl energizes an electron in an outer shell orbital thereby converting the Chl into a strong reductant energetically capable of transferring the energized electron to a nearby acceptor, the primary acceptor. Thus, the excited donor Chl, together with the nearby primary acceptor, convert into a meta-stable electron-hole pair ($\ominus - \oplus$), that becomes stabilized by further migrations of the electron and hole as they move further apart from one another. The separation of opposite charges is a form of stored energy and represents the energy captured from the initially absorbed photon. Migration of the hole away from the donor Chl corresponds to a chain of oxidations of secondary and tertiary electron donors, whereas the migration of the electron away from the donor Chl corresponds to the transfer of secondary and tertiary acceptors. From an energetic perspective, there is usually a trade-off between trapping the charge-separated state by stabilization, on the one hand, and the energetic efficiency when comparing the energy of the original photon and charge-separated state versus the redox potential of the trapped electron, on the other hand. Without sufficient loss of energy during the trapping process there is the tendency to have back-reactions. Such back-reactions are not only wasteful but also increase the risk of reactive oxygen species (ROS) formation associated with charge-recombination. Thus, photosynthetic reaction centers are tuned by evolution to capture and store as much free energy as possible in the photochemical RC yet ensure that minimal backreaction occurs. For PSII and PSI, the oxidations of the RC Chls are detected as absorbance changes at 680 nm and 700 nm, respectively. Their original spectroscopic designations of P_{680} and P_{700} in the original photosystem model of the "Z-scheme" correspond to these absorbances. These spectroscopic changes follow the excitation and for the oxidation processes with in the PSII and PSI RCs corresponding to the energization and escape of electrons from the primary donor chlorophylls with transfer to an electron acceptor that stabilizes these energized electrons. The electron transfer is vectorial since the primary electron donor is on the luminal or periplasmic side of the membrane and the primary acceptor is on the cytoplasmic side of the membrane. This contributes to and interacts with the transmembrane electrical potential, $\Delta\psi$. Meanwhile, the captured and energized electron on the acceptor side of the membrane is transferred to secondary electron acceptors to allow the electron to be utilized further, ultimately as a reductant for anabolic metabolism.

2.2.1 Light-Harvesting

Phycobilisomes

Phycobilisomes (PBS) are huge, extrinsic pigment–protein complexes attached to the cytoplasmic surface of the thylakoid membrane through protein–protein

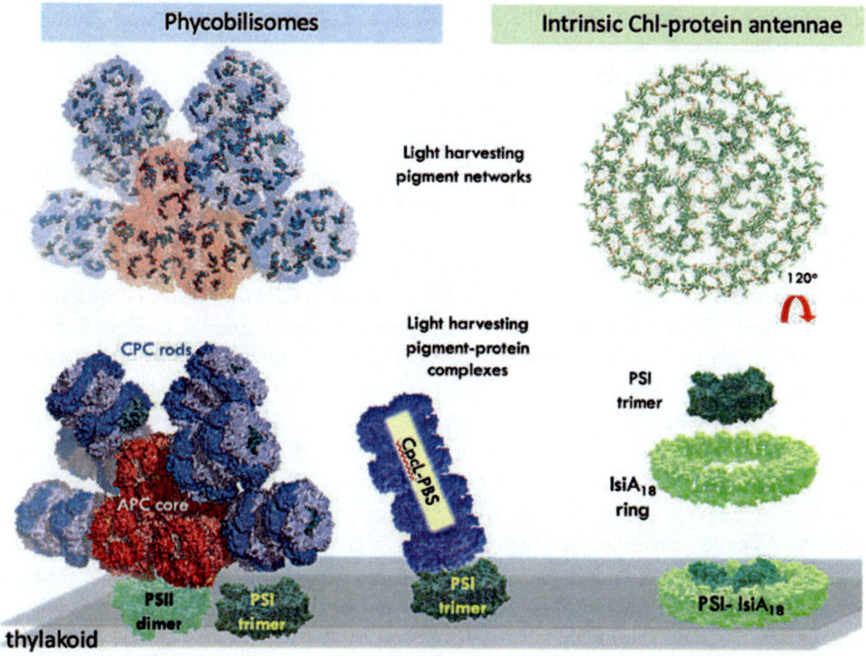

Fig. 5 Light-harvesting antenna in cyanobacteria. Phycobilisomes (PBS) are huge (5–20 MDa) extrinsic pigment–protein complexes attached to the cytoplasmic surface of the thylakoid membrane through protein–protein interactions with the photosystems. The protein matrix organizes phycocyanobilin (PCB) pigments into excitonically coupled networks that maximize energy transfer to the photosynthetic PSII and PSI reaction centers. Intrinsic Chl-protein light-harvesting complexes are exemplified by iron-stress induced intrinsic Chl membrane light-harvesting protein, IsiA, which surrounds and excitonically serves PSI. Structurally similar Chl a/b proteins are the primary antenna complexes of the highly abundant marine cyanobacteria, the *Prochlorophytes* and serve both PSI and PSII (see text for details)

interactions with the photosystems [29, 30]. PBSs absorb wavelengths of light in the spectral range between 450 and 650 nm, which is complementary to that of Chl a within PSI and PSII. Therefore, utilization of PBS for light harvesting greatly increases the usable solar energy spectrum. PBSs have a molecular mass in the range of 5–20 MDa and are situated on the cytoplasmic side of the membranes, although details of the exact connections with the photosynthetic reaction centers that they serve remain to be elucidated. The size and overall architecture of the PBS varies among species and can often be adjusted to match the light intensity and, for some species, the spectral quality of the light. The PBSs are organized as cylindrical rods-like structures with a core typically composed as a tricylindrical core, which is physically and excitonically coupled to cylindrical rods pigment–protein complexes that extend from the central tricylindrical core (Fig. 5). The cylindrical rods are comprised, in stack-like structures of adjoining disks that are themselves comprised of intensely pigmented phycobiliproteins and non-pigmented linker polypeptides that organize and bind the phycobiliproteins into the rod-like configurations. Although the overall structure-function characteristics have been known for many years and crystal structures for the individual disk-like phycobiliproteins have been resolved, only recently and with the advent of cryo-EM techniques, have high-resolution structures of the entire intact structure become available [31, 32].

The PBSs of *Synechococcus* 7002 and *Anabaena* 7120, which have hemidiscoidal architectures and share a common triangular core structure like other model cyanobacteria, was resolved using cryo-EM techniques [32]. The protein scaffolding of the PBSs organizes 288 phycocyanobilin (PCB) pigments in the case of *Synechococcus* 7002 PBS with 72 PCB organized within the triangular core. For *Anabaena* 7120, there were 348 PCB observed overall and 96 PCB in the core. Unlike the Chls of photosynthetic reaction centers and intrinsic Chl light-harvesting antenna, which are noncovalently organized within the protein matrix, PCBs are covalently attached via thioether linkages to cysteine sides chains of individual phycobiliproteins. The PCBs are arranged in networks that appear to maximize energy transfer through the pigment array and toward the core where the energy is funneled to the photosynthetic PSII and PSI reaction centers. The protein scaffolding provides an excellent example of spectral energy tuning where it is observed that the same basic PCB has specific spectral characteristics modulated by the protein environment. The light energy-harvesting and exciton transfer characteristics of the pigment–protein ensemble is made function by the variety of spectral characteristics of the individual bilin pigments and the protein environments that further tune these spectral characteristics. Generally, shorter wavelength, more energetic photons are absorbed in the more peripheral regions of the PBS and the excitation energy is transferred via a Förster resonance mechanism to progressively lower energy ("redder") bilin pigments within the antenna pigment array, which occurs within 1 ns of the initial light absorption event [33]. This results in a stochastic, yet biased diffusion of excitons down an energy gradient and then to the Chls of the photosystems where it can be trapped by the photochemical reaction center to initiate charge-separation. The central cylindrical core made of allophycocyanin (APC) joins with several peripheral rods containing phycocyanin (PC). In some species, the PC peripheral rods are further supplemented with phycoerythrocyanin (PE) on sections of the rod that are more distal from the APC core. Physically and excitonically coupled to the central trimeric core and project away from it. Excitation energy in the peripheral rods is transferred to PSI and PSII, with the PBS core APC acting as the exitonic link between PC and the Chl a in the RCs.

There is a second, more recently discovered form of the phycobilisome that consists exclusively of the phycocyanin rod structures (Fig. 5). Instead of funneling excitation energy primarily to photosystem two, these rod structures serve PSI [34]. They are linked to trimeric PSI and connect to the membrane location via a special link or polypeptide designated CpcL, (formerly CpcG2). Despite progress on the overall PBS structure, the physical connections between PBS and the RCs remain largely unresolved, although reasonable models based on analytical chemical cross-linking information have been proposed [32]. These interactions between the PBS and the RC are inherently weak, which likely relates to the functional requirement for the capacity of rapid and flexible regulation of the excitation energy transfer to the RCs under fluctuating light conditions. This regulation is multifold and includes most prominently the still poorly understood mechanism(s) of 'State I to State II' transitions which modulate the extent to which PBS excitation energy is channeled to either PSII or PSI. Overall, PBSs tend to preferentially excite PSII over PSI excitation. However, under certain conditions, a significant fraction of the excitation can be transferred to PSI in a regulatory process that occurs in the seconds time domain and is determined by the redox state of the PQ pool. In the dark, the PQ pool tends to be relatively reduced owing to respiratory electron transfer stemming from the generation of NADPH, for example from the catabolism of glycogen reserves. Under these conditions, a considerable fraction of the excitation energy is directed more to the PSI complex, which is thereby in "State I." Upon illumination, the PQ pool becomes oxidized triggering a regulatory process that results in increased excitation of PSII, corresponding to "State II," which exhibits increases in fluorescence yield since PSII is inherently more fluorescent than PSI. This is consistent with the observation that mutants in the NDH-1 complex which cannot transfer metabolic reductant to the PQ pool, which is thus more oxidized in the dark, seem to be "locked" in the high fluorescence State II in dark-adapted cells [35, 36]. While these basic features of the phenomenon are generally agreed upon, the underlying mechanism has proved difficult to resolve unequivocally and appears fundamentally different from that found in plants and algae, which have phosphorylation mechanisms that rearrange the associations between the intrinsic light-harvesting chlorophyll antenna proteins and the reaction centers [37]. A number of models ranging from mobile PBS units that shift positions between PSII and PSI, modulated spillover of excitation energy from PSII to PSI, to differential quenching processes in PSII remain under investigation.

Chlorophyll Antenna

Although PBSs are the major light-harvesting antenna of most cyanobacteria, the high abundant and ecologically significant marine cyanobacteria, *Prochlorococcus* sp., and sundry other cyanobacteria use intrinsic Chl a/b membrane light-harvesting proteins, designated prochlorophyte Chl a/b protein (pcb), which replace the classical cyanobacterial phycobiliproteins as the major antenna. Although these proteins bind both Chl a and Chl b, they are unrelated to the large LHC proteins of plants and algae [38]. Similar to the iron-stress induced intrinsic Chl membrane light-harvesting protein, IsiA [39], these proteins have six transmembrane helical segments coordinating ~12-16 Chl molecules and are homologous to the PsbC (a.k.a., CP43) protein of PSII [40]. The IsiA proteins surround PSI as shown in Fig. 5 thereby increasing its optical cross-section under stress conditions where PSI levels are reduced. The psb proteins surround and excitonically serve both PSI and PSII [41]. As with the structure of better resolved IsiA proteins [42, 43], the PCB appear to form partially or encircling arrays surrounding PSI and PSII in *Prochlorococcus* and the related *Prochlorothrix* [41, 44, 45]. Such antenna rings around PSI increase the number of light-harvesting antenna Chl to nearly 300, which brings it into the optical cross-sectional range of the PBS.

It is also important to recognize that Chl has crucial roles in light-harvesting even in PBS-containing cyanobacteria that lack Pcbs. This is because the PSII and especially for PSI complexes have so-called proximal light-harvesting antennae. These are the Chl molecules surrounding the RC Chl and provide both light-harvesting capability as well as serving as a bridge from the more peripheral antenna. The protein matrix of PSII coordinates 36 Chl, while that of PSI has 96 Chl [46]. The large number of Chl in PSI means that it provides a significant light-harvesting capacity. Furthermore, the large number of Chl in PSI combined with the facts that PSI is highly abundant in the cyanobacterial TM and the stoichiometry of PSI/PSII typically ranges from 2 to 4 depending upon the light conditions and growth stage means that PSI has a relatively large optical cross-section even without energy transfer from the PBS [23, 47, 48].

2.2.2 Photosystem II

What makes oxygenic photosynthesis so unique is that PSII utilizes water as a source of electrons to re-reduce the oxidized primary donor (Fig. 6). Thus, the remarkable evolution of oxygenic photosynthesis with its capability of oxidizing water gave rise to the oxygen-containing atmosphere of planet earth and a limitless supply of electrons that are ultimately used to reduce inorganic substrates, such a CO_2 and

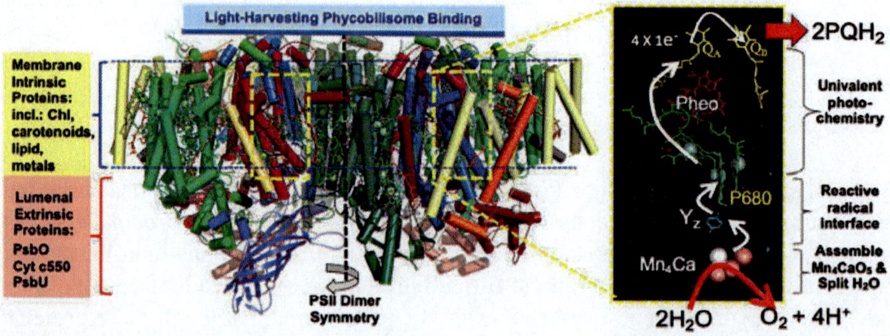

Fig. 6 Photosystem II. The 700-kDa homodimeric structure consists of over 26 different polypeptide chains organizing the cofactors involved in primary charge separation, the proximal antenna, the carotenes indispensable for photoprotection, and the metals, including the Mn_4O_5Ca [49]. The homologous D1 (PsbA) and D2 (PsbD) polypeptides each have 5 transmembrane helical segments and form the heterodimeric core. Primary charge separation (left panel) occurs within the D1/D2 heterodimer as the photooxidation of special pair Chl, P_{680}, producing the radical pair consisting of P_{680}^+ and the reduced pheophytin, $Pheo^-$. P_{680}^+ a powerful oxidant ($E_m = +1{,}100\text{–}1{,}200$) capable of extracting the tightly bound electrons of substrate H_2O via the H_2O-oxidation mechanism (see text) and producing plastoquinol that conveys the energized electron to the remainder of the electron transport chain and releasing protons on the luminal side of the TM contributing to the PMF and ATP synthesis

NO_3, into organic molecules, such sugars and amino acids that serve as the building blocks of life.

The cyanobacterial PSII complex is a 700-kDa homodimeric structure consisting of over 26 different polypeptide chains organizing the cofactors involved in primary charge separation, the proximal antenna, the carotenes indispensable for photoprotection, and the metals, including the Mn_4O_5Ca [49]. The homologous D1 (PsbA) and D2 (PsbD) polypeptides each have 5 transmembrane helical segments and form the heterodimeric core. They are largely responsible for coordinating the primary charge separation and H_2O-oxidation components, including 4 Chl, 2 Pheo, Q_A, and the Mn_4O_5Ca. The CP43 (PsbC) and CP47 (PsbB) polypeptides serve primarily as proximal antenna with each binding 13 or16 Chl, respectively. They are also the conduits of excitation energy from peripheral antenna like the PBS. There are three peripheral, hydrophilic subunits, PsbO, PsbU, and PsbV, which are situated in the lumenal side of the TM. These form a cap over the Mn_4O_5Ca and contribute to the isolation of reactive components of the H_2O-oxidation catalyst.

What is it about the mechanism of PSII H_2O-oxidation that is so bioenergetically remarkable? The evolution of PSII required the solution of two fundamental physiochemical problems, one kinetic and the other thermodynamic. The kinetic problem is that the photochemical RC works one photon and electron at a time, whereas the oxidation of water is a four-electron, four-proton process. To mechanistically solve this, the PSII H_2O-oxidation complex (WOC) functions as an oxidant accumulator during its overall catalytic cycle. This is because of a fundamental valence mismatch: the photochemistry of reaction center charge separation is univalent, H_2O oxidation is tetravalent since four electrons are required for the complete oxidation of two substrate H_2O molecules ($2H_2O$ + light energy$\rightarrow$ $O_2 + 4e^- + 4H^+$). That is, four electrons are removed from the WOC in a series of four photochemical turnovers of the PSII RC. PSII solves the multiple-oxidant-storage problem utilizing the multinuclear manganese cluster (Mn_4O_5Ca). This cluster acts as an oxidant storage device that splits two molecules of water after four flashes of light as envisioned by Bessel Kok et al. [50] based upon the experimental observations of Pierre Joliot [51]. The successive photooxidations of the photochemical RC and communication to the WOC result in the cyclic formation of a series of progressively more oxidized 'storage states' or S-states, S_0, S_1, S_2, S_3, and [S_4], with the [S_4] representing the kinetically fleeting intermediate corresponding to the formation and release of dioxygen [52]. In the process, four protons are ejected to the P-side of the membrane, contributing to the proton motive force and molecular oxygen as the by-product. The release of protons is intricately connected to the H_2O-oxidation chemistry since the kinetic mechanism is crucial in solving the thermodynamic problem.

The thermodynamic problem of H_2O-oxidation is that it requires a powerful oxidant, one that is energetically capable of removing the very tightly bound electrons of the substrate water molecules (E_m = +860 mV). The mechanistic solution was the evolutionary modification in the spatial organization and protein environment of the dimeric special pair Chl that forms the primary donor to make its oxidation energetically more intensive. Although the evolutionary pathway leading to the modern PSII reaction center is still a matter of debate [53, 54], its result is not: photooxidation of the PSII primary donor, P_{680}, produces the most powerful biological oxidant, P_{680}^+ ($E_m = +1{,}100$–$1{,}200$ mV), which is sufficient for the oxidation of water. The photoexcitation of P_{680} forms the moderately strong reductant, P_{680}^* ($E_m \approx -650$ mV) sufficient for reducing the primary electron acceptor, a pheophytin ($Pheo^-$/Pheo, $E_m = -536$ mV), which occurs within 30 ps of photoexcitation. The energized electron of $Pheo^-$ is then transferred to the bound plastoquinone, Q_A (Q_A^-/Q, $E_m = -536$ mV) within 200 ps, leading to the formation of the radical pair P_{680}^+-Q_A^-. Once formed, P_{680}^+ extracts an electron from the H_2O-oxidizing manganese cluster (Mn_4O_5Ca) via a redox-active tyrosine side chain, designated Y_z. The oxidation of Y_z restores P_{680}^+ to its initial ground state condition as P_{680} and has some interesting characteristics because the proper functioning of Y_z depends upon its close H-bonding interaction with an adjacent histidine, His190. The phenoxy -OH group of Y_z shares its proton with the imidazole nitrogen of His190 as they engage in a strong, low-barrier H-bond. The consequence of this H-bond is that the oxidation of Y_z results in the neutral radical species of the phenoxy side chain, yet instead of being lost to the surroundings, the proton "rocks" to the imidazole N of His190, thereby keeping the proton. Hence, a positive charge at that location is maintained until the $Y_z^{\cdot}$ radical is re-reduced H_2O-oxidation catalyst. For notational convenience, the oxidized Y_z is referred to as Y_z^+ with the tacit knowledge that the charge is held by the His190 imidazole.

Retention of the charge at that locality is thought to be functionally important since it facilitates the removal of other catalytically strategic protons in the vicinity. Importantly, the transfer of the electron from Y_z to P_{680}^+ is fast, but the product is relatively stable. The charge-separated state Y_z^+-P_{680}-Q_A^- forms within a couple of hundred nanoseconds, yet is stable for milliseconds. This is crucial since the subsequent H_2O-oxidation chemistry involving the Mn_4O_5Ca occurs in the µsec-msec time domain. Thus, if the energy of charge-separated state is to be exploited for H_2O-oxidation, those reactions must occur with the timeframe of the lifetime of Y_z^+-P_{680}-Q_A^- lest this state is lost to charge recombination with the energy lost primarily as heat. Importantly, proton release also accompanies the photochemical extraction of electrons from the WOC, but this is accomplished in a somewhat surprising way. It is likely that the transient positive charge at His190 formed upon the oxidation of Y_z produces an electrostatic impetus to drive proton release from the metalloprotein structure of the Mn_4O_5Ca and its immediate environment. The expulsion of a proton from the immediate vicinity of the Mn_4O_5Ca appears to provide a chemical base to receive and trap protons derived from the deprotonation of the substrate water molecules being oxidized by the Mn_4O_5Ca at catalytic intermediates of the decomposition of the substrate waters coordinated to the metals. For example, recent evidence suggests that the removal of a critical proton on the substrate waters likely triggers the extraction of the remaining electron and the concomitant formation of the dioxygen bond. Overall, deprotonation of the WOC during catalysis has been long recognized to accompany S-state advancement, and protons released are transferred from the vicinity of the Mn_4O_5Ca and its bound substrate waters through an H-bonding network leading from the WOC to the luminal side of the TM. The sequence of deprotonations occurs in a stoichiometry of 1, 0, 1, 2 proton(s) released as the WOC advances through $S_0 \rightarrow S_1$, $S_1 \rightarrow S_2$, $S_2 \rightarrow S_3$, and $S_3 \rightarrow [S_4] \rightarrow S_0$ catalytic transitions, respectively. Functionally, if protons were not coordinately released in sync with the electron transfer steps, then positive charge would build with each electron removal making the removal of the subsequent electron in the four-step process energetically more difficult for basic electrostatic reasons. Thus, a critical part of the WOC architecture is strategically located amino acids and buried water molecules that mediate the release of protons. The Y_Z moiety, the Mn_4O_5Ca cluster, and all the amino acid side chains coordinating the metals of the cluster are buried within the protein environment of the WOC. This sequestered region is formed at the interface of intrinsic and extrinsic polypeptide subunits, and despite its buried location within the protein matrix, access channels are observed to facilitate the exit of protons and O_2 and the provision of substrate H_2O. This sequestration of the metalloprotein active site optimizes the H_2O-oxidation reaction and perhaps prevents non-productive side-reactions with the catalytic intermediates of H_2O-oxidation. The electron produced upon oxidation of the primary donor Chl P_{680} is transferred to the pheophytin and then initially stabilized on the bound plastoquinone, Q_A, which accepts only one electron at a time, before transferring the electron to an exchangeable plastoquinone at the Q_B binding pocket within the PSII protein matrix. The Q_B binding pocket is located near the cytoplasmic surface of the thylakoid membrane, yet is still buried to allow exchange with a pool of free PQ that is dissolved and diffusing in the lipophilic membrane bilayer. The bound PQ with the Q_B binding pocket acts as a two-electron gate in the sense that remains firmly bound until it receives two electrons, in series, from the Q_A^-, thereby producing the doubly reduced PQH_2 which can diffuse away from the Q_B binding pocket and transfer electrons to the remainder of the electron transfer pathway.

Overall, the bioenergetics are remarkable. As a plastoquinone-water oxidoreductase PSII is able to use four quantal of light energy to split water ($E_m = +860$ mV) and reduce two plastoquinones to plastoquinol ($E_m = +80$ mV) for a redox span of 80 mV. Assuming the absorption of each of the four photons captured by the PBS is at a wavelength of 620 nm, we have an energy of 193 kJ mol^{-1} photon, that is 771 kJ mol^{-1} of light energy is used to split $2 \cdot H_2O$. The amount of energy captured in the production of PQH_2 is approximately 300 kJ mol^{-1}, plus the significant fraction that is conserved as the four protons are released to the positive side of the membrane in the TM lumen, which amounts to a total of ~80 kJ giving a net conservation of 380 kJ mol^{-1} per catalytic cycle of PSII.

2.2.3 PSI and the Production of a Strong Reductant

PSI is a light-driven plastocyanin-ferredoxin oxidoreductase. It thus serves the crucial function of producing reduced ferredoxin (Fd), the central redox energy hub that connects the light reactions of photosynthesis to the rest of cyanobacterial metabolism. Whereas the remarkable ability of PSII is its capability of generating nature's most powerful oxidant which is necessary to split water, PSI is able to generate nature's most powerful reductant, strong enough to reduce ferredoxin. As discussed below, there are multiple forms of Fd in cyanobacteria and these are observed to have a range of strongly reducing midpoint values (E_m −300-450 mV), that drive reductive nutrient assimilation, anabolic metabolism, and ROS clearance [55–58]. To satisfy this crucial function, PSI is capable of generating the strongly reducing molecular species necessary to transfer electrons, either directly or indirectly, to each of these various cyanobacterial Fds. Indeed, the excited form of the primary electron donor, the special Chl pair designated P_{700}, is the strongest known biological reductant (P_{700}*/P_{700}, $E_m' \approx -1{,}300$ mV) [59]. As shown in Fig. 7, light-induced photochemical charge separation involves the initial transfer from P_{700} through accessory Chls to the primary acceptor, a Chl designated A_0 (A_0^-/A_0, $E_m' \approx -900$–1,000 mV) to bound phylloquinone in a remarkably fast <1 ps and then to the bound phylloquinone cofactor, A_1 within ~50 ps corresponding to the formation of the radical pair state, P_{700}^+-A_1^-. Reduction of the first iron–sulfur cluster (FX), followed by FeS cluster (FA/FB) within ~200 ns and ~500 ns, respectively.

The structure of PSI complex contains nine different subunits, organized around a central core consisting of a heterodimer of two large (each ~85 kDa) homologous polypeptides (PsaA and PsaB) with 11 transmembrane α-helices (Fig. 7). This heterodimeric core contains most of the cofactors involved in the initial charge-separation process and the majority of the Chl involved in light-harvesting. Interestingly, the five c-terminal transmembrane α-helices form a heterodimeric structure that binds the electron transfer cofactors in a manner homologous to all known Type I and Type II RCs, structurally resembling the five-helix L/M and D1/D2 heterodimers of the Type II RCs. Moreover, the six transmembrane α-helices of the N-terminal portions of PsaA and PsaB adopt a similar 3D configuration as six α-helices to the light-harvesting proteins CP43 and CP47 of PSII. The PSI complex is highly abundant in the cyanobacterial TM, and recent studies show that it exists in a variety of oligomeric forms depending upon the cyanobacterial species. Moreover, the state of oligomerization can vary as a function of physiological state. Besides the originally described trimeric form of PSI, it can exist in monomeric form under stress conditions or as tetramers in some species [60–62]. Besides being an assembly intermediate in the formation of higher states of oligomerization, the abundance of PSI monomeric state appears to be regulated in response to environmental conditions [63]. For example, extreme light stress favors the monomerization of the complex [24]. The PSI trimer appears to harvest light more efficiently under low-light conditions, whereas the tetrameric form is observed in strains adapted to high-light conditions [64].

Although cyanobacteria encode and produce several paralogous ferredoxins with different midpoint potentials [55–58], the main Fd, Fd1, that intercepts electron from the acceptor side of PSI has a midpoint potential, E_m − 370 mV. Ferredoxin serves as central redox carrier in a variety of metabolic reactions that require a strong reductant. In photosynthesis, it is used to generate NADPH for CO_2 fixation via the Calvin–Benson–Basham (CBB) cycle, as well as the many other anabolic reactions the use reduced Fd (Fd^{red}) directly or require the reduced pyridine nucleotide.

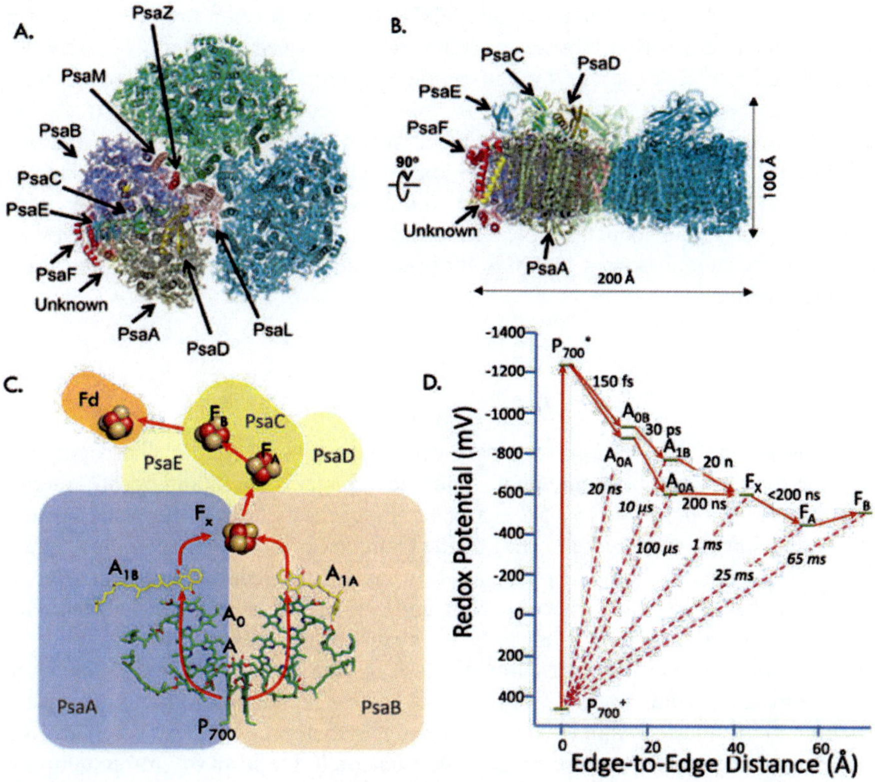

Fig. 7 Structure and electron transfer in the photosystem I complex. The trimeric form of PSI viewed looking into the plane of the thylakoid membrane (Panel **a**). Most of the pigments and redox cofactors are coordinated by the large homologous PsaA and PsaB polypeptides forming heterodimeric PSI RC core (Panels **a** and **b**). Unlike PSII which favors charge separation on only one side of the heterodimer, the electron transfer occurs up both branches of cofactors coordinated by the PsaA/B heterodimer. As a Type I reaction center, the secondary electron acceptors are the Fe-S clusters, which in PSI are designated F_X, F_A, and F_B (Panel **c**). Excitation of the primary donor Chl, P700, produces P700* which is a very strong reductant ($E_m \sim -1,200$ mV), transferring its electron with some energy loss to the primary acceptors, $A1_A$ and $A1_B$ that are phyloquinones, which have strongly negative midpoint potentials capable of reducing the bound PSI Fe-S centers and then to ferredoxin ($E_m \sim -375$ to -450 mV) (Panel **d**)

Reduced Fd^{red} drives this reductive metabolism by directly acting as a redox cofactor in enzymatic reactions or indirectly in the production of NADPH. Fd^{red} efficiently transfers its electron to $NADP^+$ via ferredoxin-NADP reductase (FNR). This transfer provides another example of valence mismatch rectification since Fd is univalent; it carries a single energized electron, whereas $NADP^+$ acts as a recipient of two such electrons. The valence mismatch is solved by the FNR cofactor, flavin adenine dinucleotide (FAD), which can exist in quinone, semi-quinone, or dihydroquinone forms. Accordingly, FNR accepts single electrons but can transfer two accumulated electrons to $NADP^+$. As discussed later, FNR also plays a central role in mediating bidirectional equilibration of low potential electrons between the pools of NADPH/$NADP^+$ and Fd^{red}/Fd. Clearly, ferredoxin plays a central role in overall metabolism by driving the reductive assimilation of inorganic substrates; however, it also is central to the regulation of autotrophic metabolism. Ferredoxin exerts its regulatory function by interacting with the thiol redox system via the transfer of electrons to thioredoxin via ferredoxin-thioredoxin reductase, which in conjunction with NADPH-thioredoxin reductase (NTR) modulates an array of regulatory processes involving the modulation of disulfide bridges in key metabolic enzymes and transcription factors.

2.2.4 Cytochrome b_6/f

The cytochrome b_6f (cyt b_6f) complex occupies a pivotal role in photosynthetic electron transfer in all oxygenic photosynthetic organisms, including cyanobacteria. It oxidizes PQH_2 produced by PSII H_2O-oxidation activity or from electrons derived from the oxidation of ferredoxin (Fd) by the Complex I-like NDH-1 complex. Thus, cyt b_6f is essential for both linear electron flow and cyclic electron flow. Moreover, it has a key regulatory role which is better understood in plants and algae, but also seems to operate, at least in part, in cyanobacteria. Enzymatically it is a plastoquinol-plastocyanin (cytochrome c_{553}) oxidoreductase. Net electron flow occurs by the oxidation of the PQH_2/PQ pool and the reduction of a soluble electron carrier proteins plastocyanin (PC) (alternatively, cytochrome c_{553} (cyt c_{553})) located in the thylakoid lumen. Energy conservation occurs in the form of proton pumping with the unusual stoichiometry of $2H^+$ transferred from the N- to P-side of the membrane per electron transferred between PQH_2 and PC (or cyt c_{553}). Like closely related cyt bc_1 protein in mitochondria and bacteria, cyt b_6f electron transfer within the complex operates by the so-called quinol (Q) cycle, which involves the oxidation of PQH_2 and the reduction of both PC and PQ at two distinct sites via electron bifurcation. The initial oxidation of PQH_2 involves the removal of both electrons PQH_2 at the Q_P catalytic site, which is located near the p-side of the membranes. This two-electron oxidation produces PQ and deposits the two protons on the p-side of the membrane contributing to Δp. The electron bifurcation reflects a mechanism whereby one electron from the two-electron oxidation of PQH_2 is irreversibly drawn to a nearby high potential Reiske Fe-S cluster, whereas the second electron is transferred internally within the complex to be recycled, seemingly counterproductively, back to a different PQ at a second PQ binding site, the Q_n site, located toward opposite N-side of the membrane. Meanwhile, the first electron transferred from PQH_2 at the Q_P site to the Reiske iron sulfur protein is further shuttled to the high potential cytochrome f heme where it can be released from the cyt b_6f complex to PC and on to PSI. It takes two cycles of oxidation of two PQH_2 molecules at the Q_P catalytic site to complete the reduction of a PQ molecule at the Q_n site along with the corresponding uptake of two protons from the N-side of the membrane to complete the overall Q-cycle of the cyt b_6f.

2.2.5 Type-1 Complexes

The cyanobacterial NDH-1, and the closely related chloroplast NDH-1 complexes, are members of the large family of Type-1 complexes (also known as Complex I), which include the familiar mitochondrial and bacterial NADH-dehydrogenases [65–70]. The cyanobacterial NDH-1 complexes are crucial for respiratory electron flow in the dark but are also important for CEF in the light. The Type I family includes the canonical mitochondrial NADH-ubiquinone oxidoreductase, known as Complex I, and its relatives in respiratory bacteria. Being the most biochemically recalcitrant complex of the respiratory chain, Complex I was the last of the respiratory membrane complexes to be crystallized. The 536-kDa bacterial respiratory complex of *Thermus thermophilus* was solved up to 3.3 Å in (Fig. 8) [71]. With the advent of high-resolution cryo-electron microscopy (cryo-EM) there has been a veritable renaissance in the structural analysis of Type-1 complexes [65, 68, 73–81]. Early genomic sequencing of cyanobacterial model organisms showed that they encode multiple paralogous copies of the *ndhD* and *ndhF* genes (*ndhD1,2,3,4,5* and *ndhF1,3,4*). This surprising abundance of multiple paralogs is now shown to be a common feature of most cyanobacteria relating to a functional diversification of NDH-1 complexes in cyanobacteria [82–85]. These variant complexes are built up in an interesting modular fashion [86–88]. Studies of high CO_2-requiring mutants revealed a new class of genes encoding proteins is essential for CO_2-uptake. These proteins, alternatively designated either Cup (CO₂ uptake) or Chp (CO_2-hydration protein), exist as two paralogous forms in most cyanobacteria: a constitutive low-affinity high flux (CupB, ChpX) and an inducible high affinity, low flux

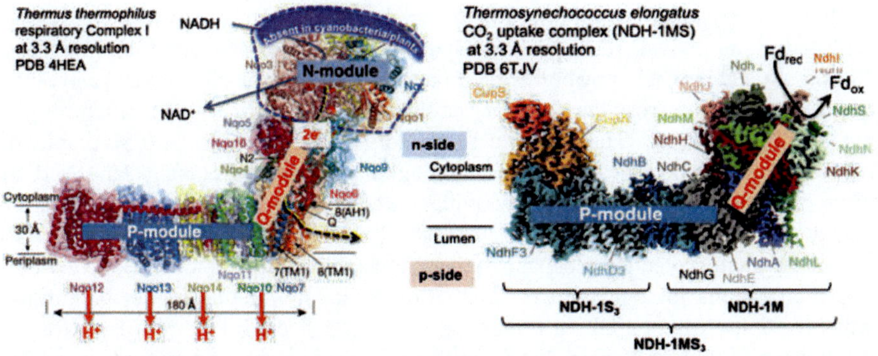

Fig. 8 Structures of the respiratory NDH-1 (Complex I) of *Thermus thermophiles* [71, 72] obtained by X-ray crystallography (left) in comparison to cyanobacterial NDH-1MS$_3$ CO_2-hydration complex (right) from *Thermosynechococcus elongatus* obtained by cryo-EM .Cyanobacteria lack the hydrophilic domain (N-module) responsible for NADH oxidation in bacteria and mitochondria (dotted blue line) but have alternative proteins involved in ferredoxin oxidation including NdhS. As noted elsewhere, some portions of the cyanobacterial complex (NDH-1S$_x$) exist in multiple forms giving different structures and functions, but all share the NDH-1M core (left panel adapted from [71, 72], right modified from [73])

(CupA, ChpY) form. These proteins are highly conserved among the cyanobacteria, yet are found nowhere else and represent a unique Pfam family PF10216 [89]. Previous studies demonstrated the NDH-1$S_{3/4}$ modules, catalyze CO_2-hydration [83–85, 90, 91]. Together the CO_2-uptake forms and the canonical respiratory forms of the NDH-1 complexes are responsible for the majority of CEF in cyanobacteria under typical conditions [35, 92, 93]. Nevertheless, contributions from other quinol reductases such as SDH also may be important for respiratory electron flow [94, 95].

2.3 *Cyclic Electron Transfer, Energy Balancing, and Photoprotection*

The juxtaposition of photosynthetic and respiratory electron complexes not only allows them to share the same set of ATP synthases to generate phosphorylating power both day and night but also leads to an intertwined and regulated network of electron pathways that is exploited for metabolic adjustments and photoprotection. Cyanobacteria accumulate energy stores in the form of glycogen during the day and then mobilize these reserves at night for respiratory metabolism that enables the myriad functions occurring during the dark phase. These range from the pre-dawn anticipatory biosynthesis of reaction center proteins to the detoxification of ROS accumulated during the preceding light phase. Each of the canonical respiratory electron transport complexes, Complexes I-IV, corresponding to the NDH-1, succinate dehydrogenase, cytochrome b_6f, and cytochrome *c* oxidase, are found in the TM (Fig. 2). Of these, the NDH-1 complex and cytochrome b_6f are especially crucial for cyanobacterial photosynthesis, as discussed above. Utilization of glycogen reserves is also bioenergetically important for energy buffering under changing conditions [96] as well as revival from quiescent states resulting from N-starvation [14]. The CM has ARTO that help power the ion fluxes necessary for nutrient uptake and other exchange reaction across the boundary of the cell and presumably operate throughout the diurnal cycle. The terminal respiratory complexes of the TM dominate respiratory O_2 uptake at night when glycogen reserves are being catabolized to drive metabolism. The respiratory activities in cyanobacteria can be broadly classified in terms of function as mediating energy production during the night. That is part of the photoprotection mechanism that allows the dissipation of excess electrons under conditions of high light or fluctuating light, or, for that matter conditions which disallow the productive utilization of reduction produced by the photosynthetic electron transport chain. In this context, the flavodiiron proteins are especially crucial for photoprotection [97, 98].

2.3.1 Cyclic Electron Flow (CEF)

Cyclic electron flow (CEF) (depicted in Fig. 9) around Photosystem I is crucial for regulating the energy outputs of the TM during photosynthesis. By adjusting the

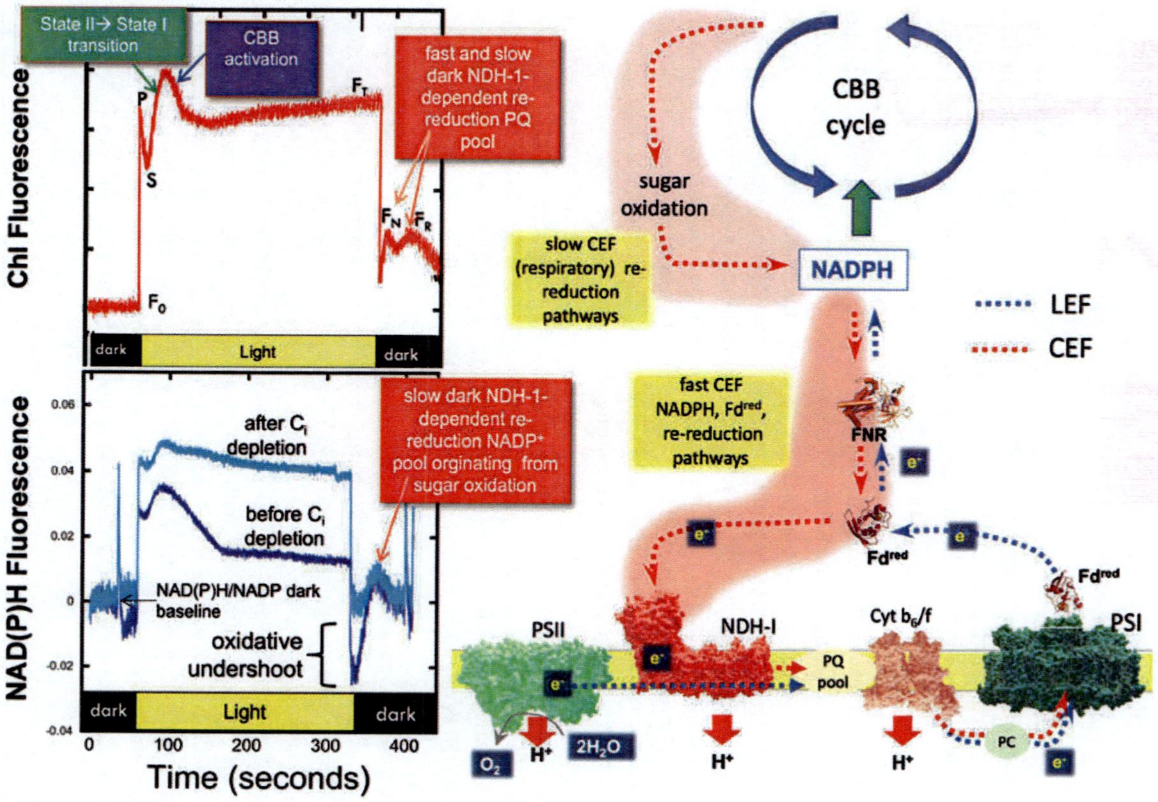

Fig. 9 Cyclic electron transport in cyanobacteria is complex. Different routes including the canonical CEF pathway involving the reduction of Fd by PSI and with Fd^{red} reducing the PQH_2/PQ pool via NDH-1 through cyt b_6f, PC and to PSI for re-energization. However, the photosynthetic generation of NADPH by LEF and reduced carbon skeletons from the activity of the CBB each constitutes a pool of reductant in the cytoplasm that can return to the PQH_2/PQ pool and is

proportions of LEF and CEF, cyanobacterial cells control the relative rates of ATP and NADPH production. Because CEF produces no net photosynthetic reductant but generates proton motive force (PMF) for the synthesis of ATP, increased CEF activity can increase the ATP/NADPH output ratio. This can become important under stress conditions, for example, exposure to high salt conditions [99]. The extent of CEF has been debated since its discovery, though most evidence suggests a considerable level [100–102], with recent studies estimating as much as 35–40% of the photosynthetic electron flow occurs via CEF [35, 103]. The main pathway for CEF in cyanobacterial is through the NDH-1 complex in a pathway that involves proton-pumping activity of both NDH-1 and cytochrome b_6f, although other pathways are not excluded. Despite its essential role in CEF, little is known about the regulation of the NDH-1 complex and its activity in cyanobacteria. The down-regulation of CEF at NDH-1 by ATP has been observed in chloroplasts and is undoubtedly critical for balancing the relative proportion of reducing and phosphorylating power by changing the balance between LEF and CEF [104]. This is one of the core bioenergetic regulatory features in oxygenic autotrophs since it enables matching energy production with the demands of metabolism. However, whether a similar mechanism operates in cyanobacteria remains to be experimentally evaluated. Thiol-modulated redox regulation may also be important, but direct tests of this hypothesis remain to be performed [22]. For longer-term energy balancing, cyanobacteria express a protein, NdhO, that acts as an inhibitor to CEF. The expression of NdhO remains constant during exposure to high light, yet there is a large increase in the expression for the NDH-1 complex components resulting increased abundance of the high-affinity CO_2 uptake form of the NDH-1 complex, NDH-1_3, consistent with the increased demands for CEF-driven CO_2 uptake [70]. Evidently, the inhibitory effects of NdhO are no longer functionally required for the NDH-1_3 complex, whereas under lower light conditions, inhibitory action is necessary.

It is also important to note that CEF is more properly considered as a set of pathways all occurring around PSI, but potentially involving very different routes with different time constants of activity. The simplified concept of cyanobacterial CEF is a pathway that starts with the reduction of Fd by PSI and with Fd^{red} recycled back into the PQH_2/PQ pool via NDH-1 on to cyt b_6f and PC (or cyt. C_{553}), and back to PSI for re-energization and with proton pumping in the process. While this appears to be largely correct, this basic route can be also extended by the fact that the photosynthetic generation of NADPH by LEF and reduced carbon skeletons from the activity of the CBB each constitutes a pool of reductant in the cytoplasm that can return to the PQH_2/PQ pool and is observable as part of the so-called post-

Fig. 9 (continued) observable as part of the so-called post-illumination rises of both Chl and NADPH fluorescence, right panel. There are two kinetically distinct peaks in fluorescence rise with one occurring ~10 s after the termination of actinic illumination and the other in the 30–40 s time range. See text for details illumination rises of both Chl and NADPH fluorescence. There are two kinetically distinct peaks in fluorescence rise with one occurring ~10 s after the termination of actinic illumination attributed to consumption of NADPH and Fd^{red}. The second peak occurs more slowly, in the range of ~30 s and is attributed to oxidation of accumulated sugars during the light period [105].

2.3.2 Control of Light Harvesting and Energy Transfer

To cope with circumstances where the bioenergetic capacity exceeds the assimilative capacity of a cell's metabolism, cyanobacteria have evolved a variety of mechanisms to safely dissipate the excess energy. The first line of defense is to downregulate the amount of light energy funneled to the photosynthetic RCs. Dissipation of excess excitation energy is, depending upon the species, most frequently accomplished in cyanobacteria by an energy-quenching process involving the orange carotenoid protein (OCP), which is a small water-soluble carotene-containing protein that interacts directly with the PBS core pigments [106–108]. OCP is activated by high-intensity white or blue light, in a process involving conformational rearrangements in the protein that are induced by the excited state of the bound carotenoid. Upon photoactivation, the OCP associates with the PBS in a manner that allows quenching of the excitation energy by the carotenoid. With the energy of incoming excitons diverted to heat, OCP drastically diminishes the flow of excitons to PSII and PSI in a process called non-photochemical quenching (NPQ). Although functionally similar, the cyanobacterial NPQ is mechanistically different than the well-studied NPQ in plants and green algae, which involves the intrinsic light-harvesting Chl proteins [109]. Mechanisms simply uncoupling the phycobilisome from the reaction center, for example by partial detachment from the membrane, are also proposed and may relate to the still poorly understood state transition mechanisms [110], but more research will be required to resolve whether such mechanisms exist and how they work.

2.3.3 Regulation of Photosynthetic Electron Flow for Photoprotection

Despite the critical importance of excitation energy dissipation as a protective mechanism, certain conditions, including high light and/or inorganic nutrient limitation, produce a situation where overexcitation of the photosystems is unavoidable, even with the activation of NPQ and other excitation energy dissipation mechanisms. Consequently, light-driven electron flow will inexorably proceed and exceed the cell's assimilative capacity to productively utilize the energized electrons. Cyanobacteria have evolved "safety valves" to cope with such situations. Electron flow that cannot be productively utilized for anabolic metabolism can be safely exhausted by a variety of mechanisms returning them to O_2, in so-called water–water cycles. The name derives from the fact that the energized electrons are derived from the PSII H_2O-oxidation reactions that are subsequently used to reduce O_2 to produce water in a protective futile cycle. There appear to be numerous mechanisms for this to occur, but the most important, at least in the model cyanobacteria, are the flavodiiron (Flv) proteins [22, 98, 111, 112] that have been demonstrated to accept electrons from NADPH [113]. Marine cyanobacteria alternatively appear to rely heavily on plastoquinol terminal oxidases (PTOX) [114]. Additionally, all oxygenic phototrophs engage in a water–water cycle via Mehler reaction where direct reduction of O_2 produces superoxide on the acceptor side of PSI with the subsequent enzymatic conversion to H_2O via dismutase and catalase activities [115].

In addition to mechanisms dissipating excess electrons, photosynthetic electron flow is regulated by changing the activity of participating electron transfer complexes. In this regard, the cyt b_6f complex appears to be very important for the regulation of electron flow to avoid light-induced damage. The operation of the Q-cycle of the cyt b_6f complex enables an unexpected regulatory feature to control electron flow when acceptor pools at limiting, such as during high light or CO_2 limitation. Under these conditions, PQH_2/PQ pool becomes over-reduced, which corresponds to a depletion of the oxidized form of PQ. The absence of available oxidized PQ prevents further enzymatic turnover of the cyt b_6f complex because the operation of the Q-cycle of the cyt b_6f complex requires the input of one PQ per completed cycle, and, as a consequence, a low amount of PQ will limit the turnover of the Q-cycle. Accordingly, electron transfer from the PQH_2/PQ pool through the cyt b_6f complex to plastocyanin will be thwarted in a process referred to as "reduction-induced suppression of electron flow (RISE)" [116]. Additionally, control of electron flow through the cyt b_6f complex may be exerted by the magnitude of ΔpH by analogy with the well-documented mechanism observed to slow electron flow through the chloroplast cyt b_6f [117], yet experimental evidence has yet to be obtained for such a mechanism in cyanobacteria. As discussed below, the RISE mechanism and/or potential ΔpH mechanisms may fit into the larger picture of how cyanobacterial cells regulate electron flow under conditions where productive reductant utilization is thwarted due to, for example, nutrient limitation. Under such limiting conditions, photosynthetic electron flow can be redirected and, in some cases, dissipated to avoid the production of damaging ROS.

2.3.4 Protection of PSI as Key to Photoacclimation

Although photodamage and repair processes involving PSII have long been recognized as part of the high light photoinhibition and response system in oxygenic autotrophs [118, 119], avoiding damage to PSI has only more recently emerged as a key feature of the overall response to excess light [120]. Although the mechanisms of photodamage remain to be identified, it is likely that damage to the Fe-S on the acceptor side is accelerated if electrons produced by photochemical charge separation are not efficiently removed. This situation arises whenever the acceptor pool of Fd^{ox} is depleted as can occur when the CBB cycle cannot proceed due to CO_2 limitation or low temperatures. As noted above, an over-reduced PQ pool may produce a block in electron flow through the cyt b_6f complex because the Q-cycle mechanism require oxidized PQ to complete its catalytic cycle. This blockage results in the oxidation of the pool of PC and the starvation of PSI for electrons and the consequent "throttling-back" of PSI reduction of Fd, which itself lowers the chances of long-lived reduced forms of the Fe-S centers on the PSI acceptor side that would be prone to damage. Acceptor-side limitation and damage of PSI is also evident in mutants lacking the NDH-1 complex, which are unable to reoxidize Fd^{red} via CEF [116], or by deletion of the flavodiiron genes [98, 121]. The downregulation of electron flow through the cyt b_6f complex would also allow the CBB cycle to begin to draw down the pools of accumulated Fd^{red} and NADPH thereby regenerating Fd^{ox} and $NADP^+$ acceptor pool, which also alleviates the electron "back-pressure" on PSI. In other words, the physiological conditions the lead to an overly reduced PQH_2/PQ pool, also usually lead to a limited pool of PSI acceptors. For example, when the availability of CO_2 lowers the sink capacity for NADPH utilization by the CBB cycle, this results in the accumulation of electrons on the Fe-S centers on the acceptor side of PSI, which then become prone to ROS damage. Similarly, an over-reduced pool of NADPH/$NADP^+$ can also be dissipated by the flavodiiron protein system, which has some functional redundancy with CEF via the NDH-1 complex [98]. This fits with observations of the Δ*glgC* mutant of *Synechocystis* which has an over-reduced PQH_2/PQ pool in the light, yet paradoxically has a slow rate of re-reduction of $P_{700}{}^+$ of PSI. Whether it is the RISE mechanisms or some other mechanism that prevents electron flow through the cyt b_6f complex with a highly reduced PQH_2/PQ pool, it is significant that it seemingly functions to prevent damage to the acceptor side of PSI. Thus, the overall picture emerging is that the acceptor side limitation of PSI can produce expensive damage that is avoided by downregulating electron flow through cyt b_6f complex to decrease the electron flow to PSI.

2.4 Energy Charge and Redox Poise

The standard free energy for the hydrolysis of ATP is ~31 kJ per mole. However, proton pumping by the photosynthetic and respiratory membranes charge the proton motive force to levels that sustain high ATP levels that reach well beyond and are typically between 50 to 60 kJ per mole. Although the energy level of the cell is sometimes expressed as the ATP/ADP ratio, that metric does not provide a complete characterization of the cellular state. Maintenance of high levels of ATP is best expressed as the adenylate "energy charge," taking into account the fact that both ATP and ADP each have phosphoanhydride bonds that can be exploited to activate the metabolic intermediates and thereby contribute to the performance of biochemical work. Energy charge (EC) can adopt values from zero to one, expressed as:

$$\mathrm{EC} = \frac{[\mathrm{ATP}] + \frac{1}{2}[\mathrm{ADP}]}{[[\mathrm{ATP}] + [\mathrm{ADP}] + [\mathrm{AMP}]]} \approx 0.6 \text{ to } 0.9 \text{ invivo}$$

Maintenance of a relatively high energy charge is a universal parameter for all living cells, typically ranging from 0.6 to 0.9 in most cell types and being affected by nutritional status [122, 123]. Measurements of energy charge in cyanobacteria are relatively scarce, but existing measurements show that the cyanobacterial EC is observed to be in the expected range of 0.6–0.9. Not surprisingly, the cyanobacterial EC is influenced by the light conditions by changing physiological demands such as the initiation of DNA replication [124, 125]. The levels of ATP were observed to exhibit diurnal fluctuations that exhibited persistent entrainment showing circadian control [126]. Deletion of the circadian regulator RpaA in *Synechococcus* sp. PCC7942 disrupts anticipatory diurnal changes in C-metabolism and causes energy charge to drop precipitously at night [125] underlining the importance of energy buffering capacity afforded in the deposition and mobilization of starch reserves. Indeed, as predicted for cells generally [122, 123], energy charge appears to regulate numerous aspects of cyanobacterial metabolism including N-metabolism [14, 127] and glycogen metabolism as part of overall energy homeostasis [96]. Mechanisms for effecting adenylate charge control range from direct interactions with ATP, ADP, and AMP mediating allosteric control of energy-utilizing and energy-yielding reactions, yet must also work in concert with other pathways. For example, involving second messengers that integrate energy metabolism with nutrient acquisition [128].

The other main form of photosynthetic power involves the flux of low-potential redox equivalents used to drive the reductive processes in metabolism. NADPH is maintained at relatively high levels in the light through the action of LEF and in the dark, through the breakdown of glycogen via the pentose-phosphate pathway. Nevertheless, the actual levels fluctuate significantly in the face of changing light and metabolic conditions. Biochemical estimates of the redox state of the NADPH/$NADP^+$ pool range [see [145] and references therein].

$$\left[\frac{\text{NADPH}}{\text{NADPH} + \text{NADP}^+}\right] \approx 0.3 \text{ to } 0.7 \text{ invivo}$$

The NADPH/$NADP^+$ redox couple has a midpoint potential of −320 mV. To estimate the fluctuations in the actual midpoint potential in the cell using the Nernst equation the potential ranges from −325 to −338 mV with the −338 mV value still sufficiently electropositive to be reduced by the main cyanobacterial ferredoxin 1 (Fd1), which has a midpoint potential of −370 mV. The actual ratio of NADPH/$NADP^+$ represents a highly dynamic steady-state balance between the production of NADPH by redox energy-releasing processes (photosynthesis, respiratory catabolism) and anabolic processes consuming NADPH. Redox homeostasis is maintained whether the cells are in the light or the dark. As expected, NADPH/$NADP^+$ values in the light are somewhat higher (~0.68) than in the dark (~0.42) indicating that the cells are poised in a slightly more oxidized state in the dark with respect to the NADPH/$NADP^+$ system according to a recent study with *Synechocystis* sp. PCC6803 [145]. Interestingly, the lower ratios are observed immediately after steady illumination is switched to darkness. This fits with spectroscopic measurements of the redox state of the NADPH/NADP$^+$ system in whole cells where a very rapid oxidation of the NADPH/NADP$^+$ pool occurs, undershooting even its dark steady-state value (illustrated in Fig. 9). This rapid consumption corresponding to the oxidative undershoot after the lights are switched off corresponds to the continued voracious consumption of NADPH by the CBB cycle until it is more gradually downregulated by redox mechanisms switching off key CBB enzymes in the dark [135, 144]. High levels of NADPH/NADP$^+$ are observed following inorganic depletion or due to inhibition of the CBB cycle, conditions that prevent the productive dissipation of redox equivalents via CO_2 fixation.

3 Integration of Bioenergetic Mechanisms into Cellular Metabolism and Growth

The bioenergetic core of cyanobacteria is the photosynthetic output of the TM membranes, which are primarily in the form of reducing and phosphorylating power. These outputs drive intermediary metabolism in support of macromolecular processes underlying biomass increase, cell growth, and proliferation. As noted, cyanobacteria have evolved mechanisms to cope with situations where there is an excess in the overall flux of reductant surpassing the assimilative capacity of the cells. Under these conditions, overall photosynthesis is sink-limited and the excess photosynthetic product beyond the sink capacity can yield potentially dangerous ROS. The photosynthetic outputs can be modulated by feedback mechanisms and intrinsic flux governors. Excess excitation energy is quenched by OCP, whereas state transitions balance the distribution of excitation energy between PSII and PSI for optimal productive photochemical quenching, meaning that the excitation energy is effectively used to drive electron flow. When rates of electron flow exceed the metabolic capacity to utilize all of the reductant, electron flow through PSI to Fd/NADPH is downregulated by slowing the flow through Cyt b_6/f by high ΔpH or RISE mechanisms. Modulation of the NADPH/ATP output ratio is regulated mainly through adjustments of CEF. The coping mechanisms matching reductant flux to metabolic demands are achieved by dissipating the excess reductant through respiratory and Flv electron flow to O_2. Each of the protection mechanisms is wasteful in the sense that a fraction of the products of photosynthetic energy capture is lost. An important set of questions to ask are to what extent can bioengineered strains exploit this waste stream of potentially valuable reductant and what sort of trade-offs might be involved?

3.1 Source-Sink Relationships: Bioengineered Product Synthesis Can Alleviate Potentially Dangerous Bioenergetic Overflows

There is now ample evidence that providing an additional sink for the primary products of photosynthesis is an effective strategy to tap into otherwise wasted photosynthetic outputs. Strains of *Synechococcus* sp. PCC7942 that express a sucrose transporter export [130, 133] is capable of exporting up to approximately 80% of total biomass in the form of sucrose. Strains of *Synechocystis* sp. PCC6803 and *Synechococcus* sp. PCC7942 have been constructed where ethylene is synthesized directly from the TCA-cycle intermediate, a-ketoglutarate, by the introduced heterologous ethylene-forming enzyme, *efe* [132, 138, 147]. Strains of *Synechococcus* sp. PCC7942 expressing genes to produce 2,3-butanediol were also highly productive with percentage of biomass produced as 2,3-butanediol ranging from 30% to 60% [140]. In each of these cases, the introduction of the pathways diverting photosynthate toward desired products increased photosynthetic activity by increasing the sink capacity. The responses of adding the additional sink included a more open electron transport chain due to greater acceptor capacity. This is evidenced by greater photochemical quenching, which is an indicator that the PQH2/PQ pool is not over-reduced, and faster oxidation of the PSI electron acceptors [141]. Moreover, engineered sinks caused the upregulation of CO_2 uptake capacity evidenced by increased levels of Rubisco [143] and the upregulation of inorganic carbon transporters [137]. Conversely, removing a sink capacity decreases overall photosynthetic performance. The ΔglgC mutant in *Synechocystis* sp. PCC6803 has a deletion of the gene encoding glucose-1-phosphate adenylyltransferase and cannot synthesize glycogen. Thus, the mutant lacks a carbon sink available to the wild-type strain. This absence of sink results in a decreased photosynthetic performance exemplified by an over-reduction of the PQH_2/PQ pool even at moderate light intensities [96, 137].

What is remarkable about the clear evidence for enhanced photosynthetic performance and upregulation of CO_2 uptake capacity in the engineered strains is that, apart from the product pathways themselves, none of the changes were explicitly engineered. Much of the experimentally observed increases in photosynthetic performance upon increasing sink capacity is readily explained by less "congestion" of the electron transfer pathways due to fast and efficient utilization of products of the light reactions. For example, it is expected, though not yet specifically tested that NADPH is more rapidly utilized so that the electron acceptor pool for PSI remains available. These physiological effects do not necessarily require increased expression of the relevant proteins such as the photosystems. Or carbon-fixing enzymes. Nevertheless, changes in protein expression can be observed in response to increased sink capacity. For example, production of ethylene was observed to increase the expression of the bicarbonate uptake transporter SbtA [137], and export of sucrose elicited higher levels of Rubisco expression and carboxysome abundance [143]. The introduction of new sinks elicited the endogenous regulatory mechanism and understanding these should provide insight into the regulatory circuits governing the expression and activity of the photosynthetic mechanism. Overall, the genetically engineered introduction of new sinks for photosynthate, at least for products that are exported from the cell, increase overall photosynthetic productivity in the cases examined so far. Thus, an important, albeit not fully tested conclusion, is that it may not be necessary to engineer alterations in the photosynthetic mechanism to accommodate new demands imposed by the introduction of new metabolic pathways for desire product production because the existing regulatory flexibility in controlling photosynthetic outputs are likely to satisfy these altered metabolic demands. One possible important exception may relate to inorganic carbon uptake since inorganic carbon often limits the overall productivity of cyanobacteria and overexpression of inorganic uptake proteins improves growth and productivity, especially under conditions of growth without CO_2 supplementation [134, 150].

3.2 *Fast-Growing Cyanobacteria and Productivity*

Another bioenergetics question regarding the engineering of product production using cyanobacteria is how the choice cyanobacterial host "chassis" with a greater bioenergetic capacity influences productivity, specifically what is the influence of more robust characteristics of the photosynthetic mechanism? A clue to this question derives from the fact that many of the same photosynthetic characteristics that are physiologically elicited upon engineering a new sink of photosynthate are the same characteristics that are associated with fast growth in cyanobacteria. The closely related *Synechococcus elongatus* UTEX 2973 and *Synechococcus* sp. PCC7942 differ at only a handful of significant genetic loci, mostly SNPs, and yet the UTEX 2973 strain grows more than twice as fast and grows optimally under much higher light intensities [148, 149]. Interestingly, the fast growth correlates with differences in the openness of photosynthetic LEF as evidenced by physiological characteristics such as faster reoxidation rates of the reduced form of the PSII acceptor, Q_A^-, and faster electron flow through PSI on both the donor and acceptor sides [146]. However, unlike the effects of engineered sinks, these are the result of what appear to be more global endogenous changes in protein expression levels, in part mediated by a different activity of the transcriptional regulator RpaA [148]. The RpaA protein has widespread effects and is "wired" into the circadian clock regulatory circuitry and part of the transcriptional program that mobilizes photosynthetic protein expression as dawn approaches [20, 131, 142]. Therefore, one can imagine that differences in RpaA activity help orchestrate a quasi-constitutive high-productivity protein expression profile. This appears to be the case since the UTEX 2973 strain contains almost twofold more Chl per cell than the PCC7942 strain and this is associated with an increase in PSI and not PSII levels [146]. Thus, the higher degree of openness of the LEF pathways is due to high levels of photosynthetic electron flux capacity downstream from PSII, notably high levels of PSI. This enables UTEX 2973 to have a higher photosynthetic capacity to harness the high light fluxes that this strain tolerates. At the same time, UTEX 2973 cells exhibit comparatively high rates of CO_2 uptake [129]. Moreover, metabolic flux analysis shows that these cells employ more carbon-efficient metabolic routes to utilize the higher output flux of the electron transport and this enables faster overall photosynthesis at the level of sugar production [136]. The metabolic flux analysis also implies increased fluxes through amino acid synthesis and other macromolecular precursor pathways so that cells are poised to match the high photosynthetic output flux with a high anabolic capacity to feed rapid cell duplication [136]. Thus, increased bioenergetic capacity appears well-matched with increased metabolic capacity. Clearly, the UTEX 2973 strain, and other fast-growing strains, are more autotrophically productive, and this is correlated with enhanced photosynthetic flux that involves changes in gene expression of photosynthetic proteins and probably other functional classes of proteins. According to optimal growth theory, fast growth depends upon optimal allocation of different functional classes of proteins within a finite cellular proteome [(i.e., a fixed total mass of protein per cell [1–3, 27, 28]]. In this context, fast-growing strains are predicted to have more optimal allocation patterns within their proteomes for growth according to specific environmental and trophic conditions. Under saturating light and inorganic nutrient conditions, this optimality is likely reflected in both a photosynthetic mechanism characterized by high flux capacity due to, for example, increased PSI levels. Moreover, the fast-growing cells under these conditions seem to be poised for greater anabolic flux capacity. How this enhanced anabolic capacity is mechanistically achieved and whether it indeed involves increased expression of key anabolic enzymes such as Rubisco remains to be determined. Answering these questions may be critical to understanding and engineering future strains. Could productivity be additively increased by using a fast-growing strain and by engineering a new sink? The answer seems to be yes at least for strains that export the product. The sucrose productivity of similarly engineered *Synechococcus elongatus* UTEX 2973 and *Synechococcus* sp. PCC7942 show a near doubling of sucrose export levels for the fast-growing UTEX 2973 compared to PCC7942 [139]. Although the positive relationship between growth rate and rates of engineered product output is intuitively satisfying, the quantitative physiological basis for this observation remains poorly understood and clearly more research needs to be performed to understand it.

References

1. Burnap RL (2015) Systems and photosystems: cellular limits of autotrophic productivity in cyanobacteria. Front Bioeng Biotechnol 3
2. Jahn M, Vialas V, Karlsen J, Maddalo G, Edfors F, Forsstrom B, Uhlen M, Kall L, Hudson EP (2018) Growth of cyanobacteria is constrained by the abundance of light and carbon assimilation proteins. Cell Rep 25:478–486.e8
3. Molenaar D, Van Berlo R, De Ridder D, Teusink B (2009) Shifts in growth strategies reflect tradeoffs in cellular economics. Mol Syst Biol 5

4. Belkin S, Mehlhorn RJ, Packer L (1987) Proton gradients in intact cyanobacteria. Plant Physiol 84:25
5. Grettenberger CL, Sumner DY, Wall K, Brown CT, Eisen JA, Mackey TJ, Hawes I, Jospin G, Jungblut AD (2020) A phylogenetically novel cyanobacterium most closely related to Gloeobacter. ISME J
6. Peschek GA, Czerny T, Schmetterer G, Nitschmann WH (1985) Transmembrane proton electrochemical gradients in dark aerobic and anaerobic cells of the cyanobacterium (blue-green alga) *Anacystis nidulans*: evidence for respiratory energy transduction in the plasma membrane. Plant Physiol 79:278–284
7. Ritchie R (1998) Bioenergetics of membrane transport in Synechococcus R-2 (Anacystis nidulans, S. leopoliensis) PCC7942. Can J Bot 76:1127–1145
8. Ritchie RJ (1991) Membrane potential and pH control in the cyanobacterium Synechococcus R-2 (Anacystis nidulans) PCC 7942. J Plant Physiol 137:409–418
9. Ritchie RJ, Christiane N, Larkum AWD (1996) Driving forces for bicarbonate transport in the cyanobacterium Synechococcus R-2 (PCC 7942). Plant Physiol 112:1573–1584
10. Milo R, Phillips R (2016) Cell biology by the numbers. Garland Science, New York
11. Baers LL, Breckels LM, Mills LA, Gatto L, Deery MJ, Stevens TJ, Howe CJ, Lilley KS, Lea-Smith DJ (2019) Proteome mapping of a cyanobacterium reveals distinct compartment organization and cell-dispersed metabolism. Plant Physiol 181:1721–1738
12. Wang J, Huang X, Ge H, Wang Y, Chen W, Zheng L, Huang C, Yang H, Li L, Sui N, Wang Y, Zhang Y, Lu D, Fang L, Xu W, Jiang Y, Huang F, Wang Y (2022) The quantitative proteome atlas of a model cyanobacterium. J Genet Genomics 49:96–108
13. Liberton M, Saha R, Jacobs JM, Nguyen AY, Gritsenko MA, Smith RD, Koppenaal DW, Pakrasi HB (2016) Global proteomic analysis reveals an exclusive role of thylakoid membranes in bioenergetics of a model cyanobacterium. Mol Cell Proteomics 15:2021–2032
14. Doello S, Burkhardt M, Forchhammer K (2021) The essential role of sodium bioenergetics and ATP homeostasis in the developmental transitions of a cyanobacterium. Curr Biol 31:1606–1615.e2
15. Wang HL, Postier BL, Burnap RL (2002) Polymerase chain reaction-based mutageneses identify key transporters belonging to multigene families involved in Na+ and pH homeostasis of Synechocystis sp. PCC 6803. Mol Microbiol 44:1493–1506
16. Mathiesen C, Hägerhäll C (2003) The 'antiporter module' of respiratory chain complex I includes the MrpC/NuoK subunit – a revision of the modular evolution scheme. FEBS Lett 549:7–13
17. Blanco-Rivero A, Leganes F, Fernandez-Valiente E, Calle P, Fernandez-Pinas F (2005) mrpA, a gene with roles in resistance to Na^+ and adaptation to alkaline pH in the cyanobacterium *Anabaena* sp. PCC7120. Microbiology 151:1671–1682
18. Wang HL, Postier BL, Burnap RL (2004) Alterations in global patterns of gene expression in *Synechocystis* sp. PCC 6803 in response to inorganic carbon limitation and the inactivation of *ndhR*, a LysR family regulator. J Biol Chem 279:5739–5751
19. Steiner J, Sazanov L (2020) Structure and mechanism of the Mrp complex, an ancient cation/proton antiporter. eLife 9:e59407
20. Diamond S, Jun D, Rubin BE, Golden SS (2015) The circadian oscillator in Synechococcus elongatus controls metabolite partitioning during diurnal growth. Proc Natl Acad Sci 112: E1916–E1925
21. Markson JS, Piechura JR, Puszynska AM, O'shea EK (2013) Circadian control of global gene expression by the cyanobacterial master regulator RpaA. Cell 155:1396–1408
22. Nikkanen L, Solymosi D, Jokel M, Allahverdiyeva Y (2021) Regulatory electron transport pathways of photosynthesis in cyanobacteria and microalgae: recent advances and biotechnological prospects. Physiol Plant 173:514–525
23. Moal G, Lagoutte B (2012) Photo-induced electron transfer from photosystem I to NADP(+): characterization and tentative simulation of the in vivo environment. Biochim Biophys Acta 1817:1635–1645

24. Huokko T, Ni T, Dykes GF, Simpson DM, Brownridge P, Conradi FD, Beynon RJ, Nixon PJ, Mullineaux CW, Zhang P, Liu L-N (2021) Probing the biogenesis pathway and dynamics of thylakoid membranes. Nat Commun 12:3475
25. Van De Meene AML, Hohmann-Marriott MF, Vermaas WFJ, Roberson RW (2006) The three-dimensional structure of the cyanobacterium Synechocystis sp PCC 6803. Arch Microbiol 184:259–270
26. Casella S, Huang F, Mason D, Zhao G-Y, Johnson GN, Mullineaux CW, Liu L-N (2017) Dissecting the native architecture and dynamics of cyanobacterial photosynthetic machinery. Mol Plant 10:1434–1448
27. Reimers A-M, Knoop H, Bockmayr A, Steuer R (2017) Cellular trade-offs and optimal resource allocation during cyanobacterial diurnal growth. Proc Natl Acad Sci 114:E6457–E6465
28. Zavrel T, Faizi M, Loureiro C, Poschmann G, Stuhler K, Sinetova M, Zorina A, Steuer R, Cerveny J (2019) Quantitative insights into the cyanobacterial cell economy. eLife 8
29. Adir N (2005) Elucidation of the molecular structures of components of the phycobilisome: reconstructing a giant. Photosynth Res 85:15–32
30. Watanabe M, Ikeuchi M (2013) Phycobilisome: architecture of a light-harvesting supercomplex. Photosynth Res 116:265–276
31. Zhang J, Ma J, Liu D, Qin S, Sun S, Zhao J, Sui S-F (2017) Structure of phycobilisome from the red alga Griffithsia pacifica. Nature 551:57–63
32. Zheng L, Zheng Z, Li X, Wang G, Zhang K, Wei P, Zhao J, Gao N (2021) Structural insight into the mechanism of energy transfer in cyanobacterial phycobilisomes. Nat Commun 12: 5497
33. Hirota Y, Serikawa H, Kawakami K, Ueno M, Kamiya N, Kosumi D (2021) Ultrafast energy transfer dynamics of phycobilisome from Thermosynechococcus vulcanus, as revealed by ps fluorescence and fs pump-probe spectroscopies. Photosynth Res 148:181–190
34. Watanabe M, Semchonok DA, Webber-Birungi MT, Ehira S, Kondo K, Narikawa R, Ohmori M, Boekema EJ, Ikeuchi M (2014) Attachment of phycobilisomes in an antenna–photosystem I supercomplex of cyanobacteria. Proc Natl Acad Sci 111:2512
35. Miller NT, Vaughn MD, Burnap RL (2021) Electron flow through NDH-1 complexes is the major driver of cyclic electron flow-dependent proton pumping in cyanobacteria. Biochim Biophys Acta Bioenerg 1862:148354
36. Ogawa T, Harada T, Ozaki H, Sonoike K (2013) Disruption of the ndhF1 gene affects Chl fluorescence through state transition in the cyanobacterium *Synechocystis* sp. PCC 6803, resulting in apparent high efficiency of photosynthesis. Plant Cell Physiol 54:1164–1171
37. Croce R (2020) Beyond 'seeing is believing': the antenna size of the photosystems in vivo. New Phytol 228:1214–1218
38. Roche JL, Van Der Staay GWM, Partensky F, Ducret A, Aebersold R, Li R, Golden SS, Hiller RG, Wrench PM, Larkum AWD, Green BR (1996) Independent evolution of the prochlorophyte and green plant chlorophyll a/b light-harvesting proteins. Proc Natl Acad Sci 93:15244–15248
39. Burnap RL, Troyan T, Sherman LA (1993) The highly abundant chlorophyll-protein complex of iron-deficient Synechococcus sp. PCC7942 (CP43') is encoded by the isiA gene. Plant Physiol 103:893–902
40. Murray JW, Duncan J, Barber J (2006) CP43-like chlorophyll binding proteins: structural and evolutionary implications. Trends Plant Sci 11:152–158
41. Boichenko VA, Pinevich AV, Stadnichuk IN (2007) Association of chlorophyll a/b-binding Pcb proteins with photosystems I and II in Prochlorothrix hollandica. Biochim Biophys Acta 1767:801–806
42. Bibby TS, Nield J, Barber J (2001) Three-dimensional model and characterization of the iron stress-induced CP43′-photosystem I supercomplex isolated from the cyanobacterium synechocystis PCC 6803. J Biol Chem 276:43246–43252

43. Toporik H, Li J, Williams D, Chiu PL, Mazor Y (2019) The structure of the stress-induced photosystem I-IsiA antenna supercomplex. Nat Struct Mol Biol 26:443–449
44. Bibby Thomas S, Nield J, Chen M, Larkum Anthony WD, Barber J (2003) Structure of a photosystem II supercomplex isolated from Prochloron didemni retaining its chlorophyll a/b light-harvesting system. Proc Natl Acad Sci 100:9050–9054
45. Bibby TS, Nield J, Partensky F, Barber J (2001) Oxyphotobacteria antenna ring around photosystem I. Nature 413:590
46. Nelson N, Yocum CF (2006) Structure and function of photosystems I and II. Annu Rev Plant Biol 57:521–565
47. Fujita Y, Murakami A, Ohki K, Hagiwara N (1988) Regulation of photosystem composition in cyanobacterial photosynthetic system: evidence indicating that photosystem I formation is controlled in response to the electron transport state. Plant Cell Physiol 29:557–564
48. Mackenzie TDB, Johnson JM, Campbell DA (2005) Dynamics of fluxes through photosynthetic complexes in response to changing light and inorganic carbon acclimation in Synechococcus elongatus. Photosynth Res 85:341–357
49. Suga M, Akita F, Hirata K, Ueno G, Murakami H, Nakajima Y, Shimizu T, Yamashita K, Yamamoto M, Ago H, Shen J-R (2015) Native structure of photosystem II at 1.95 A resolution viewed by femtosecond X-ray pulses. Nature 517:99–103
50. Kok B, Forbush B, Mcgloin M (1970) Cooperation of charges in photosynthetic oxygen evolution: I. A linear four step mechanism. Photochem Photobiol 11:457–475
51. Joliot P (1966) L'émission d'oxygène par des algues soumises à un eclairement. J Chim Phys 63:1423–1441
52. Joliot P, Kok B (1975) Oxygen evolution in photosynthesis. Bioenerg Photosynth:387–411
53. Oliver N, Avramov AP, Nürnberg DJ, Dau H, Burnap RL (2022) From manganese oxidation to water oxidation: assembly and evolution of the water-splitting complex in photosystem II. Photosynth Res
54. Oliver T, Sánchez-Baracaldo P, Larkum AW, Rutherford AW, Cardona T (2021) Time-resolved comparative molecular evolution of oxygenic photosynthesis. Biochim Biophys Acta Bioenerg 1862:148400
55. Cassier-Chauvat C, Chauvat F (2014) Function and regulation of ferredoxins in the cyanobacterium, Synechocystis PCC6803: recent advances. Life (Basel) 4:666–680
56. Motomura T, Zuccarello L, Sétif P, Boussac A, Umena Y, Lemaire D, Tripathy JN, Sugiura M, Hienerwadel R, Shen JR, Berthomieu C (2019) An alternative plant-like cyanobacterial ferredoxin with unprecedented structural and functional properties. Biochim Biophys Acta Bioenerg 1860:148084
57. Mustila H, Allahverdiyeva Y, Isojärvi J, Aro EM, Eisenhut M (2014) The bacterial-type [4Fe-4S] ferredoxin 7 has a regulatory function under photooxidative stress conditions in the cyanobacterium Synechocystis sp. PCC 6803. Biochim Biophys Acta 1837:1293–1304
58. Van Der Plas J, De Groot R, Woortman M, Cremers F, Borrias M, Van Arkel G, Weisbeek P (1988) Genes encoding ferredoxins from Anabaena sp. PCC 7937 and Synechococcus sp. PCC 7942: structure and regulation. Photosynth Res 18:179–204
59. Brettel K, Leibl W (2001) Electron transfer in photosystem I. Biochim Biophys Acta 1507: 100–114
60. Kato K, Nagao R, Jiang T-Y, Ueno Y, Yokono M, Chan SK, Watanabe M, Ikeuchi M, Shen J-R, Akimoto S, Miyazaki N, Akita F (2019) Structure of a cyanobacterial photosystem I tetramer revealed by cryo-electron microscopy. Nat Commun 10:4929
61. Semchonok DA, Mondal J, Cooper CJ, Schlum K, Li M, Amin M, Sorzano COS, Ramírez-Aportela E, Kastritis PL, Boekema EJ, Guskov A, Bruce BD (2022) Cryo-EM structure of a tetrameric photosystem I from Chroococcidiopsis TS-821, a thermophilic, unicellular, non-heterocyst-forming cyanobacterium. Plant Commun 3:100248
62. Zheng L, Li Y, Li X, Zhong Q, LI N, Zhang K, Zhang Y, Chu H, Ma C, Li G, Zhao J, Gao N (2019) Structural and functional insights into the tetrameric photosystem I from heterocyst-forming cyanobacteria. Nat Plants 5:1087–1097

63. Chen M, Liu X, He Y, Li N, He J, Zhang Y (2022) Diversity among cyanobacterial photosystem I oligomers. Front Microbiol 12
64. Dobson Z, Ahad S, Vanlandingham J, Toporik H, Vaughn N, Vaughn M, Williams D, Reppert M, Fromme P, Mazor Y (2021) The structure of photosystem I from a high-light-tolerant cyanobacteria. eLife 10
65. Laughlin TG, Bayne AN, Trempe JF, Savage DF, Davies KM (2019) Structure of the complex I-like molecule NDH of oxygenic photosynthesis. Nature 566:411–414
66. Laughlin TG, Savage DF, Davies KM (2020) Recent advances on the structure and function of NDH-1: the complex I of oxygenic photosynthesis. Biochim Biophys Acta:148254
67. Peltier G, Aro EM, Shikanai T (2016) NDH-1 and NDH-2 plastoquinone reductases in oxygenic photosynthesis. Annu Rev Plant Biol 67:55–80
68. Schuller JM, Birrell JA, Tanaka H, Konuma T, Wulfhorst H, Cox N, Schuller SK, Thiemann J, Lubitz W, Setif P, Ikegami T, Engel BD, Kurisu G, Nowaczyk MM (2019) Structural adaptations of photosynthetic compolex I enable ferredoxin-dependent electron transfer. Science 363:257–260
69. Strand DD, Fisher N, Kramer DM (2017) The higher plant plastid NAD(P)H dehydrogenase-like complex (NDH) is a high efficiency proton pump that increases ATP production by cyclic electron flow. J Biol Chem 292:11850–11860
70. Zhang C, Shuai J, Ran Z, Zhao J, Wu Z, Liao R, Wu J, Ma W, Lei M (2020) Structural insights into NDH-1 mediated cyclic electron transfer. Nat Commun 11:888
71. Efremov RG, Baradaran R, Sazanov LA (2010) The architecture of respiratory complex I. Nature 465:441–445
72. Sazanov LA (2015) A giant molecular proton pump: structure and mechanism of respiratory complex I. Nat Rev Mol Cell Biol 16:375–388
73. Schuller JM, Saura P, Thiemann J, Schuller SK, Gamiz-Hernandez AP, Kurisu G, Nowaczyk MM, Kaila VRI (2020) Redox-coupled proton pumping drives carbon concentration in the photosynthetic complex I. Nat Commun 11:494
74. Brandt U (2013) Inside view of a giant proton pump. Angew Chem Int Ed 52:7358–7360
75. Fiedorczuk K, Letts JA, Degliesposti G, Kaszuba K, Skehel M, Sazanov LA (2016) Atomic structure of the entire mammalian mitochondrial complex I. Nature 538:406–410
76. Guo R, Zong S, Wu M, Gu J, Yang M (2017) Architecture of human mitochondrial respiratory megacomplex I2III2IV2. Cell 170:1247–1257.e12
77. Kolata P, Efremov RG (2021) Structure of Escherichia coli respiratory complex I reconstituted into lipid nanodiscs reveals an uncoupled conformation. eLife 10:e68710
78. Parey K, Brandt U, Xie H, Mills DJ, Siegmund K, Vonck J, Kühlbrandt W, Zickermann V (2018) Cryo-EM structure of respiratory complex I at work. eLife 7:e39213
79. Parey K, Lasham J, Mills DJ, Djurabekova A, Haapanen O, Yoga EG, Xie H, Kühlbrandt W, Sharma V, Vonck J, Zickermann V (2021) High-resolution structure and dynamics of mitochondrial complex I-insights into the proton pumping mechanism. Sci Adv 7:eabj3221
80. Yu H, Wu C-H, Schut GJ, Haja DK, Zhao G, Peters JW, Adams MWW, Li H (2018) Structure of an ancient respiratory system. Cell 173:1636–1649.e16
81. Zhu J, Vinothkumar KR, Hirst J (2016) Structure of mammalian respiratory complex I. Nature 536:354–358
82. Klughammer B, Sultemeyer D, Badger MR, Price GD (1999) The involvement of NAD(P)H dehydrogenase subunits, NdhD3 and NdhF3, in high-affinity CO_2 uptake in *Synechococcus* sp. PCC7002 gives evidence for multiple NDH-1 complexes with specific roles in cyanobacteria. Mol Microbiol 32:1305–1315
83. Ohkawa H, Pakrasi HB, Ogawa T (2000) Two types of functionally distinct NAD(P)H dehydrogenases in *Synechocystis* sp. strain PCC6803. J Biol Chem 275:31630–31634
84. Ohkawa H, Price GD, Badger MR, Ogawa T (2000) Mutation of ndh genes leads to inhibition of CO(2) uptake rather than HCO(3)(−) uptake in Synechocystis sp. strain PCC 6803. J Bacteriol 182:2591–2596

85. Shibata M, Ohkawa H, Kaneko T, Fukuzawa H, Tabata S, Kaplan A, Ogawa T (2001) Distinct constitutive and low-CO_2-induced CO_2 uptake systems in cyanobacteria: genes involved and their phylogenetic relationship with homologous genes in other organisms. Proc Natl Acad Sci U S A 98:11789–11794
86. Battchikova N, Eisenhut M, Aro EM (2011) Cyanobacterial NDH-1 complexes: novel insights and remaining puzzles. Biochim Biophys Acta 1807:935–944
87. Ma W, Ogawa T (2015) Oxygenic photosynthesis-specific subunits of cyanobacterial NADPH dehydrogenases. IUBMB Life 67:3–8
88. Ogawa T, Mi H (2007) Cyanobacterial NADPH dehydrogenase complexes. Photosynth Res 93:69–77
89. Finn RD, Bateman A, Clements J, Coggill P, Eberhardt RY, Eddy SR, Heger A, Hetherington K, Holm L, Mistry J, Sonnhammer ELL, Tate J, Punta M (2014) Pfam: the protein families database. Nucleic Acids Res 42:D222–D230
90. Zhang P, Battchikova N, Jansen T, Appel J, Ogawa T, Aro EM (2004) Expression and functional roles of the two distinct NDH-1 complexes and the carbon acquisition complex NdhD3/NdhF3/CupA/Sll1735 in Synechocystis sp PCC 6803. Plant Cell 16:3326–3340
91. Zhang P, Battchikova N, Paakkarinen V, Katoh H, Iwai M, Ikeuchi M, Pakrasi HB, Ogawa T, Aro EM (2005) Isolation, subunit composition and interaction of the NDH-1 complexes from Thermosynechococcus elongatus BP-1. Biochem J 390:513–520
92. Artier J, Walker RM, Miller NT, Zhang M, Price GD, Burnap RL (2022) Modeling and mutagenesis of amino acid residues critical for CO2 hydration by specialized NDH-1 complexes in cyanobacteria. Biochim Biophys Acta 1863:148503
93. Bernat G, Appel J, Ogawa T, Rogner M (2011) Distinct roles of multiple NDH-1 complexes in the cyanobacterial electron transport network as revealed by kinetic analysis of P700(+) reduction in various ndh-deficient mutants of *Synechocystis* sp. strain PCC6803. J Bacteriol 193:292–295
94. Cooley JW, Vermaas WF (2001) Succinate dehydrogenase and other respiratory pathways in thylakoid membranes of Synechocystis sp. strain PCC 6803: capacity comparisons and physiological function. J Bacteriol 183:4251–4258
95. Howitt CA, Cooley JW, Wiskich JT, Vermaas WF (2001) A strain of Synechocystis sp. PCC 6803 without photosynthetic oxygen evolution and respiratory oxygen consumption: implications for the study of cyclic photosynthetic electron transport. Planta 214:46–56
96. Cano M, Holland SC, Artier J, Burnap RL, Ghirardi M, Morgan JA, Yu J (2018) Glycogen synthesis and metabolite overflow contribute to energy balancing in cyanobacteria. Cell Rep 23:667–672
97. Allahverdiyeva Y, Isojärvi J, Zhang P, Aro E-M (2015) Cyanobacterial oxygenic photosynthesis is protected by Flavodiiron proteins. Life 5:716–743
98. Nikkanen L, Santana Sanchez A, Ermakova M, Rogner M, Cournac L, Allahverdiyeva Y (2020) Functional redundancy between flavodiiron proteins and NDH-1 in Synechocystis sp. PCC 6803. Plant J 103:1460–1476
99. Jeanjean R, Matthijs HCP, Onana B, Havaux M, Joset F (1993) Exposure of the cyanobacterium Synechocystis PCC6803 to salt stress induces concerted changes in respiration and photosynthesis. Plant Cell Physiol 34:1073–1079
100. Carpentier R, Larue B, Leblanc RM (1984) Photoacoustic spectroscopy of Anacystis nidulans. III. Detection of photosynthetic activities. Arch Biochem Biophys 228:534–543
101. Herbert SK, Martin RE, Fork DC (1995) Light adaptation of cyclic electron transport through photosystem I in the cyanobacterium Synechococcus sp. PCC 7942. Photosynth Res 46:277–285
102. Mi H, Endo T, Ogawa T, Asada K (1995) Thylakoid membrane-bound, NADPH-specific pyridine nucleotide dehydrogenase complex mediates cyclic electron transport in the cyanobacterium Synechocystis sp. PCC 6803. Plant Cell Physiol 36:661–668

103. Theune ML, Hildebrandt S, Steffen-Heins A, Bilger W, Gutekunst K, Appel J (2021) In-vivo quantification of electron flow through photosystem I – cyclic electron transport makes up about 35% in a cyanobacterium. Biochim Biophys Acta 1862:148353
104. Fisher N, Bricker TM, Kramer DM (2019) Regulation of photosynthetic cyclic electron flow pathways by adenylate status in higher plant chloroplasts. Biochim Biophys Acta 1860: 148081
105. Holland SC, Kappell AD, Burnap RL (2015) Redox changes accompanying inorganic carbon limitation in *Synechocystis* sp. PCC 6803. Biochim Biophys Acta 1848:355–363
106. Bao H, Melnicki MR, Kerfeld CA (2017) Structure and functions of orange carotenoid protein homologs in cyanobacteria. Curr Opin Plant Biol 37:1–9
107. Kirilovsky D (2007) Photoprotection in cyanobacteria: the orange carotenoid protein (OCP)-related non-photochemical-quenching mechanism. Photosynth Res 93:7–16
108. Wilson A, Ajlani G, Verbavatz JM, Vass I, Kerfeld CA, Kirilovskya D (2006) A soluble carotenoid protein involved in phycobilisome-related energy dissipation in cyanobacteria. Plant Cell 18:992–1007
109. Croce R, Van Amerongen H (2020) Light harvesting in oxygenic photosynthesis: structural biology meets spectroscopy. Science 369
110. Calzadilla PI, Kirilovsky D (2020) Revisiting cyanobacterial state transitions. Photochem Photobiol Sci 19:585–603
111. Allahverdiyeva Y, Ermakova M, Eisenhut M, Zhang P, Richaud P, Hagemann M, Cournac L, Aro EM (2011) Interplay between flavodiiron proteins and photorespiration in *Synechocystis* sp. PCC 6803. J Biol Chem 286:24007–24014
112. Helman Y, Tchernov D, Reinhold L, Shibata M, Ogawa T, Schwarz R, Ohad I, Kaplan A (2003) Genes encoding a-type flavoproteins are essential for photoreduction of O_2 in cyanobacteria. Curr Biol 13:230–235
113. Brown KA, Guo Z, Tokmina-Lukaszewska M, Scott LW, Lubner CE, Smolinski S, Mulder DW, Bothner B, King PW (2019) The oxygen reduction reaction catalyzed by Synechocystis sp. PCC 6803 flavodiiron proteins. Sustain Energy Fuels 3:3191–3200
114. Bailey S, Melis A, Mackey KRM, Cardol P, Finazzi G, Van Dijken G, Berg GM, Arrigo K, Shrager J, Grossman A (2008) Alternative photosynthetic electron flow to oxygen in marine Synechococcus. Biochim Biophys Acta 1777:269–276
115. Asada K (1994) Production and action of active oxygen species in photosynthetic tissues. In: Foyer CH, Mullineaux PM (eds) Causes of photooxidative stress and amelioration of defense systems in plants. CRC Press, Boca Raton
116. Shaku K, Shimakawa G, Hashiguchi M, Miyake C (2015) Reduction-induced suppression of electron flow (RISE) in the photosynthetic electron transport system of *Synechococcus elongatus* PCC 7942. Plant Cell Physiol 57(7):1443–1453
117. Takizawa K, Cruz JA, Kanazawa A, Kramer DM (2007) The thylakoid proton motive force in vivo. Quantitative, non-invasive probes, energetics, and regulatory consequences of light-induced pmf. Biochim Biophys Acta 1767(10):1233–1244
118. Adir N, Shochat S, Ohad I (1990) Light-dependent D1 protein synthesis and translocation is regulated by reaction center II. Reaction center II serves as an acceptor for the D1 precursor. J Biol Chem 265:12563–12568
119. Nixon PJ, Michoux F, Yu J, Boehm M, Komenda J (2010) Recent advances in understanding the assembly and repair of photosystem II. Ann Bot 106:1–16
120. Lima-Melo Y, Kiliç M, Aro E-M, Gollan PJ (2021) Photosystem I inhibition, protection and signalling: knowns and unknowns. Front Plant Sci 12
121. Gerotto C, Alboresi A, Meneghesso A, Jokel M, Suorsa M, Aro E-M, Morosinotto T (2016) Flavodiiron proteins act as safety valve for electrons in Physcomitrella patens. Proc Natl Acad Sci 113:12322–12327
122. Atkinson DE (1968) Energy charge of the adenylate pool as a regulatory parameter. Interaction with feedback modifiers. Biochemistry 7:4030–4034

123. Atkinson DE (1977) Cellular energy metabolism and its regulation. Academic Press, New York
124. Jewson DH, Dokulil M (1982) Adenylate energy charge measurements in freshwater microbial studies. J Ecol 70:595–606
125. Puszynska AM, O'shea EK (2017) Switching of metabolic programs in response to light availability is an essential function of the cyanobacterial circadian output pathway. eLife 6: e23210
126. Rust MJ, Golden SS, O'shea EK (2011) Light-driven changes in energy metabolism directly entrain the cyanobacterial circadian oscillator. Science 331:220–223
127. Fokina O, Herrmann C, Forchhammer K (2011) Signal-transduction protein P(II) from Synechococcus elongatus PCC 7942 senses low adenylate energy charge in vitro. Biochem J 440:147–156
128. Selim KA, Haffner M, Burkhardt M, Mantovani O, Neumann N, Albrecht R, Seifert R, Krüger L, Stülke J, Hartmann Marcus D, Hagemann M, Forchhammer K (2021) Diurnal metabolic control in cyanobacteria requires perception of second messenger signaling molecule c-di-AMP by the carbon control protein SbtB. Sci Adv 7:eabk0568
129. Abernathy MH, Yu J, Ma F, Liberton M, Ungerer J, Hollinshead WD, Gopalakrishnan S, He L, Maranas CD, Pakrasi HB, Allen DK, Tang YJ (2017) Deciphering cyanobacterial phenotypes for fast photoautotrophic growth via isotopically nonstationary metabolic flux analysis. Biotechnol Biofuels 10:273
130. Abramson BW, Kachel B, Kramer DM, Ducat DC (2016) Increased photochemical efficiency in cyanobacteria via an engineered sucrose sink. Plant Cell Physiol 57:2451–2460
131. Ashby MK, Mullineaux CW (1999) Cyanobacterial ycf27 gene products regulate energy transfer from phycobilisomes to photosystems I and II [in process citation]. FEMS Microbiol Lett 181:253–260
132. Carbonell V, Vuorio E, Aro EM, Kallio P (2019) Enhanced stable production of ethylene in photosynthetic cyanobacterium Synechococcus elongatus PCC 7942. World J Microbiol Biotechnol 35:77
133. Ducat DC, Avelar-Rivas JA, Way JC, Silver PA (2012) Rerouting carbon flux to enhance photosynthetic productivity. Appl Environ Microbiol 78:2660–2668
134. Gupta JK, Rai P, Jain KK, Srivastava S (2020) Overexpression of bicarbonate transporters in the marine cyanobacterium Synechococcus sp. PCC 7002 increases growth rate and glycogen accumulation. Biotechnol Biofuels 13:17
135. Hagen KD, Meeks JC (2001) The unique cyanobacterial protein OpcA is an allosteric effector of glucose-6-phosphate dehydrogenase in Nostoc punctiforme ATCC 29133. J Biol Chem 276:11477–11486
136. Hendry JI, Gopalakrishnan S, Ungerer J, Pakrasi HB, Tang YJ, Maranas CD (2019) Genome-scale Fluxome of Synechococcus elongatus UTEX 2973 using transient (13)C-Labeling data. Plant Physiol 179:761–769
137. Holland SC, Artier J, Miller NT, Cano M, Yu J, Ghirardi ML, Burnap RL (2016) Impacts of genetically engineered alterations in carbon sink pathways on photosynthetic performance. Algal Res 20:87–99
138. Kallio P, Kugler A, Pyytövaara S, Stensjö K, Allahverdiyeva Y, Gao X, Lindblad P, Lindberg P (2021) Photoautotrophic production of renewable ethylene by engineered cyanobacteria: steering the cell metabolism towards biotechnological use. Physiol Plant 173:579–590
139. Lin P-C, Zhang F, Pakrasi HB (2020) Enhanced production of sucrose in the fast-growing cyanobacterium Synechococcus elongatus UTEX 2973. Sci Rep 10:390–390
140. Oliver JWK, Atsumi S (2015) A carbon sink pathway increases carbon productivity in cyanobacteria. Metab Eng 29:106–112
141. Santos-Merino M, Torrado A, Davis GA, Röttig A, Bibby TS, Kramer DM, Ducat DC (2021) Improved photosynthetic capacity and photosystem I oxidation via heterologous metabolism engineering in cyanobacteria. Proc Natl Acad Sci U S A 118

142. Scheurer NM, Rajarathinam Y, Timm S, Köbler C, Kopka J, Hagemann M, Wilde A (2021) Homologs of circadian clock proteins impact the metabolic switch between light and dark growth in the cyanobacterium Synechocystis sp. PCC 6803. Front Plant Sci 12:675227
143. Singh AK, Santos-Merino M, Sakkos JK, Walker BJ, Ducat DC (2022) Rubisco regulation in response to altered carbon status in the cyanobacterium Synechococcus elongatus PCC 7942. Plant Physiol 189:874–888
144. Tamoi M, Miyazaki T, Fukamizo T, Shigeoka S (2005) The Calvin cycle in cyanobacteria is regulated by CP12 via the NAD(H)/NADP(H) ratio under light/dark conditions. Plant J 42: 504–513
145. Tanaka K, Shimakawa G, Tabata H, Kusama S, Miyake C, Nakanishi S (2021) Quantification of NAD(P)H in cyanobacterial cells by a phenol extraction method. Photosynth Res 148:57–66
146. Ungerer J, Lin P-C, Chen H-Y, Pakrasi HB (2018) Adjustments to photosystem stoichiometry and electron transfer proteins are key to the remarkably fast growth of the cyanobacterium *Synechococcus elongatus* UTEX 2973. MBio 9
147. Ungerer J, Tao L, Davis M, Ghirardi M, Maness P-C, Yu J (2012) Sustained photosynthetic conversion of CO2 to ethylene in recombinant cyanobacterium Synechocystis 6803. Energ Environ Sci 5:8998–9006
148. Ungerer J, Wendt K, Hendry J, Maranas CD, Pakrasi HB (2018) Comparative genomics reveals the molecular determinants of rapid growth of the cyanobacterium Synechococcus elongatus UTEX 2973. Proc Natl Acad Sci 115(50):E11761–E11770
149. Yu J, Liberton M, Cliften PF, Head RD, Jacobs JM, Smith RD, Koppenaal DW, Brand JJ, Pakrasi HB (2015) Synechococcus elongatus UTEX 2973, a fast growing cyanobacterial chassis for biosynthesis using light and CO(2). Sci Rep 5:8132
150. Yu JH, Zhu LW, Xia ST, Li HM, Tang YL, Liang XH, Chen T, Tang YJ (2016) Combinatorial optimization of CO2 transport and fixation to improve succinate production by promoter engineering. Biotechnol Bioeng 113:1531–1541

Adv Biochem Eng Biotechnol (2023) 183: 65–104
https://doi.org/10.1007/10_2022_210

Published online: 28 August 2022

The Molecular Toolset and Techniques Required to Build Cyanobacterial Cell Factories

Franz Opel, Ilka M. Axmann, and Stephan Klähn

Contents

Abstract Cyanobacteria are the only prokaryotes performing oxygenic photosynthesis, a solar-driven process which allows them to obtain electrons from water to reduce and finally assimilate carbon dioxide. Consequently, they are in the spotlight

F. Opel and S. Klähn (✉)
Department of Solar Materials, Helmholtz Centre for Environmental Research – UFZ, Leipzig, Germany
e-mail: stephan.klaehn@ufz.de

I. M. Axmann
Synthetic Microbiology, Heinrich Heine University Düsseldorf, Düsseldorf, Germany

of biotechnology as photoautotrophic cell factories to generate a large variety of chemicals and biofuels in a sustainable way. Recent progress in synthetic biology has enlarged the molecular toolset to genetically engineer the metabolism of cyanobacteria, mainly targeting common model strains, such as *Synechocystis* sp. PCC 6803, *Synechococcus elongatus* PCC 7942, *Synechococcus* sp. PCC 7002, or *Anabaena* sp. PCC 7120. Nevertheless, the accessibility and flexibility of engineering cyanobacteria is still somewhat limited and less predictable compared to other biotechnologically employed microorganisms.

This chapter gives a broad overview of currently available methods for the genetic modification of cyanobacterial model strains as well as more recently discovered and promising species, such as *Synechococcus elongatus* PCC 11801. It comprises approaches based on homologous recombination, replicative broad-host-range or strain-specific plasmids, CRISPR/Cas, as well as markerless selection. Furthermore, common and newly introduced molecular tools for gene expression regulation are presented, comprising promoters, regulatory RNAs, genetic insulators like transcription terminators, ribosome binding sites, CRISPR interference, and the utilization of heterologous RNA polymerases. Additionally, potential DNA assembly strategies, like modular cloning, are described. Finally, considerations about post-translational control via protein degradation tags and heterologous proteases, as well as small proteins working as enzyme effectors are briefly discussed.

Graphical Abstract

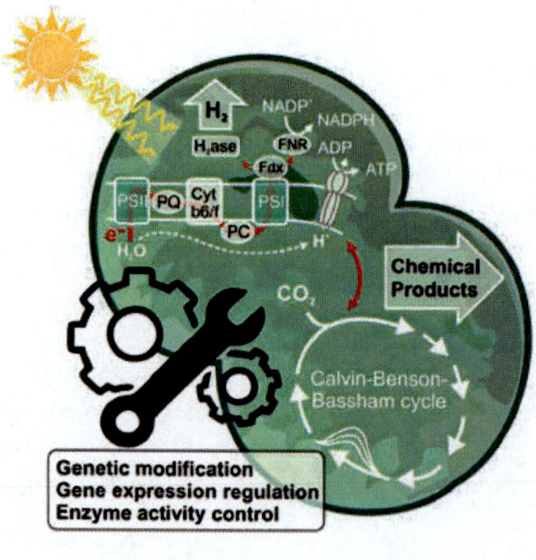

Keywords Cyanobacteria, Gene expression regulation, Genetic engineering, Molecular toolset, Synthetic biology

1 Introduction

In order to use the immense biotechnological potential of microorganisms, molecular biology is inevitable to facilitate the implementation of conducive genetic information to enable or improve a biocatalytic production process. This, for example, is required to engineer metabolic routes and to improve their yields according to a given demand. Respective molecular tools encompass standardized biological parts for the rational design and construction of artificial or at least improved biological systems. Since the early 2000s, researchers have systematically realized synthetic molecular tools and utilized them in microbial cells [1–5]. An ever since increasing toolset enabled "plug-and-play," similar to technical devices, mainly for chemoheterotrophic chassis strains like *Escherichia coli* (hereafter *E. coli*) or *Saccharomyces cerevisiae*.

Cyanobacteria are a monophyletic group of photoautotrophic bacteria. From a morphological and metabolic perspective they are quite diverse, comprising unicellular, filamentous, and multicellular strains, some of which are capable of fixing dinitrogen [6] or of synthesizing a variety of chemical compounds, e.g. toxins [7]. However, their common feature is the ability to perform oxygenic photosynthesis by combining two photosystems. Electrons are obtained by solar-driven water oxidation, thereby releasing molecular oxygen as a byproduct. The electrons are then used to drive an autotrophic metabolism, i.e. to fix carbon dioxide (CO_2). Accordingly, there is no need to provide an organic carbon source, like glucose, in contrast to chemoheterotrophic microorganisms. Consequently, cyanobacteria have raised great interest as whole-cell biocatalysts to generate various products directly from CO_2, including sugars or organic acids [8–11].

As cyanobacteria encompass a rather diverse pool of potential chassis strains, universal basic knowledge and strategies enabling their biotechnological utilization are not available so far. Strain-specific evaluations and designs of molecular tools have been performed for unicellular model strains like *Synechocystis* sp. PCC 6803 (hereafter *Synechocystis*) [12–16] and *Synechococcus elongatus* PCC 7942 (hereafter *S. elongatus* 7942) [17, 18], or the fast-growing *Synechococcus* sp. PCC 7002 (hereafter *Synechococcus* 7002) [19] and the more recently explored *Synechococcus elongatus* UTEX 2973 (hereafter *S. elongatus* UTEX 2973) [20]. Advances have been made concerning modular cloning systems, genome editing, and shuttle vectors to introduce genetic information as well as biological parts to control gene expression, and the activity of enzymes in vivo (Fig. 1). All these approaches aim to enable comparable genetic engineering of cyanobacteria across different strains, laboratories, and even industrial sites [21–31]. This chapter will give a broad overview regarding the current molecular toolset and approaches that are available to engineer frequently used model strains as well as more recently discovered promising chassis strains.

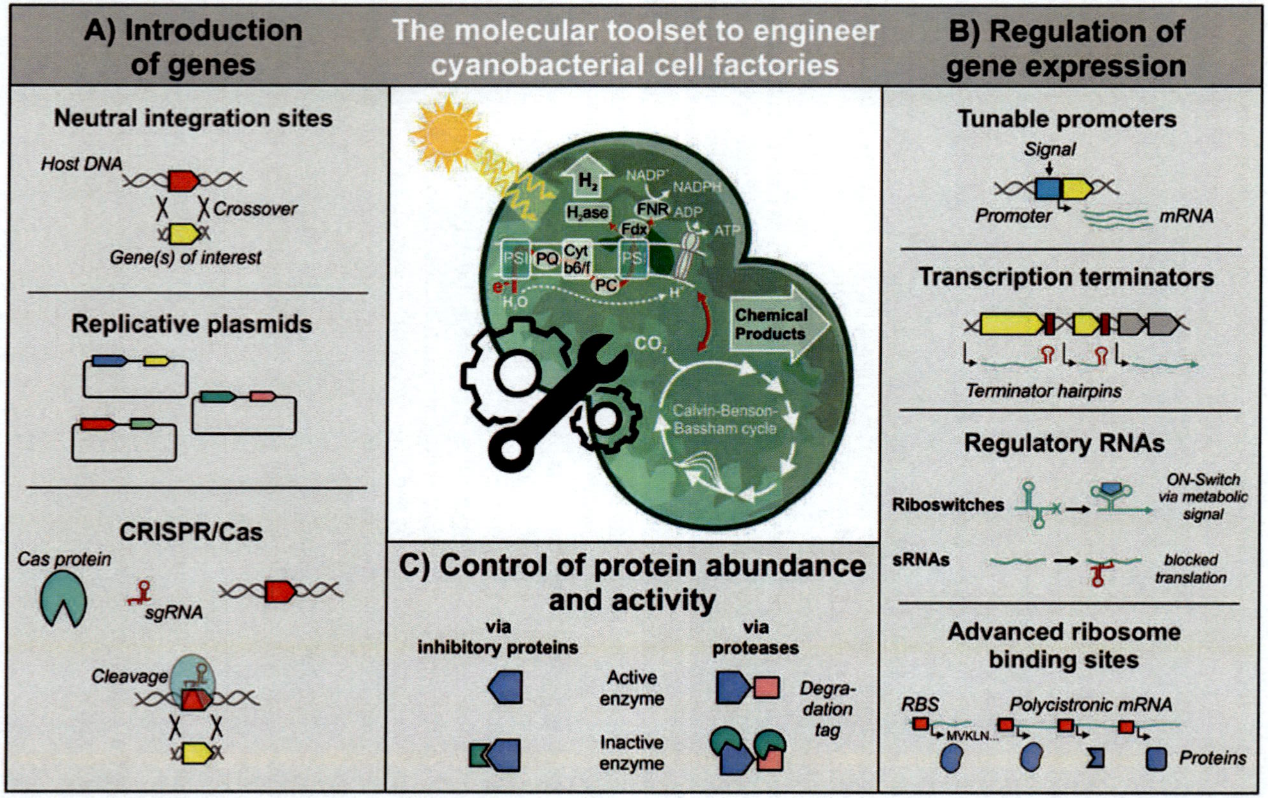

Fig. 1 The molecular toolset for genetic and metabolic engineering of cyanobacteria. (**a**) A selection of available strategies and molecular tools to introduce and express gene(s) of interest is depicted. These encompass neutral genomic integration sites for the homologous recombination (crossover) of genes into a specific locus of the host DNA. Those sites are regarded as neutral as long as their modification is not accompanied by pleiotropic effects affecting the host. Replicative plasmids are extrachromosomal elements that can also maintain genetic information. Clustered regularly interspaced short palindromic repeats (CRISPR) and

CRISPR-associated (Cas) endonucleases introduce specific double-strand breaks as a ribonucleoprotein complex. They are site directed by the single-guide RNA (sgRNA), potentially followed by homologous recombination with a donor fragment lacking the cleavage site (yellow arrow). (**b**) Expression of desired genes can be controlled at the transcriptional level using tunable promoters, which respond to a certain signal and initiate transcription of the downstream gene into messenger RNA (mRNA). Translation initiation is mediated by ribosome binding sites (RBSs) of mono- or polycistronic transcripts. Intrinsic transcription terminators based on secondary RNA structures, like the shown hairpin loop (red), block the RNA polymerase, thereby preventing transcription run-off from introduced genes into adjacent loci. Regulatory RNAs can serve for post-transcriptional or translational control, like a riboswitch via reversible conformational changes induced by a metabolic ligand (blue) or small RNAs (sRNA) that bind to a specific mRNA target sequence. (**c**) Post-translational control of recombinant protein abundance can be controlled by degradation tags that recruit the endogenous or heterologous proteases. Protein activity may also be modulated by direct interactions with other proteins. For details refer to the respective section. Further abbreviations: *PSI & II* photosystem I & II, e^- electron, *PQ* plastoquinone, *Cyt b6/f* cytochrome *b6/f* complex, *PC* plastocyanin, *Fdx* ferredoxin, *FNR* Fdx-NADP$^+$ oxidoreductase, H_2ase NiFe hydrogenase

2 Molecular Tools and Approaches to Genetically Modify Cyanobacteria

Cyanobacteria can basically be edited, i.e., by performing deletion (knockout), insertion (knockin), substitution, or alteration of (native) genes. However, it should be noted that our knowledge is restricted to few representatives so far. In particular, most abundant species that are major drivers of biogeochemical cycles in the oceans, such as *Trichodesmium* or *Prochlorococcus*, still cannot easily be manipulated by current genetic strategies [32]. Some strains, like the widely used unicellular model *Synechocystis*, are naturally competent, which enables the uptake of exogenous DNA. An integration of genetic information into the host genome can subsequently be achieved by homologous recombination at respective sites. Furthermore, genes might also be introduced via replicative plasmids and standard techniques, like conjugation or electroporation [30, 33–36]. So far reported transformation methods available for particular cyanobacterial strains and related literature may be found elsewhere [25, 37, 38], with some examples given here. This section summarizes common molecular tools for the introduction of genes (Fig. 1), including new developments concerning CRISPR/Cas as well as selection methods for recombinant cyanobacteria.

2.1 Integration of Genetic Information into Neutral Sites of the Host DNA via Homologous Recombination

Natural competence is an ecologically important process that is widely distributed among bacteria, enabling them to take up and maintain foreign DNA from the environment. The integration of genetic information into the host genome is achieved through homologous recombination (double crossover) [39]. Thus, this feature allows site-directed mutagenesis, i.e., a targeted genetic modification, which requires to add suitable DNA overhangs up- and downstream of the desired sequence to be inserted at a specific locus on the chromosome. Although many cyanobacteria seem to harbor genes known to be required, e.g., for DNA uptake and processing (type IV pili, Com proteins) as well as homologous recombination (RecA), only a few species, mostly unicellular, were shown to be transformed this way so far [40, 41]. Known strains that can be genetically edited based on natural competence are *Synechocystis* [42], *S. elongatus* 7942 [43], *Synechococcus* 7002 [44], *Thermosynechococcus elongatus* BP-1 [45, 46], as well as the more recently discovered *Thermosynechococcus elongatus* PKUAC-SCTE542 [47], *Synechococcus elongatus* PCC 11801 [48], and *Synechococcus* sp. PCC 11901 [49]. Furthermore, natural competence has been restored in mutants of *S. elongatus* UTEX 2973 (further details see below) [20].

Genetic modification based on homologous recombination requires an integration site in the host genome (Fig. 1). In general, genes of interest are introduced at presumable neutral sites, whose disruption does not cause side effects on cell metabolism while enabling ectopic gene expression from the genome, i.e. the chromosome or endogenous plasmids. Such loci are referred to as neutral, including silent genes of known and unknown function, and intergenic regions with no open reading frame (ORF) annotated. Common neutral sites are, for example, the ORF *slr0168* in *Synechocystis* [50], NSI and NSII in *S. elongatus* 7942 [51], as well as the *glpK* gene in *Synechococcus* 7002 [52]. Nevertheless, their number is rather limited, thereby fostering the search for additional sites to enable extensive genetic modifications. For example, Ng et al. identified a couple of neutral sites on the chromosome and small endogenous plasmids of *Synechocystis* by considering a global map of transcriptional start sites [53] and selecting those regions, from which no transcripts were derived under standard growth conditions, i.e. constant low light illumination (20–50 μmol photons m^{-2} s^{-1}) and ambient CO_2 supply at 30°C [54]. Furthermore, alternative integration sites for *Synechocystis*, including the four large endogenous plasmids, have been recently described by Nagy et al. [55]. These were selected genome-wide according to minimal changes in the transcriptional profile upon transitions from ambient to high CO_2 and from constant to fluctuating light [56, 57]. Presumed neutral sites have further been determined for several *Synechococcus* species, including *S. elongatus* UTEX 2973, *S. elongatus* 7942, and *S. elongatus* PCC 6301 [51, 58–62]. A comparative overview about reported chromosomal integration sites in cyanobacteria is given by Cassier-Chauvat et al. [30].

Well-characterized, dispensable genes have also been targeted by homologous recombination. For example, the *psbA2* locus in *Synechocystis* encoding the D1 protein of photosystem II has been used [63, 64]. It is strongly transcribed under standard (light) conditions but upon deletion is compensated by the functional duplicate *psbA3* [65]. Another strategy would be the substitution of endogenous genes from competing pathways by heterologous enzyme-coding sequences in the course of metabolic engineering. Liu et al. used this approach during an elaborate modular engineering strategy to re-program the carbon metabolism toward an enhanced production of 1-butanol, using a mutant of *Synechocystis* [66]. Furthermore, rational substitution of native genes like *acsA* (encoding acetyl-CoA synthetase) has been performed in *Synechococcus* sp. PCC 11901 [49] and PCC 7002 for markerless selection using acrylic acid [67].

Exogenous DNA for the transformation based on natural competence and homologous recombination can be supplied as linear fragment. Moreover, the foreign DNA can be carried by plasmids that serve for cloning purposes in other bacteria, like *E. coli*, but that are not suitable to replicate in the recipient host, as their replicon is neither self-replicative by encoding its own required factors nor supported by the endogenous DNA replication machinery. The plasmid backbone will therefore be lost during the integration of the harbored genetic information. Those constructs are often called “integrative vectors”, which is however misleading, as no integration of the entire plasmid into the host genome occurs when performing a double crossover. Plasmid-based transformation approaches have been used for various strains, including the filamentous model strain *Anabaena* sp. PCC 7120 (hereafter *Anabaena*)

Table 1 List of selected customized, non-replicative plasmids that have been used for the site-directed mutagenesis of cyanobacteria via homologous recombination

Vector or vector series	Relevant features	References
pAM1303	Sm^R, pMB1, targeting neutral site of *S. elongatus* 7942 chromosome	[59, 70–76]
pRL278	Km^R, pMB1, *sacB* gene for counter selection using sucrose, targeting *patB* gene of *Anabaena* chromosome	[69]
pSL2264-pSL2392 series	Km^R, pUC, *eyfp* reporter, targeting 15 neutral sites of *Synechocystis* chromosome as well as six neutral sites on the three small endogenous plasmids	[54]
pBS-x series	Amp^R, Km^R, pUC, targeting neutral sites 1 and 2 of *Synechocystis* chromosome	[77, 78]
pSN*n** series	Amp^R, f1, targeting 16 neutral sites of *Synechocystis* chromosome	[79]
pEERM series	Different antibiotic resistance cassettes, pMB1, targeting two neutral sites as well as *psbA2* and *sqs* sites of *Synechocystis* chromosome	[63, 66, 80, 81]
px-KO series	Km^R, pMB1, six different toxin genes for counter selection, targeting one neutral site as well as 23 small regulatory RNA sites of *S. elongatus* UTEX 2973 chromosome	[82]

Abbreviations: Amp^R ampicillin resistance, Bm^R bleomycin resistance, Cm^R chloramphenicol resistance, Km^R kanamycin resistance, Nm^R neomycin resistance, Sm^R streptomycin resistance

[68, 69], *S. elongatus* 7942 [60] or *Synechocystis*, whereby the plasmids were adapted according to the flanking homologous regions of the insert matching the DNA of the target host. Examples of such customized non-replicative vectors can be found in Table 1.

Additional factors to be considered, besides the targeted genomic site, are the strain-specific frequency for the occurrence of single and double recombination events as well as the optimal length of the flanking overhangs [36]. Homologous recombination between the supplied fragment and the wild type genomic site can occur as a single or a double crossover event, depending on whether one or both flanking regions are involved. Single recombination will thereby most likely lead to target-specific insertion of the entire sequence, whereas double crossover probably results in insertion of (or native gene replacement by) the recombinant cargo between the overhangs [36, 60, 82].

Natural competence and homologous recombination of cyanobacteria are further regulated by currently unknown factors, e.g., associated to growth phase and circadian rhythm [41]. For example, *Synechocystis* and *Synechococcus* 7002 were shown to be most competent in exponential growth phase [33]. Furthermore, *S. elongatus* 7942 shows the highest transformation efficiency at the onset of darkness during the dusk circadian peak when grown under light–dark cycles [83].

Certain genes known to be required for natural competence (see above) may be absent or altered in particular strains. In this case, they can potentially be replaced by functional copies to enable their transformation by natural competence [41]. This procedure has been done, for example, in *S. elongatus* UTEX 2973. Thereby, a second, functional copy of the *pilN* gene encoding a subunit of the pilus complex that might be involved in the translocation of exogenous double-stranded DNA into the periplasm [84] from *S. elongatus* 7942 was introduced to compensate for a mutation in the native counterpart, recovering its natural competence [20].

Altogether, transformation of cyanobacteria based on homologous recombination presents a versatile method for gene deletions and substitutions. It also enables the insertion of, e.g., large coherent sequences up to 20.8 kb that have been integrated into one neutral site of *Synechocystis* [85]. Current bottlenecks present, for example, the lack of natural competence for many species as well as the identification and availability of suitable genomic integration sites. Other obstacles for genetic modification via natural competence present physical barriers and defense mechanisms against foreign DNA [86], including, e.g., DNA degradation by exonucleases as well as restriction-modification systems involving endonucleases that may cut heterologous DNA due to a non-matching methylation pattern [41]. Both mentioned problems could potentially be circumvented by the deletion of native exo- and endonucleases [87], or pre-methylation of foreign DNA [78, 88], which were shown to improve transformation efficiency. Furthermore, polyploid strains require time-consuming segregation procedures for the selection of transformants in which finally all chromosome copies contain the modified DNA. Some cyanobacteria, like several *Prochlorococcus* strains, are known to be monoploid, whereas polyploidy, or oligoploidy, is a common trait, including, e.g., *Synechocystis*, *S. elongatus* 7942, *Synechococcus* 7002, or *Anabaena* (oligoploid) [89, 90]. Methods that are independent of homologous recombination are therefore indispensable for a fast and straight-forward insertion of genetic information in cyanobacteria.

2.2 Introduction and Maintenance of Genes via Replicative Plasmids

Replicative plasmids are extrachromosomal elements that are autonomously maintained, either by the endogenous replication machinery or in a host-independent manner (see below). Thus, no genetic segregation of polyploid cyanobacteria has to be performed after transformation. However, common cloning vectors used in *E. coli* are not replicating in cyanobacteria [30], which is why the available number of plasmids that could be used as shuttle vectors is rather low. Derivatives of the broad-host-range plasmid RSF1010 [91] are commonly used in a variety of cyanobacterial strains [37, 92]. The maintenance of RSF1010-based plasmids is host-independent as they encode their own replication factors and hence, work in a wide range of Gram-negative bacteria [93]. In 1990, the first utilization in cyanobacteria was reported for *Synechocystis* [94]. Later on, a whole library of derivatives, namely the pVZ series, was established [95], though still hampered by large size and cloning difficulties. Therefore, RSF1010-derived plasmids underwent several optimization and size-reduction steps resulting in various lineages, e.g. pPMQAK1 [12], pSL1211 [96], pFC1 [97], pDF [98], pSHDY [99, 100], and the pSEVAx5n series [14, 101]. Notably, *Synechocystis* has also been shown to maintain another replicative plasmid with broad-host range based on the replicon of RK2 [26, 102].

Only one origin of replication of the same incompatibility group can be stably maintained in one host, thus, two different plasmids based on the same replicon, e.g. RSF1010, are not suitable for genetic modification of bacteria. Due to the lack of such alternative plasmids for cyanobacteria, several studies have been performed aiming at the generation of synthetic shuttle vectors derived from a strain-specific endogenous plasmid. These are composed of two fused replicons: one mediating replication in certain cyanobacteria and another for the maintenance in *E. coli*. Similarly, new vectors of choice could be designed. Many of such representatives were designed for filamentous strains that are difficult to be transformed by homologous recombination, like *Anabaena*. However, some also exist for *Synechocystis*, *S. elongatus* 7942, and *Synechococcus* 7002. A selection of cyanobacterial shuttle vectors with their encountered host range is listed in Table 2.

Replicative plasmids can be introduced into the host cell via conjugation, if they contain the genetic information necessary for DNA transfer (see below), or via electroporation of electro-competent cells, whereby the latter spares the need for subsequent microbial segregation [33, 35, 36, 123]. Conjugal transfer does not rely on natural competence but is rather carried out by transient co-cultivation of the recipient cyanobacterial strain and usually one or two *E. coli* strains. The *E. coli* cells carry the vector to be transferred (named cargo or donor) as well as another vector (named helper; the plasmid or its features are either provided by the same or by a separate cell) to facilitate its mobilization via the RP4 conjugative transfer apparatus. Further, the helper provides DNA methylases protecting the target DNA against degradation by the recipient organisms' endonucleases. The cargo vector therefore needs to harbor a vegetative origin of replication (*oriV*) for *E. coli* as well as an origin of transfer (*oriT*), and either the helper or the cargo plasmid need to encode mobilization (*mob*) genes [35, 124–126]. Conjugal transfer has been described for various cyanobacteria, including, for example, *Anabaena* [103], *Synechocystis* strains PCC 6803 [127] and PCC 6714 as well as *S. elongatus* PCC 7942, PCC 6301 [125] or UTEX 2973 [127, 128].

2.3 Modifications Based on CRISPR/Cas

Clustered regularly interspaced short palindromic repeats (CRISPR) and CRISPR-associated (Cas) proteins are part of a natural prokaryotic RNA-based defense mechanisms against viruses and foreign DNA and have been described for many know prokaryotes, including cyanobacteria [129–131]. Meanwhile, molecular procedures based on the CRISPR/Cas mechanism have been established and became an important technique for precise editing of prokaryotic and eukaryotic genomes

Table 2 Examples of cyanobacterial shuttle vectors derived from endogenous plasmids and their reported host range (in addition to *E. coli*)

Shuttle vector or shuttle vector series	Derived from endogenous plasmid/host strain	Strains for which plasmid maintenance was reported	References
pRL series	pDU1/*Nostoc* sp. PCC 7524	*Anabaena*, *Nostoc* sp. PCC 7524, and *Anabaena* sp. M-131	[103]
pRL2697	pDU1/*Nostoc* sp. PCC 7524	*Anabaena*	[104]
pRL25C	pDU1/*Nostoc* sp. PCC 7524	*Anabaena*, *Fischerella muscicola* PCC 7414, *Chlorogloeopsis fritschii* PCC 6912, and *Fischerella thermalis* (also known as *Mastigocladus laminosus*) SAG 4.84	[105–107]
pDUCA7	pDU1/*Nostoc* sp. PCC 7524	*Anabaena*; *Chroococcidiopsis* spp. str. 029, 057, and 123	[108, 109]
pRL489	pDU1/*Nostoc* sp. PCC 7524	*Anabaena*; *Chroococcidiopsis* spp. str. 029, 057, and 123; *Oscillatoria* MKU 277	[109–111]
pRL1050	pDU1/*Nostoc* sp. PCC 7524	*Anabaena*	[112]
pSCR119/202	pDC1/*Nostoc* sp. MAC PCC 8009	*Nostoc punctiforme*	[113]
pSUN119/202	pDC1/*Nostoc* sp. MAC PCC 8009	*Nostoc punctiforme*	[114]
pPBH201	pGL3/*Plectonema boryanum* sp. PCC 6306	*Anabaena* and *Plectonema boryanum* sp. PCC 6306	[115]
pAQE series	pAQ1/ *Synechococcus* 7002	*Synechococcus* 7002	[116]
pJCF series	pFDA/*Fremyella diplosiphon*	*Fremyella diplosiphon*	[117]
pAM series	pANS (also called pUH24)/ *S. elongatus* 7942	*S. elongatus* 7942 and *Anabaena*	[75]
pLS103	pANS/*S. elongatus* 7942	*S. elongatus* 7942	[118]
pSG111	pANS/*S. elongatus* 7942	*S. elongatus* 7942	[119]
pCA-UC118	pCA2.4/ *Synechocystis*	*Synechocystis*	[15]
pSOMA10–14	pCA2.4/ *Synechocystis*	*Synechocystis* and *Anabaena*	[120]
pCB-SC101	pCB2.4/ *Synechocystis*	*Synechocystis*	[15]
pSOMA15–19	pCB2.4/ *Synechocystis*	*Synechocystis*	[120]

(continued)

Table 2 (continued)

Shuttle vector or shuttle vector series	Derived from endogenous plasmid/host strain	Strains for which plasmid maintenance was reported	References
pSCB series	pCC5.2/ *Synechocystis*	*Synechocystis*	[121]
pCCM1-FbFP	pCC5.2/ *Synechocystis*	*Synechocystis*	[122]

Table 3 Examples of markerless selection methods in cyanobacteria

Method	Applied strains	References
Sucrose-mediated counter selection by applying the *sacB* gene in combination with an antibiotic resistance	*Anabaena*, *Synechocystis,* and *S. elongatus* 7942	[26, 60, 68, 136–139]
Acrylate tolerance upon endogenous *acsA* gene loss of function disruption	*Synechococcus* 7002 and *Synechococcus* sp. PCC 11901	[49, 67]
Inducible expression of toxin genes for counter selection in combination with an antibiotic resistance	*Synechocystis* and *S. elongatus* UTEX 2973	[82, 140–142]
Free fatty acid tolerance upon endogenous *aas* gene loss of function disruption	*Synechocystis* and *Synechococcus* 7002	[143]
Inducible expression of Flp/FRT recombination system for counter selection in combination with an antibiotic resistance	*Synechocystis* and *S. elongatus* 7942	[144]
Inducible expression of Cre/Lox recombination system for counter selection in combination with an antibiotic resistance	*Synechocystis* and *Anabaena*	[145]
CRISPR/Cas	*Synechocystis*, *S. elongatus* UTEX 2973, *S. elongatus* 7942, *Anabaena*, and *S. elongatus* PCC 11801	[62, 122, 146–150]

Abbreviations: *sacB* gene: gene encoding levansucrase, *acsA* gene: gene encoding acetyl-CoA synthetase, *aas* gene: gene encoding acyl-ACP synthetase, Flp/FRT: flippase recombinase and flippase recognition target sites, Cre/Lox: Cre recombinase and Lox target sites, CRISPR/Cas: clustered regularly interspaced short palindromic repeats and CRISPR-associated protein

[132]. Synthetic CRISPR-based techniques, including CRISPR/Cas9 and CRISPR/Cas12, have been applied in cyanobacteria for markerless selection, introduction of point mutations, gene deletions and insertions, and simultaneous manipulation of multiple genes in some strains (see Table 3) [28, 133]. Generally, a single-guide RNA (sgRNA) needs to be designed to specifically target a DNA locus, where it forms a ribonucleoprotein complex with the cognate Cas endonuclease (Fig. 1). Upon expression of both components, the system introduces a double-strand break of the genomic DNA, which needs to be repaired and edited in a way that the sgRNA does no longer bind to the former cleavage site. The cyanobacterial DNA repair mechanisms are homologous recombination and non-homologous end joining. Non-homologous end joining may be sufficient to gain inactivating random mutations, whereas specific alterations can be achieved by homologous recombination in

the case of a supplied donor fragment that harbors the desired sequence flanked by homology arms [23, 28, 133], see Sect. 2.1. The field of CRISPR-mediated genetic modification in cyanobacteria is still emerging and will likely become more fundamental during the next years. Limitations that have to be overcome are the significant occurrence of false positive transformants without the desired genetic modification, toxicity of the foreign Cas proteins, low specificity of the cleavage complex leading to genomic off-targets, and the lack of utilizable endogenous CRISPR/Cas systems [23, 28, 133].

2.4 Marker-Based and Marker-Free Selection of Recombinant Strains

The classical microbiological workflow to genetically engineer bacteria includes the combination of the desired sequence(s) to be introduced; for example, a gene of interest, with a suitable selection marker, most often genes encoding the resistance against certain antibiotics. Thus, recombinant strains that are able to grow on media supplemented with the respective antibiotics can be discriminated against those not carrying the desired modification. Final confirmation of genetic modifications can be performed by PCR and Sanger sequencing, or a classical Southern Blot. Common antibiotics for effective selection in cyanobacteria are erythromycin, neomycin/kanamycin, spectinomycin/streptomycin, chloramphenicol, and gentamicin with varying concentrations [36, 134].

Markerless selection of cyanobacteria might be of interest for multiple consecutive gene insertions, selection marker recycling, and concerning stable maintenance of an inserted construct in the genome without selection pressure. In the course of genetic modification via homologous recombination, a lethal "killer" gene (negative selection marker) may be used in addition to the antibiotic resistance (positive selection marker) to remove the undesired regions of the introduced sequence from the host genome, including the markers themselves. For this, initial transformation via the positive selection marker can be followed by a second step employing the lethal gene [23, 24]. Thereby, only those transformants that lost the negative selection marker are able to grow upon supplementation of a certain substrate (see Sect. 2.1.). A frequently reported example of such two-step selection in Gram-negative bacteria resulting in unmarked transformants is the employment of the *sacB* gene encoding the levansucrase that converts sucrose into toxic products [135]. Thereby, removal of both markers, i.e. the antibiotic resistance and the *sacB* gene, requires first the selection of clones on the respective antibiotic followed by supplementation of sucrose that leads to a consecutive recombination step, whereby only the desired DNA is retained [135]. This counter-selection procedure has been successfully applied in various cyanobacterial strains [36, 51, 60, 136]. Saar et al., for example, used this sucrose-mediated approach to introduce up to five genomic modifications in one strain of *Synechocystis* [137]. Moreover, Dietsch et al.

introduced markerless genomic deletions in *Synechocystis* to boost isoprene-derived production in cyanobacteria [138].

Selection based on the phenotype may be achieved without antibiotics by direct insertions of heterologous genetic information without any selection marker into a specific genomic site, like the *ascA* locus, whose disruption abolishes sensitivity toward acrylate [67]. Most recently, recombinant CRISPR-based modification methods (see Sect. 2.3.) have been applied in a range of cyanobacterial species. They rely on cell death due to specifically cut genomic DNA, while the respective recognition sites are altered and thereby protected in positive transformants. Although genome editing can be done this way without the need of additional selection and is therefore considered markerless, initial implementation of the heterologous CRISPR/Cas components may require a selection marker [23, 28, 133]. A list of examples for markerless selection tools can be found in Table 3.

3 Customized Biological Parts to Control the Expression of Introduced Genes

Biological parts (also genetic parts) enable biological functions, such as a tightly controllable gene expression. In this regard, several biological parts have to be assembled on DNA level in close proximity to the gene(s) of interest to mediate transcription and translation in a defined, i.e., regulated manner. Genetic organization of the so-called expression cassettes in bacteria mainly encompasses regulatory elements located upstream of the gene of interest. These include the promoter, a DNA sequence that recruits the RNA polymerase and hence, mediates the transcription of the downstream region starting at the transcriptional start site (TSS). Further, a ribosome binding site (RBS) is localized in the 5' untranslated region (5' UTR) of the mRNA to initiate protein biosynthesis at the start codon of an ORF. A transcription terminator can be included downstream of the gene(s) of interest, completing the insulated synthetic expression unit. Genes may be organized as monocistronic expression unit or polycistronic cluster in an operon, with an RBS in front of each single gene. Moreover, additional linkers are potentially needed for the spatial interconnection and optimal spacing of the respective biological parts.

The expression level for a gene of interest, which may correlate to the strength of a specific phenotype, e.g., the abundance and overall activity of an enzyme, is an important aspect with respect to biotechnological processes. In parts this depends on the gene copy number (gene dosage) and thus, whether the genes are located on the chromosome or on replicative plasmids [55]. However, gene expression is mainly determined by regulatory elements that control the molecular processes of mRNA and/or protein accumulation, depending on the synthesis and the half-live of both molecules [151, 152]. In addition, it is also of importance to control and fine-tune a biocatalytic process, and not only switch an introduced metabolic route entirely on or off. For this, molecular tools enabling high as well as tunable gene expression are required.

3.1 Promoters

Bacterial promoters contain core elements of conserved sequences around positions −35 and −10 upstream of the TSS (usually referred to as +1). These are recognized by sigma (σ) factors, mainly σ^{70} (also called Rpo^D or σ^D) that are required to recruit the RNA polymerase and together form the active holoenzyme. Tunable promoters that can be controlled via environmental or metabolic stimuli, commonly contain further sequence motifs that are recognized by transcriptional regulators acting either as repressors or as activators of transcription. Promoters are the most commonly used and best-studied biological parts for regulating gene expression, and there have been numerous studies on them. Promoters can be derived from the host, a different organism, or even be synthetic. Furthermore, promoters can be divided into constitutive promoters, which are constantly active, and inducible promoters that respond to a stimulus and hence, need further regulatory factors. With respect to biocatalysis, inducible promoters are preferred, as they enable a defined time point of expression and might be also titratable for a tight as well as a dynamic dose response. However, such tightly regulated promoters enabling high tunability are rare in cyanobacteria, or inherit downsides, like the need for toxic or light-sensitive inducers. Furthermore, heterologous promoters frequently used in *E. coli* are often impaired by leakiness (high basal expression when supposed to be off) or low expression output.

A comprehensive overview of currently available promoters for cyanobacteria can be found elsewhere, e.g. in Zhang and Song [22] and Wang et al. [134]. Expression strength, induction control, and the dynamic range of promoters have also been recently summarized [29]. Alternatively, cyanobacterial sigma factors may also be considered for biotechnological applications employing native promoters [153, 154]. Particularities of fundamental cyanobacterial transcription and RNA polymerases are reviewed elsewhere [155, 156]. This section focuses on commonly used and broadly applicable promoters for genetic engineering of a variety of different cyanobacterial species given in Table 4.

3.1.1 Homologous Promoters Derived from Cyanobacteria

Several native, predominantly strong promoters which are often derivatives of P_{cpc} or P_{psbA} have been used in cyanobacteria [29]. Those promoters natively drive the transcription of genes for highly abundant components of the photosynthetic apparatus such as the phycobilisomes (P_{cpc} [196]) or the D1 protein of photosystem II (P_{psbA} [65]). For *Synechocystis*, even a "super strong" derivative, P_{cpc560}, was described that yielded a protein with a concentration of 15% of total protein content [164]. The same promoter has also been recently utilized in *S. elongatus* UTEX 2973 [26, 166]. P_{psbA} promoter derivatives are regulated by light intensity, although being constitutive under constant light illumination. P_{psbA} variants have been applied in several species, including *Synechocystis* [13, 16, 173–175] and *S. elongatus* 7942 [92, 169–172]. Further common examples for *Synechocystis* are P_{rbc} variants

Table 4 Commonly used promoters and their application in different cyanobacterial strains

Promoter name	Source	Mode of transcription	Applied strains	References
BioBrick BBa_J23101	Synthetic	Constitutive	*Synechocystis*, *S. elongatus* 7942, and *Anabaena*	[13, 120, 157–160]
BioBrick BBa_J23119	Synthetic	Constitutive	*Synechocystis*, *S. elongatus* 7942, *Synechococcus* 7002, *Anabaena*, and *S. elongatus* UTEX 2973	[13, 19, 26, 99, 146, 160, 161]
P_{conII}	*E. coli*, semisynthetic	Constitutive	*Synechocystis*, *S. elongatus* 7942, *Synechocystis* sp. strain WHSyn, *Anabaena*, and *Leptolyngbya* BL0902	[37, 92, 160, 162]
P_{isiAB}	*Synechocystis*	Inducible with Fe^{3+}	*Synechocystis*; *S. elongatus* 7942, *Anabaena*, and *Leptolyngbya* BL0902	[50, 92, 163]
P_{phoA}	*S. elongatus* 7942	Inducible with inorganic phosphorus	*Synechocystis*, *Synechococcus elongatus* sp. PCC 7942, *Anabaena*, and *Leptolyngbya* BL0902	[92]
P_{cpc560}	*Synechocystis*	Constitutive	*Synechocystis* and *S. elongatus* UTEX 2973	[16, 26, 54, 99, 164–166]
P_{psbA}	*Amaranthus hybridus*	Constitutively active in the light	*Synechocystis*, *S. elongatus* 7942, *Anabaena*, and *Synechococcus* 7002	[16, 110, 167, 168]
P_{psbA1}	*S. elongatus* 7942	Constitutive, inducible with low light, repressed at high light	*S. elongatus* 7942, *Synechocystis*, *Anabaena*, and *Leptolyngbya* BL0902	[92, 169–172]
P_{psbA2}	*Synechocystis*	Constitutively active in the light	*Synechocystis*	[13, 16, 173–175]
P_{rbc} different variants	*Synechocystis*	Constitutive	*Synechocystis*	[12, 13, 174, 176]
P_{rnpB}	*Synechocystis*	Constitutive	*Synechocystis*	[12, 13, 158]
P_{nrsB}	*Synechocystis*	Inducible with Ni^{2+} or Co^{2+}	*Synechocystis*	[63, 120, 177–180]
P_{petE}	*Synechocystis*	Inducible with Cu^{2+}	*Synechocystis*	[13, 98, 158, 176, 181–184]
P_{petE}	*Anabaena*	Inducible with Cu^{2+}	*Anabaena*	[161, 185–189]
P_{trc1O}	Synthetic hybrid promoter from *E. coli*	Constitutive	*Synechocystis*	[12, 16, 98, 157, 158]

(continued)

Table 4 (continued)

Promoter name	Source	Mode of transcription	Applied strains	References
P_{rhaBAD}	*E. coli*	Inducible with rhamnose	*Synechocystis*	[99, 190–192]
P_{L03}	*E. coli* synthetic derivative	Inducible with anhydrotetracycline	*Synechocystis* and *Anabaena*	[99, 157, 161, 189, 193–195]

[12, 13, 174, 176] and P_{rnpB} [12, 13, 158], which are responsible for the transcription of the genes encoding likewise highly abundant ribulose-1,5-bisphosphate carboxylase/oxygenase (RubisCO) and the RNA subunit of RNase P, respectively. However, both promoters, P_{rbc} and P_{rnpB}, showed only weak or moderate strength. Recently, Sengupta et al. generated a library of different P_{rbc} and P_{cpc} versions that function, apart from *E. coli,* in *S. elongatus* 7942 and fast-growing *Synechoccocus* strains, suggesting a broad host range [197].

Inducible promoters preferably allow tight regulation with no basal gene expression, fast response upon induction, and increasing expression levels correlating with the applied inducer amounts (dose response). On the one hand, native promoters are compatible with the host transcription machinery without the need of recombinant transcriptional regulators. On the other hand, the molecular mechanisms that are responsible for their inducibility are often not known and the inducers interfere with the host metabolism, which limits the orthogonal regulation ability [12, 29, 98]. Frequently, heavy metal-ion responsive promoters are used in cyanobacteria, which are natively related to metal homoeostasis. They are often tightly regulated, but impaired by inducer toxicity at higher concentrations [13, 29, 96]. For example, the *nrsb* promoter of *Synechocystis* is inducible by nickel (Ni^{2+}) and cobalt (Co^{2+}) ions, as it drives transcription of genes coding for associated efflux pumps [198]. P_{nrsB} has been proven to be the most versatile, because it allows tight control as well a good tunability while facilitating rather high expression levels when induced with Ni^{2+} and Co^{2+} or only Ni^{2+}, compared to other native promoters from *Synechocystis* [13]. It has been applied in several studies in *Synechocystis* [63, 120, 177–180]. Another common example is the copper ion (Cu^{2+})-inducible *petE* promoter, showing moderate expression strength but also leakiness. It has been widely used in *Synechocystis* [13, 98, 158, 176, 181–184] and *Anabaena* [161, 185–189]. More examples for native inducible promoters that have been applied in different strains can be found in Table 4.

3.1.2 Heterologous Promoters Derived from Other Sources

Heterologous promoters include synthetic, semisynthetic, and newly assembled hybrid parts or those derived from a different organism, like *E. coli*. They can be used in a constitutive mode or in combination with the required recombinant transcriptional regulators. Widely used heterologous constitutive promoters in cyanobacteria are representatives of the synthetic J23-series from the Anderson collection that confer different expression strengths [199] or the semisynthetic P_{conII} [92, 162] (see Table 4 for details). Commonly used lactose-, or rather isopropyl β-d-1-thiogalactopyranoside (IPTG)-inducible *lac* promoters derived from *E. coli* often perform poorly or display high leakiness due to inefficient repression in cyanobacteria [29, 158]. However, some *lac* promoters, such as the synthetic hybrid promoter P_{trc1O} *of E. coli*, have been used for constitutive expression without the LacI repressor in *Synechocystis* [12, 16, 98, 157, 158]. Another heterologous, sugar-responsive promoter that performs well in cyanobacteria is the rhamnose-inducible P_{rhaBAD}, which additionally requires activation by the recombinant transcriptional regulator RhaS. P_{rhaBAD} has displayed low leakiness and no inducer cross-reactions, while allowing good tunability in *Synechocystis* [99, 190–192]. De-repression of recombinant TetR through orthogonal induction by anhydrotetracycline may be performed with the synthetic *L03* promoter derivative of the *E. coli* P_{tet} that has been used in *Synechocystis* [99, 157] and *Anabaena* [161, 189, 193] for tight expression control, but suffers from photolability of the inducer, anhydrotetracycline.

Non-native RNA polymerases, such as those originating from the bacteriophage T7, and their cognate promoters (P_{T7}) are another option. This approach offers the advantage of lower crosstalk with the host transcription machinery, as the endogenous RNA polymerase is thereby only needed for the synthesis of the recombinant polymerase. The transcription of the gene(s) of interest that are fused to P_{T7} is then specifically performed by the orthogonal RNA polymerase. Furthermore, bacteriophage-transcription systems feature robust, rapid, and strong performance. For these reasons, T7 RNA polymerases have been previously introduced in various microbes including *E. coli*, where they even have become a routine application for strong gene (over)expression [200]. Examples for the usage of T7 RNA polymerases in cyanobacteria are given for *Synechocystis* [165, 201], *Anabaena* [112, 202], *Synechococcus* 7002 [203], *S. elongatus* 7942 [17, 204], and *S. elongatus* UTEX 2973 [20]. Thereby, it proved to be a strong transcriptional system that caused high reporter signals comparable to other strong promoters that instead utilize the endogenous RNA polymerase.

3.2 Transcription Terminators

In addition to transcription initiation mediated by the promoter, termination of transcription is also crucial for sufficient expression of genes of interest. In particular, transcription terminators ensure RNA polymerase detachment after transcript elongation and are hence located downstream of genes. Termination is either dependent on the endogenous Rho transcription termination factor or by intrinsic secondary structures, like hairpin loops. The latter is believed to be the predominant mechanism in cyanobacteria, as no Rho homologs could be identified [205]. Customized transcription terminators can be used in synthetic expression units to prevent read-through into adjacent loci and thus, stabilize the inserted sequence while insulating it from the downstream regions [191, 206]. Native, synthetic, and heterologous transcription terminators have been evaluated in *Synechocystis* with respect to efficiencies and final gene expression output [15, 191, 207]. A common example is the synthetic hybrid BioBrick BBa_B0015 double terminator (also referred to as T_{rrnB}) [12, 199]. Gale et al. recently compared 34 Rho-independent intrinsic terminators. Therein, five parts with a termination efficiencies of >96% were identified for *E. coli*, *Synechocystis*, and *S. elongatus* UTEX 2973, including BBa_B0015 [207].

3.3 Ribosome Binding Sites

Ribosome binding sites (RBSs) are commonly required for the translation of an mRNA into an amino acid sequence. They are spatially situated in the 5' UTR of mRNAs, i.e. between the TSS and the start codon of ORFs. In case of operons, the polycistronic mRNAs contain RBSs upstream of each ORF (Fig. 1). Classical bacterial RBSs contain a conserved Shine–Dalgarno (SD) core sequence (5'-GGAGG-3') that is complementary to the anti-SD sequence of the 16S ribosomal RNA, which is required to direct the start codon into the correct position in the ribosome. Regulatory promoter regions, including the associated RBSs, can be utilized upstream of protein-coding sequences. For example, the native 5' UTR of P_{cpc560} was shown to promote the highest translation in *Synechocystis* [26]. RBSs can also be separately inserted and artificially combined with different promoters, flanked by suitable linkers. For *Synechocystis*, the optimal spacing between the centered A of the SD core sequence (underlined) and the first base of the start codon (5'-ATG-3') was evaluated to be 9–11 bp [36, 208].

Moreover, customized RBSs can be employed to build artificial operons that are transcriptionally regulated by a single inducible promoter, which in turn results in polycistronic co-transcription of several genes, e.g. encoding whole pathways triggered by one stimulus. A minimal RBS that was associated with high protein synthesis is the synthetic RBS* with a size of only 10 nt, which contains the consensus SD sequence [36]. Another example for a strong synthetic RBS is BioBrick BBa_0034 [199], which yielded high expression levels for two different fluorescent reporter genes in *Synechocystis* [13]. Synthetic and native RBSs have been characterized, e.g., in *Synechocystis* [13, 15, 16, 36, 209, 210], and *S. elongatus* 7942 [211]. Thus, RBSs of different strengths are available to control or at least determine gene expression at the translational level.

However, the translation efficiency mediated by the RBS generally depends largely on the respective sequence and cellular context, which is, compared to general model bacteria such as *E. coli*, less predictable in cyanobacteria [19, 209, 212]. Nevertheless, in silico tools exist and have also been rarely applied to cyanobacterial strains [22]. For example, the RBS calculator [3] was used to optimize RBSs for the expression of two different fluorescent reporter genes, whereby the output signal was enhanced in each case by several folds in *Synechocystis*, *S. elongatus* 7942, *Anabaena*, and *Leptolyngbya* BL0902 [92]. The RBS calculator completed with UTR Designer [213] and RBS Designer [4] modeling software was furthermore used in *Synechococcus* 7002 for the design and evaluation of a synthetic RBS library, which allowed the prediction of gross changes in the final gene product [19]. Thereby, the correlation between actual reporter signals and predicted strengths was rather weak, suggesting that fine control of translation may not be achieved using these in silico tools.

3.4 Regulatory RNAs

In addition to transcriptional regulators, bacteria possess numerous and diverse means of RNA-mediated gene regulation. These regulatory RNA elements do not encode proteins and thus are non-coding. RNA-based molecular tools display great potential for metabolic engineering, as they, for example, pose only a minor metabolic burden on the host [214]. Those systems may be used in combination with other biological parts, like inducible promoters, to enhance tunability of gene expression and/or to lower the problem of leakiness in cyanobacteria.

3.4.1 Small RNAs

A major group of those non-coding RNAs, called small regulatory RNAs (sRNAs), can activate or repress gene expression at post-transcriptional level by complementary base pairing with mRNAs and contribute to the specific and customized synthesis of the respective proteins [215]. These elements can be divided into two classes, *cis*- and *trans*-encoded sRNAs [216]. With respect to their target genes the *cis*-encoded sRNAs are usually encoded at the same DNA locus but on the opposite strand and hence show long and perfect complementarity with the targeted mRNA. These elements are also called antisense RNAs (asRNAs). In contrast, the *trans*-encoded sRNAs show short, imperfect base pairing interactions but frequently overlap with sequences required for translation initiation [215]. The application of sRNAs as molecular tools to engineer gene expression in cyanobacteria has been recently reviewed [21, 217]. For example, a distinct down-regulation (knockdown) of a gene of interest could be achieved by expressing an artificial asRNA complementary to the desired target mRNA. This strategy has been applied, e.g., in *Anabaena* with, however, rather low regulatory efficiency [193, 218].

A specific situation combining RNA secondary structures with sRNA functions are the so-called riboregulators, which have been used as molecular tools for the genetic engineering of cyanobacteria, reviewed by Ueno and Tsukakoshi et al. [219]. For example, the synthetic riboregulatory crR*2/taR*2 was described in *E. coli*. It represses translation via a *cis*-repressing RNA by stem-loop formation upstream of the RBS, which is reversed by the distinct binding of a *trans*-activating

RNA [220]. Variants of this translation ON riboregulator have been employed in *Synechocystis* [221–224] and *Anabaena* [194], enabling low leakiness and up to 78-fold dynamic range. Studies in *Synechococcus* 7002 were conducted with another *E. coli*-derived sRNA-based tool, IS10 [225], which led to 70% effective translation repression [226]. Versatile, yet still leaky regulation was achieved in a study by Sun et al., using two synthetic sRNA-based tools in *Synechocystis* [227]. The applied paired termini asRNAs (PTRNAs) [228] as well as Hfq chaperone and MicC scaffold (Hfq-MicC) [229] have been previously introduced in *E. coli.* Thereby, Hfq-MicC, which in combination with an asRNA leads to a site-directed mRNA degradation, enabled simultaneous multiple gene regulation affecting fatty acid synthesis. Additionally, a re-direction of carbon flux from competing pathways toward malonyl-CoA was achieved, leading to a 41% increased production compared to the wild type [227]. Very recently, Sun et al. introduced a genetic switch inducible by *N*-acetylneuraminic acid that displayed orthogonality in combination with a theophylline-responsive riboregulator for a defined binary gene transcription regulation [230].

3.4.2 Riboswitches

Riboswitches are another major group of regulatory RNAs, which are part of untranslated regions (UTRs) of mRNAs and can regulate gene expression by ligand-induced structural modulation [231]. Riboswitches are composed of an aptamer, which specifically binds the ligand (e.g., metabolites, inorganic ions), and an expression platform, which determines the read out of genetic information by interfering with the transcriptional or translational machinery in response to the structural modulation induced by the respective aptamer [232]. In cyanobacteria, for example, glutamine-responsive riboswitches are involved in controlling nitrogen assimilation [159]. They appear to be unique to cyanobacteria as the glutamine-binding aptamers are not present in other bacterial genomes. Besides natural riboswitches that respond to metabolic signals in the host, synthetic riboswitches can be fused upstream to the gene of interest. Most riboswitches that have been used to trigger heterologous gene expression in cyanobacteria are synthetic, e.g. those that activate translation in the presence of theophylline [29]. For example, representatives of the theophylline-responsive translation ON riboswitches A-F [233] have been used in *Synechocystis* [192, 227, 234], *S. elongatus* 7942 [162, 235], *S. elongatus* UTEX 2973 [20, 82], and *Anabaena* [161, 162], whereby mainly variant F allowed tight control and a high dynamic range. An example for a cyanobacterial-derived riboswitch is the transcriptional OFF cobalamin-repressible riboswitch from *Synechococcus* sp. PCC 73109, applied in *Synechococcus* 7002, which showed low leakiness with an induction fold of ~6 [236].

3.5 *Knockdown via CRISPR Interference*

CRISPR interference (CRISPRi) is a rather new branch of genetic engineering that, opposed to CRISPR/Cas, relies on a catalytically inactive deficient endonuclease protein, like dead or deactivated Cas9 (dCas9), which is guided to a specific locus via the sgRNA, but does not introduce a double-stand break of the DNA. Instead, this ribonucleoprotein complex simply binds its target sequence and thus, reversible blocks the transcription of a downstream gene. Therefore, the design of a target-specific sgRNA in combination with the controlled expression of both proteins, dCas9 and target protein, as well as the RNA component may be used as knockdown strategy [237]. CRISPRi has been applied in *Synechocystis* [138, 192, 195, 238, 239], *S. elongatus* 7942 [160, 240, 241], *Synechococcus* 7002 [242], *Anabaena* [161], and *S. elongatus* UTEX 2973 [243]. Thereby, effective repression of target genes has been achieved, but requires a tight promoter for the expression of the CRISPRi components for full activity under non-repressing conditions. Liu et al., for example, combined a theophylline-responsive riboswitch and a rhamnose-inducible promoter to activate the biosynthesis of a CRISPR/dCas12a complex. Therein, they were able to achieve 95% repression of a target gene, which was completely reverted upon removal of both inducers [192]. Recently, Dietsch et al. downregulated carotenoid formation using CRISPRi on *crtE* down to <10% of the wild type level, which enabled heterologous valencene production in *Synechocystis* [138].

4 Post-Translational Control of Proteins and Their Activities

Post-translational control of protein accumulation and activity represents another level of cyanobacterial engineering (Fig. 1). Recombinant proteins, especially of non-prokaryotic origin, may be unstable, which might be circumvented by fusing them to a leader peptide sequence. This was shown to enhance recombinant protein stability and attain up to ~20% of total cellular protein content in *Synechocystis* [64, 244, 245]. In contrast, protein abundance and activity could be specifically regulated by adding protein degradation tags that are targeted by native or heterologous proteases for a defined turnover (see below). Protein activities can also be directly modulated by interaction factors, like small proteins. Small proteins with a size of ~100 amino acids are coming into focus – a large family, whose representatives exhibit numerous functions in bacterial cells [246]. They are likewise believed to play diverse roles in the regulation of cyanobacterial metabolism and therefore, may exhibit great potential for biotechnological applications [247]. Some metabolic engineering studies have been performed in this regard. The deletion of inhibitory (small) proteins controlling enzymes with desired activities, like CP12 in *S. elongatus* 7942 [248] or PirC in *Synechocystis* [249], led to an increase in product yields.

4.1 Protein Degradation Tags and Heterologous Proteases

Native protein degradation tags recruit the endogenous proteasome, the regulatory machinery of the cell's proteome. They may be included in the ORF of a gene of interest for a defined protein abundance. Thereby, protein degradation tags can also minimize potential toxicity associated with the recombinant protein, e.g. metabolic burden or toxic reaction products. They usually involve a C-terminal *ssrA* RNA-encoded peptide variant that continuously recruits inter alia ClpX protease homologs at different strengths for an orchestrated protein turnover. This native quality control system has been described in bacteria as co-translational modification of mRNAs from stalled proteins for ribosome release and subsequent degradation [250]. It has furthermore been extensively used as molecular tool for various bacteria [250], including cyanobacteria [152]. Heterologous as well as native candidates, which both were processed by the endogenous proteasome, were characterized in *Synechocystis* [12, 251], and enabled different turnover rates from 1–50% of the level of an untagged reporter protein. A common tag for rapid protein turnover is LVA [252] that has been used, for example, in *Anabaena* [189]. Moderate degradation may be achieved through the ASV tag [99, 252].

An example for an orthogonal, heterologous degradation mechanism is the LON protease derived from the transfer-messenger RNA system of *Mesoplasma florum* that has been initially introduced into *E. coli* employing specific tags [253]. Its inducible synthesis enabled reversible and rapid degradation of native as well as recombinant proteins in *S. elongatus* 7942 [254].

Furthermore, the tobacco etch virus (TEV) protease may be implemented for the defined orthogonal in vivo cleavage of recombinant proteins at distinct internal sites. This might be of interest for purification, dynamics evaluation, and the reversible regulation of essential proteins. For example, Zhang et al. created a genetically encoded system for the inducible separation of leader peptide and target segment from fusion proteins by the TEV protease in *Synechocystis* [64].

5 Assembly of Biological Parts

Standard molecular cloning techniques, for example, BioBrick assembly methods [199, 255, 256], joining or removing DNA fragments that do not rely on restriction digest and ligation, are usually based on overhang sequences. This enables the interconnection of two or more adjacent biological parts with suitable linkers, e.g., BioBrick scars [199], for their scarless cloning, usually via Gibson Assembly [257] or AQUA cloning [258]. Furthermore, Golden Gate cloning employs a DNA ligase and Type IIs restriction enzymes that cut outside their recognition sites and hence, can create non-palindromic overhangs of usually four required bases for a directed, stepwise assembly [259, 260], enabling the generation of plasmids in a modular fashion [261]. DNA molecules harboring suitable extensions are therefore chemically synthesized or amplified with a high-fidelity DNA polymerase, such as Q5, and applied for several modifications, including insertion, deletion, mutagenesis, and combinatorial cloning as simultaneous one pot de novo assembly. In vivo cloning steps are usually carried out in *E. coli* and the generated plasmids afterwards isolated for the transformation of cyanobacteria. Notably, the development of the CyanoGate modular cloning (MoClo) system, which is based on the assembly library and syntax of the Plant Golden Gate MoClo kit, by Vasudevan et al. could rapidly speed up genetic modifications of cyanobacteria. This customized, commercially available Golden Gate cloning suite encompasses biological parts as well as non-replicative and replicative vectors for the generation of marked and unmarked mutants, and multigene regulation systems [26, 127].

6 Conclusion and Future Perspectives

Synthetic biology of cyanobacteria is a dynamic field with many advances and highly promising applications. The molecular tools we reviewed here enable genetic and metabolic engineering, for the construction of cyanobacterial cell factories. Sophisticated genetic circuits, composed of multiple regulatory elements, have been constructed for well-defined gene expression [29, 152], and to create genetic logic gates [241, 262]. Furthermore, strategies like modular engineering [66] and modifications using small proteins [248, 249] might facilitate metabolic optimization of biocatalytically relevant reactions in the future. Bottlenecks at the moment represent the lacking orthogonality and compatibility of systems transferred from established industrial work horses like *E. coli*, which often do not function similarly in cyanobacteria, i.e., they lack sufficient control and strength, and thus require further attention. Moreover, the less considered RNA degradation in cyanobacteria that is highly depended on light might become key for future production hosts [151]. Apart from rational engineering, adaptive laboratory evolution (ALE) presents an alternative route to attain desired phenotypes by serial exposure of cells to defined conditions. Generated mutants were shown to be stable while being classified to be non-genetically modified, which makes them especially attractive for industry as it facilitates the licensing procedure of a given process [263, 264].

Altogether, recent progress in synthetic biology has largely extended the molecular toolset to engineer cyanobacteria. However, rational utilization of in silico design and prediction tools, modular cloning techniques and faster genetic modifications combined with lab automation, will replace traditional time-consuming testing on a limited number of mutants in the future. Cyanobacteria are in the spotlight of biotechnology as photoautotrophic cell factories for sustainable production, and recent advances will help deliver on this promise.

References

1. Knight T (2003) Idempotent vector design for standard assembly of Biobricks. MIT Artificial Intelligence Laboratory
2. Voigt CA (2006) Genetic parts to program bacteria. Curr Opin Biotechnol 17:548–557. https://doi.org/10.1016/j.copbio.2006.09.001
3. Salis HM, Mirsky EA, Voigt CA (2009) Automated design of synthetic ribosome binding sites to control protein expression. Nat Biotechnol 27:946–950. https://doi.org/10.1038/nbt.1568
4. Na D, Lee D (2010) RBSDesigner: software for designing synthetic ribosome binding sites that yields a desired level of protein expression. Bioinformatics 26:2633–2634. https://doi.org/10.1093/bioinformatics/btq458
5. Seo SW, Yang J-S, Cho H-S et al (2014) Predictive combinatorial design of mRNA translation initiation regions for systematic optimization of gene expression levels. Sci Rep 4:4515. https://doi.org/10.1038/srep04515
6. Rippka R, Deruelles J, Waterbury JB et al (1979) Generic assignments, strain histories and properties of pure cultures of cyanobacteria. Microbiology 111:1–61. https://doi.org/10.1099/00221287-111-1-1
7. Bláha L, Babica P, Maršálek B (2009) Toxins produced in cyanobacterial water blooms – toxicity and risks. Interdiscip Toxicol 2:36–41. https://doi.org/10.2478/v10102-009-0006-2
8. Ducat DC, Way JC, Silver PA (2011) Engineering cyanobacteria to generate high-value products. Trends Biotechnol 29:95–103. https://doi.org/10.1016/j.tibtech.2010.12.003
9. Wijffels RH, Kruse O, Hellingwerf KJ (2013) Potential of industrial biotechnology with cyanobacteria and eukaryotic microalgae. Curr Opin Biotechnol 24:405–413. https://doi.org/10.1016/j.copbio.2013.04.004
10. Al-Haj L, Lui Y, Abed R et al (2016) Cyanobacteria as chassis for industrial biotechnology: progress and prospects. Life 6:42. https://doi.org/10.3390/life6040042
11. Jodlbauer J, Rohr T, Spadiut O et al (2021) Biocatalysis in green and blue: cyanobacteria. Trends Biotechnol 39:875–889. https://doi.org/10.1016/j.tibtech.2020.12.009
12. Huang H-H, Camsund D, Lindblad P et al (2010) Design and characterization of molecular tools for a synthetic biology approach towards developing cyanobacterial biotechnology. Nucleic Acids Res 38:2577–2593. https://doi.org/10.1093/nar/gkq164
13. Englund E, Liang F, Lindberg P (2016) Evaluation of promoters and ribosome binding sites for biotechnological applications in the unicellular cyanobacterium *Synechocystis* sp. PCC 6803. Sci Rep 6:36640. https://doi.org/10.1038/srep36640
14. Ferreira EA, Pacheco CC, Pinto F et al (2018) Expanding the toolbox for *Synechocystis* sp. PCC 6803: validation of replicative vectors and characterization of a novel set of promoters. Synth Biol (Oxf) 3:ysy014. https://doi.org/10.1093/synbio/ysy014
15. Liu D, Pakrasi HB (2018) Exploring native genetic elements as plug-in tools for synthetic biology in the cyanobacterium *Synechocystis* sp. PCC 6803. Microb Cell Factories 17:48. https://doi.org/10.1186/s12934-018-0897-8
16. Wang B, Eckert C, Maness P-C et al (2018) A genetic toolbox for modulating the expression of heterologous genes in the cyanobacterium *Synechocystis* sp. PCC 6803. ACS Synth Biol 7:276–286. https://doi.org/10.1021/acssynbio.7b00297
17. Kim WJ, Lee S-M, Um Y et al (2017) Development of SyneBrick vectors as a synthetic biology platform for gene expression in *Synechococcus elongatus* PCC 7942. Front Plant Sci 8:293. https://doi.org/10.3389/fpls.2017.00293
18. Sengupta A, Sunder AV, Sohoni SV et al (2019) Fine-tuning native promoters of *Synechococcus elongatus* PCC 7942 to develop a synthetic toolbox for heterologous protein expression. ACS Synth Biol 8:1219–1223. https://doi.org/10.1021/acssynbio.9b00066
19. Markley AL, Begemann MB, Clarke RE et al (2015) Synthetic biology toolbox for controlling gene expression in the cyanobacterium *Synechococcus* sp. strain PCC 7002. ACS Synth Biol 4:595–603. https://doi.org/10.1021/sb500260k

20. Li S, Sun T, Xu C et al (2018) Development and optimization of genetic toolboxes for a fast-growing cyanobacterium *Synechococcus elongatus* UTEX 2973. Metab Eng 48:163–174. https://doi.org/10.1016/j.ymben.2018.06.002
21. Sun T, Li S, Song X et al (2018) Toolboxes for cyanobacteria: recent advances and future direction. Biotechnol Adv 36:1293–1307. https://doi.org/10.1016/j.biotechadv.2018.04.007
22. Zhang W, Song X (2018) Synthetic biology of cyanobacteria, vol 1080. Springer, Singapore
23. Santos-Merino M, Singh AK, Ducat DC (2019) New applications of synthetic biology tools for cyanobacterial metabolic engineering. Front Bioeng Biotechnol 7:33. https://doi.org/10.3389/fbioe.2019.00033
24. Xia P-F, Ling H, Foo JL et al (2019) Synthetic biology toolkits for metabolic engineering of cyanobacteria. Biotechnol J 14:e1800496. https://doi.org/10.1002/biot.201800496
25. Gale GAR, Schiavon Osorio AA, Mills LA et al (2019) Emerging species and genome editing tools: future prospects in cyanobacterial synthetic biology. Microorganisms 7:409. https://doi.org/10.3390/microorganisms7100409
26. Vasudevan R, Gale GAR, Schiavon AA et al (2019) CyanoGate: a modular cloning suite for engineering cyanobacteria based on the plant MoClo syntax. Plant Physiol 180:39–55. https://doi.org/10.1104/pp.18.01401
27. Ng I-S, Keskin BB, Tan S-I (2020) A critical review of genome editing and synthetic biology applications in metabolic engineering of microalgae and cyanobacteria. Biotechnol J 15: e1900228. https://doi.org/10.1002/biot.201900228
28. Pattharaprachayakul N, Lee M, Incharoensakdi A et al (2020) Current understanding of the cyanobacterial CRISPR-Cas systems and development of the synthetic CRISPR-Cas systems for cyanobacteria. Enzym Microb Technol 140:109619. https://doi.org/10.1016/j.enzmictec.2020.109619
29. Till P, Toepel J, Bühler B et al (2020) Regulatory systems for gene expression control in cyanobacteria. Appl Microbiol Biotechnol 104:1977–1991. https://doi.org/10.1007/s00253-019-10344-w
30. Cassier-Chauvat C, Blanc-Garin V, Chauvat F (2021) Genetic, genomics, and responses to stresses in cyanobacteria: biotechnological implications. Genes (Basel) 12:500. https://doi.org/10.3390/genes12040500
31. Srivastava A, Shukla P (2021) Emerging tools and strategies in cyanobacterial omics. Trends Biotechnol 40:4–7. https://doi.org/10.1016/j.tibtech.2021.05.004
32. Laurenceau R, Bliem C, Osburne MS et al (2020) Toward a genetic system in the marine cyanobacterium *Prochlorococcus*. Access Microbiol 2:acmi000107. https://doi.org/10.1099/acmi.0.000107
33. Porter RD (1986) Transformation in cyanobacteria. Crit Rev Microbiol 13:111–132. https://doi.org/10.3109/10408418609108736
34. Thiel T (1995) Genetic analysis of cyanobacteria. In: Bryant DA (ed) The molecular biology of cyanobacteria. Kluwer Academic Publ, Dordrecht, pp 581–611
35. Koksharova OA, Wolk CP (2002) Genetic tools for cyanobacteria. Appl Microbiol Biotechnol 58:123–137. https://doi.org/10.1007/s00253-001-0864-9
36. Heidorn T, Camsund D, Huang H-H et al (2011) Synthetic biology in cyanobacteria engineering and analyzing novel functions. Methods Enzymol 497:539–579. https://doi.org/10.1016/B978-0-12-385075-1.00024-X
37. Bishé B, Taton A, Golden JW (2019) Modification of RSF1010-based broad-host-range plasmids for improved conjugation and cyanobacterial bioprospecting. iScience 20:216–228. https://doi.org/10.1016/j.isci.2019.09.002
38. Gutiérrez S, Lauersen KJ (2021) Gene delivery technologies with applications in microalgal genetic engineering. Biology (Basel) 10:265. https://doi.org/10.3390/biology10040265
39. Johnsborg O, Eldholm V, Håvarstein LS (2007) Natural genetic transformation: prevalence, mechanisms and function. Res Microbiol 158:767–778. https://doi.org/10.1016/j.resmic.2007.09.004

40. Wendt KE, Pakrasi HB (2019) Genomics approaches to deciphering natural transformation in cyanobacteria. Front Microbiol 10:1259. https://doi.org/10.3389/fmicb.2019.01259
41. Schirmacher AM, Hanamghar SS, Zedler JAZ (2020) Function and benefits of natural competence in cyanobacteria: from ecology to targeted manipulation. Life (Basel) 10:249. https://doi.org/10.3390/life10110249
42. Grigorieva G, Shestakov S (1982) Transformation in the cyanobacterium *Synechocystis* sp. 6803. FEMS Microbiol Lett 13:367–370. https://doi.org/10.1111/j.1574-6968.1982.tb08289.x
43. Shestakov SV, Khyen NT (1970) Evidence for genetic transformation in blue-green alga *Anacystis nidulans*. Mol Gen Genet 107:372–375. https://doi.org/10.1007/BF00441199
44. Stevens SE, Porter RD (1980) Transformation in *Agmenellum quadruplicatum*. Proc Natl Acad Sci U S A 77:6052–6056. https://doi.org/10.1073/pnas.77.10.6052
45. Iwai M, Katoh H, Katayama M et al (2004) Improved genetic transformation of the thermophilic cyanobacterium, *Thermosynechococcus elongatus* BP-1. Plant Cell Physiol 45:171–175. https://doi.org/10.1093/pcp/pch015
46. Onai K, Morishita M, Kaneko T et al (2004) Natural transformation of the thermophilic cyanobacterium *Thermosynechococcus elongatus* BP-1: a simple and efficient method for gene transfer. Mol Gen Genomics 271:50–59. https://doi.org/10.1007/s00438-003-0953-9
47. Liang Y, Tang J, Luo Y et al (2019) *Thermosynechococcus* as a thermophilic photosynthetic microbial cell factory for CO2 utilisation. Bioresour Technol 278:255–265. https://doi.org/10.1016/j.biortech.2019.01.089
48. Jaiswal D, Sengupta A, Sohoni S et al (2018) Genome features and biochemical characteristics of a robust, fast growing and naturally transformable cyanobacterium *Synechococcus elongatus* PCC 11801 isolated from India. Sci Rep 8:16632. https://doi.org/10.1038/s41598-018-34872-z
49. Włodarczyk A, Selão TT, Norling B et al (2020) Newly discovered *Synechococcus* sp. PCC 11901 is a robust cyanobacterial strain for high biomass production. Commun Biol 3:215. https://doi.org/10.1038/s42003-020-0910-8
50. Kunert A, Hagemann M, Erdmann N (2000) Construction of promoter probe vectors for *Synechocystis* sp. PCC 6803 using the light-emitting reporter systems Gfp and LuxAB. J Microbiol Methods 41:185–194. https://doi.org/10.1016/S0167-7012(00)00162-7
51. Andersson CR, Tsinoremas NF, Shelton J et al (2000) Application of bioluminescence to the study of circadian rhythms in cyanobacteria. In: Bioluminescence and chemiluminescence part C, vol 305. Elsevier, pp 527–542
52. Peterson ES, McCue LA, Schrimpe-Rutledge AC et al (2012) VESPA: software to facilitate genomic annotation of prokaryotic organisms through integration of proteomic and transcriptomic data. BMC Genomics 13:131. https://doi.org/10.1186/1471-2164-13-131
53. Mitschke J, Georg J, Scholz I et al (2011) An experimentally anchored map of transcriptional start sites in the model cyanobacterium *Synechocystis* sp. PCC6803. Proc Natl Acad Sci U S A 108:2124–2129. https://doi.org/10.1073/pnas.1015154108
54. Ng AH, Berla BM, Pakrasi HB (2015) Fine-tuning of photoautotrophic protein production by combining promoters and neutral sites in the cyanobacterium *Synechocystis* sp. strain PCC 6803. Appl Environ Microbiol 81:6857–6863. https://doi.org/10.1128/AEM.01349-15
55. Nagy C, Thiel K, Mulaku E et al (2021) Comparison of alternative integration sites in the chromosome and the native plasmids of the cyanobacterium *Synechocystis* sp. PCC 6803 in respect to expression efficiency and copy number. Microb Cell Factories 20:539. https://doi.org/10.1186/s12934-021-01622-2
56. Mustila H, Allahverdiyeva Y, Isojärvi J et al (2014) The bacterial-type 4Fe-4S ferredoxin 7 has a regulatory function under photooxidative stress conditions in the cyanobacterium *Synechocystis* sp. PCC 6803. Biochim Biophys Acta 1837:1293–1304. https://doi.org/10.1016/j.bbabio.2014.04.006
57. Mustila H, Paananen P, Battchikova N et al (2016) The flavodiiron protein Flv3 functions as a homo-oligomer during stress acclimation and is distinct from the Flv1/Flv3 hetero-oligomer

specific to the O_2 photoreduction pathway. Plant Cell Physiol 57:pcw047. https://doi.org/10.1093/pcp/pcw047
58. Bustos SA, Golden SS (1992) Light-regulated expression of the *psbD* gene family in *Synechococcus* sp. strain PCC 7942: evidence for the role of duplicated *psbD* genes in cyanobacteria. Mol Gen Genet 232:221–230. https://doi.org/10.1007/BF00280000
59. Kulkarni RD, Golden SS (1997) mRNA stability is regulated by a coding-region element and the unique 5′ untranslated leader sequences of the three *Synechococcus psbA* transcripts. Mol Microbiol 24:1131–1142. https://doi.org/10.1046/j.1365-2958.1997.4201768.x
60. Clerico EM, Ditty JL, Golden SS (2007) Specialized techniques for site-directed mutagenesis in cyanobacteria. Methods Mol Biol 362:155–171. https://doi.org/10.1007/978-1-59745-257-1_11
61. Niederholtmeyer H, Wolfstädter BT, Savage DF et al (2010) Engineering cyanobacteria to synthesize and export hydrophilic products. Appl Environ Microbiol 76:3462–3466. https://doi.org/10.1128/AEM.00202-10
62. Ungerer J, Wendt KE, Hendry JI et al (2018) Comparative genomics reveals the molecular determinants of rapid growth of the cyanobacterium *Synechococcus elongatus* UTEX 2973. Proc Natl Acad Sci U S A 115:E11761–E11770. https://doi.org/10.1073/pnas.1814912115
63. Englund E, Andersen-Ranberg J, Miao R et al (2015) Metabolic engineering of *Synechocystis* sp. PCC 6803 for production of the plant diterpenoid manoyl oxide. ACS Synth Biol 4:1270–1278. https://doi.org/10.1021/acssynbio.5b00070
64. Zhang X, Betterle N, Hidalgo Martinez D et al (2021) Recombinant protein stability in cyanobacteria. ACS Synth Biol 10:810–825. https://doi.org/10.1021/acssynbio.0c00610
65. Mohamed A, Jansson C (1989) Influence of light on accumulation of photosynthesis-specific transcripts in the cyanobacterium *Synechocystis* 6803. Plant Mol Biol 13:693–700. https://doi.org/10.1007/BF00016024
66. Liu X, Miao R, Lindberg P et al (2019) Modular engineering for efficient photosynthetic biosynthesis of 1-butanol from CO_2 in cyanobacteria. Energy Environ Sci 12:2765–2777. https://doi.org/10.1039/c9ee01214a
67. Begemann MB, Zess EK, Walters EM et al (2013) An organic acid based counter selection system for cyanobacteria. PLoS One 8:e76594. https://doi.org/10.1371/journal.pone.0076594
68. Cai YP, Wolk CP (1990) Use of a conditionally lethal gene in *Anabaena* sp. strain PCC 7120 to select for double recombinants and to entrap insertion sequences. J Bacteriol 172:3138–3145. https://doi.org/10.1128/jb.172.6.3138-3145.1990
69. Jones KM, Buikema WJ, Haselkorn R (2003) Heterocyst-specific expression of *patB*, a gene required for nitrogen fixation in *Anabaena* sp. strain PCC 7120. J Bacteriol 185:2306–2314. https://doi.org/10.1128/JB.185.7.2306-2314.2003
70. Lee H-M, Flores E, Herrero A et al (1998) A role for the signal transduction protein PII in the control of nitrate/nitrite uptake in a cyanobacterium. FEBS Lett 427:291–295. https://doi.org/10.1016/S0014-5793(98)00451-7
71. Lee HM, Flores E, Forchhammer K et al (2000) Phosphorylation of the signal transducer PII protein and an additional effector are required for the PII-mediated regulation of nitrate and nitrite uptake in the Cyanobacterium *Synechococcus* sp. PCC 7942. Eur J Biochem 267:591–600. https://doi.org/10.1046/j.1432-1327.2000.01043.x
72. Ditty JL, Canales SR, Anderson BE et al (2005) Stability of the *Synechococcus elongatus* PCC 7942 circadian clock under directed anti-phase expression of the kai genes. Microbiology (Reading) 151:2605–2613. https://doi.org/10.1099/mic.0.28030-0
73. Holtman CK, Chen Y, Sandoval P et al (2005) High-throughput functional analysis of the *Synechococcus elongatus* PCC 7942 genome. DNA Res 12:103–115. https://doi.org/10.1093/dnares/12.2.103
74. Chen Y, Holtman CK, Magnuson RD et al (2008) The complete sequence and functional analysis of pANL, the large plasmid of the unicellular freshwater cyanobacterium *Synechococcus elongatus* PCC 7942. Plasmid 59:176–192. https://doi.org/10.1016/j.plasmid.2008.01.005

75. Chen Y, Taton A, Go M et al (2016) Self-replicating shuttle vectors based on pANS, a small endogenous plasmid of the unicellular cyanobacterium *Synechococcus elongatus* PCC 7942. Microbiology (Reading) 162:2029–2041. https://doi.org/10.1099/mic.0.000377
76. Puszynska AM, O'Shea EK (2017) Switching of metabolic programs in response to light availability is an essential function of the cyanobacterial circadian output pathway. elife 6: e23210. https://doi.org/10.7554/eLife.23210
77. Wang B, Pugh S, Nielsen DR et al (2013) Engineering cyanobacteria for photosynthetic production of 3-hydroxybutyrate directly from CO_2. Metab Eng 16:68–77. https://doi.org/10.1016/j.ymben.2013.01.001
78. Wang B, Yu J, Zhang W et al (2015) Premethylation of foreign DNA improves integrative transformation efficiency in *Synechocystis* sp. strain PCC 6803. Appl Environ Microbiol 81: 8500–8506. https://doi.org/10.1128/AEM.02575-15
79. Pinto F, Pacheco CC, Oliveira P et al (2015) Improving a *Synechocystis*-based photoautotrophic chassis through systematic genome mapping and validation of neutral sites. DNA Res 22: 425–437. https://doi.org/10.1093/dnares/dsv024
80. Durall C, Lindberg P, Yu J et al (2020) Increased ethylene production by overexpressing phosphoenolpyruvate carboxylase in the cyanobacterium *Synechocystis* PCC 6803. Biotechnol Biofuels 13:16. https://doi.org/10.1186/s13068-020-1653-y
81. Eungrasamee K, Incharoensakdi A, Lindblad P et al (2020) *Synechocystis* sp. PCC 6803 overexpressing genes involved in CBB cycle and free fatty acid cycling enhances the significant levels of intracellular lipids and secreted free fatty acids. Sci Rep 10:4515. https://doi.org/10.1038/s41598-020-61100-4
82. Chen L, Liu H, Wang L et al (2021) Synthetic counter-selection markers and their application in genetic modification of *Synechococcus elongatus* UTEX2973. Appl Microbiol Biotechnol 105:5077–5086. https://doi.org/10.1007/s00253-021-11391-y
83. Taton A, Erikson C, Yang Y et al (2020) The circadian clock and darkness control natural competence in cyanobacteria. Nat Commun 11:1688. https://doi.org/10.1038/s41467-020-15384-9
84. Yoshihara S, Geng X, Okamoto S et al (2001) Mutational analysis of genes involved in pilus structure, motility and transformation competency in the unicellular motile cyanobacterium *Synechocystis* sp. PCC 6803. Plant Cell Physiol 42:63–73. https://doi.org/10.1093/pcp/pce007
85. Tsujimoto R, Kotani H, Yokomizo K et al (2018) Functional expression of an oxygen-labile nitrogenase in an oxygenic photosynthetic organism. Sci Rep 8:7380. https://doi.org/10.1038/s41598-018-25396-7
86. Stucken K, Koch R, Dagan T (2013) Cyanobacterial defense mechanisms against foreign DNA transfer and their impact on genetic engineering. Biol Res 46:373–382. https://doi.org/10.4067/S0716-97602013000400009
87. Kufryk GI, Sachet M, Schmetterer G et al (2002) Transformation of the cyanobacterium *Synechocystis* sp. PCC 6803 as a tool for genetic mapping: optimization of efficiency. FEMS Microbiol Lett 206:215–219. https://doi.org/10.1111/j.1574-6968.2002.tb11012.x
88. Elhai J, Vepritskiy A, Muro-Pastor AM et al (1997) Reduction of conjugal transfer efficiency by three restriction activities of *Anabaena* sp. strain PCC 7120. J Bacteriol 179:1998–2005. https://doi.org/10.1128/jb.179.6.1998-2005.1997
89. Griese M, Lange C, Soppa J (2011) Ploidy in cyanobacteria. FEMS Microbiol Lett 323:124–131. https://doi.org/10.1111/j.1574-6968.2011.02368.x
90. Watanabe S (2020) Cyanobacterial multi-copy chromosomes and their replication. Biosci Biotechnol Biochem 84:1309–1321. https://doi.org/10.1080/09168451.2020.1736983
91. Guerry P, van Embden J, Falkow S (1974) Molecular nature of two nonconjugative plasmids carrying drug resistance genes. J Bacteriol 117:619–630. https://doi.org/10.1128/jb.117.2.619-630.1974
92. Taton A, Unglaub F, Wright NE et al (2014) Broad-host-range vector system for synthetic biology and biotechnology in cyanobacteria. Nucleic Acids Res 42:e136. https://doi.org/10.1093/nar/gku673

93. Meyer R (2009) Replication and conjugative mobilization of broad host-range IncQ plasmids. Plasmid 62:57–70. https://doi.org/10.1016/j.plasmid.2009.05.001
94. Kreps S, Ferino F, Mosrin C et al (1990) Conjugative transfer and autonomous replication of a promiscuous IncQ plasmid in the cyanobacterium *Synechocystis* PCC 6803. Mol Gen Genet 221:129–133. https://doi.org/10.1007/BF00280378
95. Zinchenko V, Piven IV, Melnik VA, Shestakov SV (1999) Vectors for the complementation analysis of cyanobacterial mutants. Russ J Genet 35:228–232
96. Berla BM, Saha R, Immethun CM et al (2013) Synthetic biology of cyanobacteria: unique challenges and opportunities. Front Microbiol 4:246. https://doi.org/10.3389/fmicb.2013.00246
97. Wang B, Wang J, Zhang W et al (2012) Application of synthetic biology in cyanobacteria and algae. Front Microbiol 3:344. https://doi.org/10.3389/fmicb.2012.00344
98. Guerrero F, Carbonell V, Cossu M et al (2012) Ethylene synthesis and regulated expression of recombinant protein in *Synechocystis* sp. PCC 6803. PLoS One 7:e50470. https://doi.org/10.1371/journal.pone.0050470
99. Behle A, Saake P, Germann AT et al (2020) Comparative dose-response analysis of inducible promoters in cyanobacteria. ACS Synth Biol 9:843–855. https://doi.org/10.1021/acssynbio.9b00505
100. Behle A, Axmann IM (2022) pSHDY: a new tool for genetic engineering of cyanobacteria. Methods Mol Biol 2379:67–79. https://doi.org/10.1007/978-1-0716-1791-5_4
101. Martínez-García E, Goñi-Moreno A, Bartley B et al (2020) SEVA 3.0: an update of the standard European vector architecture for enabling portability of genetic constructs among diverse bacterial hosts. Nucleic Acids Res 48:D1164–D1170. https://doi.org/10.1093/nar/gkz1024
102. Ingram LC, Richmond MH, Sykes RB (1973) Molecular characterization of the R factors implicated in the carbenicillin resistance of a sequence of *Pseudomonas aeruginosa* strains isolated from burns. Antimicrob Agents Chemother 3:279–288. https://doi.org/10.1128/aac.3.2.279
103. Wolk CP, Vonshak A, Kehoe P et al (1984) Construction of shuttle vectors capable of conjugative transfer from *Escherichia coli* to nitrogen-fixing filamentous cyanobacteria. Proc Natl Acad Sci U S A 81:1561–1565. https://doi.org/10.1073/pnas.81.5.1561
104. Wolk CP, Fan Q, Zhou R et al (2007) Paired cloning vectors for complementation of mutations in the cyanobacterium *Anabaena* sp. strain PCC 7120. Arch Microbiol 188:551–563. https://doi.org/10.1007/s00203-007-0276-z
105. Wolk CP, Cai Y, Cardemil L et al (1988) Isolation and complementation of mutants of *Anabaena* sp. strain PCC 7120 unable to grow aerobically on dinitrogen. J Bacteriol 170:1239–1244. https://doi.org/10.1128/jb.170.3.1239-1244.1988
106. Stucken K, Ilhan J, Roettger M et al (2012) Transformation and conjugal transfer of foreign genes into the filamentous multicellular cyanobacteria (subsection V) *Fischerella* and *Chlorogloeopsis*. Curr Microbiol 65:552–560. https://doi.org/10.1007/s00284-012-0193-5
107. Antonaru LA, Nürnberg DJ (2017) Role of PatS and cell type on the heterocyst spacing pattern in a filamentous branching cyanobacterium. FEMS Microbiol Lett 364:fnx154. https://doi.org/10.1093/femsle/fnx154
108. Buikema WJ, Haselkorn R (1991) Isolation and complementation of nitrogen fixation mutants of the cyanobacterium *Anabaena* sp. strain PCC 7120. J Bacteriol 173:1879–1885. https://doi.org/10.1128/jb.173.6.1879-1885.1991
109. Billi D, Friedmann EI, Helm RF et al (2001) Gene transfer to the desiccation-tolerant cyanobacterium *Chroococcidiopsis*. J Bacteriol 183:2298–2305. https://doi.org/10.1128/JB.183.7.2298-2305.2001
110. Elhai J (1993) Strong and regulated promoters in the cyanobacterium *Anabaena* PCC 7120. FEMS Microbiol Lett 114:179–184. https://doi.org/10.1111/j.1574-6968.1993.tb06570.x

111. Ravindran CRM, Suguna S, Shanmugasundaram S (2006) Electroporation as a tool to transfer the plasmid pRL489 in *Oscillatoria* MKU 277. J Microbiol Methods 66:174–176. https://doi.org/10.1016/j.mimet.2005.11.011
112. Wolk CP, Elhai J, Kuritz T et al (1993) Amplified expression of a transcriptional pattern formed during development of *Anabaena*. Mol Microbiol 7:441–445. https://doi.org/10.1111/j.1365-2958.1993.tb01135.x
113. Summers ML, Wallis JG, Campbell EL et al (1995) Genetic evidence of a major role for glucose-6-phosphate dehydrogenase in nitrogen fixation and dark growth of the cyanobacterium *Nostoc* sp. strain ATCC 29133. J Bacteriol 177:6184–6194. https://doi.org/10.1128/jb.177.21.6184-6194.1995
114. Argueta C, Yuksek K, Summers M (2004) Construction and use of GFP reporter vectors for analysis of cell-type-specific gene expression in *Nostoc punctiforme*. J Microbiol Methods 59: 181–188. https://doi.org/10.1016/j.mimet.2004.06.009
115. Walton DK, Gendel SM, Atherly AG (1993) DNA sequence and shuttle vector construction of plasmid pGL3 from *Plectonema boryanum* PCC 6306. Nucleic Acids Res 21:746. https://doi.org/10.1093/nar/21.3.746
116. Buzby JS, Porter RD, Stevens SE (1985) Expression of the *Escherichia coli lacZ* gene on a plasmid vector in a cyanobacterium. Science 230:805–807. https://doi.org/10.1126/science.2997920
117. Cobley JG, Zerweck E, Reyes R et al (1993) Construction of shuttle plasmids which can be efficiently mobilized from *Escherichia coli* into the chromatically adapting cyanobacterium, *Fremyella diplosiphon*. Plasmid 30:90–105. https://doi.org/10.1006/plas.1993.1037
118. Sherman LA, van de Putte P (1982) Construction of a hybrid plasmid capable of replication in the bacterium *Escherichia coli* and the cyanobacterium *Anacystis nidulans*. J Bacteriol 150: 410–413. https://doi.org/10.1128/jb.150.1.410-413.1982
119. Golden SS, Sherman LA (1983) A hybrid plasmid is a stable cloning vector for the cyanobacterium *Anacystis nidulans* R2. J Bacteriol 155:966–972. https://doi.org/10.1128/jb.155.3.966-972.1983
120. Opel F, Siebert NA, Klatt S et al (2022) Generation of synthetic shuttle vectors enabling modular genetic engineering of cyanobacteria. ACS Synth Biol 11:1758–1771. https://doi.org/10.1021/acssynbio.1c00605
121. Jin H, Wang Y, Idoine A et al (2018) Construction of a shuttle vector using an endogenous plasmid from the cyanobacterium *Synechocystis* sp. PCC6803. Front Microbiol 9:1662. https://doi.org/10.3389/fmicb.2018.01662
122. Xiao Y, Wang S, Rommelfanger S et al (2018) Developing a Cas9-based tool to engineer native plasmids in *Synechocystis* sp. PCC 6803. Biotechnol Bioeng 115:2305–2314. https://doi.org/10.1002/bit.26747
123. Thiel T, Poo H (1989) Transformation of a filamentous cyanobacterium by electroporation. J Bacteriol 171:5743–5746. https://doi.org/10.1128/jb.171.10.5743-5746.1989
124. Elhai J, Wolk C (1988) Conjugal transfer of DNA to cyanobacteria. In: Cyanobacteria, vol 167. Elsevier, pp 747–754
125. Marraccini P, Bulteau S, Cassier-Chauvat C et al (1993) A conjugative plasmid vector for promoter analysis in several cyanobacteria of the genera *Synechococcus* and *Synechocystis*. Plant Mol Biol 23:905–909. https://doi.org/10.1007/BF00021546
126. Waters VL, Guiney DG (1993) Processes at the nick region link conjugation, T-DNA transfer and rolling circle replication. Mol Microbiol 9:1123–1130. https://doi.org/10.1111/j.1365-2958.1993.tb01242.x
127. Gale GAR, Schiavon Osorio AA, Puzorjov A et al (2019) Genetic modification of cyanobacteria by conjugation using the CyanoGate modular cloning toolkit. J Vis Exp. https://doi.org/10.3791/60451
128. Yu J, Liberton M, Cliften PF et al (2015) *Synechococcus elongatus* UTEX 2973, a fast growing cyanobacterial chassis for biosynthesis using light and CO_2. Sci Rep 5:8132. https://doi.org/10.1038/srep08132

129. Cai F, Axen SD, Kerfeld CA (2013) Evidence for the widespread distribution of CRISPR-Cas system in the phylum cyanobacteria. RNA Biol 10:687–693. https://doi.org/10.4161/rna.24571
130. Makarova KS, Wolf YI, Alkhnbashi OS et al (2015) An updated evolutionary classification of CRISPR-Cas systems. Nat Rev Microbiol 13:722–736. https://doi.org/10.1038/nrmicro3569
131. Koonin EV, Makarova KS (2019) Origins and evolution of CRISPR-Cas systems. Philos Trans R Soc Lond Ser B Biol Sci 374:20180087. https://doi.org/10.1098/rstb.2018.0087
132. Hashemi A (2018) CRISPR-cas system as a genome engineering platform: applications in biomedicine and biotechnology. CGT 18:115–124. https://doi.org/10.2174/1566523218666180221110627
133. Behler J, Vijay D, Hess WR et al (2018) CRISPR-based technologies for metabolic engineering in cyanobacteria. Trends Biotechnol 36:996–1010. https://doi.org/10.1016/j.tibtech.2018.05.011
134. Wang F, Gao Y, Yang G (2020) Recent advances in synthetic biology of cyanobacteria for improved chemicals production. Bioengineered 11:1208–1220. https://doi.org/10.1080/21655979.2020.1837458
135. Gay P, Le Coq D, Steinmetz M et al (1985) Positive selection procedure for entrapment of insertion sequence elements in gram-negative bacteria. J Bacteriol 164:918–921. https://doi.org/10.1128/jb.164.2.918-921.1985
136. Lea-Smith DJ, Vasudevan R, Howe CJ (2016) Generation of marked and markerless mutants in model cyanobacterial species. J Vis Exp. https://doi.org/10.3791/54001
137. Saar KL, Bombelli P, Lea-Smith DJ et al (2018) Enhancing power density of biophotovoltaics by decoupling storage and power delivery. Nat Energy 3:75–81. https://doi.org/10.1038/s41560-017-0073-0
138. Dietsch M, Behle A, Westhoff P et al (2021) Metabolic engineering of *Synechocystis* sp. PCC 6803 for the photoproduction of the sesquiterpene valencene. Metab Eng Commun 13:e00178. https://doi.org/10.1016/j.mec.2021.e00178
139. Viola S, Rühle T, Leister D (2014) A single vector-based strategy for marker-less gene replacement in *Synechocystis* sp. PCC 6803. Microb Cell Factories 13:4. https://doi.org/10.1186/1475-2859-13-4
140. Cheah YE, Albers SC, Peebles CAM (2013) A novel counter-selection method for markerless genetic modification in *Synechocystis* sp. PCC 6803. Biotechnol Prog 29:23–30. https://doi.org/10.1002/btpr.1661
141. Sebesta J, Werner A, Peebles CAM (2019) Genetic engineering of cyanobacteria: design, implementation, and characterization of recombinant *Synechocystis* sp. PCC 6803. Methods Mol Biol 1927:139–154. https://doi.org/10.1007/978-1-4939-9142-6_10
142. Caicedo-Burbano P, Smit T, Pineda Hernández H et al (2020) Construction of fully segregated genomic libraries in polyploid organisms such as *Synechocystis* sp. PCC 6803. ACS Synth Biol 9:2632–2638. https://doi.org/10.1021/acssynbio.0c00353
143. Kojima K, Keta S, Uesaka K et al (2016) A simple method for isolation and construction of markerless cyanobacterial mutants defective in acyl-acyl carrier protein synthetase. Appl Microbiol Biotechnol 100:10107–10113. https://doi.org/10.1007/s00253-016-7850-8
144. Tan X, Liang F, Cai K et al (2013) Application of the FLP/FRT recombination system in cyanobacteria for construction of markerless mutants. Appl Microbiol Biotechnol 97:6373–6382. https://doi.org/10.1007/s00253-013-4837-6
145. Zhang Y, Pu H, Wang Q et al (2007) PII is important in regulation of nitrogen metabolism but not required for heterocyst formation in the cyanobacterium *Anabaena* sp. PCC 7120. J Biol Chem 282:33641–33648. https://doi.org/10.1074/jbc.M706500200
146. Ungerer J, Pakrasi HB (2016) Cpf1 is a versatile tool for CRISPR genome editing across diverse species of cyanobacteria. Sci Rep 6:39681. https://doi.org/10.1038/srep39681
147. Wendt KE, Ungerer J, Cobb RE et al (2016) CRISPR/Cas9 mediated targeted mutagenesis of the fast growing cyanobacterium *Synechococcus elongatus* UTEX 2973. Microb Cell Factories 15:115. https://doi.org/10.1186/s12934-016-0514-7

148. Li H, Shen CR, Huang C-H et al (2016) CRISPR-Cas9 for the genome engineering of cyanobacteria and succinate production. Metab Eng 38:293–302. https://doi.org/10.1016/j.ymben.2016.09.006
149. Niu T-C, Lin G-M, Xie L-R et al (2019) Expanding the potential of CRISPR-Cpf1-based genome editing technology in the cyanobacterium *Anabaena* PCC 7120. ACS Synth Biol 8: 170–180. https://doi.org/10.1021/acssynbio.8b00437
150. Sengupta A, Pritam P, Jaiswal D et al (2020) Photosynthetic co-production of succinate and ethylene in a fast-growing cyanobacterium, *Synechococcus elongatus* PCC 11801. Meta 10: 250. https://doi.org/10.3390/metabo10060250
151. Gordon GC, Pfleger BF (2018) Regulatory tools for controlling gene expression in cyanobacteria. In: Zhang W, Song X (eds) Synthetic biology of cyanobacteria, vol 1080. Springer, Singapore, pp 281–315
152. Immethun CM, Moon TS (2018) Synthetic gene regulation in cyanobacteria. In: Zhang W, Song X (eds) Synthetic biology of cyanobacteria, vol 1080. Springer, Singapore, pp 317–355
153. Srivastava A, Summers ML, Sobotka R (2020) Cyanobacterial sigma factors: current and future applications for biotechnological advances. Biotechnol Adv 40:107517. https://doi.org/10.1016/j.biotechadv.2020.107517
154. Srivastava A, Varshney RK, Shukla P (2021) Sigma factor modulation for cyanobacterial metabolic engineering. Trends Microbiol 29:266–277. https://doi.org/10.1016/j.tim.2020.10.012
155. Stensjö K, Vavitsas K, Tyystjärvi T (2018) Harnessing transcription for bioproduction in cyanobacteria. Physiol Plant 162:148–155. https://doi.org/10.1111/ppl.12606
156. Riaz-Bradley A (2019) Transcription in cyanobacteria: a distinctive machinery and putative mechanisms. Biochem Soc Trans 47:679–689. https://doi.org/10.1042/BST20180508
157. Huang H-H, Lindblad P (2013) Wide-dynamic-range promoters engineered for cyanobacteria. J Biol Eng 7:10. https://doi.org/10.1186/1754-1611-7-10
158. Camsund D, Heidorn T, Lindblad P (2014) Design and analysis of LacI-repressed promoters and DNA-looping in a cyanobacterium. J Biol Eng 8:4. https://doi.org/10.1186/1754-1611-8-4
159. Klähn S, Bolay P, Wright PR et al (2018) A glutamine riboswitch is a key element for the regulation of glutamine synthetase in cyanobacteria. Nucleic Acids Res 46:10082–10094. https://doi.org/10.1093/nar/gky709
160. Huang C-H, Shen CR, Li H et al (2016) CRISPR interference (CRISPRi) for gene regulation and succinate production in cyanobacterium *S. elongatus* PCC 7942. Microb Cell Factories 15: 196. https://doi.org/10.1186/s12934-016-0595-3
161. Higo A, Isu A, Fukaya Y et al (2018) Application of CRISPR interference for metabolic engineering of the heterocyst-forming multicellular cyanobacterium *Anabaena* sp. PCC 7120. Plant Cell Physiol 59:119–127. https://doi.org/10.1093/pcp/pcx166
162. Ma AT, Schmidt CM, Golden JW (2014) Regulation of gene expression in diverse cyanobacterial species by using theophylline-responsive riboswitches. Appl Environ Microbiol 80:6704–6713. https://doi.org/10.1128/AEM.01697-14
163. Kunert A, Vinnemeier J, Erdmann N et al (2003) Repression by Fur is not the main mechanism controlling the iron-inducible *isiAB* operon in the cyanobacterium *Synechocystis* sp. PCC 6803. FEMS Microbiol Lett 227:255–262. https://doi.org/10.1016/S0378-1097(03)00689-X
164. Zhou J, Zhang H, Meng H et al (2014) Discovery of a super-strong promoter enables efficient production of heterologous proteins in cyanobacteria. Sci Rep 4:4500. https://doi.org/10.1038/srep04500
165. Shono C, Ariyanti D, Abe K et al (2021) A green light-regulated T7 RNA polymerase gene expression system for cyanobacteria. Mar Biotechnol (NY) 23:31–38. https://doi.org/10.1007/s10126-020-09997-w
166. Zhang L, Chen L, Diao J et al (2020) Construction and analysis of an artificial consortium based on the fast-growing cyanobacterium *Synechococcus elongatus* UTEX 2973 to produce the platform chemical 3-hydroxypropionic acid from CO_2. Biotechnol Biofuels 13:82. https://doi.org/10.1186/s13068-020-01720-0

167. Dzelzkalns VA, Owens GC, Bogorad L (1984) Chloroplast promoter driven expression of the chloramphenicol acetyl transferase gene in a cyanobacterium. Nucleic Acids Res 12:8917–8925. https://doi.org/10.1093/nar/12.23.8917
168. Jacobsen JH, Frigaard N-U (2014) Engineering of photosynthetic mannitol biosynthesis from CO_2 in a cyanobacterium. Metab Eng 21:60–70. https://doi.org/10.1016/j.ymben.2013.11.004
169. Sakai M, Ogawa T, Matsuoka M et al (1997) Photosynthetic conversion of carbon dioxide to ethylene by the recombinant cyanobacterium, *Synechococcus* sp. PCC 7942, which harbors a gene for the ethylene-forming enzyme of *Pseudomonas syringae*. J Ferment Bioeng 84:434–443. https://doi.org/10.1016/S0922-338X(97)82004-1
170. Nair U, Thomas C, Golden SS (2001) Functional elements of the strong *psbAI* promoter of *Synechococcus elongatus* PCC 7942. J Bacteriol 183:1740–1747. https://doi.org/10.1128/JB.183.5.1740-1747.2001
171. Takahama K, Matsuoka M, Nagahama K et al (2003) Construction and analysis of a recombinant cyanobacterium expressing a chromosomally inserted gene for an ethylene-forming enzyme at the *psbAI* locus. J Biosci Bioeng 95:302–305. https://doi.org/10.1016/S1389-1723(03)80034-8
172. Ducat DC, Sachdeva G, Silver PA (2011) Rewiring hydrogenase-dependent redox circuits in cyanobacteria. Proc Natl Acad Sci U S A 108:3941–3946. https://doi.org/10.1073/pnas.1016026108
173. Lindberg P, Park S, Melis A (2010) Engineering a platform for photosynthetic isoprene production in cyanobacteria, using *Synechocystis* as the model organism. Metab Eng 12:70–79. https://doi.org/10.1016/j.ymben.2009.10.001
174. Liu X, Sheng J, Curtiss R (2011) Fatty acid production in genetically modified cyanobacteria. Proc Natl Acad Sci U S A 108:6899–6904. https://doi.org/10.1073/pnas.1103014108
175. Albers SC, Peebles CAM (2017) Evaluating light-induced promoters for the control of heterologous gene expression in *Synechocystis* sp. PCC 6803. Biotechnol Prog 33:45–53. https://doi.org/10.1002/btpr.2396
176. Tan X, Yao L, Gao Q et al (2011) Photosynthesis driven conversion of carbon dioxide to fatty alcohols and hydrocarbons in cyanobacteria. Metab Eng 13:169–176. https://doi.org/10.1016/j.ymben.2011.01.001
177. Peca L, Kós PB, Vass I (2007) Characterization of the activity of heavy metal-responsive promoters in the cyanobacterium *Synechocystis* PCC 6803. Acta Biol Hung 58(Suppl):11–22. https://doi.org/10.1556/ABiol.58.2007.Suppl.2
178. Peca L, Kós PB, Máté Z et al (2008) Construction of bioluminescent cyanobacterial reporter strains for detection of nickel, cobalt and zinc. FEMS Microbiol Lett 289:258–264. https://doi.org/10.1111/j.1574-6968.2008.01393.x
179. Blasi B, Peca L, Vass I et al (2012) Characterization of stress responses of heavy metal and metalloid inducible promoters in *Synechocystis* PCC6803. J Microbiol Biotechnol 22:166–169. https://doi.org/10.4014/jmb.1106.06050
180. Tüllinghoff A, Uhl MB, Nintzel FEH et al (2022) Maximizing photosynthesis-driven Baeyer-Villiger oxidation efficiency in recombinant *Synechocystis* sp. PCC6803. Front Catal. https://doi.org/10.3389/fctls.2021.780474
181. Oliveira P, Lindblad P (2008) An AbrB-like protein regulates the expression of the bidirectional hydrogenase in *Synechocystis* sp. strain PCC 6803. J Bacteriol 190:1011–1019. https://doi.org/10.1128/jb.01605-07
182. Klähn S, Schaal C, Georg J et al (2015) The sRNA *NsiR4* is involved in nitrogen assimilation control in cyanobacteria by targeting glutamine synthetase inactivating factor IF7. Proc Natl Acad Sci U S A 112:E6243–E6252. https://doi.org/10.1073/pnas.1508412112
183. Bolay P, Rozbeh R, Muro-Pastor MI et al (2021) The novel PII-interacting protein PirA controls flux into the cyanobacterial ornithine-ammonia cycle. MBio 12:e00229-21. https://doi.org/10.1128/mBio.00229-21

184. Brandenburg F, Theodosiou E, Bertelmann C et al (2021) Trans-4-hydroxy-L-proline production by the cyanobacterium *Synechocystis* sp. PCC 6803. Metab Eng Commun 12:e00155. https://doi.org/10.1016/j.mec.2020.e00155
185. Buikema WJ, Haselkorn R (2001) Expression of the *Anabaena hetR* gene from a copper-regulated promoter leads to heterocyst differentiation under repressing conditions. Proc Natl Acad Sci U S A 98:2729–2734. https://doi.org/10.1073/pnas.051624898
186. Callahan SM, Buikema WJ (2001) The role of HetN in maintenance of the heterocyst pattern in *Anabaena* sp. PCC 7120. Mol Microbiol 40:941–950. https://doi.org/10.1046/j.1365-2958.2001.02437.x
187. Higa KC, Callahan SM (2010) Ectopic expression of *hetP* can partially bypass the need for *hetR* in heterocyst differentiation by *Anabaena* sp. strain PCC 7120. Mol Microbiol 77:562–574. https://doi.org/10.1111/j.1365-2958.2010.07257.x
188. Saha SK, Golden JW (2011) Overexpression of *pknE* blocks heterocyst development in *Anabaena* sp. strain PCC 7120. J Bacteriol 193:2619–2629. https://doi.org/10.1128/jb.00120-11
189. Higo A, Isu A, Fukaya Y et al (2016) Efficient gene induction and endogenous gene repression systems for the filamentous cyanobacterium *Anabaena* sp. PCC 7120. Plant Cell Physiol 57:387–396. https://doi.org/10.1093/pcp/pcv202
190. Kelly CL, Taylor GM, Hitchcock A et al (2018) A rhamnose-inducible system for precise and temporal control of gene expression in cyanobacteria. ACS Synth Biol 7:1056–1066. https://doi.org/10.1021/acssynbio.7b00435
191. Kelly CL, Taylor GM, Šatkutė A et al (2019) Transcriptional terminators allow leak-free chromosomal integration of genetic constructs in cyanobacteria. Microorganisms 7:263. https://doi.org/10.3390/microorganisms7080263
192. Liu D, Johnson VM, Pakrasi HB (2020) A reversibly induced CRISPRi system targeting photosystem II in the cyanobacterium *Synechocystis* sp. PCC 6803. ACS Synth Biol 9:1441–1449. https://doi.org/10.1021/acssynbio.0c00106
193. Higo A, Isu A, Fukaya Y et al (2017) Designing synthetic flexible gene regulation networks using RNA devices in cyanobacteria. ACS Synth Biol 6:55–61. https://doi.org/10.1021/acssynbio.6b00201
194. Higo A, Isu A, Fukaya Y et al (2018) Spatio-temporal gene induction systems in the heterocyst-forming multicellular cyanobacterium *Anabaena* sp. PCC 7120. Plant Cell Physiol 59:82–89. https://doi.org/10.1093/pcp/pcx163
195. Yao L, Cengic I, Anfelt J et al (2016) Multiple gene repression in cyanobacteria using CRISPRi. ACS Synth Biol 5:207–212. https://doi.org/10.1021/acssynbio.5b00264
196. Johnson TR, Haynes JI, Wealand JL et al (1988) Structure and regulation of genes encoding phycocyanin and allophycocyanin from *Anabaena variabilis* ATCC 29413. J Bacteriol 170:1858–1865. https://doi.org/10.1128/jb.170.4.1858-1865.1988
197. Sengupta A, Madhu S, Wangikar PP (2020) A library of tunable, portable, and inducer-free promoters derived from cyanobacteria. ACS Synth Biol 9:1790–1801. https://doi.org/10.1021/acssynbio.0c00152
198. López-Maury L, García-Domínguez M, Florencio FJ et al (2002) A two-component signal transduction system involved in nickel sensing in the cyanobacterium *Synechocystis* sp. PCC 6803. Mol Microbiol 43:247–256. https://doi.org/10.1046/j.1365-2958.2002.02741.x
199. Registry of Standard Biological Parts. http://parts.igem.org
200. Wang W, Li Y, Wang Y et al (2018) Bacteriophage T7 transcription system: an enabling tool in synthetic biology. Biotechnol Adv 36:2129–2137. https://doi.org/10.1016/j.biotechadv.2018.10.001
201. Jin H, Lindblad P, Bhaya D (2019) Building an inducible T7 RNA polymerase/T7 promoter circuit in *Synechocystis* sp. PCC6803. ACS Synth Biol 8:655–660. https://doi.org/10.1021/acssynbio.8b00515

202. Xie X, Tian Y, Tian J et al (2020) Construction of T7 RNA polymerase gene expression system in *Anabaena* sp. PCC 7120 for the expression of *hG-CSF*. Sheng Wu Gong Cheng Xue Bao 36:2467–2477. https://doi.org/10.13345/j.cjb.200173
203. Jones CM, Korosh TC, Nielsen DR et al (2021) Optimization of a T7-RNA polymerase system in *Synechococcus* sp. PCC 7002 mirrors the protein overproduction phenotype from *E. coli* BL21(DE3). Appl Microbiol Biotechnol 105:1147–1158. https://doi.org/10.1007/s00253-020-11085-x
204. Azevedo R, Lopes JL, de Souza MM et al (2019) Synechococcus elongatus as a model of photosynthetic bioreactor for expression of recombinant β-glucosidases. Biotechnol Biofuels 12:174. https://doi.org/10.1186/s13068-019-1505-9
205. Vijayan V, Jain IH, O'Shea EK (2011) A high resolution map of a cyanobacterial transcriptome. Genome Biol 12:R47. https://doi.org/10.1186/gb-2011-12-5-r47
206. Stueber D, Bujard H (1982) Transcription from efficient promoters can interfere with plasmid replication and diminish expression of plasmid specified genes. EMBO J 1(11):1399–1404
207. Gale GAR, Wang B, McCormick AJ (2020) Evaluation and comparison of the efficiency of transcription terminators in different cyanobacterial species. Front Microbiol 11:624011. https://doi.org/10.3389/fmicb.2020.624011
208. Ma J, Campbell A, Karlin S (2002) Correlations between Shine-Dalgarno sequences and gene features such as predicted expression levels and operon structures. J Bacteriol 184:5733–5745. https://doi.org/10.1128/JB.184.20.5733-5745.2002
209. Thiel K, Mulaku E, Dandapani H et al (2018) Translation efficiency of heterologous proteins is significantly affected by the genetic context of RBS sequences in engineered cyanobacterium *Synechocystis* sp. PCC 6803. Microb Cell Factories 17:34. https://doi.org/10.1186/s12934-018-0882-2
210. Sebesta J, Peebles CA (2020) Improving heterologous protein expression in *Synechocystis* sp. PCC 6803 for alpha-bisabolene production. Metab Eng Commun 10:e00117. https://doi.org/10.1016/j.mec.2019.e00117
211. Oliver JWK, Machado IMP, Yoneda H et al (2013) Cyanobacterial conversion of carbon dioxide to 2,3-butanediol. Proc Natl Acad Sci U S A 110:1249–1254. https://doi.org/10.1073/pnas.1213024110
212. Oliver JWK, Machado IMP, Yoneda H et al (2014) Combinatorial optimization of cyanobacterial 2,3-butanediol production. Metab Eng 22:76–82. https://doi.org/10.1016/j.ymben.2014.01.001
213. Seo SW, Yang J-S, Kim I et al (2013) Predictive design of mRNA translation initiation region to control prokaryotic translation efficiency. Metab Eng 15:67–74. https://doi.org/10.1016/j.ymben.2012.10.006
214. Gaida SM, Al-Hinai MA, Indurthi DC et al (2013) Synthetic tolerance: three noncoding small RNAs, *DsrA*, *ArcZ* and *RprA*, acting supra-additively against acid stress. Nucleic Acids Res 41:8726–8737. https://doi.org/10.1093/nar/gkt651
215. Waters LS, Storz G (2009) Regulatory RNAs in bacteria. Cell 136:615–628. https://doi.org/10.1016/j.cell.2009.01.043
216. Brantl S (2007) Regulatory mechanisms employed by cis-encoded antisense RNAs. Curr Opin Microbiol 10:102–109. https://doi.org/10.1016/j.mib.2007.03.012
217. Li S, Sun T, Chen L et al (2021) Designing and constructing artificial small RNAs for gene regulation and carbon flux redirection in photosynthetic cyanobacteria. Methods Mol Biol 2290:229–252. https://doi.org/10.1007/978-1-0716-1323-8_16
218. Srivastava A, Brilisauer K, Rai AK et al (2017) Down-regulation of the alternative sigma factor SigJ confers a photoprotective phenotype to *Anabaena* PCC 7120. Plant Cell Physiol 58:287–297. https://doi.org/10.1093/pcp/pcw188
219. Ueno K, Tsukakoshi K, Ikebukuro K (2018) Riboregulator elements as tools to engineer gene expression in cyanobacteria. Appl Microbiol Biotechnol 102:7717–7723. https://doi.org/10.1007/s00253-018-9221-0

220. Isaacs FJ, Dwyer DJ, Ding C et al (2004) Engineered riboregulators enable post-transcriptional control of gene expression. Nat Biotechnol 22:841–847. https://doi.org/10.1038/nbt986
221. Abe K, Miyake K, Nakamura M et al (2014) Engineering of a green-light inducible gene expression system in *Synechocystis* sp. PCC 6803. Microb Biotechnol 7:177–183. https://doi.org/10.1111/1751-7915.12098
222. Sakai Y, Abe K, Nakashima S et al (2015) Scaffold-fused riboregulators for enhanced gene activation in *Synechocystis* sp. PCC 6803. Microbiology 4:533–540. https://doi.org/10.1002/mbo3.257
223. Sakamoto I, Abe K, Kawai S et al (2018) Improving the induction fold of riboregulators for cyanobacteria. RNA Biol 15:353–358. https://doi.org/10.1080/15476286.2017.1422470
224. Ueno K, Sakai Y, Shono C et al (2017) Applying a riboregulator as a new chromosomal gene regulation tool for higher glycogen production in *Synechocystis* sp. PCC 6803. Appl Microbiol Biotechnol 101:8465–8474. https://doi.org/10.1007/s00253-017-8570-4
225. Kittle JD, Simons RW, Lee J et al (1989) Insertion sequence IS10 anti-sense pairing initiates by an interaction between the 5′ end of the target RNA and a loop in the anti-sense RNA. J Mol Biol 210:561–572. https://doi.org/10.1016/0022-2836(89)90132-0
226. Zess EK, Begemann MB, Pfleger BF (2016) Construction of new synthetic biology tools for the control of gene expression in the cyanobacterium *Synechococcus* sp. strain PCC 7002. Biotechnol Bioeng 113:424–432. https://doi.org/10.1002/bit.25713
227. Sun T, Li S, Song X et al (2018) Re-direction of carbon flux to key precursor malonyl-CoA via artificial small RNAs in photosynthetic *Synechocystis* sp. PCC 6803. Biotechnol Biofuels 11:26. https://doi.org/10.1186/s13068-018-1032-0
228. Nakashima N, Tamura T, Good L (2006) Paired termini stabilize antisense RNAs and enhance conditional gene silencing in *Escherichia coli*. Nucleic Acids Res 34:e138. https://doi.org/10.1093/nar/gkl697
229. Na D, Yoo SM, Chung H et al (2013) Metabolic engineering of *Escherichia coli* using synthetic small regulatory RNAs. Nat Biotechnol 31:170–174. https://doi.org/10.1038/nbt.2461
230. Sun X, Li S, Zhang F et al (2021) Development of a N-acetylneuraminic acid-based sensing and responding switch for orthogonal gene regulation in cyanobacterial *Synechococcus* strains. ACS Synth Biol 10:1920–1930. https://doi.org/10.1021/acssynbio.1c00139
231. Mandal M, Breaker RR (2004) Gene regulation by riboswitches. Nat Rev Mol Cell Biol 5:451–463. https://doi.org/10.1038/nrm1403
232. Garst AD, Edwards AL, Batey RT (2011) Riboswitches: structures and mechanisms. Cold Spring Harb Perspect Biol 3:a003533. https://doi.org/10.1101/cshperspect.a003533
233. Topp S, Reynoso CMK, Seeliger JC et al (2010) Synthetic riboswitches that induce gene expression in diverse bacterial species. Appl Environ Microbiol 76:7881–7884. https://doi.org/10.1128/AEM.01537-10
234. Ohbayashi R, Akai H, Yoshikawa H et al (2016) A tightly inducible riboswitch system in *Synechocystis* sp. PCC 6803. J Gen Appl Microbiol 62:154–159. https://doi.org/10.2323/jgam.2016.02.002
235. Nakahira Y, Ogawa A, Asano H et al (2013) Theophylline-dependent riboswitch as a novel genetic tool for strict regulation of protein expression in cyanobacterium *Synechococcus elongatus* PCC 7942. Plant Cell Physiol 54:1724–1735. https://doi.org/10.1093/pcp/pct115
236. Pérez AA, Liu Z, Rodionov DA et al (2016) Complementation of cobalamin auxotrophy in *Synechococcus* sp. strain PCC 7002 and validation of a putative cobalamin riboswitch *in vivo*. J Bacteriol 198:2743–2752. https://doi.org/10.1128/JB.00475-16
237. Qi LS, Larson MH, Gilbert LA et al (2013) Repurposing CRISPR as an RNA-guided platform for sequence-specific control of gene expression. Cell 152:1173–1183. https://doi.org/10.1016/j.cell.2013.02.022
238. Kirtania P, Hódi B, Mallick I et al (2019) A single plasmid based CRISPR interference in *Synechocystis* 6803 – a proof of concept. PLoS One 14:e0225375. https://doi.org/10.1371/journal.pone.0225375

239. Yao L, Shabestary K, Björk SM et al (2020) Pooled CRISPRi screening of the cyanobacterium *Synechocystis* sp PCC 6803 for enhanced industrial phenotypes. Nat Commun 11:1666. https://doi.org/10.1038/s41467-020-15491-7
240. Choi SY, Woo HM (2020) CRISPRi-dCas12a: a dCas12a-mediated CRISPR interference for repression of multiple genes and metabolic engineering in cyanobacteria. ACS Synth Biol 9: 2351–2361. https://doi.org/10.1021/acssynbio.0c00091
241. Lee M, Woo HM (2020) A logic NAND gate for controlling gene expression in a circadian rhythm in cyanobacteria. ACS Synth Biol 9:3210–3216. https://doi.org/10.1021/acssynbio.0c00455
242. Gordon GC, Korosh TC, Cameron JC et al (2016) CRISPR interference as a titratable, trans-acting regulatory tool for metabolic engineering in the cyanobacterium *Synechococcus* sp. strain PCC 7002. Metab Eng 38:170–179. https://doi.org/10.1016/j.ymben.2016.07.007
243. Knoot CJ, Biswas S, Pakrasi HB (2020) Tunable repression of key photosynthetic processes using Cas12a CRISPR interference in the fast-growing cyanobacterium *Synechococcus* sp. UTEX 2973. ACS Synth Biol 9:132–143. https://doi.org/10.1021/acssynbio.9b00417
244. Chaves JE, Rueda-Romero P, Kirst H et al (2017) Engineering isoprene synthase expression and activity in cyanobacteria. ACS Synth Biol 6:2281–2292. https://doi.org/10.1021/acssynbio.7b00214
245. Chaves JE, Melis A (2018) Biotechnology of cyanobacterial isoprene production. Appl Microbiol Biotechnol 102:6451–6458. https://doi.org/10.1007/s00253-018-9093-3
246. Storz G, Wolf YI, Ramamurthi KS (2014) Small proteins can no longer be ignored. Annu Rev Biochem 83:753–777. https://doi.org/10.1146/annurev-biochem-070611-102400
247. Brandenburg F, Klähn S (2020) Small but smart: on the diverse role of small proteins in the regulation of cyanobacterial metabolism. Life 10:322. https://doi.org/10.3390/life10120322
248. Kanno M, Carroll AL, Atsumi S (2017) Global metabolic rewiring for improved CO2 fixation and chemical production in cyanobacteria. Nat Commun 8:14724. https://doi.org/10.1038/ncomms14724
249. Koch M, Bruckmoser J, Scholl J et al (2020) Maximizing PHB content in *Synechocystis* sp. PCC 6803: a new metabolic engineering strategy based on the regulator PirC. Microb Cell Factories 19:231. https://doi.org/10.1186/s12934-020-01491-1
250. Fritze J, Zhang M, Luo Q et al (2020) An overview of the bacterial SsrA system modulating intracellular protein levels and activities. Appl Microbiol Biotechnol 104:5229–5241. https://doi.org/10.1007/s00253-020-10623-x
251. Landry BP, Stöckel J, Pakrasi HB (2013) Use of degradation tags to control protein levels in the cyanobacterium *Synechocystis* sp. strain PCC 6803. Appl Environ Microbiol 79:2833–2835. https://doi.org/10.1128/aem.03741-12
252. Andersen JB, Sternberg C, Poulsen LK et al (1998) New unstable variants of green fluorescent protein for studies of transient gene expression in bacteria. Appl Environ Microbiol 64:2240–2246. https://doi.org/10.1128/AEM.64.6.2240-2246.1998
253. Gur E, Sauer RT (2008) Evolution of the ssrA degradation tag in mycoplasma: specificity switch to a different protease. Proc Natl Acad Sci U S A 105:16113–16118. https://doi.org/10.1073/pnas.0808802105
254. Sakkos JK, Hernandez-Ortiz S, Osteryoung KW et al (2021) Orthogonal degron system for controlled protein degradation in cyanobacteria. ACS Synth Biol 10:1667–1681. https://doi.org/10.1021/acssynbio.1c00035
255. Røkke G, Korvald E, Pahr J et al (2014) BioBrick assembly standards and techniques and associated software tools. Methods Mol Biol 1116:1–24. https://doi.org/10.1007/978-1-62703-764-8_1
256. Matsumura I (2020) Methylase-assisted subcloning for high throughput BioBrick assembly. PeerJ 8:e9841. https://doi.org/10.7717/peerj.9841
257. Gibson DG, Young L, Chuang R-Y et al (2009) Enzymatic assembly of DNA molecules up to several hundred kilobases. Nat Methods 6:343–345. https://doi.org/10.1038/NMETH.1318

258. Beyer HM, Gonschorek P, Samodelov SL et al (2015) AQUA cloning: a versatile and simple enzyme-free cloning approach. PLoS One 10:e0137652. https://doi.org/10.1371/journal.pone.0137652
259. Engler C, Kandzia R, Marillonnet S (2008) A one pot, one step, precision cloning method with high throughput capability. PLoS One 3:e3647. https://doi.org/10.1371/journal.pone.0003647
260. Engler C, Gruetzner R, Kandzia R et al (2009) Golden gate shuffling: a one-pot DNA shuffling method based on type IIs restriction enzymes. PLoS One 4:e5553. https://doi.org/10.1371/journal.pone.0005553
261. Weber E, Engler C, Gruetzner R et al (2011) A modular cloning system for standardized assembly of multigene constructs. PLoS One 6:e16765. https://doi.org/10.1371/journal.pone.0016765
262. Immethun CM, Ng KM, DeLorenzo DM et al (2016) Oxygen-responsive genetic circuits constructed in *Synechocystis* sp. PCC 6803. Biotechnol Bioeng 113:433–442. https://doi.org/10.1002/bit.25722
263. Mavrommati M, Daskalaki A, Papanikolaou S et al (2022) Adaptive laboratory evolution principles and applications in industrial biotechnology. Biotechnol Adv 54:107795. https://doi.org/10.1016/j.biotechadv.2021.107795
264. Portnoy VA, Bezdan D, Zengler K (2011) Adaptive laboratory evolution--harnessing the power of biology for metabolic engineering. Curr Opin Biotechnol 22:590–594. https://doi.org/10.1016/j.copbio.2011.03.007

Adv Biochem Eng Biotechnol (2023) 183: 105–144
https://doi.org/10.1007/10_2023_218

Published online: 25 April 2023

Metabolic Engineering Design Strategies for Increasing Carbon Fluxes Relevant for Biosynthesis in Cyanobacteria

Arvin Y. Chen, Jason T. Ku, Teresa P. Tsai, Jenny J. Hung, Billy C. Hung, and Ethan I. Lan

Contents

A. Y. Chen, J. T. Ku, J. J. Hung, and B. C. Hung
Institute of Molecular Medicine and Bioengineering, National Yang Ming Chiao Tung University, Hsinchu City, Taiwan

T. P. Tsai
Institute of Biomedical Engineering, National Yang Ming Chiao Tung University, Hsinchu City, Taiwan

E. I. Lan (✉)
Department of Biological Science and Technology, National Yang Ming Chiao Tung University, Hsinchu City, Taiwan
e-mail: ethanilan@nycu.edu.tw

Abstract Cyanobacteria are promising microbial cell factories for the direct production of biochemicals and biofuels from CO_2. Through genetic and metabolic engineering, they can be modified to produce a variety of both natural and non-natural compounds. To enhance the yield of these products, various design strategies have been developed. In this chapter, strategies used to enhance metabolic fluxes towards common precursors used in biosynthesis, including pyruvate, acetyl-CoA, malonyl-CoA, TCA cycle intermediates, and aromatics, are discussed. Additionally, strategies related to cofactor availability and mixotrophic conditions for bioproduction are also summarize.

Graphical Abstract

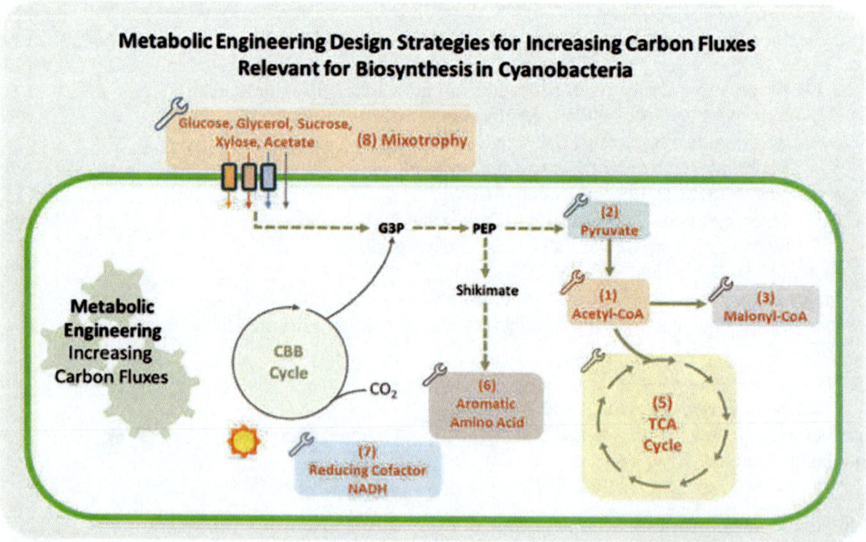

Keywords Cyanobacteria, Metabolic engineering, Design strategies, Carbon flux, Central metabolites

1 Introduction

Development of cyanobacterial cell factories has emerged as an attractive approach to simultaneously address sustainability and environmental challenges. In recent decades, numerous studies have been published using cyanobacteria as hosts for biochemical and biofuel production. Model cyanobacterium *Synechococcus elongatus* PCC 7942, *Synechococcus* sp. PCC 7002, and *Synechocystis* sp. PCC 6803 were the most commonly employed for such studies. In these works, biochemicals such as ethanol [1–10], ethylene [11], lactate [12–20], 3-hydroxypropionic acid [21, 22], 1,3-propandiol [23, 24], 1,2-propandiol [25], glycerol [23, 26, 27], acetone [28], butanol [4, 29–34], butyrate [35], succinate [36], 2,3-butandiol [37–41], 3-hydroxybutyrate [42, 43], isopropanol [44, 45], isobutanol [46–49], isobutene [50], isobutaldehyde [48, 51, 52], 2-methylbutanol [53], isoprene [54], itaconic acid [55], *p*-coumaric acid [56], limonene [57], squalene [58], lysine [59], sucrose [60, 61],mannitol [62], fatty acids [59, 63], fatty alcohols [64, 65], and alka(e)nes [66] have been produced directly from CO_2. While numerous compounds have been demonstrated for production in these metabolically engineered cyanobacteria, some compounds are produced with higher fluxes than others.

The directions of natural carbon fluxes in cyanobacterial metabolism have been obtained through steady-state ^{13}C metabolic flux analysis (SS-MFA) and isotopically nonstationary metabolic flux analysis (INST-MFA) [67, 68] data. The major carbon fluxes in cyanobacteria occur through reactions in the Calvin–Benson–Bassham (CBB) cycle for CO_2 assimilation by ribulose-1,5-bisphosphate carboxylase/oxygenase (RuBisCO) under photoautotrophic condition. High levels of carbon flux were also observed during mixotrophic growth, during which carbon flux in the CBB cycle is about twofold higher than that of glycolysis [69]. In contrast to terrestrial plants, photorespiration as a result of oxygenation by RuBisCO is generally insignificant with negligible glycolate secreted [67]. Relevant to the CBB cycle is the oxidative pentose phosphate pathway (PPP) exhibiting CO_2 loss about 13% of total fixed carbon [67]. However, the PPP is important for the synthesis of five and four carbon sugar phosphate intermediates used for the nucleotides and aromatic amino acid biosynthesis.

A notable feature in cyanobacterial flux different from heterotrophs is the conversion of PEP to pyruvate via malate. Most organisms typically convert PEP to pyruvate using pyruvate kinase. However, many cyanobacteria achieve this conversion through PEP carboxylase (Ppc), malate dehydrogenase (Mdh), and malic enzyme (ME) [67, 68, 70]. Another unique metabolic difference between cyanobacteria and heterotrophs is their tricarboxylic acid (TCA) cycle. TCA cycle in many model cyanobacteria was commonly recognized as a non-complete cycle due to lack of α-ketoglutarate dehydrogenase (OGDH). Although there are OGDH bypasses such as via succinic semialdehyde and γ-aminobutyrate shunt, these fluxes are low. Nonetheless, the oxidative branch of TCA cycle up to the synthesis of α-ketoglutarate accounted for approximately 12–14% of total flux [70] of the fixed carbon as α-ketoglutarate is required for nitrogen assimilation. Given these natural fluxes, metabolic engineering strategies are often needed to support redirecting carbon fluxes toward the desired chemical products.

This chapter dives into cyanobacterial metabolic engineering design strategies used to re-route carbon flux for bioproduction. Relevant to this chapter are the strategies used to increase carbon fluxes to acetyl-CoA, pyruvate, malonyl-CoA, TCA cycle intermediates, aromatic amino acid biosynthesis, and terpenoids. This chapter also discusses strategies used to enhance carbon fixation, coupling production pathways to proper reducing cofactor, and uses of mixotrophic production mode. To guide readers through each type of strategies used, Tables 1, 2, 3, 4, 5, 6, and 7 summarize each strategy by their primary goal. Above all, this chapter aims to guide metabolic engineers interested in developing cyanobacterial cell factories through these useful design strategies.

2 Increasing Carbon Flux Toward Acetyl-CoA and Its Derivatives

Acetyl-CoA is a versatile central metabolite that serves as the precursor for numerous downstream metabolic products, including fatty acids, terpenoids, alcohols, and TCA cycle intermediates. However, cyanobacteria generally have lower acetyl-CoA availability compared to commonly used heterotrophs. For instance, intracellular acetyl-CoA concentration in cyanobacteria is about 50 μM, while in *Escherichia coli* it is about 1,000 μM [54]. Therefore, increasing acetyl-CoA availability is an important engineering target for establishing efficient acetyl-CoA-derived compound production pathways in cyanobacteria. To address the low acetyl-CoA availability, metabolic engineering strategies (Table 1) such as competing pathway removal and upstream enzyme activity improvement were frequently applied.

2.1 Dark Anoxic Cultivation Increases Acetyl-CoA Flux by Degrading Carbon Storage Compounds

Early demonstration of bioproduction in cyanobacteria for acetyl-CoA derivatives commonly applied dark anoxic conditions to increase the intracellular acetyl-CoA concentration by degrading endogenous carbon storage compounds such as glycogen. This approach is simple and easy to apply as no genetic engineering is required. For instance, isopropanol titer was increased from around none to 10 mg/L when

Table 1 Examples for increasing carbon flux via acetyl-CoA for biosynthesis

Approach	Bioproduct	Host	Effectiveness	Reference
Dark/anoxic incubation increases acetyl-CoA flux by degrading carbon storage compounds	1-butanol	PCC 7942	Butanol titer increased from 0.5 to 7 mg/L in 14 days (under dark anoxic condition)	[30]
	Isopropanol	PCC 7942	Isopropanol titer increased from around 1 to 26.5 mg/L in 9 days (under dark anoxic condition)	[45]
	Isopropanol	PCC 7942	Isopropanol titer increased from 22.8 to 114 mg/L in 10 days (from dark anoxic to light aerobic condition)	[71]
Increasing upstream enzyme activity	Acetate	PCC 7942	Overexpressing *pdh* increased acetate titer from 84 to 595 mg/L in 21 days with a twofold increased intracellular acetyl-CoA concentration	[44]
	1-butanol	PCC 6803	Overexpressing *xfpk* increased butanol titer from 20 to 35 mg/L in 8 days with a sixfold increased intracellular acetyl-CoA concentration	[32]
	Isopropanol	PCC 7942	Overexpressing *pdh* increased isopropanol titer from 30 to 114 mg/L in 10 days with a twofold increased intracellular acetyl-CoA concentration	[44]
	PHB	PCC 6803	Overexpressing *xfpk* increased PHB titer from 50 to 150 mg/L in 25 days	[75]
	PHB	PCC 6803	Removal of *pirC* potentially increased the activity of 2,3-bisphosphoglycerate phosphoglycerate mutase, leading to twofold intracellular PHB content	[78]
Competing pathway removal/repression	Acetone	PCC 6803	Deleting *pta* increased acetone titer from 5 to 36 mg/L in 4 days	[28]
	1-butanol	PCC 6803	Using CRISPRi system to partially inhibit *gltA* increased butanol titer from 75 to 120 mg/L in 12 days	[34]
	Ethanol	PCC 6803	Deleting *phaAB* increased ethanol titer from 1.12 to 1.49 g/L in 26 days	[83]

(continued)

Table 1 (continued)

Approach	Bioproduct	Host	Effectiveness	Reference
	Fatty acid	PCC 6803	Replacing *pta* by *tesA137* increased fatty concentration from 146 to 197 mg/L	[63]
	3-hydroxybutyrate	PCC 6803	Deleting PHB polymerase *phaEC* increased 3-hydroxybutyrate titer from 45 to 94 mg/L in 5 days	[82]
	PHB	PCC 6803	Deleting *pta* and *ach* increased PHB titer from 150 to 232 mg/L in 32 days	[75]
	Succinate	PCC 7942	Deleting *glgC* increased succinate titer from 41 to 182 mg/L	[81]
Constructing metabolic bypasses for enhancing carbon flux to acetyl-CoA	Acetone	PCC 7942	Expressing *pdc*, *ald*, *acs*, and *mmc* (AAM bypass) enabled acetone production at 411 mg/L in 8 days with 9.5-fold increased intracellular acetyl-CoA concentration.	[85]
	Ketoisocaproate	PCC 7942	Expressing *ppc, mdh, mtk, mcl, gcl, glxR*, and *garK* (MCG pathway) increased ketoisocaproate titer from around 100 to 433 mg/L in 6 days with twofold increased intracellular acetyl-CoA concentration	[84]

simply switching the cultivation environment into dark anoxic condition [45]. Isopropanol production required two acetyl-CoA per each isopropanol produced. Two acetyl-CoA first condense into acetoacetyl-CoA, followed by transfer of the CoA moiety to acetate via the activity of a CoA-transferase resulting in acetoacetate. Acetoacetate then decarboxylates into acetone that can be reduced to isopropanol. Such pathway requires a large amount of acetyl-CoA flux. The dark anoxic incubation allows cyanobacteria to behave similar to heterotrophs, as the cellular stock of glycogen is processed through glycolysis to acetyl-CoA. Anoxic condition diminishes respiratory activity, causing a reduced carbon flux into the TCA cycle. This simple cultivation approach was further applied for acetate production. Another study used a two-step cultivation approach to increase isopropanol production [71]. In this case, acetate was produced under dark anoxic conditions in the first step, followed by light aerobic cultivation conditions to produce isopropanol. The resulting isopropanol titer increased from 0.38 mM (22.8 mg/L) to 1.9 mM (114 mg/L). Acetate produced in the first dark anoxic stage was

Table 2 Examples for increasing carbon flux toward pyruvate

Strategy		Product	Host	Effectiveness	Reference
Add pyruvate		Acetoin	PCC 7942	Carbon partitioning[a] to acetoin increased from 54% (ambient CO_2) to 73% and acetoin production was increased from 263 to 354 mg/L in first day	[41]
		Ethanol	PCC 7942	Carbon partitioning[a] to ethanol increased from 23% (50 mM $NaHCO_3$) to 26% and ethanol productivity was increased from 36.3 to 64.5 mg/L/day	[9]
Elimination of glycogen biosynthesis	Knockout *glgC* and under nitrogen deficiency (or limited) condition	Pyruvate	PCC 7942	Around 17.6 mg/L pyruvate secretion in 4 days	[89]
			PCC 6803	Around 100 mg/L pyruvate secretion in 4 days	[87]
			PCC 6803	Around 35.2 mg/L pyruvate secretion in 2 days	[88]
			PCC 6803	Around 528 mg/L pyruvate secretion in 7 days (with 3.5 mM nitrate adding)	[91]
			PCC 7002	Around 12.3 mg/L pyruvate secretion in 2 days	[90]
		L-lactate	PCC 6803	Productivity increased from 8.9 mg/L/OD/day to 21.6 mg/L/OD/day under nitrogen starvation on the first day	[92]
	Knockout *glgC* only	Branched chain alcohols	PCC 7942	Higher alcohol productivity (by 1.6- to 2.4-fold increasing) during 6 days (maxima was 76.9 mg/L/OD/day). Total fixed carbon increased from 22% to 52%	[46]
	Knockout *glgA* (*sll0945* and *sll1393*)	Pyruvate	PCC 6803	Around 123 mg/L pyruvate secretion in 2 days by photomixotrophic condition	[88]
Mimic nitrogen starvation	Knockdown *glnA*	L-lactate	PCC 7002	Higher lactate productivity from 360 to 810 mg/L/day in 4 days	[95]
	Knockout *gdc*	–	PCC 6803	Intracellular level of pyruvate was 1.2-fold higher	[96]

(continued)

Table 2 (continued)

Strategy		Product	Host	Effectiveness	Reference
Improve flux from 3PG toward pyruvate	Overexpression of *pyk*	2,3-BDO	PCC 7942	Productivity was increased from 84 to 132 mg/L/OD/day. Total carbon productivity was 4.3 mmol/gDW/h, which was equal to 117% of the theoretical maximum for carbon fixation rate (3.7 mmol/gDW/h)	[41]
		L-lactate	PCC 6803	Heterologous expression of *pyk* from *Enterococcus faecalis* (or *E. coli*) increased titer from 540 to 810 mg/L in 17 days	[12]
		IBA	PCC 7942	Increase Isobutyraldehyde (IBA) productivity from 51.9 to 77.8 mg/L/gCDW/day	[68]
	Knockout *pirC* (*sll0944*)	–	PCC 6803	Intracellular pyruvate level was 14-fold higher than the wild-type. (after 48 h of nitrogen starvation)	[138]
Partition the flux from TCA cycle intermediate to pyruvate	Expression of *pyk* and *pck* from *E. coli*	IBA	PCC 7942	Increase productivity approximately from 155 to 224 mg/L/gCDW/day (compared to strain with *pyk* overexpression already [68]) and accumulate intracellular pyruvate	[51]
	Expression of endogenous *mae* (PCC 7942) and *mdh* (PCC 6803)	IBA	PCC 7942	Increase IBA productivity from 51.9 to 86.5 mg/L/gCDW/day and accumulate intracellular pyruvate	[68]
	Overexpress *mae*	D-lactate	PCC 6803	Productivity improved from 50 to 87.5 mg/L/day in 4 days under anoxic fermentation	[20]
Prevent competing pathway	Knockdown *pdh* via antisense RNA	IBA	PCC 7942	Productivity approximately from 155 mg/L/gCDW/day to 207 mg/L/gCDW/day	[51]
	Knockout *ppsA* (slr0301)	Ethanol	PCC 6803	Titer increased from 591 to 878 mg/L after 9 days	[97]

(continued)

Table 2 (continued)

Strategy		Product	Host	Effectiveness	Reference
	Partial knockout of *ppc*	L-lactate	PCC 6803	Increase titer approximately from 153 to 205 mg/L in 14 days	[12]

[a] Partitioning is calculated by dividing the carbon productivity of the product by the sum of carbon productivities for product and biomass during the same time period

assimilated for isopropanol production during the second light aerobic stage with a yield of 50–67% of theoretical, indicating the presence of competing fluxes for acetate. Similar to isopropanol, 1-butanol production also benefited from this cultivation condition switch strategy [30]. Upon dark anoxic incubation of metabolically engineered cyanobacteria containing the *Clostridial* butanol synthesis pathway, over 7 mg/L of butanol were achieved, which was a 10-fold increase compared to the same strain cultivated under light aerobic condition. Nevertheless, dark anoxic incubation is not a sustainable approach for long-term chemical production as the internal carbon storage is limited in supply and only replenished during photoautotrophic growth. Therefore, in order to replicate the effects of increasing acetyl-CoA flux by dark anoxic condition for sustainable production, directly genetic and corresponding metabolic manipulation should be applied.

2.2 *Enhancing Upstream Enzyme Activity Increases Acetyl-CoA Flux for Biosynthesis*

Since pyruvate dehydrogenase, converting pyruvate to acetyl-CoA, is the primary enzyme responsible for acetyl-CoA synthesis, increasing the pyruvate dehydrogenase activity is a straightforward metabolic strategy for enhancing the carbon flux toward bioproducts via acetyl-CoA as intermediate. Moreover, pyruvate is relatively abundant in cyanobacteria, and a 20- to 500-fold higher intracellular concentration compared to acetyl-CoA has been reported [54, 72], indicating a promising possibility of increasing acetyl-CoA flux by pyruvate dehydrogenase overexpression. However, pyruvate dehydrogenase overexpression is challenging due to its complicated regulation by endogenous regulon, allosteric effectors, and covalent modifications [73]. Although very few studies reported pyruvate dehydrogenase overexpression in cyanobacteria, it was recently applied to generate hosts for acetate and isopropanol production with improved product titers [44]. Using the strong promoter P_{trc} to drive the overexpression of pyruvate dehydrogenase in *Synechococcus elongatus* strain PCC 7942 (hereafter PCC 7942), its activity was increased significantly, leading to a twofold increase in intracellular acetyl-CoA content. Correlating with the higher acetyl-CoA availability, acetate and isopropanol production also improved by 7-fold and 3.8-fold, respectively.

Table 3 Examples for increasing carbon flux toward malonyl-CoA-derived compound

Strategy	Product	Hosts	Effectiveness	Reference
Overexpression of acetyl-CoA carboxylase (ACC)	Free fatty acid (FFA)	PCC 6803	Overexpression of native ACC in cyanobacteria did not significantly increase titer	[63]
		PCC 7942	Heterologous expression of ACC from *Chlamydomonas reinhardtii* increased FFA titer from 370 to 550 mmol/gDCW on the 18th day (compared to without ACC expression)	[101]
	Lipids	PCC 6803	Overexpressing native ACC and knocking out phospholipid hydrolysis (lipids to free fatty acids) led to increased titer, from 87.94 to 126.03 mg/L on the fourth day (compared to wild-type)	[102]
	Alka(e)nes Heptedecane Heptedecene	PCC 6803	Overexpressing native ACC drove total hydrocarbons' increase from approximately 100 to 161.6 μg/L/OD730 in 12 days	[64]
		PCC 6803	Overexpressing ACC and expressing acyl-acyl carrier protein reductase as well as aldehyde-deformylating oxygenase led to increased alka(e)nes production up to 1,100 μg/L/OD730 in 14 days	[66]
Overexpression of ACC + *birA*	3HP	PCC 6803	Expressing malonyl-CoA reductase, ACC, and *birA* led to 31.62–130.40 mg/L 4.5-fold increase in 6 days	[22]
Inhibition of fatty acid synthesis	Intracellular malonyl-CoA	PCC 6803	Using artificial small RNAs Hfq-MicC tools to inhibit fatty acid synthesis pathways *fabDFH* led to intracellular malonyl-CoA increase from approximately 0.49 to 0.69 μg/L/OD_{730} at 48 h	[104]
Feed cerulenin	Resveratrol, naringenin, bisdemethoxycurcumin	PCC 7942	Adding cerulenin limited the amount of malonyl-CoA channeling to fatty acid synthesis. It increased resveratrol production from 3.9 to 4.6 mg/L, naringenin from 4.3 to 7.1 mg/L, and bisdemethoxycurcumin from 2.9 to 4.1 mg/L	[105]

Table 4 Examples for increasing carbon flux toward TCA cycle

Strategy			Product	Host	Effectiveness	Reference
Enhance carbon flux into TCA cycle		Express *gltA* and *ppc*	Succinate	PCC 7942	Productivity increased from around 23 to 30 mg/L/day with *gltA* expression and further to 80 mg/L/day with additional *ppc* expression	[36]
		Express *pps*A and *ppc*	Ethylene	PCC 6803	Productivity increased from 6.4 to 10.5 mg/L/OD/day with *ppc* expression and further increased to 16.8 mg/L/OD/day with additional *ppsA* expression	[11]
		Express *aceAB*	Succinate	PCC 6803	It further increased production titer from around 40 to 400 mg/L/OD for 4 days production compared to a strain with 2 copies of *ppc* expression	[108]
Increase TCA cycle intermediate level	Through GABA shunt	Knock out *gabdh*	Succinate, glutamate, GABA	PCC 6803	It increased intracellular metabolites of around 2 μg/gDW succinate, 8 ng/gDW glutamate, and 15 ng/gDW GABA, which were 5-, 2-, 15-folds, respectively, compared to wild-type strain	[111]
	Through reductive TCA pathway	Express *citH*	Malate, fumarate, succinate	PCC 6803	Excreted level of malate, fumarate, and succinate was increased around 2.8-, 2.1-, and 1.2-folds, reached highest titer of 441, 637, 677 mg/L, respectively in 3 days	[112]

Due to the complex regulation of pyruvate dehydrogenase, alternative approaches were investigated to increase the carbon flux toward acetyl-CoA. For instance, phosphoketolase catalyzes the conversion of fructose-6-phsophate to erythrose-4-phosphate and acetyl-phosphate. Acetyl-phosphate can be further converted to acetyl-CoA via phosphotransacetylase [74]. Expression of phosphoketolase,

Table 5 Strategies of improving carbon flux toward aromatic amino acids to increase derived compound titer

Strategies	Products	Hosts	Effectiveness	Reference
Knockout putative laccase gene *slr1573*	*p*-coumaric acid	PCC 6803	Increased titer from non-detectable to 82.6 mg/L in 7 days	[56]
Overexpress feedback-resistant (fbr) genes	2-Phenylethanol	PCC 7942	Overexpression of DAHP synthase (aroG^{fbr}) and (pheA^{fbr}) from *E. coli* caused phenylethanol titer to increase from 0.4 to 23 mg/L/OD_{730nm} by KivD route[a], 1.8 to 14.07 mg/L/OD_{730nm} by Paas route[b] in 10 days	[139]
	Tryptophan	PCC 6803	Overexpression of DAHP synthase (*aroG*$^{fbr\ (L175D)}$) and anthranilate synthase (*trpE*$^{fbr\ (S40F)}$) increased tryptophan titer from non-detectable to approximately 13 mg/L in 7 days	[140]
	p-coumaric acid	PCC 7942	Overexpression of DAHP synthase (encoded by aroG^{fbr}) increased *p*-coumaric acid titer from 0.3 to 22 mg/L/OD_{730nm} in 10 days	[141]
	Phenylalanine, tyrosine, tryptophan	PCC 6803	Overexpression of DAHP synthase (aroG^{fbr}) and arogenate dehydrogenases (tyrA^{fbr}) in a Δ*slr1573* strain increased phenylalanine titer from non-detectable to 903 mg/gDW. Tyrosine titer increased from non-detectable to 64 mg/gDW. Tryptophan titer was improved from non-detectable to 19.3 mg/gDW in 10 days	[142]

[a] KivD route: overexpressing 2-keto acid decarboxylase (*kivD*) gene to convert phenylpyruvate to phenylacetaldehyde in cyanobacteria

[b] Paas route: overexpressing phenylacetaldehyde synthase (*paas*) gene to convert phenylalanine to phenylacetaldehyde in cyanobacteria

encoded by *xfpk*, led to a sixfold increased acetyl-CoA concentration, which benefited butanol production titer by twofold [32]. Continuing studies showed a comparable behavior by combining expression of phosphoketolase and phosphotransacetylase [31]. An analogous strategy has been applied for poly-3-hydroxybutyrate (PHB) production in *Synechocystis* sp. strain PCC 6803 (hereafter PCC 6803) [75]. Since PHB is synthesized from two acetyl-CoA condensations, acetyl-CoA availability is important for efficient PHB production. With the expression of phosphoketolase, the PHB content increased up to 11.5-fold compared to the strain without phosphoketolase expression. Moreover, a small regulatory protein PirC (encoded by *sll0944*) is known for triggering glycogen accumulation [76] by inhibiting phosphoglycerate mutase [77]. In strain PCC 6803, *sll0944* was removed to increase the carbon flux for PHB synthesis. Although no direct evidence of the

Table 6 Selected examples for increasing cofactor availability for enhanced biosynthesis

Approaches	Products	Host	Effectiveness	Reference
Using NADPH-dependent enzymes	1-butanol	PCC 7942	Endogenous NADPH-dependent alcohol dehydrogenase Slr1192 used	[123]
	1-butanol	PCC 6803	NADPH-dependent sADH, YqhD used	[31]
	2,3-butanediol	PCC 7942	NADPH-dependent PhaB, Bldh, YqhD used	[125]
	Ethanol	PCC 6803	NADPH-dependent PhaB, Ccr, Slr1192 used; achieving the highest titer of ethanol so far	[83]
	Isobutanol	PCC 7942	5.5-fold increase in titer by replacing NADH-dependent AdhA with NADPH-dependent YqhD	[48]
	1,2-propanediol	PCC 7942	Tested 4 NADPH-dependent alcohol dehydrogenase to synthesize different chirality of 1,2-PDO	[126]
	Succinate	PCC 7942	NADP-dependent succinate semialdehyde dehydrogenase GabD used	[36]
Changing cofactor preference by protein engineering	Lactate	PCC 6803	Expressing NADPH-dependent Ldh mutant (V39R) instead of wild-type Ldh increased lactate titer from 140 to 345 mg/L in 14 days	[12]
		PCC 7942	Expressing engineered NADPH-dependent Ldh^{ARSdR} instead of wild-type Ldh increased lactate titer from 101 to 798 mg/L in 10 days although the total activity of wild-type Ldh is higher than that of Ldh^{ARSdR}	[14]
Manipulating intracellular NAD(P)H concentration	Acetate, lactate	PCC 7002	Expression of *gap1* increased glycolytic rate, leading to increased lactate and acetate production under dark anoxic condition. NADH and NADPH were increased by 4-fold and 2.9-fold, respectively	[131]
	2,3-butanediol	PCC 6803	Expressing *sth* gene from *P. aeruginosa* increased 2,3-butanediol to acetoin ration	[129]
	Ethanol	PCC 6803	Expressing *zwf* increased ethanol titer from 440 to 590 mg/L in 14 days	[130]
	Lactate	PCC 6803	Expressing transhydrogenase *sth* from *P. aeruginosa* increased lactate titer from 58 to 288 mg/L in 17 days	[18]

change in intracellular acetyl-CoA flux and concentration was reported, the pyruvate concentration [77] and PHB content [78] were increased upon removal of *sll0944*, indicating the increased glycolysis and acetyl-CoA flux.

Table 7 Examples for increasing carbon flux by mixotrophic growth

Feeding	Products	Host	Effectiveness	Reference
Glucose	Biomass (growth)	PCC 7942	Growth rate increased from 0.16 to 0.24 OD_{700}/day. Overexpressing glucose uptake gene *galP* from *E. coli* increased growth rate from 0.24 to 0.54 OD_{700}/day	[133]
	2,3-butanediol	PCC 7942	Overexpression of *galP* and glucose feeding increased titer from 0.5 to 3.0 g/L in a fed-batch fermentation	[134]
		PCC 7942	Overexpression of *galP* and glucose feeding increased titer from 70 mg/L/OD to 270 mg/L/OD	[38]
Xylose	Biomass (growth)	PCC 7942	Xylose feeding increased growth rate from 0.16 to 0.35 OD_{700}/day. Expression of *xylEAB* further increased growth rate to 0.47 OD_{700}/day in light condition and 0.29 OD_{700}/day in dark	[133]
	2,3-butanediol	PCC 7942	Overexpression of *xylEAB* and xylose feeding increased titer from 39 to 75 mg/L	[134]
	Ethylene	PCC 6803	Overexpression of *xylAB* and xylose feeding increased ethylene titer from 8.5 to 11.3 mg/L	[135]
Sucrose	Biomass (growth)	PCC 7942	Sucrose feeding increased growth rate from 0.16 to 0.31 OD_{700}/day. Overexpression of *cscB* and *cscK* further improved growth rate to 0.38 OD_{700}/day in light condition and 0.13 OD_{700}/day in dark	[133]
Glycerol	2,3-butanediol	PCC 7942	Overexpression of *gldFKD* and *tpiA* with glycerol feeding increased titer from 33 to 59 mg/L/OD	[136]
Acetate	Isopropanol	PCC 7942	By acetate feeding, titer increased from non-detectable to 21.7 mg/L at seventh day	[45]
	D-lactate	PCC 6803	By acetate feeding, titer increased from 57 mg/L/OD to 110 mg/L/OD at 24 h	[15]
	Polyhydroxybutyrate	PCC 6803	By acetate feeding, titer increased from 26% of DCW to 35% of DCW at ninth day	[137]

2.3 *Reducing Competing Pathway Activities Concentrates Acetyl-CoA Flux Toward Targeted Bioproducts*

In addition to increasing the influx to acetyl-CoA, removal or blocking of competing pathways that potentially consume acetyl-CoA is another conventional strategy for increasing acetyl-CoA availability [79]. Acetate is a common by-product diverging carbon flux from acetyl-CoA. It was observed in some biosynthesis, such as butyrate [35], 3-hydroxybutyrate [80], and acetone [28]. Removal of acetate synthesis usually led to improved production titer of acetyl-CoA-based bioproducts. For instance, deletion of *pta* encoding for phosphotransacetylase improved acetone titer by 7.2-fold with 15% residual acetate produced [28]. Another study showed the deletion of *pta* combined with the heterologous expression of *tesA'* encoding for an acyl-ACP thioesterase increased fatty acid titer by 35% [63]. Moreover, PHB production was also reported with a 1.5-fold improvement by deletion of *pta* and *ach* encoding an acetyl-CoA hydrolase [75].

Cyanobacteria naturally accumulate carbon storage compounds such as glycogen and poly-3-hydroxybutyrate (PHB) that potentially diverge carbon flux from acetyl-CoA. Therefore, removal of such carbon storage pathways is a promising strategy for increasing carbon flux toward target bioproducts. Inactivation of the glycogen synthesis pathway by the deletion of *glgC* enabled succinate production [81]. Directly relevant to acetyl-CoA flux consumption is the native biosynthesis of PHB, which uses acetyl-CoA as its biosynthetic precursor. Wild-type strain PCC 6803 naturally accumulates around 10% PHB/cell dry weight (CDW) [78], representing a significant acetyl-CoA drain. Therefore, inactivation of PHB synthesis was used for constructing strain PCC 6803 for valuable chemical synthesis. A straightforward target for PHB synthesis pathway inactivation is its monomer, 3-hydroxybutyrate. Deletion of *phaEC* encoding PHB polymerase increased 3-hydroxybutyrate production by 2.1-fold [82]. With further improvement, a 3HB titer of 1.84 g/L was later on reported [42]. Ethanol titer, on the other hand, was increased by 33% with the deletion of *phaAB*, encoding PHA-specific β-ketothiolase and acetoacetyl-CoA reductase, respectively, which catalyze the first two steps of PHB synthesis [83]. The optimized strain achieved high production titers of 5.5 g/L.

In addition to the competing genes and pathways described above, which are generally regarded as non-essential to the cell growth upon removal, there are competing pathways that are essential for cell growth. As such, where these essential genes have been targets of engineering, they were generally partially repressed instead of deleted to concentrate the carbon flux toward the desired synthesis pathways. Citrate synthase (*gltA*) catalyzes the first step in the TCA cycle, which is an important route for synthesizing α-ketoglutarate-derived biomass precursors. Therefore, deletion of *gltA* usually prohibits the cell proliferation. As an alternative, CRISPRi system was used for inhibiting *gltA* for butanol production [34]. Although repression of 90% activity of GltA led to fivefold increased carbon partition in butanol, indicating the effective redirecting carbon flux to acetyl-CoA derivative, the butanol titer was not increased due to poor cell growth. By further modulating *gltA* repression level, butanol titer was improved by 1.6-fold with minor growth inhibition effect caused by 11% gltA repression, indicating the importance of tunable genetic repression tools.

2.4 *Constructing Metabolic Bypasses for Enhancing Carbon Flux to Acetyl-CoA*

In addition to conventional metabolic engineering strategies, synthetic pathway construction for enhancing carbon flux was reported recently. A malyl-CoA–glycerate pathway (MCG pathway) constructed by Yu et al. was designed to convert phosphoenolpyruvate (PEP) to two molecules of acetyl-CoA with additional carbon fixed [84]. The MCG pathway was demonstrated to effectively improve intracellular acetyl-CoA concentration by twofold in strain PCC 7942. In combination with the expression of ketoisocaproate production pathway, 433 mg/L of ketoisocaproate was observed, which was more than fivefold improvement compared to the strain without MCG pathway.

In another study, an acetate-acetyl-CoA/malonyl-CoA (AAM) bypass was constructed to circumvent the highly regulated pyruvate dehydrogenase (Pdh) for acetyl-CoA synthesis [85]. The pathway comprises an acetate-acetyl-CoA module (AA-module) and a malonyl-CoA module (M-module). The AA-module was composed by pyruvate decarboxylase (*pdc*), aldehyde dehydrogenase (*ald6*), and acetyl-CoA synthetase (*acs*), which together allowed acetyl-CoA synthesis from pyruvate independent of the Pdh-catalyzed reaction. Acetyl-CoA concentration was around eightfold in the strain with AA-module expression compared to that without. While solely expressing genes encoding the acetone synthesis pathway did not lead to acetone production, additional expression of the respective genes for the AA-module resulted in 378 mg/L acetone, showing the effectiveness of AA-module on redirecting carbon flux to acetyl-CoA [85]. Since cyanobacteria synthesizes significant amounts of lipid as required for the thylakoid membranes, acetyl-CoA carboxylation into malonyl-CoA, the precursor to lipids, may be a major competing pathway for acetyl-CoA flux. Therefore, removal of the carboxyl group from malonyl-CoA to regenerate acetyl-CoA was considered as a viable approach to increase acetyl-CoA pool for biosynthesis. To remove the carboxyl group from malonyl-CoA without releasing it as a CO_2, methylmalonyl-CoA carboxyltransferase, designated as the M-module, was expressed together with the AA-module. Methylmalonyl-CoA carboxyltransferase catalyzes the transfer of carboxyl group from malonyl-CoA to pyruvate, forming oxaloacetate and acetyl-CoA. This final strain produced 411 mg/L of acetone. While acetone production was only marginally improved, the intracellular acetyl-CoA pool increased from 129 to 165 μg/gDCW when comparing expression of AA-module and the entire AAM bypass, indicating that manipulation of acetyl-CoA pool was successful, but downstream product acetone titer was not proportionally increased, possibly due to limitation of acetone synthesis pathway.

In summary, strategies (Fig. 1) including culture condition optimization, upstream enzyme activity improvement, competing pathway removal, and synthetic pathway design and constriction were shown to effectively increase acetyl-CoA flux, concentration, and biosynthesis of acetyl-CoA-derived compounds.

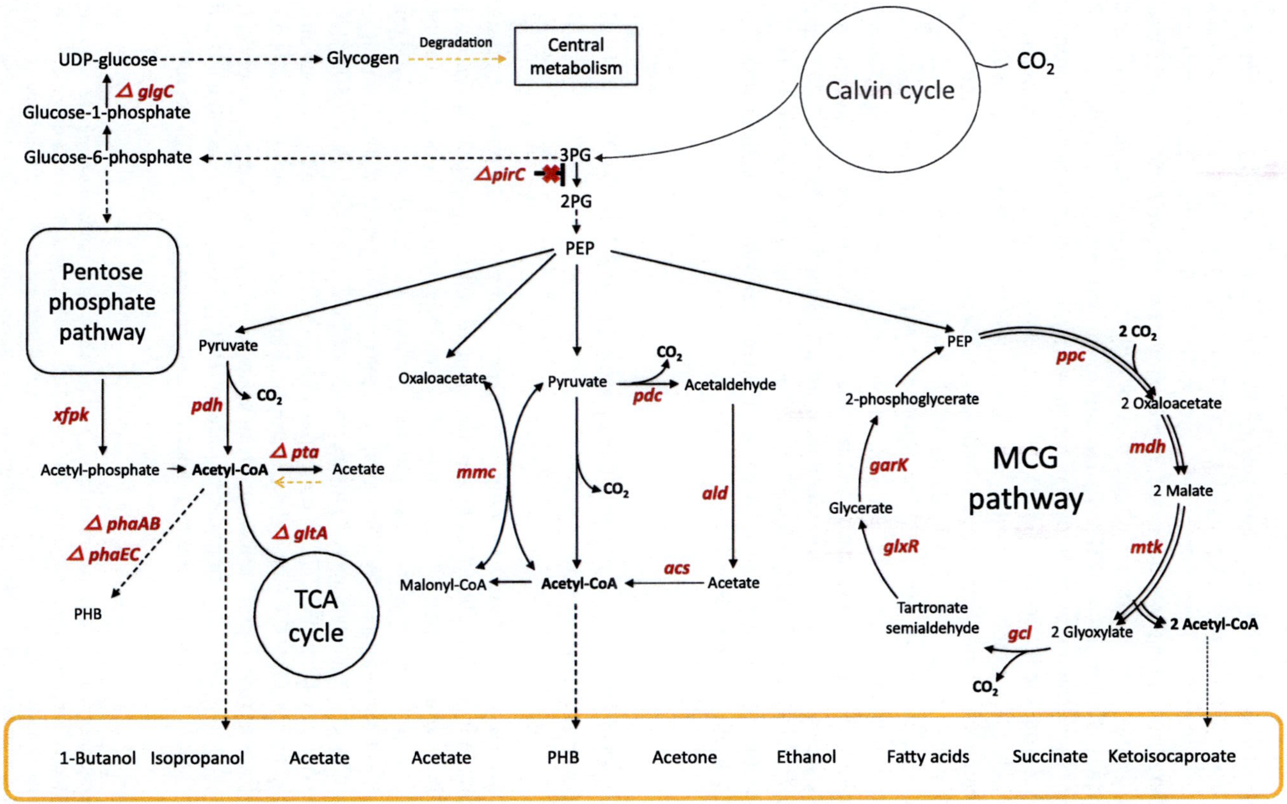

Fig. 1 Schematics of strategies for increasing carbon flux for acetyl-CoA-derived chemical synthesis in cyanobacteria. Genetic modifications are labeled in red. The chemical production improved by applying strategies for increasing acetyl-CoA carbon flux is listed in yellow box. Dashed lines indicated multistep

3 Strategies to Increase Flux of Pyruvate

Pyruvate is a key central metabolite for cell to grow. It is essential for amino acid biosynthesis and precursor to acetyl-CoA. Furthermore, pyruvate is also one of the starting substrates for isoprenoid biosynthesis through MEP pathway (methylerythritol phosphate pathway), condensing one pyruvate and one glyceraldehyde 3-phosphate to form building blocks isopentenyl diphosphate (IPP) and dimethylallyl diphosphate (DMAPP) leading to the synthesis of a wide array of terpenes. Therefore, pyruvate widely participates in numerous biochemical productions. According to a previous report [54], intracellular pyruvate concentration of strain PCC 7942 was about 1,851 μM, which is 11-fold higher than *E. coli* (164 μM). Using commercial kits, intracellular pyruvate concentration was measured to be about 0.4–0.8 mg/gDCW [85] in strain PCC 7942 and 25 nmol/g fresh weight [86] (approximately 50 μM [18]) in strain PCC 6803. To show that the native pyruvate availability may be insufficient for biosynthesis, studies have used exogenous pyruvate supplementation in culture medium. In doing so, the downstream products acetoin and ethanol were increased from 263 mg/L to 354 mg/L [41] and 40.2 mg/L to 71.3 mg/L [9], respectively. Therefore, numerous metabolic design strategies for increasing pyruvate availability have been proposed, summarized in Table 2 and illustrated in Fig. 2.

3.1 Elimination of Glycogen Biosynthesis and Mimic Nitrogen Starvation

Cyanobacteria use glycogen to store assimilated carbon during photosynthesis. Therefore, to increase carbon flux toward desirable direction, glycogen biosynthesis is an often addressed target. Disruption of glycogen biosynthesis can be achieved by deletion of *glgC* or *glgA* encoding glucose-1-phosphate transferase or glycogen synthase, respectively [87–89]. In doing so, pyruvate is secreted under nitrogen starvation by glycogen synthesis disrupted strains of PCC 7942 [89], PCC 6803 [87], and *Synechococcus* sp. PCC 7002 (hereafter PCC 7002) [90] or under mixotrophic condition for PCC 6803 [88]. Upon further nitrate concentration optimization, 528 mg/L of pyruvate production had been observed [91]. Following this concept, knocking out *glgC* in engineered strain PCC 7942 for isobutanol and 3-methylbutanol production led to higher per cell productivity by 1.6- to 2.4-fold

---

Fig. 1 (continued) conversions. Abbreviations: *pta,* phosphate acetyltransferase*; phaAB,* thiolase and acetoacetyl-CoA reductase; *phaEC,* PHA synthase*; pdh,* pyruvate dehydrogenase; *xfpk,* phosphotransketolase; *glgC,* glucose-1-phosphate adenylyltransferase; *pirC,* PII-interacting regulator; *pdc,* pyruvate decarboxylase; *ald,* aldehyde dehydrogenase; *acs,* acetyl-CoA synthetase; *ppc,* PEP carboxylase; *mdh,* malate dehydrogenase; *mtk,* malate thiokinase; *mcl,* malyl-CoA lyase; *gcl,* glyoxylate carboligase; *glxR,* tartronate semialdehyde reductase; *garK,* Glycerate kinase

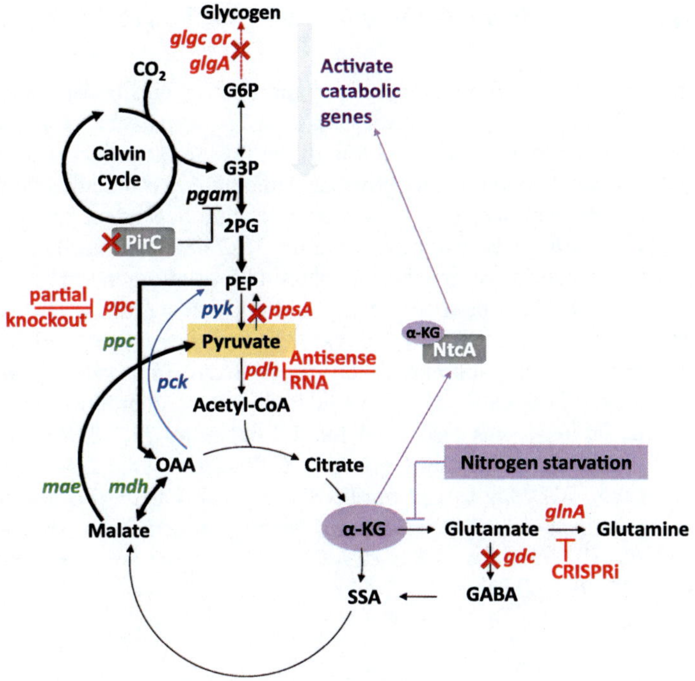

Fig. 2 Schematics of strategies for increasing pyruvate flux in cyanobacteria. Enhance pyruvate flux by knocking out or knocking down genes as shown in red, overexpressing genes as shown in green and blue, which are two modules, and nitrogen starvation shown in purple. Nitrogen starvation leads to α-ketoglutarate, abbreviated as α-KG, accumulation and triggers downstream regulation by NctA. Abbreviations: *glgC*, glucose-1-phosphate adenylyltransferase; *glgA*, glycogen synthases; *ppsA*, phosphoenolpyruvate synthase; *gdc*, glutamate decarboxylase; *ppc*, phosphoenolpyruvate carboxylase; *glnA*, glutamine synthase; *pirC*, PII-interacting regulator; *pdh*, pyruvate dehydrogenase; *mdh*, malate dehydrogenase; *mae*, malic enzyme; *pck*, phosphoenolpyruvate carboxykinase; *pyk*, pyruvate kinase

improvement with 52% of total fixed carbon directed toward alcohol production [46]. Similarly, *glgC* knockout was used to improve lactate production as lactate productivity increased from 8.9 mg/L/OD/day to 21.6 mg/L/OD/day [92]. In many cases, disruption of glycogen synthesis is coupled to nitrogen starvation to direct carbon fluxes for biosynthesis. Nitrogen starvation contributed to intracellular α-ketoglutarate accumulation as glutamate dehydrogenase is lacking ammonia to react. High level of α-ketoglutarate activates global nitrogen transcriptional regulators, such as NctA [93], which turns on sugar catabolic genes [94] and increases carbon flux down toward pyruvate. As an alternative to nitrogen starvation for achieving α-ketoglutarate accumulation, repression of glutamine synthetase I gene (*glnA*) by CRISPRi [95] and deletion of glutamate decarboxylase gene (*gdc*) [96] have also been shown to be effective.

3.2 *Improve Flux from 3-Phosphoglycerate Toward Pyruvate*

Since the majority of the carbon flux within cyanobacterial metabolism runs through the CBB cycle, 3-phosphoglycerate represents a vital node between CO_2 fixation and biosynthesis of pyruvate. Recently, deletion of the PII-interactor PirC was used as a strategy to direct more carbon flux to pyruvate. This strategy is based on the rationale that PirC inhibits the activity of 2,3-phosphoglycerate-independent phosphoglycerate mutase. When PirC was disrupted in strain PCC 6803, intracellular pyruvate level was 14-fold higher than the wild type after 48 h of nitrogen starvation [77].

Pyruvate flux could also be increased by higher pyruvate kinase (Pyk) activity, which irreversibly converts PEP to pyruvate. This strategy was used for lactate production where heterologous expression of *pyk* from *Enterococcus faecalis* or *E. coli* in engineered PCC 6803 led to 1.5-fold higher titer and productivity of lactate [12]. The same strategy was also used for 2,3-butandiol (2,3-BDO) production where the expression of endogenous *pyk* in PCC 7942 increased 2,3-BDO productivity from 84 mg/L/OD/day to 132 mg/L/OD/day [41]. Lastly, this approach was used for isobutyraldehyde (IBA) production where productivity of IBA increased from 51.9 mg/L/gCDW/day to 77.8 mg/L/gCDW/day by overexpression of endogenous *pyk* in PCC 7942 [68].

3.3 *Partitioning Flux from TCA Cycle to Pyruvate*

In addition to increasing pyruvate flux from upper glycolysis, flux may also be drawn from TCA cycle intermediates. Some cyanobacteria naturally synthesize pyruvate through oxaloacetate and malate. Therefore, flux of pyruvate biosynthesis may be increased by enhancing the expression of malate dehydrogenase (Mdh) and malic enzyme (ME). Using this approach, the productivity of isobutyraldehyde was increased from 51.9 to 86.5 mg/L/gCDW/day upon overexpression of the genes encoding endogenous Mdh and ME [68]. This approach was also used to increase lactate production where overexpression of ME increased lactate production from 50 to 87.5 mg/L/day [20]. In another study, PEP carboxykinase was overexpressed together with pyruvate kinase to increase isobutyraldehyde production. The resulting strain increased isobutyraldehyde production from 155 to 224 mg/L/gCDW/day [51], indicating the synergistic effect of drawing flux from upper glycolysis and the TCA cycle.

3.4 *Reducing Native Pyruvate Consumption Fluxes*

Inhibition of competing pathways is generally an effective strategy for increasing desirable carbon flux. PEP carboxylase is a major competitive route for pyruvate biosynthesis. Therefore, reducing PEP carboxylase activity in conjunction with strategies mentioned above is an effective approach to increase pyruvate availability. However, because PEP carboxylase is essential for growth as it is required for the biosynthesis of oxaloacetate used for the synthesis of numerous amino acids, it could only be repressed instead of completely deleted. Taking advantage of the multiploid nature of cyanobacteria, it was possible to repress the PEP carboxylase encoding gene using conventional homologous recombination in an incomplete chromosomal segregated strain [12]. This strain showed an improved lactate production from 153 to 205 mg/L.

In addition to PEP carboxylase, other reactions such as those catalyzed by PEP synthase and pyruvate dehydrogenase directly consume pyruvate, lowering the productivity of desired chemical targets. PEP synthase is non-essential and the corresponding genes can therefore be deleted. Applying this strategy in PCC 6803 engineered for ethanol production led to increased product titers from 591 to 878 mg/L [97]. Pyruvate dehydrogenase, on the other hand, is essential for growth. Using a strategy based on antisense RNA, the respective genes have been repressed and thereby isobutyraldehyde production rate was improved from 155 to 207 mg/L/gCDW/day. Together, these strategies for increasing pyruvate availability have positive effects on the production of downstream chemicals and should be considered when designing cyanobacterial cell factories for the production of pyruvate-based chemicals.

4 Strategies to Increase Flux of Malonyl-CoA

Malonyl-coenzyme A (malonyl-CoA) is an important intermediate involved in the biosynthesis of polyketides, flavonoids, and fatty acids, which have wide applications in biofuels, food additives, and pharmaceutical industries. Naturally, malonyl-CoA is derived from acetyl-CoA through acetyl-CoA carboxylase (ACC) in a multistep reaction. Intracellular malonyl-CoA, which is limited due to the extensive regulation, is often believed to be the key bottleneck of downstream products synthesis [98]. As a result, multiple metabolic engineering strategies have been demonstrated successfully in *E. coli* and yeast to increase intracellular malonyl-CoA carbon flux [99, 100]. Although cyanobacteria have not been as extensively studied for the biosynthesis of malonyl-CoA-derived compounds as commonly used heterotrophs, similar strategies for increasing malonyl-CoA flux have been applied and shown to be effective (Table 3). These strategies primarily involve the overexpression of acetyl-CoA carboxylase and inhibition of competitive pathways such as fatty acid biosynthesis (Fig. 3).

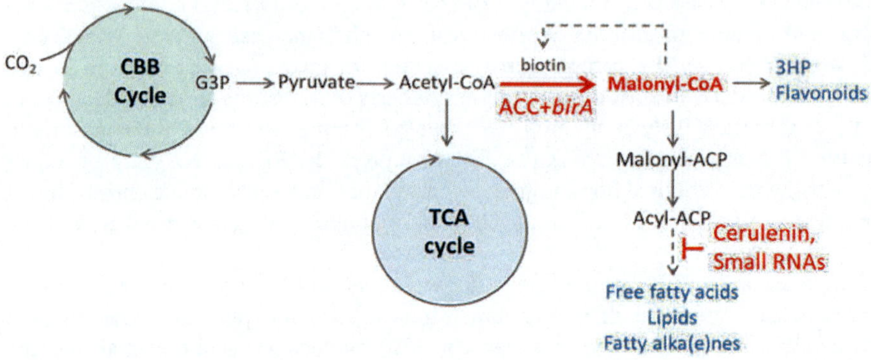

Fig. 3 Schematics of strategies for increasing malonyl-CoA carbon flux in cyanobacteria. Malonyl-CoA could be used to provide flux toward 3-hydroxypropionic acid, 3HP, and flavonoids production, as well as native fatty acid biosynthesis. Besides, malonyl-CoA flux could be increased by acetyl-CoA carboxylase, ACC, overexpression (ACC with *birA*), as well as downregulation of fatty acid biosynthesis flux, including adding inhibitor cerulenin and expressing small RNAs

4.1 Overexpression of Acetyl-CoA Carboxylase

Cyanobacteria, like most other prokaryotes, use heteromeric ACC, which is a biotin- and ATP-dependent multi-subunit protein, to catalyze the carboxylation of acetyl-CoA to malonyl-CoA. ACC is conventionally regarded as the major limiting step for downstream product synthesis. Therefore, overexpression of ACC coding genes *accABCD* and biotin ligase (*birA*) to increase malonyl-CoA carbon flux has been used to increase malonyl-CoA flux. This approach has been used in the biosynthesis of free fatty acids [63], but with minimal success. This work showed the overexpression optimization of acyl-acyl carrier protein (Acyl-ACP) thioesterase was more important for the production of free fatty acid, indicating the limiting step was thioesterase. Furthermore, by expressing ACC from *Chlamydomonas reinhardtii* [101], a model microalgae known for producing triacylglycerides, as well as RuBisCO and deleting the gene coding for acyl-ACP synthetase, led to an increase in fatty acid production from 370 to 550 mmol/gDCW. In another study that aimed to improve cyanobacterial lipid production, ACC overexpression led to an increase in lipid titer of from 88 to 126 mg/L [102]. Relevant to fatty acid production, hydrocarbon biosynthesis is based on the same pathway and therefore also benefits from ACC overexpression. With only ACC overexpressed in strain PCC 6803, total hydrocarbons increased from approximately 100 to 161.6 μg/L/OD730 [64]. Further improvement by overexpressing alka(e)nes biosynthetic enzymes acyl-acyl carrier protein reductase and aldehyde-deformylating oxygenase led to an increase in hydrocarbon titer to approximately 1,100 μg/L/OD730. In addition to ACC overexpression, overexpression of biotin ligase (*birA*) may also be necessary to increase malonyl-CoA flux. Overexpression of ACC alone may result in the increased translation of ACC polypeptides. However, functional ACC requires biotinylation by BirA. BirA activates biotin to form the BirA-bio-5'-AMP complex and transfers biotin to AccB subunit of acetyl-CoA carboxylase, resulting in functional and biotinylated AccB. Combined overexpression of BirA and ACC was shown effective for increasing 3-hydroxypropionic acid (3HP) production, which is a reduction product of malonyl-CoA. Using this strategy, 3HP titer increased from 31.62 to 130.40 mg/L [103]. These studies show that ACC is a promising target to improve carbon flux to malonyl-CoA.

4.2 *Inhibiting Downstream Fatty Acid Biosynthesis*

Inhibition of malonyl-CoA downstream consumption pathways is also a useful strategy. Naturally, most of the malonyl-CoA carbon flux goes to fatty acids biosynthesis. Fatty acids are the building blocks of cell membranes and essential components to ensure cell viability. Conventional gene-knockout strategies on fatty acid biosynthesis would be lethal. Therefore, downregulating their gene expression allows for reduced carbon flux entering fatty acid biosynthesis but maintaining sufficient flux for growth.

In strain PCC 6803, the artificial small RNAs Hfq-MicC tools has been used to manipulate the transcriptional level of fatty acid synthesis pathways genes *fabDFH*. The resulting strain had an increased malonyl-CoA level from 0.49 to 0.69 μg/L/OD730, a 41% improvement compared to wild-type strains [104]. Alternatively, cerulenin, an antifungal antibiotic that inhibits keto-acyl-ACP synthase, can be used to reduce carbon flux into fatty acid biosynthesis. This approach was applied for flavonoid biosynthesis which requires malonyl-CoA as a precursor. Addition of cerulenin increased resveratrol, naringenin, and bisdemethoxycurcumin titers by approximately 1.5-fold [105].

In addition to these two strategies, previous studies used other methods to improve the intracellular synthesis of malonyl-CoA in *E. coli* and yeast. For example, introducing malonate assimilation pathway from *Rhizobium trifolii* coupled to malonate supplementation has significantly increased intracellular malonyl-CoA [100]. Increased intracellular free CoA levels through manipulation of CoA biosynthesis also showed improvement in *E. coli* [106]. While these strategies have yet to be tested in cyanobacteria, they may serve as effective ways to increase malonyl-CoA carbon flux in cyanobacteria as well.

5 Strategies to Drive Flux Toward TCA Cycle

Photoautotrophs primarily produce reducing cofactors and ATP through the photosynthetic electron transport chain (PET), resulting in a lower demand for carbon flux entering the TCA cycle as compared to conventional heterotrophs used in biochemical productions. Therefore, it is important to increase carbon flux through TCA cycle if the desired target compound is downstream of TCA cycle intermediates. Summarized in Table 4 are the strategies used to increase flux into the TCA cycle, especially toward the biosynthesis of TCA cycle diacids.

The key enzymes controlling carbon flux into the TCA cycle are citrate synthase (encoded by *gltA*) and phosphoenolpyruvate (PEP) carboxylase (encoded by *ppc*). Citrate synthase catalyzes the condensation reaction of oxaloacetate and acetyl-CoA to form citric acid. PEP carboxylase catalyzes the carboxylation of PEP to oxaloacetate. The overexpression of these two genes enhanced succinate productivity in an engineered succinate producing strain of PCC 7942 [36]. In this study, *gltA* and *ppc* were cloned from *Corynebacterium glutamicum* due to the organism's ability to overproduce glutamic acid, a derivative of α-ketoglutarate. When only *gltA* was overexpressed, the succinate production rate only increased slightly from 23 to 30 mg/L/day. When both *gltA* and *ppc* were overexpressed, the rate was further enhanced to 80 mg/L/day resulting in a succinate titer of 430 mg/L succinate in 8 days. This result indicated that the native carbon flux from PEP to oxaloacetate is insufficient. The overexpression of *ppc* is a necessary step to increase carbon flux into the TCA cycle.

Increasing PEP biosynthesis and availability further increases carbon flux toward oxaloacetate biosynthesis and therefore throughout the TCA cycle. This strategy was applied in ethylene production where ethylene is derived from α-ketoglutarate through ethylene forming enzyme. The expression of PEP synthase (encoded by *ppsa*), converting pyruvate to PEP, together with *ppc* increased ethylene productivity to 16.8 mg/L/OD/day, whereas solely overexpressing *ppc* only achieved productivity of 10.5 mg/L/OD/day [11].

Cyanobacteria were previously thought to have an incomplete TCA cycle due to the lack of α-ketoglutarate dehydrogenase [107]. However, it was later revealed that there are three TCA cycle variants in cyanobacteria (Fig. 4), namely glyoxylate shunt, γ-aminobutyric acid (GABA) shunt, and α-ketoglutarate decarboxylase/succinate semialdehyde dehydrogenase route to connect α-ketoglutarate with succinate and downstream intermediates. Overexpression of these pathways is useful to produce the C4 diacids in the TCA cycle. For example, overexpression of the glyoxylate shunt was used to produce succinate. When isocitrate lyase (encoded by *aceA*) and malate synthase (encoded by *aceB*) were overexpressed in PCC 6803 containing two copies of *ppc*, 400 mg/L/OD of succinate accumulated, while the control strain only produced around 40 mg/L/OD in 4 days of incubation in the dark [108]. Similarly, photosynthetic succinate production in strain PCC 7942 [36] and *Synechococcus elongatus* PCC 11801 (hereafter PCC 11801) [109] was enabled by engineering a α-ketoglutarate decarboxylase/succinate semialdehyde dehydrogenase route. In the absence of this pathway, succinate production from a strain of PCC 7942 engineered by inhibiting succinate dehydrogenase with CRISPRi yielded only 0.63 mg/L [110]. When α-ketoglutarate decarboxylase and succinate semialdehyde dehydrogenase were introduced into this host, product titer was enhanced 900-fold to 576 mg/L, indicating the effectiveness of this bypass for directing carbon flux from α-ketoglutarate to succinate.

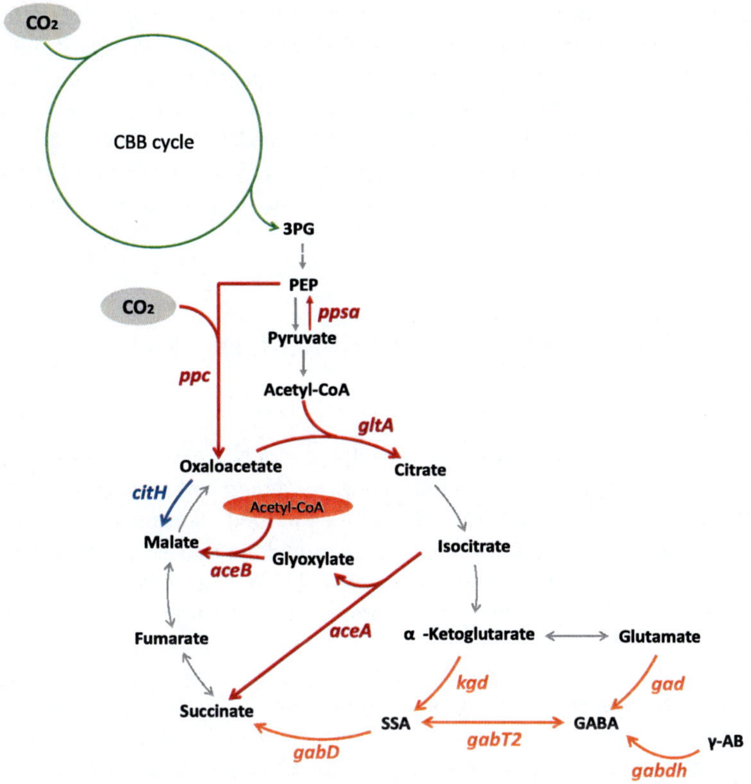

Fig. 4 Schematics of pathways related to redirection of TCA cycle fluxes in cyanobacteria. Enhance carbon flux into TCA cycle by overexpression of *ppsA*, *PEP synthase, ppc*, PEP carboxylase, and *gltA,* citrate synthase, or *aceBA, the* glyoxylate shunt, as indicated by the red arrows. Increase carbon flux via the oxidative branch of TCA cycle by overexpressing *kgd*, 2-ketoglutarate decarboxylase, and *gabD*, succinate semialdehyde dehydrogenase, or the GABA shunt composed of *gad,* glutamate decarboxylase, *gabT2* 4-aminobutyrate transaminase, and *gabD*, succinate semialdehyde (SSA) dehydrogenase, or via the gamma-aminobutanal (γAB) degradation pathway composed of *gabdh,* aminobutanal dehydrogenase, *gabT2*, and *gabD* as indicated by the orange arrows. Increase carbon flux via the reductive TCA branch by overexpression of *citH*, malate dehydrogenase, as indicated by the blue arrow

For strains naturally using the GABA shunt for converting α-ketoglutarate to succinate, the respective flux may be modulated by competing pathway removal. For example, it has been shown in PCC 6803 that upon knockout of aminobutanal dehydrogenase (*gabdh*), transcriptional levels of α-ketoglutarate decarboxylase, glutamate decarboxylase, and GABA amino transferase were upregulated, leading to increased intracellular succinate, glutamate, and GABA concentration by 5-, 2-, and 15-folds, respectively, compared to the wild-type strain [111].

In addition to increasing carbon flux from α-ketoglutarate to succinate, the C4 diacids in the TCA cycle may also be produced via the reductive branch of TCA cycle through auto-fermentation. During auto-fermentation cyanobacteria ferment intracellular glycogen to obtain energy during dark anoxic conditions and in the process secrete fermentative products. To enhance the flux entering the reductive branch of TCA cycle, malate dehydrogenase (encoded by *citH*) was overexpressed in strain PCC 6803. The resulting strain excreted malate, fumarate, and succinate with 2.8-, 2.1-, and 1.2-folds increase, respectively [112]. This result demonstrated that malate dehydrogenase is a key enzyme for the activation of reductive TCA branch.

6 Increasing Carbon Flux Toward Aromatic Amino Acid and Its Derivatives

Numerous natural products such as phenylpropanoids are derived from aromatic amino acids. However, amino acid biosynthesis is generally highly controlled in the cell. As such, metabolic flux engineering is often needed to channel carbon flux into the biosynthesis of aromatic amino acids. Aromatic amino acids are synthesized through the strictly regulated shikimate pathway [113]. Currently adopted strategies (Fig. 5) for increasing the respective carbon flux rely primarily on the use of feedback-inhibition resistant enzymes.

Naturally, phenylalanine and tyrosine feedback inhibit chorismate mutase/prephenate dehydrogenase, encoded by *pheA* or *tyrA*, and 3-deoxy-d-arabino-heptulonate-7-phosphate synthetase, encoded by *aroG* [114, 115]. Feedback-inhibition resistant mutants of these enzymes have been well studied. Baez-Viveros et al. [116] relived feedback inhibition by phenylalanine via deletion of the R-domain of PheA. Zhou et al. [117] reported on a T326P mutation in the R-domain of PheA gene after random mutagenesis with the resulting enzyme tolerating up to 200 mM phenylalanine, while native PheA could only tolerate less than 10 mM phenylalanine. Similarly, Eversloh and Stephanopoulos et al. [118] applied random mutagenesis to obtain a TyrA variant resistant to feedback inhibition. Furthermore, feedback-resistant mutant of AroG has also been generated through random mutagenesis [119]. Application of these feedback-resistant mutants has been used in the production of 2-phenylethanol, coumaric acid, and aromatic amino acids. Table 5 summarizes the effect of using these feedback-resistant mutants in these studies.

In addition to the use of feedback-inhibition resistant enzymes for driving the carbon flux into the shikimate pathway, removal of native laccase is an additional strategy used to redirect flux. Laccases oxidize phenolic compounds and thus are responsible for flux loss in the biosynthesis of aromatic compounds. Therefore, deletion of the gene responsible for laccase is a viable approach. Xue et al. [56] identified a putative laccase gene *slr1573* in strain PCC 6803 that codes for laccase degrading *p*-coumaric acid into 4-vinylphenol. Deleting this gene resulted in a PCC 6803 variant accumulating 82.6 mg/L of *p*-coumaric acid, whereas no *p*-coumaric

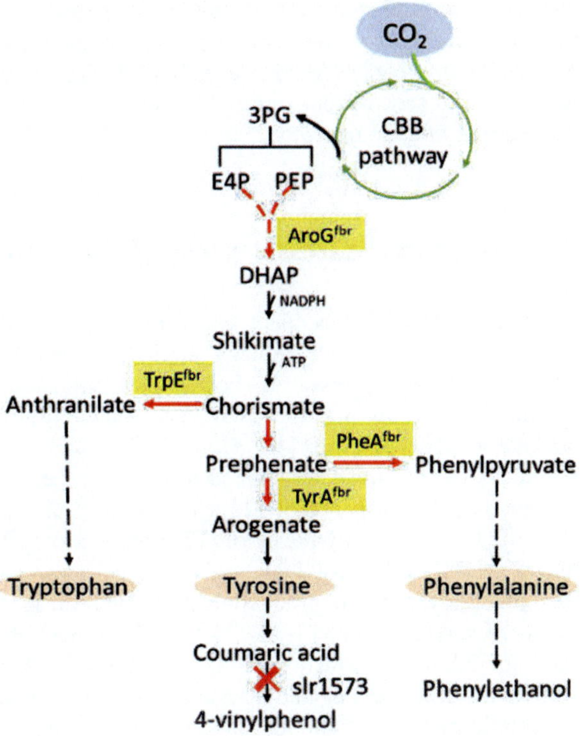

Fig. 5 Strategies for increasing carbon flux toward aromatic amino acids. Overexpression of feedback-resistant 3-deoxy-7-phosphoheptulonate synthase (aroG^{fbr}), anthranilate synthase (TrpE^{fbr}), and chorismate mutase/prephenate dehydrogenase (PheA^{fbr} and TyrA^{fbr}) bypasses the natural intracellular flux control. Knockout of gene slr1573 has been shown to be effective in PCC 6803 for reducing flux loss

acid was detected applying the parent strain. Together, these efforts show that the conventional metabolic engineering strategies used for driving carbon flux into the shikimate pathway are also applicable to cyanobacteria.

7 Increasing Reducing Cofactor Availability Drives Carbon Flux for Biosynthesis

In addition to directly enhancing carbon flux by manipulating enzyme activities and pathway re-routing, optimizing reducing cofactor availability is also a useful strategy for enhancing carbon flux. NADH is mostly produced in catabolism, while NADPH is primarily consumed in anabolic pathways, leading to distinct physiological availability dependent on various parameters, such as strain background [120] and growth condition [121]. Since $NADP^+$ is the terminal electron acceptor of photosynthesis, its relative abundance in cyanobacteria is higher than NAD+. The NADP (H) concentration is 6.4 ± 0.1 μmol/mg chlorophyll whereas the concentration of NAD(H) is 0.6 ± 0.2 μmol/mg chlorophyll [122]. This cofactor availability is differing significantly from conventional heterotrophic microbes and may pose a problem when transferring metabolic engineering strategies from heterotrophic to cyanobacterial host strains. Therefore, harmonizing cofactor needs and availability by introducing, e.g., NADPH-dependent enzyme and/or trans-hydrogenases, may increase the respective enzyme activity and thus the overall carbon flux toward the desired biosynthetic pathways.

7.1 Adopting NADPH-Dependent Enzymes for Biosynthesis

Substituting NADH-dependent enzymes with NADPH-dependent enzymes was shown to be an effective strategy for increasing carbon flux for biosynthesis in cyanobacteria. Early research showed that isobutanol titer increased from 11 to 60 mg/L by replacing the NADH-dependent AdhA with a broad substrate NADPH-dependent YqhD [48]. Given its NADPH dependency and broad substrate specificity, YqhD was later applied to various other biosynthetic pathways such as ethanol [4], 1,3-propanediol [23], 1-butanol [123], isobutanol [47], 2-methyl-1-butanol [124], and 2,3-butanediol [125] production pathway. In addition to YqhD, other NADPH-dependent dehydrogenases such as secondary alcohol dehydrogenase used for 1,2-propanediol production have also been shown to be effective [126]. Furthermore, an endogenous NADPH-dependent alcohol dehydrogenase *slr1192* was used for ethanol production in PCC 6803 [83]. Using Slr1192 instead of the commonly used alcohol dehydrogenase AdhII from *Zymomonas mobilis* led to a 1.5-fold increase in ethanol production. The activity of Slr1192 is similar to that of AdhII, indicating that carbon flux is higher when coupled to NADPH oxidation. This approach has also been applied for coenzyme A-dependent butanol production pathway. Lan et al. [123] showed an improved n-butanol production through replacing three NADH-dependent enzymes to those utilizing NADPH. Later studies using strain PCC 6803 for butanol production also showed NADPH-dependent enzymes outperforming NADH-dependent enzymes for butanol production, achieving the highest butanol cumulative titer so far of 4.8 g/L [31]. The pathway was later on modified for efficient butyrate production [35].

Lactate synthesis generally uses lactate dehydrogenase (Ldh) to convert pyruvate to lactate. However, as most of the natural Ldh are NADH-dependent [127], recruiting NADPH-dependent Ldh thus requires further protein engineering efforts. A NADH-dependent Ldh from *Bacillus subtilis* was engineered through structure-based site-directed mutagenesis [127]. A single-point mutation V39R on the Ldh resulted in a 250-fold increased catalytic efficiency for NADPH-dependent reduction but no significant decrease in NADH-dependent Ldh activity. Both wild-type Ldh and Ldh (V39R) were compared for their ability to produce lactate using strain PCC 6803 as host. The strain using Ldh (V39R) produced 4 mM of lactate while the strain using original Ldh produced 1.5 mM [12], indicating the ability to use NADPH drives increased pyruvate reduction. Another study shifted the cofactor preference of a Ldh from *Lactobacillus bulgaricus* from NADH to NADPH [14]. With 4 mutations, D176A, I177R, F178S, and N180R, the catalytic efficiency for the reaction using NADPH became similar to that of NADH. By introducing this mutant Ldh into strain PCC 7942, lactate production was improved by 4.5-fold compared to the strain with wild-type Ldh, reaching 453 mg/L. The same study also showed a slight decrease in NADPH:NADP$^+$ ratio when comparing wild-type and the engineered strain expressing mutant Ldh, indicating that Ldh is efficient to consume NADPH to drive lactate biosynthesis. In addition to recruiting NADPH-dependent enzymes for efficient reduction such as the case for ethanol and lactate production, biosynthetic pathway requiring oxidation steps using NADP-dependent enzyme was also shown to be effective for driving carbon flux as demonstrated in the case of succinate production [36]. Altogether, these results indicate that selecting NADPH-dependent dehydrogenases is an important design strategy for directing carbon flux for biosynthesis.

7.2 *Manipulating Intracellular Reducing Cofactor Availability*

Although recruiting NADP(H)-dependent enzymes was shown to be an effective strategy for cyanobacterial host improvement, NADP(H)-dependent enzymes may be unavailable or changing the cofactor specificity via protein engineering may not be possible. Therefore, as an alternative strategy, trans-hydrogenases that interconvert NADPH to NADH are used to shift the intracellular ratio of NADH to NADPH and thereby provide more NADH. Despite the development of NADPH-dependent Ldh mutant for lactate production, this variant was not available before the year 2011. Prior studies reported successful lactate production in cyanobacteria through the expression of NADH-dependent Ldh together with a transhydrogenase. Expression of transhydrogenase together with NADH-dependent Ldh led to a five- to sixfold higher lactate production compared to the strain expressing solely the Ldh [18, 128]. Having shown that Ldh activity was the same between the strain with and without transhydrogenase expression, these studies supported the idea that expressing transhydrogenase may aid driving carbon flux to lactate production by providing proper cofactor availability. Expression of transhydrogenase was also used for 2,3-butanediol synthesis [129]. Upon expression of transhydrogenase, the increase of NADPH drives the reduction of acetoin into 2,3-butanediol. The resulting strain showed reduced acetoin secretion and increased 2,3-butanediol production.

In addition to expressing trans-hydrogenases which exchange reducing cofactors but do not significantly affect the overall ratio of the reduced cofactor to its oxidized form, increasing the overall NAD(P)H:NAD(P)$^+$ ratio may drive biosynthesis of products that are highly reduced. To increase NADPH levels, overexpressing *zwf* encoding for glucose-6-phosphate dehydrogenase increased intracellular NADPH concentration and thereby ethanol titers were improved from 440 to 590 mg/L [130] in strain PCC 6803. In another study, expression of *gap1* encoding for glyceraldehyde-3-phosphate dehydrogenase increased glycolytic rate, leading to increased

NADH and NADPH levels by 4-fold and 2.9-fold, respectively. The resulting strain showed higher level of acetate and lactate secretion [131].

Studies discussed above and summarized in Table 6 showed promising results on increasing carbon flux relevant for biosynthesis in cyanobacteria through manipulating cofactor availability. Dependent on the availability of enzymes with proper cofactor preference, strategies including direct recruiting of NADPH-dependent enzymes, protein engineering for altering cofactor preference, and manipulating intracellular cofactor content could be applied for efficient channeling of carbon flux for biosynthesis.

8 Engineered Mixotrophy for Increased Biosynthetic Fluxes

While numerous metabolic design strategies have been developed to increase carbon fluxes toward important central metabolites, overall carbon flux may be further increased by exogenous addition of organic substrates to facilitate higher biosynthetic fluxes. Renewable substrates such as glucose, xylose, glycerol and acetate have been used to augment cyanobacterial chemical productivity, especially for diurnal cycles. While photomixotrophic chemical production had been reviewed in detail by the previous review article [132], this section briefly discusses the development of photomixotrophic chemical production hosts (Table 7).

8.1 Glucose

Cyanobacteria naturally use the Calvin–Benson–Bassham (CBB) cycle to assimilate CO_2 into biomass and glycogen in the presence of light. Because of its natural presence, glycogen breakdown and glucose utilization pathway already exist in wild-type strains. However, most cyanobacterial strains show none or limited glucose uptake ability, indicating the lack of or low expression of glucose transporters. Several glucose transporters have been shown to enhance glucose uptake [133]. Overexpression of hexose transporter *galP* (Fig. 6) from *E. coli* with glucose supplementation in media increased PCC 7942 growth rate to 0.54 OD700/day, whereas the wild type achieved a growth rate of 0.24 OD700/day. Without glucose feeding, the wild type grew at a rate of 0.16 OD700/day. This result indicated that PCC 7942 may have endogenous sugar transporters, but these are low in either activity or expression. Expression of the heterologous sugar transporter significantly increased biosynthetic rates. Based on *galP* overexpression, 2,3-butanediol titer has been successfully increased by providing glucose in the respective production host PCC 7942 and product titers increased from 57 to 95 mg/L [134]. Mixotrophic

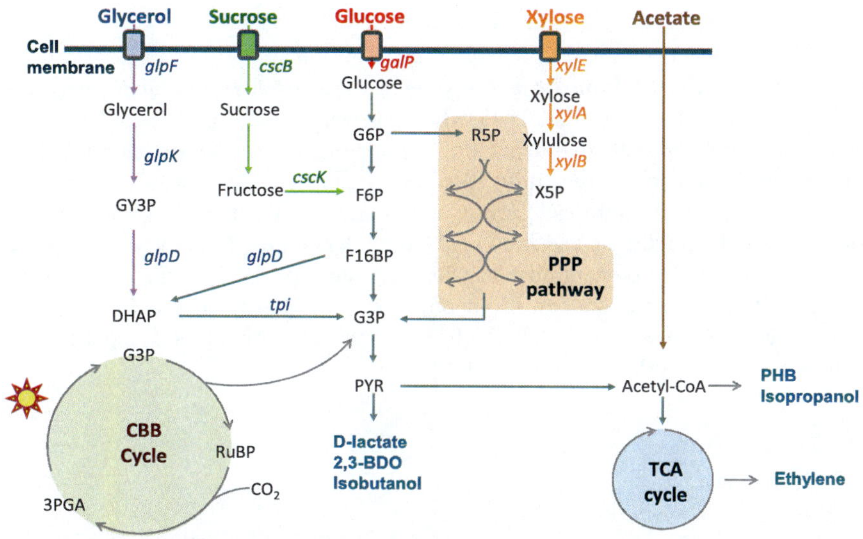

Fig. 6 Increased carbon fluxes via engineered mixotrophic metabolism and cultivation. Engineered cyanobacteria could utilize glycerol, sucrose, glucose, xylose, or acetate with CO_2. Glucose utilization requires the expression of *galP*, a glucose transporter gene. Sucrose utilization requires the expression of *cscB*, a sucrose transporter gene, and *cscK*, a fructokinase gene. Xylose utilization requires the expression of *xylE*, a xylose transporter gene, *xylA*, a xylose isomerase gene, and *xylB*, a xylulose kinase gene. Glycerol utilization requires the expression of *glpF*, a glycerol transporter gene, *glpK*, a glycerol kinase gene, and *glpD*, a glycerol-3-phosphate dehydrogenase gene. Acetate utilization utilizes native acetyl-CoA synthetase

growth also allowed for continuous production and growth in light–dark diurnal cycle conditions [38].

8.2 *Xylose*

Xylose is abundant in hemicellulose and is a promising renewable feedstock for microbial production. Native *S. elongatus* strain PCC 7942 can consume xylose slowly. Overexpression of a xylose transporter, however, showed no improved cell growth. This result indicated none or little activity in xylose downstream pathway enzymes. Introducing *xylE, xylA*, and *xylB* encoding a xylose transporter, xylose isomerase, and xylulokinase, respectively, from *E. coli* finally enabled xylose assimilation [133], and it showed improved growth rates on xylose for continuous light and diurnal condition, by 35% and 480%, respectively. Based on this strategy, xylose feeding increased 2,3-butanediol titers from 39 to 75 mg/L under continuous light cultivation and from 25 to 120 mg/L under diurnal cycles [134]. A similar effect was reported for mixotrophic ethylene production [135].

8.3 *Sucrose*

Sucrose is synthesized from storage glycogen as an osmoprotectant to resist the osmotic pressure by strain PCC7942 when cultivated under high saline environments. Similar to glucose, this behavior indicates that cyanobacteria naturally have catabolic pathways for sucrose utilization. For strain PCC 7942, it was shown that supplying sucrose as additional carbon source led to an enhanced growth rate in continuous light conditions and even enabled moderate growth in the dark, indicating that this strain is naturally capable of sucrose uptake and metabolism. Overexpression of *cscB* and *cscK,* encoding sucrose permease and fructokinase, led to an increase in sucrose uptake, utilization rate, and growth rates in continuous light conditions as well as in the dark [133].

8.4 *Glycerol*

Glycerol is currently a by-product from biodiesel production and a relatively inexpensive carbon source used for heterotrophic chemical production. Also cyanobacteria are capable of utilizing this carbon source in mixotrophic growth as shown for PCC 7942. By introducing heterologous genes *glpFKD* and *tpiA,* encoding glycerol facilitator, glycerol kinase, glycerol 3-phosphate dehydrogenase, and triosephosphate isomerase to apply for mixotrophic growth on glycerol, it was possible to increase 2,3-butanediol production from 33 mg/L/OD to 59 mg/L/OD in the respective host [136], indicating an effective assimilation of glycerol in the central glycolytic metabolism.

8.5 *Acetate*

Acetate enters cellular metabolism through acetyl-CoA. However, most model cyanobacteria lack the glyoxylate shunt to convert acetyl-CoA into glycolytic intermediates. Therefore, instead of redirecting carbon flux toward pyruvate and glycolytic intermediates, co-feed of acetate was shown to enhance acetyl-CoA availability, which is useful for pathways requiring acetyl-CoA. A prominent example is isopropanol production, which requires two molecules of acetyl-CoA per molecule of isopropanol produced in addition to acetate as a co-substrate for CoA-transferase, the latter being a key step in the pathway. Exogenous addition of acetate in the culture media increased isopropanol titers from non-detectable to 21.7 mg/L [45]. Similarly, polyhydroxybutyrate (PHB) is a polymer composed of 3-hydroxybutyrate which is synthesized from two molecules of acetyl-CoA. Exogenous addition of acetate to the respective cyanobacterial host increased intracellular PHB content from 26% to 35% DCW [137]. Interestingly, it has been previously shown that supplementation of acetate also aids the production of lactate [15]. This result was surprising as the lack of the glyoxylate shunt prevents acetyl-CoA from entering glycolysis. ^{13}C labeled flux analysis revealed that acetate was not directly used for the synthesis of lactate, but was used for the biosynthesis of amino acids such as glutamate. In turn, less glycolytic flux is required to provide acetyl-CoA, resulting in more pyruvate for lactate production, and the product titer was doubled [15].

9 Conclusion

Metabolic engineering of cyanobacteria has enabled the direct production of numerous compounds using carbon dioxide as carbon source. However, due to inherent difference in the metabolic context between cyanobacteria and conventional heterotrophic workhorse, metabolic engineering approaches often cannot simply be transferred from one to the other host. This chapter discussed several cyanobacterial metabolic design strategies for increasing carbon flux toward important central metabolites, enabling higher productivity of their downstream products. While effective, overall carbon flux available may be the next limiting step as suggested by increased biosynthetic fluxes from engineered photomixotrophy. Therefore, as a future perspective, it is important to further evolve and or engineer cyanobacterial hosts toward faster carbon assimilation rates.

References

1. Velmurugan R, Incharoensakdi A (2020) Heterologous expression of ethanol synthesis pathway in glycogen deficient Synechococcus elongatus PCC 7942 resulted in enhanced production of ethanol and exopolysaccharides. Front Plant Sci 11
2. Gao ZX, Zhao H, Li ZM, Tan XM, Lu XF (2016) Photosynthetic production of ethanol from carbon dioxide in genetically engineered cyanobacteria (vol 5, p. 9857, 2012). Energ Environ Sci 9:1113
3. Dexter J, Fu PC (2009) Metabolic engineering of cyanobacteria for ethanol production. Energ Environ Sci 2:857–864
4. Lan EI, Ro SY, Liao JC (2013) Oxygen-tolerant coenzyme A-acylating aldehyde dehydrogenase facilitates efficient photosynthetic n-butanol biosynthesis in cyanobacteria. Energ Environ Sci 6:2672–2681
5. Deng MD, Coleman JR (1999) Ethanol synthesis by genetic engineering in cyanobacteria. Appl Environ Microbiol 65:523–528
6. Quinn L et al (2019) Zymobacter palmae pyruvate decarboxylase is less effective than that of zymomonas mobilis for ethanol production in metabolically engineered Synechocystis sp. PCC6803. Microorganisms 7
7. Testa RL, Delpino C, Estrada V, Diaz SM (2019) In silico strategies to couple production of bioethanol with growth in cyanobacteria. Biotechnol Bioeng 116:2061–2073
8. Bartasun P et al (2019) The effect of modulating the quantity of enzymes in a model ethanol pathway on metabolic flux in Synechocystis sp. PCC 6803. PeerJ 7

9. Luan GD et al (2015) Combinatory strategy for characterizing and understanding the ethanol synthesis pathway in cyanobacteria cell factories. Biotechnol Biofuels 8
10. Gao EB et al (2022) Res Sq
11. Durall C, Lindberg P, Yu J, Lindblad P (2020) Increased ethylene production by overexpressing phosphoenolpyruvate carboxylase in the cyanobacterium Synechocystis PCC 6803. Biotechnol Biofuels 13:16
12. Angermayr SA et al (2014) Exploring metabolic engineering design principles for the photosynthetic production of lactic acid by Synechocystis sp. PCC6803. Biotechnol Biofuels 7:99
13. Angermayr SA, Hellingwerf KJ (2013) On the use of metabolic control analysis in the optimization of cyanobacterial biosolar cell factories. J Phys Chem B 117:11169–11175
14. Li C et al (2015) Enhancing the light-driven production of D-lactate by engineering cyanobacterium using a combinational strategy. Sci Rep 5
15. Varman AM, Yu Y, You L, Tang YJ (2013) Photoautotrophic production of D-lactic acid in an engineered cyanobacterium. Microb Cell Fact 12:117
16. Zhou J, Zhang HF, Meng HK, Zhang YP, Li Y (2014) Production of optically pure D-lactate from CO2 by blocking the PHB and acetate pathways and expressing D-lactate dehydrogenase in cyanobacterium Synechocystis sp. PCC 6803. Process Biochem 49:2071–2077
17. Niederholtmeyer H, Wolfstadter BT, Savage DF, Silver PA, Way JC (2010) Engineering cyanobacteria to synthesize and export hydrophilic products (vol 76, p. 3462, 2010). Appl Environ Microbiol 76:6023
18. Angermayr SA, Paszota M, Hellingwerf KJ (2012) Engineering a cyanobacterial cell factory for production of lactic acid. Appl Environ Microbiol 78:7098–7106
19. Joseph A et al (2013) Utilization of lactic acid bacterial genes in Synechocystis sp. PCC 6803 in the production of lactic acid. Biosci Biotech Bioch 77:966–970
20. Hidese R, Matsuda M, Osanai T, Hasunuma T, Kondo A (2020) Malic enzyme facilitates D-lactate production through increased pyruvate supply during anoxic dark fermentation in Synechocystis sp. PCC 6803. ACS Synth Biol 9:260–268
21. Lan EI et al (2015) Metabolic engineering of cyanobacteria for photosynthetic 3-hydroxypropionic acid production from CO2 using Synechococcus elongatus PCC 7942. Metab Eng 31:163–170
22. Wang YP et al (2016) Biosynthesis of platform chemical 3-hydroxypropionic acid (3-HP) directly from CO2 in cyanobacterium Synechocystis sp. PCC 6803. Metab Eng 34:60–70
23. Hirokawa Y, Maki Y, Tatsuke T, Hanai T (2016) Cyanobacterial production of 1,3-propanediol directly from carbon dioxide using a synthetic metabolic pathway. Metab Eng 34:97–103
24. Hirokawa Y, Maki Y, Hanai T (2017) Improvement of 1,3-propanediol production using an engineered cyanobacterium, Synechococcus elongatus by optimization of the gene expression level of a synthetic metabolic pathway and production conditions. Metab Eng 39:192–199
25. Li H, Liao JC (2013) Engineering a cyanobacterium as the catalyst for the photosynthetic conversion of CO2 to 1,2-propanediol. Microb Cell Fact 12
26. Savakis P et al (2015) Photosynthetic production of glycerol by a recombinant cyanobacterium. J Biotechnol 195:46–51
27. Wang Y, Tao F, Ni J, Li C, Xu P (2015) Production of C3 platform chemicals from CO2 by genetically engineered cyanobacteria. Green Chem 17:3100–3110
28. Zhou J, Zhang H, Zhang Y, Li Y, Ma Y (2012) Designing and creating a modularized synthetic pathway in cyanobacterium Synechocystis enables production of acetone from carbon dioxide. Metab Eng 14:394–400
29. Lan EI, Liao JC (2012) ATP drives direct photosynthetic production of 1-butanol in cyanobacteria. Proc Natl Acad Sci U S A 109:6018–6023
30. Lan EI, Liao JC (2011) Metabolic engineering of cyanobacteria for 1-butanol production from carbon dioxide. Metab Eng 13:353–363
31. Liu XF, Miao R, Lindberg P, Lindblad P (2019) Modular engineering for efficient photosynthetic biosynthesis of 1-butanol from CO2 in cyanobacteria. Energ Environ Sci 12:2765–2777

32. Anfelt J et al (2015) Genetic and nutrient modulation of acetyl-CoA levels in synechocystis for n-butanol production. Microb Cell Fact 14:167
33. Fathima AM et al (2018) Iterative cycle of widely targeted metabolic profiling for the improvement of 1-butanol titer and productivity in Synechococcus elongatus. Biotechnol Biofuels 11:188
34. Shabestary K et al (2018) Targeted repression of essential genes to arrest growth and increase carbon partitioning and biofuel titers in cyanobacteria. ACS Synth Biol 7:1669–1675
35. Lai MJ, Lan EI (2019) Photoautotrophic synthesis of butyrate by metabolically engineered cyanobacteria. Biotechnol Bioeng 116:893–903
36. Lan EI, Wei CT (2016) Metabolic engineering of cyanobacteria for the photosynthetic production of succinate. Metab Eng 38:483–493
37. Nozzi NE, Case AE, Carroll AL, Atsumi S (2017) Systematic approaches to efficiently produce 2,3-butanediol in a marine cyanobacterium. ACS synthetic biology 6:2136–2144
38. Kanno M, Carroll AL, Atsumi S (2017) Global metabolic rewiring for improved CO2 fixation and chemical production in cyanobacteria. Nat Commun 8:14724
39. Savakis PE, Angermayr SA, Hellingwerf KJ (2013) Synthesis of 2,3-butanediol by Synechocystis sp. PCC6803 via heterologous expression of a catabolic pathway from lactic acid- and enterobacteria. Metab Eng 20:121–130
40. Atsumi S, Oliver JWK, Machado IMP, Yoneda H (2014) Cyanobacterial conversion of carbon dioxide to 2,3-butanediol. Abstr Pap Am Chem S 247
41. Oliver JWK, Atsumi S (2015) A carbon sink pathway increases carbon productivity in cyanobacteria. Metab Eng 29:106–112
42. Wang B, Xiong W, Yu JP, Maness PC, Meldrum DR (2018) Unlocking the photobiological conversion of CO2 to (R)-3-hydroxybutyrate in cyanobacteria. Green Chem 20:3772–3782
43. Putri SP et al (2018) Identifying metabolic elements that contribute to productivity of 1-propanol bioproduction using metabolomic analysis. Metabolomics 14
44. Hirokawa Y, Kubo T, Soma Y, Saruta F, Hanai T (2020) Enhancement of acetyl-CoA flux for photosynthetic chemical production by pyruvate dehydrogenase complex overexpression in Synechococcus elongatus PCC 7942. Metab Eng 57:23–30
45. Kusakabe T et al (2013) Engineering a synthetic pathway in cyanobacteria for isopropanol production directly from carbon dioxide and light. Metab Eng 20:101–108
46. Li XQ, Shen CR, Liao JC (2014) Isobutanol production as an alternative metabolic sink to rescue the growth deficiency of the glycogen mutant of Synechococcus elongatus PCC 7942. Photosynth Res 120:301–310
47. Miao R, Liu X, Englund E, Lindberg P, Lindblad P (2017) Isobutanol production in Synechocystis PCC 6803 using heterologous and endogenous alcohol dehydrogenases. Metab Eng Commun 5:45–53
48. Atsumi S, Higashide W, Liao JC (2009) Direct photosynthetic recycling of carbon dioxide to isobutyraldehyde. Nat Biotechnol 27:1177–1180
49. Varman AM, Xiao Y, Pakrasi HB, Tang YJJ (2013) Metabolic engineering of Synechocystis sp. Strain PCC 6803 for isobutanol production. Appl Environ Microbiol 79:908–914
50. Mustila H, Kugler A, Stensjo K (2021) Isobutene production in Synechocystis sp. PCC 6803 by introducing alpha-ketoisocaproate dioxygenase from Rattus norvegicus. Metab Eng Commun 12:e00163
51. Cheah YE et al (2020) Systematic identification and elimination of flux bottlenecks in the aldehyde production pathway of Synechococcus elongatus PCC 7942. Metab Eng 60:56–65
52. Jazmin LJ et al (2017) Isotopically nonstationary C-13 flux analysis of cyanobacterial isobutyraldehyde production. Metab Eng 42:9–18
53. Shen CR, Liao JC (2012) Photosynthetic production of 2-methyl-1-butanol from CO2 in cyanobacterium Synechococcus elongatus PCC7942 and characterization of the native acetohydroxyacid synthase. Energ Environ Sci 5:9574–9583
54. Gao X et al (2016) Engineering the methylerythritol phosphate pathway in cyanobacteria for photosynthetic isoprene production from CO2. Energ Environ Sci 9:1400–1411

55. Chin T, Sano M, Takahashi T, Ohara H, Aso Y (2015) Photosynthetic production of itaconic acid in Synechocystis sp. PCC6803. J Biotechnol 195:43–45
56. Xue Y, Zhang Y, Cheng D, Daddy S, He QF (2014) Genetically engineering Synechocystis sp. Pasteur culture collection 6803 for the sustainable production of the plant secondary metabolite p-coumaric acid. Proc Natl Acad Sci U S A 111:9449–9454
57. Lin PC, Saha R, Zhang F, Pakrasi HB (2017) Metabolic engineering of the pentose phosphate pathway for enhanced limonene production in the cyanobacterium Synechocysti s sp. PCC 6803. Sci Rep 7:17503
58. Choi SY et al (2017) Improvement of squalene production from CO2 in Synechococcus elongatus PCC 7942 by metabolic engineering and scalable production in a photobioreactor. ACS synthetic biology 6:1289–1295
59. Ruffing AM (2014) Improved free fatty acid production in cyanobacteria with Synechococcus sp. PCC 7002 as host. Front Bioeng Biotechnol 2:17
60. Ducat DC, Avelar-Rivas JA, Way JC, Silver PA (2012) Rerouting carbon flux to enhance photosynthetic productivity. Appl Environ Microbiol 78:2660–2668
61. Qiao CC et al (2018) Effects of reduced and enhanced glycogen pools on salt-induced sucrose production in a sucrose-secreting strain of Synechococcus elongatus PCC 7942. Appl Environ Microbiol 84
62. Jacobsen JH, Frigaard NU (2014) Engineering of photosynthetic mannitol biosynthesis from CO2 in a cyanobacterium. Metab Eng 21:60–70
63. Liu XY, Sheng J, Curtiss R (2011) Fatty acid production in genetically modified cyanobacteria. Proc Natl Acad Sci U S A 108:6899–6904
64. Tan XM et al (2011) Photosynthesis driven conversion of carbon dioxide to fatty alcohols and hydrocarbons in cyanobacteria. Metab Eng 13:169–176
65. Gao QQ, Wang WH, Zhao H, Lu XF (2012) Effects of fatty acid activation on photosynthetic production of fatty acid-based biofuels in Synechocystis sp. PCC6803. Biotechnol Biofuels 5
66. Wang W, Liu X, Lu X (2013) Engineering cyanobacteria to improve photosynthetic production of alka(e)nes. Biotechnol Biofuels 6:69
67. Young JD, Shastri AA, Stephanopoulos G, Morgan JA (2011) Mapping photoautotrophic metabolism with isotopically nonstationary (13)C flux analysis. Metab Eng 13:656–665
68. Jazmin LJ et al (2017) Isotopically nonstationary (13)C flux analysis of cyanobacterial isobutyraldehyde production. Metab Eng 42:9–18
69. Yang C, Hua Q, Shimizu K (2002) Metabolic flux analysis in Synechocystis using isotope distribution from 13C-labeled glucose. Metab Eng 4:202–216
70. Abernathy MH et al (2017) Deciphering cyanobacterial phenotypes for fast photoautotrophic growth via isotopically nonstationary metabolic flux analysis. Biotechnol Biofuels 10:273
71. Hirokawa Y, Suzuki I, Hanai T (2015) Optimization of isopropanol production by engineered cyanobacteria with a synthetic metabolic pathway. J Biosci Bioeng 119:585–590
72. Dempo Y, Ohta E, Nakayama Y, Bamba T, Fukusaki E (2014) Molar-based targeted metabolic profiling of cyanobacterial strains with potential for biological production. Metabolites 4:499–516
73. Patel MS, Nemeria NS, Furey W, Jordan F (2014) The pyruvate dehydrogenase complexes: structure-based function and regulation. J Biol Chem 289:16615–16623
74. Chuang DSW, Liao JC (2021) Role of cyanobacterial phosphoketolase in energy regulation and glucose secretion under dark anaerobic and osmotic stress conditions. Metab Eng 65:255–262
75. Carpine R et al (2017) Genetic engineering of Synechocystis sp. PCC6803 for poly-β-hydroxybutyrate overproduction. Algal Res 25:117–127
76. Muro-Pastor MI et al (2020) CfrA, a novel carbon flow regulator, adapts carbon metabolism to nitrogen deficiency in cyanobacteria. Plant Physiol 184:1792–1810
77. Orthwein T et al (2021) The novel P-II-interactor PirC identifies phosphoglycerate mutase as key control point of carbon storage metabolism in cyanobacteria. Proc Natl Acad Sci U S A 118

78. Koch M et al (2020) Maximizing PHB content in Synechocystis sp. PCC 6803: a new metabolic engineering strategy based on the regulator PirC. Microb Cell Fact 19:231
79. Ku JST, Chen AVY, Lan EI (2020) Metabolic engineering design strategies for increasing acetyl-CoA flux. Metabolites 10
80. Ku JT, Lan EI (2018) A balanced ATP driving force module for enhancing photosynthetic biosynthesis of 3-hydroxybutyrate from CO2. Metab Eng 46:35–42
81. Li H et al (2016) CRISPR-Cas9 for the genome engineering of cyanobacteria and succinate production. Metab Eng 38:293–302
82. Wang B, Pugh S, Nielsen DR, Zhang W, Meldrum DR (2013) Engineering cyanobacteria for photosynthetic production of 3-hydroxybutyrate directly from CO2. Metab Eng 16:68–77
83. Gao Z, Zhao H, Li Z, Tan X, Lu X (2012) Photosynthetic production of ethanol from carbon dioxide in genetically engineered cyanobacteria. Energ Environ Sci 5:9857–9865
84. Yu H, Li X, Duchoud F, Chuang DS, Liao JC (2018) Augmenting the Calvin–Benson–Bassham cycle by a synthetic malyl-CoA-glycerate carbon fixation pathway. Nat Commun 9:2008
85. Lee HJ, Son J, Sim SJ, Woo HM (2020) Metabolic rewiring of synthetic pyruvate dehydrogenase bypasses for acetone production in cyanobacteria. Plant Biotechnol J 18:1860–1868
86. Takahashi H, Uchimiya H, Hihara Y (2008) Difference in metabolite levels between photoautotrophic and photomixotrophic cultures of Synechocystis sp. PCC 6803 examined by capillary electrophoresis electrospray ionization mass spectrometry. J Exp Bot 59:3009–3018
87. Carrieri D, Paddock T, Maness PC, Seibert M, Yu JP (2012) Photo-catalytic conversion of carbon dioxide to organic acids by a recombinant cyanobacterium incapable of glycogen storage. Energ Environ Sci 5:9457–9461
88. Grundel M, Scheunemann R, Lockau W, Zilliges Y (2012) Impaired glycogen synthesis causes metabolic overflow reactions and affects stress responses in the cyanobacterium Synechocystis sp. PCC 6803. Microbiology (Reading) 158:3032–3043
89. Benson PJ et al (2016) Factors altering pyruvate excretion in a glycogen storage mutant of the cyanobacterium, Synechococcus PCC7942. Front Microbiol 7:475
90. Davies FK, Work VH, Beliaev AS, Posewitz MC (2014) Engineering limonene and bisabolene production in wild type and a glycogen-deficient mutant of Synechococcus sp. PCC 7002. Front Bioeng Biotechnol 2:21
91. Carrieri D et al (2015) Enhancing photo-catalytic production of organic acids in the cyanobacterium Synechocystis sp. PCC 6803 DeltaglgC, a strain incapable of glycogen storage. J Microbial Biotechnol 8:275–280
92. van der Woude AD, Angermayr SA, Puthan Veetil V, Osnato A, Hellingwerf KJ (2014) Carbon sink removal: increased photosynthetic production of lactic acid by Synechocystis sp. PCC6803 in a glycogen storage mutant. J Biotechnol 184:100–102
93. Herrero A, Muro-Pastor AM, Flores E (2001) Nitrogen control in cyanobacteria. J Bacteriol 183:411–425
94. Osanai T et al (2006) Nitrogen induction of sugar catabolic gene expression in Synechocystis sp. PCC 6803. DNA Res 13:185–195
95. Gordon GC et al (2016) CRISPR interference as a titratable, trans-acting regulatory tool for metabolic engineering in the cyanobacterium Synechococcus sp. strain PCC 7002. Metab Eng 38:170–179
96. Monshupanee T, Chairattanawat C, Incharoensakdi A (2019) Disruption of cyanobacterial gamma-aminobutyric acid shunt pathway reduces metabolites levels in tricarboxylic acid cycle, but enhances pyruvate and poly(3-hydroxybutyrate) accumulation. Sci Rep 9:8184
97. Gao E-B et al (2020) Increased ethanol production by disrupting the competitive phosphoenolpyruvate synthesis pathway and enhancing the expression of ethanol-producing genes in Synechocystis Sp PCC6803. Res Sq
98. Takamura Y, Nomura G (1988) Changes in the intracellular concentration of acetyl-CoA and malonyl-CoA in relation to the carbon and energy metabolism of Escherichia coli K12. J Gen Microbiol 134:2249–2253

99. Leonard E, Lim K-H, Saw P-N, Koffas MA (2007) Engineering central metabolic pathways for high-level flavonoid production in Escherichia coli. Appl Environ Microbiol 73:3877–3886
100. Leonard E et al (2008) Strain improvement of recombinant Escherichia coli for efficient production of plant flavonoids. Mol Pharm 5:257–265
101. Ruffing AM (2013) Borrowing genes from Chlamydomonas reinhardtii for free fatty acid production in engineered cyanobacteria. J Appl Phycol 25:1495–1507
102. Eungrasamee K, Miao R, Incharoensakdi A, Lindblad P, Jantaro S (2019) Improved lipid production via fatty acid biosynthesis and free fatty acid recycling in engineered Synechocystis sp. PCC 6803. Biotechnol Biofuels 12:8
103. Hu B et al (2016) Comprehensive molecular characterization of methylobacterium extorquens AM1 adapted for 1-butanol tolerance. Biotechnol Biofuels 9:84
104. Sun T et al (2018) Re-direction of carbon flux to key precursor malonyl-CoA via artificial small RNAs in photosynthetic Synechocystis sp. PCC 6803. Biotechnol Biofuels 11:1–16
105. Ni J, Tao F, Wang Y, Yao F, Xu P (2016) A photoautotrophic platform for the sustainable production of valuable plant natural products from CO 2. Green Chem 18:3537–3548
106. Satoh S et al (2020) Enhancement of fatty acid biosynthesis by exogenous acetyl-CoA carboxylase and pantothenate kinase in Escherichia coli. Biotechnol Lett 42:2595–2605
107. Zhang S, Bryant DA (2011) The tricarboxylic acid cycle in cyanobacteria. Science 334:1551–1553
108. Durall C et al (2021) Production of succinate by engineered strains of Synechocystis PCC 6803 overexpressing phosphoenolpyruvate carboxylase and a glyoxylate shunt. Microb Cell Fact 20:39
109. Sengupta S et al (2020) Metabolic engineering of a fast-growing cyanobacterium Synechococcus elongatus PCC 11801 for photoautotrophic production of succinic acid. Biotechnol Biofuels 13
110. Huang CH et al (2016) CRISPR interference (CRISPRi) for gene regulation and succinate production in cyanobacterium S. elongatus PCC 7942. Microb Cell Fact 15:196
111. Kanwal S, Incharoensakdi A (2020) GABA synthesis mediated by gamma-aminobutanal dehydrogenase in Synechocystis sp. PCC6803 with disrupted glutamate and alpha-ketoglutarate decarboxylase genes. Plant Sci 290:110287
112. Iijima H et al (2021) Four-carbon dicarboxylic acid production through the reductive branch of the open cyanobacterial tricarboxylic acid cycle in Synechocystis sp. PCC 6803. Metab Eng 65:88–98
113. Agarwal AS, Zhai Y, Hill D, Sridhar N (2011) The electrochemical reduction of carbon dioxide to formate/formic acid: engineering and economic feasibility. ChemSusChem 4:1301–1310
114. Brown K, Doy C (1966) Control of three isoenzymic 7-phospho-2-oxo-3-deoxy-d-arabino-heptonate-d-erythrose-4-phosphate lyases of Escherichia coli W and derived mutants by repressive and “inductive” effects of the aromatic amino acids. Biochim Biophys Acta 118:157–172
115. Hall G, Flick M, Gherna R, Jensen R (1982) Biochemical diversity for biosynthesis of aromatic amino acids among the cyanobacteria. J Bacteriol 149:65–78
116. Baez-Viveros JL et al (2004) Metabolic engineering and protein directed evolution increase the yield of L-phenylalanine synthesized from glucose in Escherichia coli. Biotechnol Bioeng 87:516–524
117. Zhou H, Liao X, Wang T, Du G, Chen J (2010) Enhanced l-phenylalanine biosynthesis by co-expression of pheA(fbr) and aroF(wt). Bioresour Technol 101:4151–4156
118. Lütke-Eversloh T, Stephanopoulos G (2005) Feedback inhibition of chorismate mutase/prephenate dehydrogenase (TyrA) of Escherichia coli: generation and characterization of tyrosine-insensitive mutants. Appl Environ Microbiol 71:7224–7228

119. Kikuchi Y, Tsujimoto K, Kurahashi O (1997) Mutational analysis of the feedback sites of phenylalanine-sensitive 3-deoxy-D-arabino-heptulosonate-7-phosphate synthase of Escherichia coli. Appl Environ Microbiol 63:761–762
120. Tanaka K et al (2021) Quantification of NAD(P)H in cyanobacterial cells by a phenol extraction method. Photosynth Res 148:57–66
121. Tamoi M, Miyazaki T, Fukamizo T, Shigeoka S (2005) The Calvin cycle in cyanobacteria is regulated by CP12 via the NAD (H)/NADP (H) ratio under light/dark conditions. Plant J 42: 504–513
122. Cooley JW, Vermaas WF (2001) Succinate dehydrogenase and other respiratory pathways in thylakoid membranes of Synechocystis sp. strain PCC 6803: capacity comparisons and physiological function. J Bacteriol 183:4251–4258
123. Lan EI, Liao JC (2012) ATP drives direct photosynthetic production of 1-butanol in cyanobacteria. Proc Natl Acad Sci 109:6018–6023
124. Shen CR, Liao JC (2012) Photosynthetic production of 2-methyl-1-butanol from CO 2 in cyanobacterium Synechococcus elongatus PCC7942 and characterization of the native acetohydroxyacid synthase. Energ Environ Sci 5:9574–9583
125. Oliver JW, Machado IM, Yoneda H, Atsumi S (2013) Cyanobacterial conversion of carbon dioxide to 2, 3-butanediol. Proc Natl Acad Sci 110:1249–1254
126. Li H, Liao JC (2013) Engineering a cyanobacterium as the catalyst for the photosynthetic conversion of CO 2 to 1, 2-propanediol. Microb Cell Fact 12:1
127. Richter N, Zienert A, Hummel W (2011) A single-point mutation enables lactate dehydrogenase from Bacillus subtilis to utilize NAD+ and NADP+ as cofactor. Eng Life Sci 11:26–36
128. Niederholtmeyer H, Wolfstädter BT, Savage DF, Silver PA, Way JC (2010) Engineering cyanobacteria to synthesize and export hydrophilic products. Appl Environ Microbiol 76: 3462–3466
129. Savakis PE, Angermayr SA, Hellingwerf KJ (2013) Synthesis of 2, 3-butanediol by Synechocystis sp. PCC6803 via heterologous expression of a catabolic pathway from lactic acid-and enterobacteria. Metab Eng 20:121–130
130. Choi YN, Park JM (2016) Enhancing biomass and ethanol production by increasing NADPH production in Synechocystis sp. PCC 6803. Bioresour Technol 213:54–57
131. Kumaraswamy GK et al (2013) Reprogramming the glycolytic pathway for increased hydrogen production in cyanobacteria: metabolic engineering of NAD(+)-dependent GAPDH. Energ Environ Sci 6:3722–3731
132. Matson MM, Atsumi S (2018) Photomixotrophic chemical production in cyanobacteria. Curr Opin Biotechnol 50:65–71
133. McEwen JT, Machado IM, Connor MR, Atsumi S (2013) Engineering Synechococcus elongatus PCC 7942 for continuous growth under diurnal conditions. Appl Environ Microbiol 79:1668–1675
134. McEwen JT, Kanno M, Atsumi S (2016) 2,3 Butanediol production in an obligate photoautotrophic cyanobacterium in dark conditions via diverse sugar consumption. Metab Eng 36: 28–36
135. Lee TC et al (2015) Engineered xylose utilization enhances bio-products productivity in the cyanobacterium Synechocystis sp. PCC 6803. Metab Eng 30:179–189
136. Kanno M, Atsumi S (2017) Engineering an obligate photoautotrophic cyanobacterium to utilize glycerol for growth and chemical production. ACS Synth Biol 6:69–75
137. Khetkorn W, Incharoensakdi A, Lindblad P, Jantaro S (2016) Enhancement of poly-3-hydroxybutyrate production in Synechocystis sp. PCC 6803 by overexpression of its native biosynthetic genes. Bioresour Technol 214:761–768
138. Orthwein T et al (2021) The novel PII-interactor PirC identifies phosphoglycerate mutase as key control point of carbon storage metabolism in cyanobacteria. Proc Natl Acad Sci U S A 118

139. Ni J, Liu HY, Tao F, Wu YT, Xu P (2018) Remodeling of the photosynthetic chain promotes direct CO2 conversion into valuable aromatic compounds. Angew Chem Int Ed 57:15990–15994
140. Deshpande A, Vue J, Morgan J (2020) Combining random mutagenesis and metabolic engineering for enhanced tryptophan production in Synechocystis sp. Strain PCC 6803. Appl Environ Microbiol 86
141. Ni J, Tao F, Wang Y, Yao F, Xu P (2016) A photoautotrophic platform for the sustainable production of valuable plant natural products from CO2. Green Chem 18:3537–3548
142. Brey LF et al (2020) Metabolic engineering of Synechocystis sp. PCC 6803 for the production of aromatic amino acids and derived phenylpropanoids. Metab Eng 57:129–139

Adv Biochem Eng Biotechnol (2023) 183: 145–170
https://doi.org/10.1007/10_2022_213

Published online: 11 February 2023

Production of Fatty Acids and Derivatives Using Cyanobacteria

Pachara Sattayawat, Ian S. Yunus, and Patrik R. Jones

Contents

P. Sattayawat
Department of Biology, Faculty of Science, Chiang Mai University, Chiang Mai, Thailand
e-mail: pachara.sattayawat@cmu.ac.th

I. S. Yunus
Joint BioEnergy Institute, Emeryville, CA, USA

Biological Systems and Engineering Division, Lawrence Berkeley National Laboratory, Berkeley, CA, USA

P. R. Jones (✉)
Department of Life Sciences, Imperial College London, London, UK
e-mail: p.jones@imperial.ac.uk

Abstract Fatty acids and their derivatives are highly valuable chemicals that can be produced through chemical or enzymatic processes using plant lipids. This may compete with human food sources. Therefore, there has been an urge to create a new method for synthesizing these chemicals. One approach is to use microbial cells, specifically cyanobacteria, as a factory platform. Engineering may need to be implemented in order to allow a cost-competitive production and to enable a production of a variety of different fatty acids and derivatives. In this chapter, we explain in details the importance of fatty acids and their derivatives, including fatty aldehydes, fatty alcohols, hydrocarbons, fatty acid methyl esters, and hydroxy fatty acids. The production of these chemicals using cyanobacterial native metabolisms together with strategies to engineer them are also explained. Moreover, recent examples of fatty acid and fatty acid derivative production from engineered cyanobacteria are gathered and reported. Commercial opportunities to manufacture fatty acids and derivatives are also discussed in this chapter. Altogether, it is clear that fatty acids and their derivatives are important chemicals, and with recent advancements in genetic engineering, a cyanobacterial platform for bio-based production is feasible. However, there are regulations and guidelines in place for the use of genetically modified organisms (GMOs) and some further developments are still needed before commercialization can be reached.

Graphical Abstract

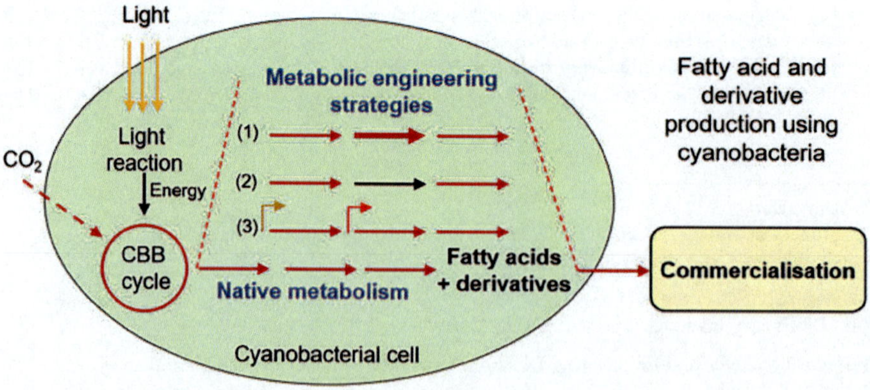

Keywords Fatty acids, Derivatives, Cyanobacteria, Metabolic engineering, Commercial opportunity

1 Introduction

Fatty acids and their derivatives are important chemical feedstocks used in several industries. Currently, they are mainly obtained by chemical or enzymatic transformation of plant lipids [1]. However, plant-based production of chemicals competes with agriculture, which already has a substantial environmental footprint [2]. In addition, environmental impacts and health risks caused during lipid extraction are further reasons why more sustainable production methods should urgently be developed [3, 4]. Once fatty acids have been obtained, they can be altered to generate functional derivatives with required properties [5, 6]. The process of fatty acid derivatization is to transform the carboxyl group on fatty acids to other functional groups such as hydroxyl or aldehyde in one or more steps.

Recently, there has been great interest in utilizing oleaginous microalgae for fatty acid manufacturing [7]. By definition, these microorganisms can synthesize relatively high amounts of lipids as an internal storage compound [8]. Another alternative is to utilize cyanobacteria, prokaryotic microalgae, though these need to first be engineered in order to biosynthesise fatty acids or derivatives thereof [9–12]. In general, microalgae are attractive as they can directly utilize CO_2 as their sole carbon source and thereby consume rather than produce greenhouse gas during manufacturing, thereby avoiding the need for a multi-step process in which terrestrial plants first accumulate sugars. Since cyanobacteria are relatively easy to engineer there are many examples in which the direct conversion of CO_2 into a wide range of chemicals has been demonstrated. A simplified diagram depicting indirect and direct approaches for renewable production of biofuels and biochemicals is presented in Fig. 1.

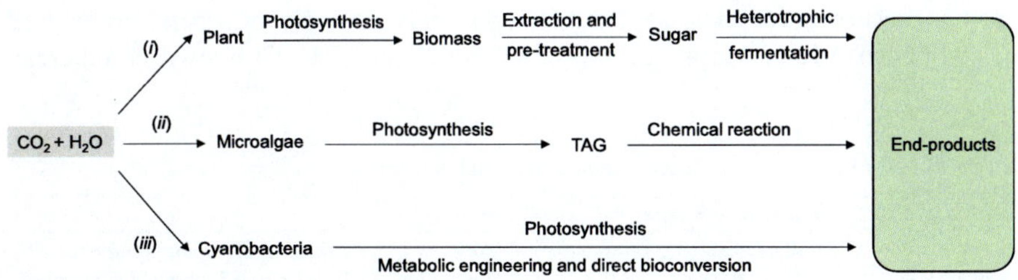

Fig. 1 Simplified diagram comparing indirect and direct approaches for biotechnology-manufactured biofuels and biochemicals from CO_2 adapted from Wang et al. [13]. Renewable production of bio-based chemicals via (*i*) heterotrophic fermentation typically relies on plant-derived biomass, which often requires complicated extraction and pre-treatment processes to obtain fermentable sugars. (*ii*) Microalgae accumulate triacylglycerol (TAG) which can be converted into fatty alcohols by ex vivo esterification and hydrogenation. (*iii*) Engineered cyanobacteria can directly convert atmospheric CO_2 into biofuels and biochemicals

2 Fatty Acids, Fatty Acid Derivatives, and Their Importance

Fatty acids are compounds that consist of two parts: (1) a terminal carboxyl group (-COOH) and (2) an aliphatic chain (saturated or unsaturated). With such a broad definition, there is a wide range of diverse structures, depending on the number of carbon atoms and unsaturation. Derivatives of fatty acids are compounds where the carboxyl group has been modified to other functional groups such as aldehydes, alcohols, alkanes, alkenes, and esters.

Fatty acids are obtained from biotechnology through three different routes summarized in Fig. 1. However, currently the main plant sources include soybean, palm, peanut, olive, rapeseed, coconut, where free fatty acids are obtained by hydrolysis of extracted oils [5]. Naturally existing fatty acids from plants or animals are typically even chains with 20–24 carbon atoms. They are further grouped into saturated aliphatic fatty acids that are short (C2-C6), medium (C8-C12), long (C14-C18), and very long chain (C20-C24) [14]. Anneken and colleagues [5] have summarized how the industrial use corresponds to the chain length of fatty acids and their derivatives. Medium-chain fatty acids and derivatives are usually used for coatings, inks, and plastics. Longer-chain molecules (C14-C22) are used for soaps and detergents. Notably, the utilities often overlap, for example, fatty acids and derivatives ranging from C8 to C22 (apart from C14) can all be used as emulsifiers.

Switching our attention to derivatives (Table 1), *fatty aldehydes* are mainly used as precursors for the synthesis of rubbers and plastics [15], but are also used as ingredients in flavour and fragrance industries. Commercially, they are synthesized from fatty acids using substrates, solvents, and catalysts, some of which can be toxic [16].

Fatty alcohols have high reactivity due to their hydroxyl group [17] and are often used as precursors for chemical synthesis. Saturated fatty alcohols can also be used as components, for example, in household detergents [18]. Currently, commercial

Table 1 List of fatty acid derivatives included in this chapter

Derivative	Commercial source	Common use
Fatty aldehydes	Synthesized from natural fatty acids	– Precursors for rubber and plastic synthesis – Ingredients in flavour and fragrance industries
Fatty alcohols	Synthesized from natural fatty acids	– Precursors in industrial chemical synthesis – Components in household products
Hydrocarbons	Natural gas and petroleum refineries	– Fuels – Precursors for industrial chemical synthesis
Fatty acid esters	Synthesized from natural fatty acids	– Fuels – Chemical components in several industries such as in flavour and fragrance
Hydroxy fatty acids	Synthesized from natural fatty acids	– Precursors used for pharmaceutical and polymer applications

fatty alcohols are mainly produced by chemical reactions using plant-derived free fatty acids as precursors [19, 20].

Hydrocarbons are classified as alkanes (saturated) or alkenes (unsaturated). They are produced from natural gas and petroleum refineries by direct extraction or chemical conversion from natural gas/petroleum-based precursors [21]. They are predominantly used as fuels and lubricants. Alkenes are also used as precursors for chemical synthesis. For example, the synthesis of oxo-alcohols for the plasticizer industry utilizes linear 1-alkenes as a substrate [21].

Fatty acid esters are currently synthesized through transesterification of oils/fats (such as vegetable oils and animal fats) and alcohols (mostly methanol and ethanol) to form fatty acid methyl esters or fatty acid ethyl esters. However, the production of these esters is limited by the lipid sources and often requires toxic chemicals for the alkyl donor such as methanol [22–25]. Apart from short methyl or ethyl esters as biodiesel derived from methanol or ethanol as substrates, esters with a longer side chain are also important in the industries. Octyl acetate, as an example, has been used in flavour and fragrance industries and has currently been synthesized via chemical reactions [26, 27].

Another derivative of fatty acids included in this chapter does not contain modified carboxyl group but contains an extra hydroxyl group on the alkyl chain – *hydroxy fatty acids*. They are often long-chain fatty acids with at least one hydroxyl group on the alkyl saturated or unsaturated chain [28]. The presence of an additional hydroxyl group renders then attractive as precursors for pharmaceutical and polymer applications [28]. Hydroxy fatty acids are commonly produced by introducing hydroxy groups in hydrocarbon chains of plant oils using a chemical catalyst.

3 Fatty Acid Metabolism in Cyanobacteria

Fatty acids are the building blocks of cell membranes of all organisms with phospholipid bilayer membranes, including cyanobacteria [29]. Cyanobacteria synthesize fatty acids via fatty acid elongation pathway. This pathway is similar to the pathway found in other bacteria. In the first step of fatty acid ßbiosynthesis, acetyl-CoA generated from the CBB cycle is converted to malonyl-CoA and then to malonyl-ACP by acetyl-CoA carboxylase (Acc) and malonyl-CoA:ACP transacylase (FabD), respectively. Then, the elongation cycle starts by generation of acetoacetyl-ACP via Claisen condensation of acetyl-CoA and malonyl-ACP catalysed by a β-ketoacyl-ACP synthase (KAS). Three KASs are present in microorganisms including *E. coli* and cyanobacteria – KASI (FabB), KASII (FabF), and KASIII (FabH). Importantly, KASI specifically converts acetyl-CoA rather than other acyl-ACPs. Hence, KASI plays a role in the initiation of the elongation cycle as it catalyses the first step by converting acetyl-CoA to acetoacetyl-CoA. KASII and KASIII, on the other hand, play a role in later stages of fatty acid elongation pathway as they can terminate the elongation steps by catalysing longer-chain acyl-CoAs.

3-Hydroxyacyl-ACP (or 3-hydroxybutyryl-ACP in the first cycle) is the next intermediate converted from acetoacetyl-ACP by β-ketoacyl-ACP reductase (FabG) and then to enoyl-ACP (or 2-butenoyl-ACP in the first cycle) by a 3-hydroxyacyl-ACP (FabA and FabZ). Following this step, enoyl-ACP is reduced to acyl-ACP (or butyryl-ACP in the first cycle) by an enoyl-ACP reductase (FabI). Butyryl-ACP from the first round enters the cycle again for the second round of elongation by which 2 carbons are added to the intermediate for each cycle until acyl-ACPs of 16 and18 carbons are generated. Respective fatty acids can be synthesized from these long-chain acyl-ACPs by a native thioesterase (Tes) or acyl-ACPs are transformed to phosphatidic acid by a set of enzymes (phosphate acetyltransferase (PlsX), acylglycerol-phosphate acyltransferase (PlsY), and lysophosphatidic acid acyltransferase (PlsC)) for cell membrane synthesis. However, cyanobacteria do not possess the fatty acid degradation pathway (also known as β-oxidation pathway). Therefore, fatty acid utilization relies on a set of enzymes including acyl-acyl carrier protein synthetase (Aas), which converts fatty acids to respective fatty acyl-ACPs [30, 31]. The overall fatty acid metabolism in cyanobacteria is illustrated in Fig. 2.

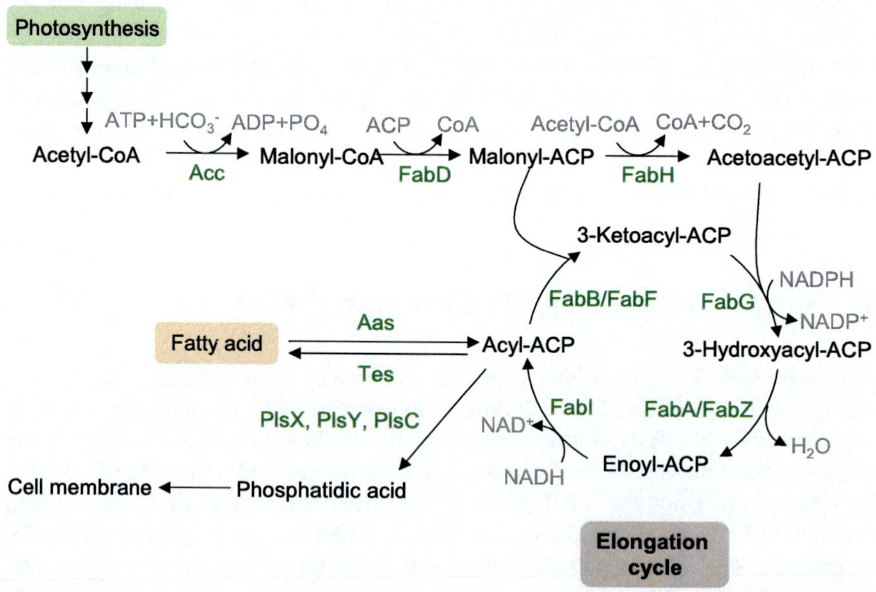

Fig. 2 Native fatty acid biosynthesis in cyanobacteria. The diagram is adapted from Wang et al. [13]. Fatty acid biosynthesis in cyanobacteria is similar to in *E. coli* in the early steps but lacks β-oxidation, the degradation pathway of fatty acids. Acc, acetyl-coA carboxylase; FabD, malonyl-CoA:ACP transacylase; FabH, β-Ketoacyl-ACP synthase III; FabG, β-Ketoacyl-ACP reductase; FabA/FabZ, β-hydroxyacyl-ACP dehydratase/isomerase; FabI, enoyl-ACP reductase I; FabB/FabF, β-Ketoacyl-ACP synthase I or II; Tes, thioesterase; Aas, acyl-ACP synthetase; PlsX, phosphate acetyltransferase; PlsY, acylglycerol-phosphate acyltransferase; PlsC, lysophosphatidic acid acyltransferase. Pathway enzymes are indicated in green

4 The Use of Metabolic Engineering to Implement Pathways for Fatty Acid Derivatives

Cyanobacteria have been genetically engineered for bioproduction of fatty acid derivatives by extending or redirecting the native fatty acid pathway towards targets of interest. The engineered pathways are illustrated in Fig. 3.

4.1 Fatty Aldehydes

Fatty aldehydes are important intermediates for synthesizing other fatty acid-derived molecules including fatty alcohols and fatty alkanes ([32]). To date, there are two biosynthetic routes for fatty aldehyde production in cyanobacteria: (1) acyl-ACP-dependent pathway which utilizes acyl-ACP, and (2) CAR-dependent pathway which utilizes fatty acids as precursors. In the first route, acyl-ACPs are converted to respective fatty aldehydes by an enzyme, namely acyl-ACP reductase (AAR) (Fig. 3). Notably, if the cells utilize this route, aldehydes can be produced without the

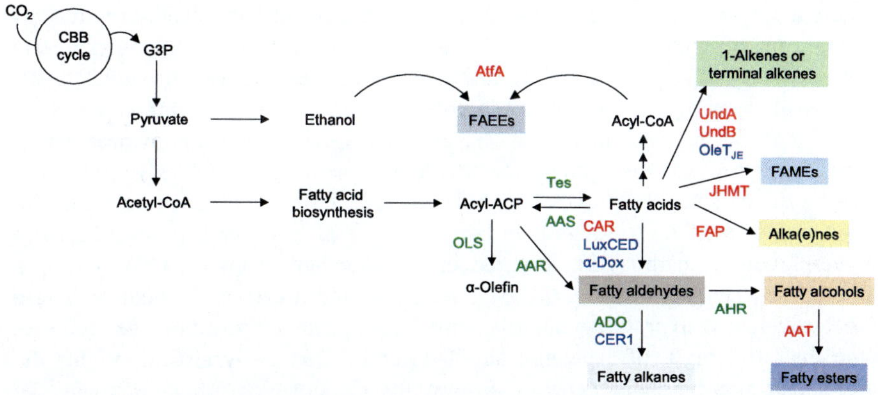

Fig. 3 Overview of synthetic pathways for the production of free fatty acids and derivatives in cyanobacteria. This diagram illustrates simplified fatty acid biosynthesis in cyanobacteria and derived pathways for fatty acid derivatives; fatty aldehydes, fatty alcohols, fatty alka(e)nes, and fatty acid esters (FAMEs, FAEEs, and other fatty esters). Enzymes catalysing each reaction step are illustrated, red indicates enzymes that have been reported to be heterologously expressed in cyanobacteria, blue indicates the enzymes that could catalyse the step but have never been reported in cyanobacteria and green indicates endogenous enzymes in cyanobacteria. Tes, thioesterase; AAS, acyl-ACP synthetase; OLS, olefin synthase; AAR, acyl-ACP reductase; ADO, aldehyde deformylating oxygenase; CER1, fatty aldehyde decarboxylase; CAR, carboxylic acid reductase; LuxCED, a combination of enzymes fatty acid reductase (LuxC), a fatty acyl synthetase (LuxE), and a fatty acyl transferase (LuxD); α-Dox, alpha-dioxygenase; AHR, aldehyde reductase; FAP, fatty acid photodecarboxylase; JHMT, juvenile hormone *O*-methyl transferase; UndA, oxidative decarboxylase; UndB, desaturase-like enzyme; OleTJE, cytochrome P450 fatty acid decarboxylase; AtfA, A:diacylglycerol acyltransferase; AAT, alcohol *O*-acyltransferase

need to generate fatty acids as an intermediate. In the latter route, fatty acids are transformed into fatty aldehydes by a carboxylic acid reductase (CAR) enzyme (Fig. 3). CAR is a versatile enzyme with broad substrate specificity [33] that allows the production of fatty aldehydes from various acids including fatty acids (C4-C18). Production of fatty aldehydes in cyanobacteria, however, has been a challenge. Cyanobacteria, such as *Synechocystis* sp. PCC 6803, carry multiple endogenous aldehyde reductases (AHRs), which convert fatty aldehydes to fatty alcohols. Although inactivation of endogenous AHRs has been demonstrated to enhance the accumulation of fatty aldehydes in *E. coli* [15], such strategies remain to be tested in cyanobacteria.

4.2 Fatty Alcohols

Several studies have reported production of fatty alcohols in cyanobacteria. Yao and colleagues first reported the production of long-chain fatty alcohols in *Synechocystis* sp. PCC 6803 by expressing fatty acyl-Coenzyme A (acyl-CoA) reductase gene from *Marinobacter aquaeolei* VT8. Together with aldehyde deformylating oxygenase (ADO) and acyl-ACP reductase (AAR) knock-outs, a total of 2.87 mg/g dry weight of fatty alcohols was produced [34]. Acyl-ACP-dependent fatty alcohol production suffers from competition of acyl-ACP precursors with the membrane biosynthesis pathway. To overcome this issue, Kaczmarzyk and colleagues established CRISPRi systems to downregulate six genes involved in acyl-ACP consuming pathways [35]. The results indicated that strains carrying CRISPRi systems for downregulation of *slr1510* (gene encoding phosphate acyltransferase enzyme or PlsX) produced a higher titre of long-chain fatty alcohols. In a different study, we reported the production of medium to long-chain fatty alcohols in *Synechocystis* sp. PCC 6803 by employing a carboxylic acid reductase-dependent pathway [10]. Using a thioesterase (Tes3) from *Anaerococcus tetradius* and a carboxylic acid reductase (CAR) together with promoter and ribosome-binding site optimization, we produced more than 100 mg/L of 1-octanol and 1-decanol. This study also highlights the thioesterase as a bottleneck enzyme in the fatty alcohol biosynthetic pathway. To further improve the titre, we screened several plant thioesterases with enhanced catalytic activity and produced approximately 526 mg/L 1-octanol after optimizing the cultivation condition [11]. In the same study, we also tested the strain stability and productivity for over 180 days of continuous cultivation. The tested strain reached an accumulative titre of >3.5 g/L 1-octanol in 180 days, the highest titre reported to date from cyanobacteria.

4.3 Hydrocarbons

As for hydrocarbon production in cyanobacteria, at present, there are at least two routes that have been reported for hydrocarbon (alkanes and/or alkenes) production via fatty acid biosynthesis: (1) acyl-ACP- and (2) free fatty acid-dependent pathways. In the former route, 2 enzymes are key players. First, acyl-ACPs from the fatty acid elongation cycle are converted directly to fatty aldehydes by an acyl-ACP reductase (AAR) and then fatty aldehydes can be converted further to fatty alkanes by a native cyanobacterial aldehyde deformylating oxygenase (ADO) [36] (Fig. 3). Several efforts have been made to enhance alkane production in cyanobacteria by overexpression of the native or non-native AAR and ADO enzyme couple [37–40]. This strategy relies on acyl-ACP as the precursor. Although the amounts of naturally accumulating alkanes have been enhanced through genetic manipulation achieving high carbon partitioning in a lipid-accumulating cyanobacterium, *Nostoc punctiforme* (up to 12.9% (g/g) CDW) [39], similar efforts in a non-lipid accumulating model cyanobacterium *Synechocystis sp.* PCC 6803 only yielded 1.1% (g/g_{CDW}) at best [37, 40]. Another route of acyl-ACP-dependent alkene production is via a native enzyme, namely olefin synthase (OLS), which is one of the two hydrocarbon biosynthetic pathways in cyanobacteria [41, 42]. This enzyme extends the carbon chain of acyl-ACP by generating a covalently bound C_{n+2} acyl chain via a decarboxylative condensation reaction with malonyl-CoA as a carbon donor (Fig. 3). Sequentially, these intermediates are decarboxylated to generate C_{n+1} alkenes [41, 42]. The cyanobacterium *Synechococcus* sp. PCC 7002 uses the olefin synthase (OLS) pathway to produce 1-alkenes, such as 1-nonadecene (C19:1) and 1,14-nonadecadiene (C19:2). To increase the production of α-olefin, Mendez-Perez and co-workers [43] replaced the upstream region (i.e., transcriptional regulation region) of *ols* in *Synechococcus* sp. PCC 7002 with a strong P_{psbA} promoter [44] from *Amaranthus hybridus,* a species of annual flowering plant. Although the resulting strain showed a twofold increase in 1-nonadecene and fivefold increase in 1,14-nonadecadiene titres compared to the wild-type strain, the mutant strain only produced 4.2 mg/L/OD_{730} [43]. In addition, the hydrocarbon species was restricted to 1-nonadecene and 1,14-nonadecadiene due to a narrow substrate preferential of olefin synthase. As these pathways do not need fatty acids as an intermediate, in principle, this route may not be considered as a fatty acid-derived pathway. Unlike acyl-ACP-dependent pathway, free fatty acid-dependent pathway requires a conversion of fatty acyl-ACP to fatty acids by a thioesterase (Tes), similar to the fatty alcohol pathway. The produced free fatty acids are then catalysed further by different enzymes to yield alka(e)nes (Fig. 3). Metabolically, three different sets of enzymes have been reported to play a role in converting fatty acids to fatty alkanes or alkenes: via (1) $OleT_{JE}$, UndA, UndB (2) CAR, LuxCED, α-Dox and ADO, CER1, and (3) FAP. In the case of terminal alkenes (or 1-alkenes), a handful number of enzymes have been reported to convert free fatty acids to fatty alkenes in one single step. $OleT_{JE}$ (a cytochrome P450 fatty acid decarboxylase belonging to the CYP152 family from a gram-positive bacterium *Jeotgalicoccus* sp. ATCC 8456) [45],

UndA (a non-heme iron oxygenase from *Pseudomonas fluorescens* Pf-5), and UndB (a family of desaturase-like enzymes from *Pseudomonas mendocina*) have been reported to be responsible for such reactions [46–48]. Secondly, fatty acids yielded from thioesterases can be converted to fatty aldehydes first either by a carboxylic acid reductase (CAR) [33], LuxCED (a combination of enzymes fatty acid reductase (LuxC), a fatty acyl synthetase (LuxE), and a fatty acyl transferase (LuxD) [49], or α-DOX (an alpha-dioxygenase) [50]. Once fatty aldehydes are provided, aldehyde deformylating oxygenase (ADO) or fatty aldehyde decarboxylase (CER1) catalyses the formation of fatty alkanes. Thirdly, fatty acid photodecarboxylase (FAP) converts these fatty acids directly to fatty alkanes or alkenes [51, 52]. Based on these studies, Yunus and colleagues engineered *Synechocystis* sp. PCC 6803 for production of hydrocarbons by employing a thioesterase with a wide substrate specificity, 'TesA, in combination with a desaturase-like enzyme (UndB) for terminal alkene production, a carboxylic acid reductase (CAR) and a native aldehyde deformylating oxygenase (ADO) for fatty alkane production, and a fatty acid photodecarboxylase (FAP) for fatty alka(e)ne production [12]. Strains overexpressing CAR did not accumulate hydrocarbons but, instead, produced long-chain fatty alcohols. In contrast, both strains overexpressing UndB or FAP produced higher hydrocarbon titres in comparison with the wild-type strain, demonstrating the first successful implementation of fatty acid-dependent hydrocarbon synthetic metabolic pathway in cyanobacteria. However, all hydrocarbons produced from this study accumulated intracellularly, which will likely increase the cost of downstream processing (e.g. cell harvesting, extraction, etc.). To engineer hydrocarbon-producing cyanobacteria that do not require cell harvesting or extraction, Amer and colleagues co-expressed a butyric acid-specific thioesterase (Tes4) and mutated FAP_{G462V} for production of propane gas [53]. The study showed a modest titre of propane (25 mg/L) from the cyanobacterial engineered strain. In a different study, recently, Yunus and colleagues also reported the development of cyanobacterial systems for production of short to medium-hydrocarbon chain lengths by coexpressing UndB or FAP or ADO with a short-medium acyl-ACP-specific thioesterase. The resulting engineered strains produced terminal alkenes or fatty alkanes extracellularly which did not require cell lysis for product isolation [9].

4.4 Fatty Acid Methyl Esters

Fatty acid methyl esters (FAMEs) can be produced from a fatty acid biosynthesis-derived pathway without the need to add methanol ([54] a). A thioesterase first converts acyl-ACPs to fatty acids and then these fatty acids are converted to FAMEs by an enzyme called juvenile hormone *O*-methyl transferase (JHMT) or fatty acid *O*-methyltransferase (FAMT) [54]. In the second reaction, *S*-adenosylmethionine (SAM) is needed as a methyl donor instead of methanol. SAM is a native metabolite inside the cell. It is used as a co-substrate in several processes such as methylation of DNA, RNA, and proteins [55]. In the case of fatty acid ethyl esters (FAEEs), the pathway is depending on a key enzyme in a group of acyltransferases (A: diacylglycerol acyltransferase (AtfA)) that generates FAEEs from ethanol and acyl-CoA substrates [56]. The role of this enzyme is to convert acyl-CoAs and ethanol to ethyl esters. However, to provide ethanol to the system, enzymes involved in ethanol synthesis may be needed. The most common enzyme is pyruvate dehydrogenase complex (Pdc), which converts pyruvate to acetaldehyde and then acetaldehyde is converted to ethanol by an alcohol dehydrogenase (Adh) [56]. Another example of ester production derived from the fatty acid biosynthetic pathway by cyanobacteria is the production of octyl acetate. As fatty alcohols are toxic to cyanobacterial cells, our group recently reported the concept of bioderivatization to reduce the toxicity of 1-octanol by converting 1-octanol into its corresponding ester compound, octyl acetate [57]. The strain producing octyl acetate exhibited better growth and accumulated higher molarity of the final product.

4.5 *Hydroxy Fatty Acids*

Recently, ($\omega - 1$)-hydroxy fatty acids were produced by *Synechocystis* sp. PCC 6803 with a novel synthetic pathway [58]. The pathway was extended from the polyhydroxybutyrate (PHB) native biosynthetic pathway by expressing aaKASIII and a fatty acid synthase (FAS). The engineered strain produced 2.1 mol% of ($\omega - 1$)-hydroxy fatty acids [58].

5 Metabolic Engineering Strategies for Enhancing Bioproduction of Fatty Acids and Fatty Acid Derivatives in Cyanobacteria

Fatty acids are the basic components of cell membranes. In natural settings, however, evolved metabolic networks of cyanobacteria are not genetically optimized to produce metabolites over the minimal levels for cellular needs. Thus, metabolic engineering approaches are implemented to enhance bioproduction of target metabolites. Here, we describe and summarize recent metabolic engineering efforts to produce fatty acids and fatty acid-derived compounds in cyanobacteria.

5.1 *Knocking Out Acyl-ACP Synthetase* (aas) *Gene*

The acyl-ACP synthetase (*aas*) gene encodes an enzyme that recycles free fatty acids back to the elongation pathway (Fig. 2). Hence, knocking out *aas* is the first step to overproduce free fatty acids [31]. This strategy has been previously reported to enhance the fatty acid accumulation and secretion in multiple studies [10, 59–61]. Interestingly, a recent publication demonstrated that overexpression of *aas* together with *rbcLXS* and *glpD* both encoding enzymes involved in CBB cycle (*rbcLXS* encodes RuBisCo large, small and chaperone subunits, and *glpD* encodes glycerol-3-phosphate dehydrogenase (GPD)) increased free fatty acid secretion in *Synechocystis* sp. PCC 6803 in comparison with the wild type [62]. RuBisCo or (ribulose-1,5-bisphosphate carboxylase/oxygenase) converts CO_2 and ribulose-1,5-bisphosphate to 3-phosphoglycerate (3-PG), which is the first step in CBB cycle. GPD enzyme converts dihydroxyacetone phosphate (DHAP) – one to the intermediates in the CBB cycle – to glycerol-3-phosphate, which is later used in cellular metabolism. The overexpression of *aas* in this work was in contrast to previous reports mentioned above where *aas* was often inactivated. An explanation could be that in addition to *aas,* genes encoding enzymes involved in the CBB cycle (*rbcLXS* and *glpD*) have been overexpressed concurrently; thus, these observations are the result of the sum impact from overall expression of 3 genes.

5.2 Expression of Genes Involved in Fatty Acids Biosynthesis (FAB)

To overproduce fatty acids, acyl-ACP thioesterases (Tes) is critical as this enzyme converts acyl-ACPs – intermediates in the fatty acid elongation cycle – to free fatty acids [63]. Insertion of *'tesA*, the gene encoding Tes, into *Synechocystis* sp. PCC 6803 wild type in combination with *aas* inactivation resulted in 83.6 mg/L of secreted fatty acids (compared with 1.8 mg/L produced from the corresponding wild type) [60]. 'TesA is an *E. coli* acyl-acyl carrier protein (ACP) thioesterase I that has been modified to remove the leader sequence involved in protein maturation. This modification allows the protein to be expressed and remain in the cytoplasm [64]. *E. coli* TesA catalyses the formation of fatty acids from acyl-ACPs and was reported to be specific for longer-chain substrates; thus, free fatty acids produced from 'TesA-overexpressing strains are mainly C16:0 and C18:0 [60]. A similar study found that deletion of *aas* and overexpression of *'tesA* could also improve the production of fatty acids in another model cyanobacterium, *Synechococcus elongatus* PCC 7942, by almost 80-fold compared to the wild type [65]. The *E. coli* TesA is arguably one of the most well-studied thioesterases, however, to diversify the product chain length, other thioesterases have also been investigated. For example, a plant thioesterase, FatB1, from *Umbellularia californica* (UcFatB1) was expressed in the *Synechococcus* sp. PCC 7002 *aas* deletion strain and showed C12:0 fatty acids accumulation unlike the wild type where C12:0 could not be detected [66]. This is in contrast to the specificity of TesA, as the UcFatB1 thioesterase showed specificity towards a C12 substrate. This C12-thioesterase was also expressed in *Synechocystis* sp. PCC 6803 *aas* deletion strain and yielded 6.67 mg/L of fatty acids [67]. Another example is a thioesterase from *Cinnamomum camphora* that showed to have specificity towards a C14 substrate when expressed in *E. coli* [68]. This variation of enzyme-specificity that nature has to offer allows us to handpick genetic parts for the most optimized systems. In addition to this, we recently characterized three medium to long-chain acyl-ACP-specific plant thioesterases (*Ca*FatB3 from *Cuphea avigera var. pulcherrima*, *Cho*FatB2 from *Cuphea hookeriana*, and *Cp*FatB1 from *Cuphea palustris*) and improved their activity in *Synechocystis* sp. PCC 6803 by truncating the N-terminal signal peptides [11].

In addition, FabH is an enzyme catalysing the reaction step prior to fatty acid elongation cycle (Fig. 2) and the expression of heterologous FabH from a planktonic diatom, *Chaetoceros* GSL56, as a replacement of FabH in *Synechococcus elongatus* PCC 7942 increased C12:0 fatty acid production by fivefold [69]. To provide more examples of how expression of key enzymes can contribute to lipid production, heterologous expression of glycerol-3-phosphate dehydrogenase (GPD) and diacylglycerol acyltransferase (DGAT) in *Synechocystis* sp. PCC 6803 showed enhanced lipid production without growth reduction [70]. Both of these enzymes play a role in lipid biosynthesis. Overexpression of native *plsX* and *plsC* genes responsible for phosphatidic acids synthesis which are consequently used as membrane lipids showed to enhance lipid accumulation in *Synechocystis* sp. PCC 6803 [71].

5.3 Enhance Cellular Productivity

Other than the expression of genes involved directly to fatty acid biosynthesis (FAB), heterologous expression of RuBisCo subunits (encoded by *rbcLS*) was found to increase free fatty acid titres in *Synechococcus* sp. PCC 7002 by threefold [61]. This indicates that indirect strategies could also be implemented as this did not involve a direct modification of the genes for fatty acid biosynthesis to enhance the production.

5.4 Alleviation of Free Fatty Acid Toxicity

Free fatty acids at deleterious concentrations can cause damages to cyanobacterial plasma membrane. RNA-Seq analysis was used to investigate the stress response to free fatty acid production in *Synechococcus elongatus* PCC 7942 and 15 genes were discovered to be involved in detoxification of free fatty acids. Consequently, single-gene mutations via gene knockout (for gene downregulation) and overexpression (for gene upregulation) were investigated and led to an increase in free fatty acid production [72], suggesting that alleviation of the product toxicity is another strategy to enhance free fatty acid production in cyanobacteria. This has also been demonstrated in another bio-product, 1-octanol [57]. Removal of metabolites from liquid cultures using solvent overlay has also been beneficial for product titre improvement, especially when the products are toxic or volatile. Yunus and colleagues used isopropyl myristate overlay to alleviate toxicity of fatty alcohols [10]. Kato and colleagues also used isopropyl myristate to capture free fatty acids and demonstrated a significant improvement in the product titre [73].

5.5 Manipulation of the Transcriptional Regulators

In 2017, Kizawa and colleagues discovered a transcription factor, namely LexA, that controls a set of genes in cyanobacterial fatty acid biosynthesis (*fabD*, *fabH*, *fabF*, *fabG*, *fabZ*, and *fabI*). Deletion of this transcription factor resulted in upregulation of *fabD*, *fabH*, *fabF*, and *fabG* genes and consequently resulted in higher accumulation of fatty acids [74]. Apart from transcription factors that are involved directly in fatty acid metabolism, deletion of transcription factors for other cellular processes also showed to increase fatty acid production. For example, cyAbrB2 is a transcription factor involved in carbon and nitrogen metabolism in response to the environment and was shown that deletion of this transcription factor increased fatty acid production by ~2-fold [75].

5.6 Increasing Free Fatty Acid Secretion

Transporters are an important factor for microbial cell factories as they play a role in exporting the products and have been shown to alleviate possible toxic effects caused by the products [76]. It has been shown that overexpression of *rndA1* and *rndB1* – genes responsible for resistance nodulation division (RND)-type porin/transporter in cyanobacteria – showed ~3-fold higher excretion of fatty acids compared to the control strain [77]. Moreover, a recent work showed that *sll0180* and *slr2131* (genes involved in AcrAB-TolC multidrug efflux system) knock-out strains showed a higher intracellular accumulation of free fatty acids in *Synechocystis* sp. PCC 6803 [78]. A recent study, however, demonstrated that overexpressing RND-like efflux pumps such as AcrA (*sll0180*), AcrB (*slr2131*), or TolC (*slr1270*) did not improve 1-octanol production in *Synechocystis* sp. PCC 6803 [11]. This suggests that these transporters might play different roles in transportation of fatty acid-derived products and open opportunities for future discoveries.

5.7 Free Fatty Acid Recovery from Lipid Membrane

An interesting strategy to release fatty acids from the lipid membrane was implemented by expressing lipolytic enzymes induced by CO_2 limitation, which resulted in the release of fatty acids to culture media when CO_2 was limited [79]. This strategy enables the recovery of fatty acids when the cells are in stationary phase or when extraction of free fatty acids from biomass is needed.

5.8 *Increasing the Precursor Pool*

One of the commonly used strategies to improve bioproduction is to increase the availability of central precursors that can then be redirected towards the targets. Pathways are basically defined as connecting chemical reactions in which the pool of substrates or intermediates in each step could impact the reaction rate of the following steps. In this sense, high availability of key intermediates could lead to high reaction rates of the following steps. To give an example, acetyl-CoA as a central metabolite in cellular metabolism can be a target for engineering to increase the flux towards the synthesis pathways of several target compounds [80]. In cyanobacteria, poly-3-hydroxybutyrate (PHB) production is one of the pathways competing for acetyl-CoA with fatty acid biosynthesis. In one study on engineered *Synechocystis* sp. PCC 6803 for fatty acid production, deletion of genes involved in PHB production in combination with *acc* overexpression resulted in ~46-fold increase in free fatty acid secretion yield [60]. Apart from direct modification of early reaction steps leading to the accumulation of precursors, manipulation of regulatory systems that can increase the amount of central metabolites could be an alternative. P_{II} signal transduction proteins have been shown to play a role in regulating fatty acid metabolism. Interestingly, in cyanobacteria, it has been shown that *Synechocystis* sp. PCC 6803 lacking a P_{II} protein showed decreased acetyl-CoA levels but slightly increased fatty acid levels [81].

Fatty acid derivatives are derived from fatty acids; hence, any engineering made to increase fatty acid yield are, in theory, beneficial to produce these derivatives as well. Brief explanations of biosynthetic pathways of these derivatives in cyanobacteria were explained in the previous section, here, strategies used to enhance such production are collectively discussed.

5.9 *Optimization of Metabolic Engineering*

Genetic engineering, especially when it involves non-native genetic parts, is rarely optimal the first time it is introduced into an organism. Optimized promoters and ribosome-binding sites (RBSs) used in synthetic pathways have been reported in multiple studies to contribute to enhanced production of fatty acid derivatives. For example, optimization of promoters and genetic insulators has shown to increase 1-butanol production by 2.3-fold [82]. Methanol-free biosynthesis of methyl laurate in cyanobacteria was reported previously by our group. The pathway was designed and assembled as explained above to obtain the first-generation strain producing

6.9 mg/L methyl laurate. Then, optimization of promoter sequences was shown to enhance the production up to 120 mg/L [54]. Moreover, as mentioned above, olefin synthase (OLS) is a native enzyme in cyanobacteria responsible for the production of terminal alkenes [43]. A study showed that by replacing the promoter in front of the native *ols* gene with a strong P_{psbA} promoter resulted in twofold increase of 1-nonadecene and fivefold increase of 1,14-nonadecadiene titres [43]. However, it is worth noting that protein expression and resultant production titre are not always correlating with the strength of these genetic parts as they are rather context-dependent [83]. Meaning that a set of parts shown to be optimized for the production of one compound does not guarantee to show the same effects when used to produce another compound. Moreover, tuning the expression of proteins via optimization of inducer concentrations is another strategy that relies on the same principle with genetic part refactoring. 1-Octanol has been produced by *Synechocystis* sp. PCC 6803 for the first time in 2018 via CAR-dependent pathway [10]. Recently, concentrations of the chemical inducer have been optimized and, together with thioesterase selection and light intensity optimization, 526 mg/L of 1-octanol was produced after 12 days [11].

5.10 Protein Engineering

Protein engineering of FAP from *Chlorella variabilis* (hereafter CvFAP) has been investigated to alter the specificity towards propane production in a few model chassis including in *Synechocystis* sp. PCC 6803. It has been shown that mutation of CvFAP at G462A position resulted in ~16-fold increase in propane production when compared with the wild-type strain [53]. All strategies used to enhance the production of free fatty acids and their derivatives from cyanobacteria are illustrated in Fig. 4. Despite such achievements, it should be noted that the production of fatty acid-derived chemicals through engineered cyanobacteria is not well-established yet, compared to other model microorganisms.

Recently, CRISPR technology has been applied to the production of fatty acid derivatives from cyanobacteria. CRISPRi was used to repress PlsX activity in order to eliminate the competitive pathway and increase the acyl-ACP pool for fatty alcohols. In combination with the expression of fatty acyl-CoA/ACP reductase (FAR), it resulted in 10.3 mg/g CDW of octadecanol. Though it should be noted that the use of FAR bypasses the need to synthesize fatty acid intermediate as FAR directly converts acyl-ACP to fatty alcohols [35].

Other strategies similar to strategies used to overproduce fatty acids have also been implemented for the production of fatty acid derivatives. Here, we elucidate a few examples. As mentioned in Sect. 1.3.3, production of odd chain-length alkanes has been reported in cyanobacteria via a synthetic pathway expressing 'TesA and a truncated light-dependent fatty acid photodecarboxylase (FAP) in an *aas* deletion strain with 111 mg/L (77 mg/g CDW) total alkanes being produced [12]. Moreover, the use of light-dependent FAP to generate alkenes was investigated by expressing

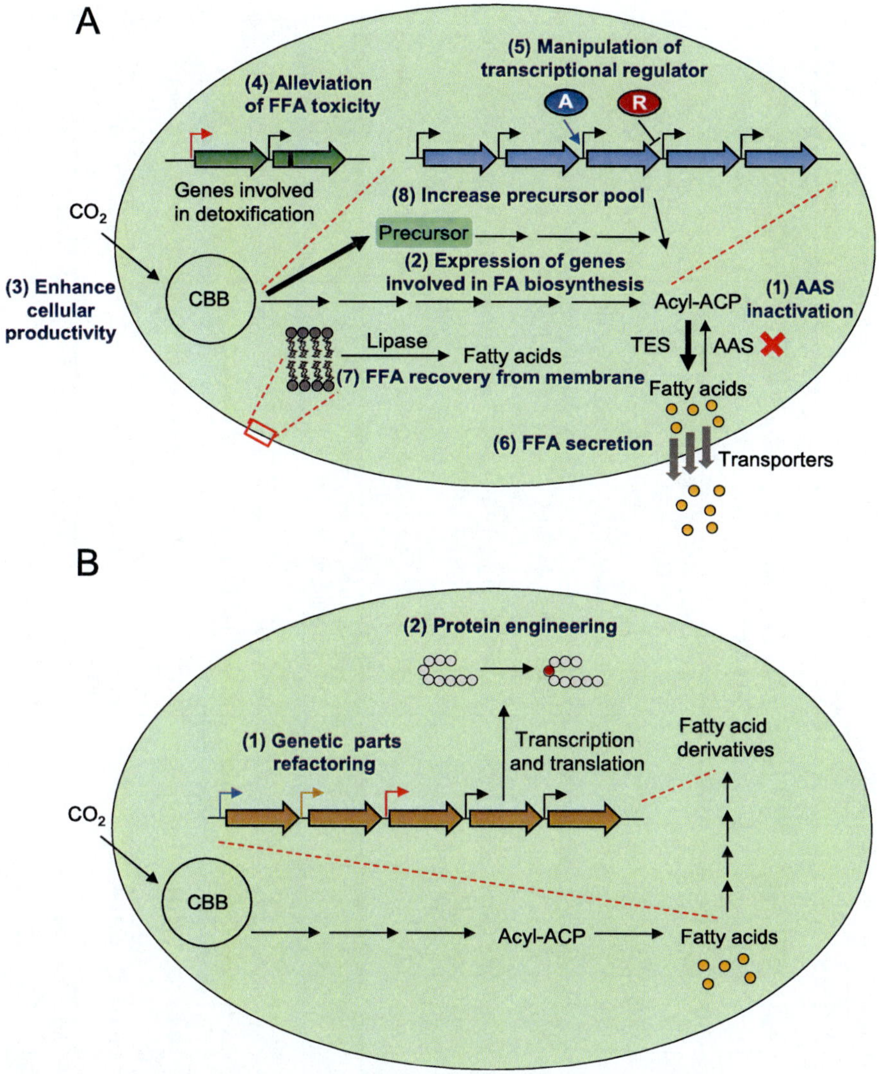

Fig. 4 Simplified diagrams depicting strategies to enhance free fatty acid and derivative production. (**a**) Strategies used for free fatty acid overproduction; (**b**) Strategies used to enhance the production of fatty acid-derived chemicals

FAP from *Chlamydomonas reinhardtii* (*CrFAP*) together with '*tesA* for fatty acid accumulation. This resulted in ~7-fold improvement of 7-heptadecene production compared to the unoptimized system [12]. In addition, removal of chloroplast transit peptides from *Cr*FAP was also shown to contribute to a higher yield [12]. Recent reports on the production of fatty acids and their derivatives are collectively listed in Table 2.

Table 2 Recent examples of fatty acid and fatty acid derivative production from engineered cyanobacteria (2015–2022)

Compounds	Strain	Modification	Cultivation condition	Yield/titre	Reference
Fatty acids					
Extracellular free fatty acids (total)	*Synechocystis* sp. PCC 6803	Overexpression of *aas*, *glpD*, and *rbcLXS*	BG11, 28°C, 50 μmol photons m^{-2} s^{-1}, 10 days	8.2 mg/L	Eungrasamee et al. [62]
Free fatty acids (total)	*Synechocystis* sp. PCC 6803	Δ*cyabrB2* Δ*aas::UcTE*	BG-11 with 5 mM TES-KOH, 40 μmol photons m^{-2} s^{-1} with bubbling of air, 20 days	22.58 mg/L	Kawahara et al. [67]
Free fatty acids (total)	*Synechocystis* sp. PCC 6803	Δ*cyabrB2* Δ*aas::UcTE*	BG-11 with 20 mM HEPES-NaOH, 30°C, 40 μmol photons m^{-2} s^{-1} with bubbling of air, 14 days	~20 mg/L	Kodama et al. [75]
Free fatty acids (total)	*Synechocystis* sp. PCC 6803	ΔP_{II}	BG11 with 5 mM $NaHCO_3$, 27°C, 50–80 μmol photons m^{-2} s^{-1}, 8 h after N-starvation	21.15 nmol/ 1×10^8 cells	Hauf et al. [81]
C12:0	*Synechococcus elongatus* PCC 7942	Δ*FabH:: KASIII* (FabH from *Chaetoceros* GSL56)	A+, RT, ~200 μmol PAR m^{-2} s^{-1} and aerated with atmospheric CO_2, 20 days	80 mg/L	Gu et al. [69]
C8:0-C18:0	*Synechocystis* sp. PCC 6803	Δ*aas-tes3*	BG11-Co, 30°C, 60 μmol photons m^{-2} s^{-1}, 180 rpm, 12 days	43 mg/L	Yunus et al. [12]
C12:0-C18:0	*Synechocystis* sp. PCC 6803	Δ*aas-'tesA*	BG11-Co, 30°C, 60 μmol photons m^{-2} s^{-1}, 180 rpm, 12 days	220 mg/L	Yunus et al. [12]
Fatty alcohols					
1-Butanol	*Synechocystis* sp. PCC 6803	Δ*slr0168-nphT7-phaB-ptaBs*, Δ*phaEC-pduP-slr1192*, Δ*ach-ccr-phaJ-pkPa*	BG-11, 30°C, 50 μmol photons m^{-2} s^{-1} with feeding $NaHCO_3$ every day, 30 days	4.8 g/L	Liu et al. [82]
1-Octanol	*Synechocystis* sp. PCC 6803	Δ*aas-'CpFatB1-4*, *car*, *sfp*	BG11-Co, 30°C, 60 μmol photons m^{-2} s^{-1}, 180 rpm, 12 days	526 mg/L	Yunus et al. [11]
1-Decanol	*Synechocystis* sp. PCC 6803	Δ*aas-'CpFatB1-4*, *car*, *sfp*	BG11-Co, 30°C, 60 μmol photons m^{-2} s^{-1}, 180 rpm, 12 days	54 mg/L	Yunus et al. [11]
C12-OH – C18-OH	*Synechocystis* sp. PCC 6803	Δ*aas-'tesA*, *car*, *sfp*	BG11-Co, 30°C, 60 μmol photons m^{-2} s^{-1}, 180 rpm, 12 days	70 mg/L	Yunus et al. [12]

Hydrocarbons					
Propane	*Synechocystis* sp. PCC 6803	Δ*aas-CvFAP*	BG11$^+$, 30°C with maximal stirring, airflow of 1.21 L/min, 30 μE, automated pH maintenance, 45 days	3.5 mg/g cells	Amer et al. [53]
Fatty alkanes (C11-C17)	*Synechocystis* sp. PCC 6803	Δ*aas-'tesA-'FAP*	BG11-Co, 30°C, 60 μmol photons m^{-2} s^{-1}, 180 rpm, 10 days	77 mg/g CDW	Yunus et al. [12]
1-Alkenes (C15-C17)	*Synechocystis* sp. PCC 6803	Δ*aas-'tesA-undB*	BG11-Co, 30°C, 60 μmol photons m^{-2} s^{-1}, 180 rpm, 10 days	20 mg/g CDW	Yunus et al. [12]
Fatty alkanes (C11-C13)	*Synechocystis* sp. PCC 6803	Δ*aas-'UcFatB1-'FAP*	BG11-Co, 30°C, 60 μmol photons m^{-2} s^{-1}, 180 rpm, 10 days and blue light LEDs (100–150 μE)	25 mg/L	Yunus et al. [9]
Fatty alkanes (C7-C13)	*Synechocystis* sp. PCC 6803	Δ*aas-'CpFatB1.4-'FAP*	BG11-Co, 30°C, 60 μmol photons m^{-2} s^{-1}, 180 rpm, 10 days and blue light LEDs (100–150 μE)	10 mg/L	Yunus et al. [9]
1-Alkenes (C7-13)	*Synechocystis* sp. PCC 6803	Δ*aas-'ChoFatB2.2-undB*	BG11-Co, 30°C, 60 μmol photons m^{-2} s^{-1}, 180 rpm, 10 days	6 mg/L	Yunus et al. [9]
1-Alkenes (C11-13)	*Synechocystis* sp. PCC 6803	Δ*aas-'UcFatB1-undB*	BG11-Co, 30°C, 60 μmol photons m^{-2} s^{-1}, 180 rpm, 10 days	240 mg/L	Yunus et al. [9]
Fatty esters					
Methyl laurate	*Synechocystis* sp. PCC 6803	Δ*aas-'UcFatB1, DmJHAMT*	BG11-Co, 30°C, 60 μmol photons m^{-2} s^{-1}, 180 rpm, 10 days	120 mg/L	Yunus et al. [54]
Octyl acetate	*Synechocystis* sp. PCC 6803	Δ*aas-'CpFatB1-4, car, sfp, atf1*	BG11-Co, 30°C, 60 μmol photons m^{-2} s^{-1}, 180 rpm	2.4 mM	Sattayawat et al. [57]
FAEEs (Total)	*Synechococcus elongatus* PCC 7942	*NSI::Bb1s-atfA-xpkA-pta NSII::Bb1k-pdc-adh*	30°C, 100 μmol photons/m^2/s)	10.0 mg/L/ OD_{730}	Lee et al. [56]
Hydroxy fatty acids					
(ω − 1)-hydroxy fatty acids	*Synechocystis* sp. PCC 6803	*aaKASIII, phaAB,* Δ*phaC*	BG11 with 20 mM HEPES, 34°C, white fluorescent lamps and aerated with 1% (v/v) CO_2-enriched air	2.1 mol% of total fatty acids	Inada et al. [58]

6 Commercial Opportunities to Manufacture Fatty Acids and Derivatives Thereof

Free fatty acids are currently produced via chemical or enzymatic splitting of terrestrial plant-derived fats and oils. As discussed throughout this chapter, recent advances in synthetic biology provide an alternative and possibly more sustainable method for manufacturing fatty acids and their derivatives using cyanobacteria. However, this concept is relatively new, and we are not aware of any current commercial manufacturing of fatty acids or their derivatives using cyanobacteria. Eukaryotic algae-based poly-unsaturated fatty acids, however, can already be found on the nutraceuticals market, although this only utilizes native strains that have not been improved through genetic engineering. Originally, the aim of using microorganisms as a source for fatty acids and derivatives was to reduce the use of palm and coconut oils as the use of these oils is competing with food and land use, not to mention certain environmental problems caused by the production processes [84]. The goals have expanded over time especially since a wider variety of bioproducts are being synthesized from microorganisms with the help of synthetic biology and metabolic engineering. Nevertheless, as promising as this sounds, differing views by various consumer groups on genetically modified organisms (GMOs) have impacted commercialisation of precision-engineered biological species. The experiences of Ecover illustrate how conflicting views could halt the development of GMO-based technology. In 2014, Ecover announced a new range of commercial laundry detergents produced from algae oil as a replacement of palm oil. Shortly after that, however, the activist group ‘ETC’ raised concerns about the use by Ecover of ingredients from precision-engineered organisms. Following this publicity, Ecover decided to withdraw the use of algae oil in their products [85]. This suggests that synthetic biology-based approaches to produce chemicals are still considered a sensitive matter, despite the potential environmental benefits it may bring.

7 Conclusion

Substantial progress in engineering biological manufacturing systems for the production of a wide range of fatty acids and derived products has been demonstrated in research laboratories across the world. As commercial biotechnology employing engineered cyanobacteria is still not a common practice, there are still plenty of challenges ahead before it is possible to switch from unsustainable tropical plant agriculture to theoretically more sustainable manufacturing using cyanobacteria.

References

1. Kenar JA, Moser BR, List GR (2017) Naturally occurring fatty acids. Elsevier Inc. https://doi.org/10.1016/b978-0-12-809521-8.00002-7
2. Sutton MA (2011) Too much of a good thing? Nature 472:5–7. https://doi.org/10.1126/scitranslmed.3008071
3. Kumar RR, Rao PH, Arumugam M (2015) Lipid extraction methods from microalgae: a comprehensive review. Front Energy Res 3:1–9. https://doi.org/10.3389/fenrg.2014.00061
4. Saini RK, Prasad P, Shang X, Keum YS (2021) Advances in lipid extraction methods – a review. Int J Mol Sci 22:1–19. https://doi.org/10.3390/ijms222413643
5. Anneken DJ, Both S, Christoph R, Fieg G, Steinberner U, Wesfechtel A (2012) Fatty acids. Comp Biotechnol:66–78. https://doi.org/10.1016/B978-0-444-64046-8.00151-8
6. Yu AQ, Pratomo Juwono NK, Leong SSJ, Chang MW (2014) Production of fatty acid-derived valuable chemicals in synthetic microbes. Front Bioeng Biotechnol 2:1–12. https://doi.org/10.3389/fbioe.2014.00078
7. Patel A, Karageorgou D, Rova E, Katapodis P, Rova U, Christakopoulos P et al (2020) An overview of potential oleaginous microorganisms and their role in biodiesel and omega-3 fatty acid-based industries. Microorganisms 8. https://doi.org/10.3390/microorganisms8030434
8. Cho IJ, Choi KR, Lee SY (2020) Microbial production of fatty acids and derivative chemicals. Curr Opin Biotechnol 65:129–141. https://doi.org/10.1016/j.copbio.2020.02.006
9. Yunus IS, Anfelt J, Sporre E, Miao R, Hudson EP, Jones PR (2022) Synthetic metabolic pathways for conversion of CO2 into secreted short-to medium-chain hydrocarbons using cyanobacteria. Metab Eng
10. Yunus IS, Jones PR (2018) Photosynthesis-dependent biosynthesis of medium chain-length fatty acids and alcohols. Metab Eng 49:59–68. https://doi.org/10.1016/j.ymben.2018.07.015
11. Yunus IS, Wang Z, Sattayawat P, Muller J, Zemichael FW, Hellgardt K et al (2021) Improved bioproduction of 1-octanol using engineered Synechocystis sp. PCC 6803. ACS Synth Biol. https://doi.org/10.1021/acssynbio.1c00029
12. Yunus IS, Wichmann J, Wördenweber R, Lauersen KJ, Kruse O, Jones PR (2018) Synthetic metabolic pathways for photobiological conversion of CO2 into hydrocarbon fuel. Metab Eng 49:201–211. https://doi.org/10.1016/j.ymben.2018.08.008
13. Wang L, Chen L, Yang S, Tan X (2020) Photosynthetic conversion of carbon dioxide to oleochemicals by cyanobacteria: recent advances and future perspectives. Front Microbiol 11. https://doi.org/10.3389/fmicb.2020.00634
14. Clarke SD, Nakamura MT (2013) Fatty acid structure and synthesis. Elsevier Inc. https://doi.org/10.1016/B978-0-12-378630-2.00038-4
15. Rodriguez GM, Atsumi S (2014) Toward aldehyde and alkane production by removing aldehyde reductase activity in Escherichia coli. Metab Eng 25:227–237. https://doi.org/10.1016/j.ymben.2014.07.012
16. Johannsen J, Baek G, Fieg G, Waluga T (2021) An innovative approach for fatty acid reduction to fatty aldehydes. Green Chem Lett Rev 14:454–460. https://doi.org/10.1080/17518253.2021.1943006
17. Noweck K, Grafahrend W (2012) Fatty alcohols. J Am Oil Chem Soc 31:564–568. https://doi.org/10.1007/BF02638573
18. Hill EF, Wilson GR, Steinle EC (1954) Production, properties, and uses of fatty alcohols. Ind Eng Chem:1917–1921
19. Monick JA (1979) Fatty alcohols. J Am Oil Chem Soc 56:853–860. https://doi.org/10.1007/BF02667462
20. Sánchez MA, Torres GC, Mazzieri VA, Pieck CL (2017) Selective hydrogenation of fatty acids and methyl esters of fatty acids to obtain fatty alcohols – a review. J Chem Technol Biotechnol 92:27–42. https://doi.org/10.1002/jctb.5039
21. Cubo E (2010) Hydrocarbons. Encycl Mov Disord:49–51. https://doi.org/10.1016/B978-0-12-374105-9.00035-6

22. Singh SP, Singh D (2010) Biodiesel production through the use of different sources and characterization of oils and their esters as the substitute of diesel: a review. Renew Sustain Energy Rev 14:200–216. https://doi.org/10.1016/j.rser.2009.07.017
23. U.S. Department of Energy (2009) Biodiesel handling and use guide, fourth edition. National Renewable Energy Laboratory, U.S. Department of Energy, pp 1–56
24. Wang X, Xu X, Wang Q, Huang Z, He J, Qiu T (2020) Fatty acid methyl Ester synthesis through transesterification of palm oil with methanol in microchannels: flow pattern and reaction kinetics. Energy Fuel 34:3628–3639. https://doi.org/10.1021/acs.energyfuels.9b03365
25. Yu A, Zhao Y, Li J, Li S, Pang Y, Zhao Y et al (2020) Sustainable production of FAEE biodiesel using the oleaginous yeast Yarrowia lipolytica. Microbiology 9:1–14. https://doi.org/10.1002/mbo3.1051
26. Chandane VS, Rathod AP, Wasewar KL, Sonawane SS (2017) Efficient cenosphere supported catalyst for the esterification of n-octanol with acetic acid. C R Chim 20:818–826. https://doi.org/10.1016/j.crci.2017.03.007
27. Tomke PD, Rathod VK (2016) Enzyme as biocatalyst for synthesis of octyl ethanoate using acoustic cavitation: optimization and kinetic study. Biocatal Agric Biotechnol 7:145–153. https://doi.org/10.1016/j.bcab.2016.04.010
28. Cao Y, Zhang X (2013) Production of long-chain hydroxy fatty acids by microbial conversion. Appl Microbiol Biotechnol 97:3323–3331. https://doi.org/10.1007/s00253-013-4815-z
29. De Carvalho CCCR, Caramujo MJ (2018) The various roles of fatty acids. Molecules 23. https://doi.org/10.3390/molecules23102583
30. Beld J, Finzel K, Burkart MD (2014) Versatility of acyl-acyl carrier protein synthetases. Chem Biol 21:1293–1299. https://doi.org/10.1016/j.chembiol.2014.08.015
31. Kaczmarzyk D, Fulda M (2010) Fatty acid activation in cyanobacteria mediated by acyl-acyl carrier protein synthetase enables fatty acid recycling. Plant Physiol 152:1598–1610. https://doi.org/10.1104/pp.109.148007
32. Kaiser BK, Carleton M, Hickman JW, Miller C, Lawson D, Budde M et al (2013) Fatty aldehydes in cyanobacteria are a metabolically flexible precursor for a diversity of biofuel products. PLoS One 8:e58307. https://doi.org/10.1371/journal.pone.0058307
33. Akhtar MK, Turner NJ, Jones PR (2013) Carboxylic acid reductase is a versatile enzyme for the conversion of fatty acids into fuels and chemical commodities. Proc Natl Acad Sci U S A 110: 87–92. https://doi.org/10.1073/pnas.1216516110
34. Yao L, Qi F, Tan X, Lu X (2014) Improved production of fatty alcohols in cyanobacteria by metabolic engineering. Biotechnol Biofuels 7:1–9. https://doi.org/10.1186/1754-6834-7-94
35. Kaczmarzyk D, Cengic I, Yao L, Hudson EP (2018) Diversion of the long-chain acyl-ACP pool in Synechocystis to fatty alcohols through CRISPRi repression of the essential phosphate acyltransferase PlsX. Metab Eng 45:59–66. https://doi.org/10.1016/j.ymben.2017.11.014
36. Syuhada R, Noor R, Raja Z, Rahman A, Ha N, Kamarudin A et al (2020) Cyanobacterial aldehyde deformylating oxygenase: structure, function, and potential in biofuels production. Int J Biol Macromol 164:3155–3162. https://doi.org/10.1016/j.ijbiomac.2020.08.162
37. Hu P, Borglin S, Kamennaya NA, Chen L, Park H, Mahoney L et al (2013) Metabolic phenotyping of the cyanobacterium Synechocystis 6803 engineered for production of alkanes and free fatty acids. Appl Energy 102:850–859. https://doi.org/10.1016/j.apenergy.2012.08.047
38. Kageyama H, Waditee-Sirisattha R, Sirisattha S, Tanaka Y, Mahakhant A, Takabe T (2015) Improved alkane production in nitrogen-fixing and halotolerant cyanobacteria via abiotic stresses and genetic manipulation of alkane synthetic genes. Curr Microbiol 71:115–120. https://doi.org/10.1007/s00284-015-0833-7
39. Peramuna A, Morton R, Summers ML (2015) Enhancing alkane production in cyanobacterial lipid droplets: a model platform for industrially relevant compound production. Life 5:1111–1126. https://doi.org/10.3390/life5021111
40. Wang W, Liu X, Lu X (2013) Engineering cyanobacteria to improve photosynthetic production of alka(e)nes. Biotechnol Biofuels 6:1–9. https://doi.org/10.1186/1754-6834-6-69

41. Knoot CJ, Pakrasi HB (2019) Diverse hydrocarbon biosynthetic enzymes can substitute for olefin synthase in the cyanobacterium Synechococcus sp. PCC 7002. Sci Rep:1–12. https://doi.org/10.1038/s41598-018-38124-y
42. Knoot CJ, Pakrasi HB (2019) Diverse hydrocarbon biosynthetic enzymes can substitute for olefin synthase in the cyanobacterium Synechococcus sp. PCC 7002. Sci Rep 9. https://doi.org/10.1038/s41598-018-38124-y
43. Mendez-perez D, Begemann MB, Pfleger BF (2011) Modular synthase-encoding gene involved in α-olefin biosynthesis in Synechococcus sp. strain PCC 7002. Appl Environ Microbiol 77: 4264–4267. https://doi.org/10.1128/AEM.00467-11
44. Elhai J (1993) Strong and regulated promoters in the cyanobacterium Anabaena PCC. FEMS Microbiol Lett:7120
45. Rude MA, Baron TS, Brubaker S, Alibhai M, Del Cardayre SB, Schirmer A (2011) Terminal olefin (1-alkene) biosynthesis by a novel P450 fatty acid decarboxylase from Jeotgalicoccus species. Appl Environ Microbiol 77:1718–1727. https://doi.org/10.1128/AEM.02580-10
46. Rui Z, Harris NC, Zhu X, Huang W, Zhang W (2015) Discovery of a family of desaturase-like enzymes for 1-alkene biosynthesis. ACS Catal 5:7091–7094. https://doi.org/10.1021/acscatal.5b01842
47. Rui Z, Li X, Zhu X, Liu J, Domigan B, Barr I et al (2014) Microbial biosynthesis of medium-chain 1-alkenes by a nonheme iron oxidase. Proc Natl Acad Sci U S A 111:18237–18242. https://doi.org/10.1073/pnas.1419701112
48. Zhou YJ, Hu Y, Zhu Z, Siewers V, Nielsen J (2018) Engineering 1-alkene biosynthesis and secretion by dynamic regulation in yeast. ACS Synth Biol 7:584–590. https://doi.org/10.1021/acssynbio.7b00338
49. Howard TP, Middelhaufe S, Moore K, Edner C, Kolak DM, Taylor GN et al (2013) Synthesis of customized petroleum-replica fuel molecules by targeted modification of free fatty acid pools in Escherichia coli. Proc Natl Acad Sci U S A 110:7636–7641. https://doi.org/10.1073/pnas.1215966110
50. Koeduka T, Matsui K, Akakabe Y, Kajiwara T (2002) Catalytic properties of rice α-oxygenase. A comparison with mammalian prostaglandin H synthases. J Biol Chem 277:22648–22655. https://doi.org/10.1074/jbc.M110420200
51. Huijbers MME, Zhang W, Tonin F, Hollmann F (2018) Light-driven enzymatic decarboxylation of fatty acids. 57:13648–13651. https://doi.org/10.1002/anie.201807119
52. Sorigué D, Légeret B, Cuiné S, Blangy S, Moulin S, Billon E et al (2017) An algal photoenzyme converts fatty acids to hydrocarbons. Science 907:903–907
53. Amer M, Wojcik EZ, Sun C, Hoeven R, Hoeven R, Hughes JMX et al (2020) Low carbon strategies for sustainable bio-alkane gas production and renewable energy. Energ Environ Sci 13:1818–1831. https://doi.org/10.1039/d0ee00095g
54. Yunus IS, Palma A, Trudeau DL, Tawfik DS, Jones PR (2020) Methanol-free biosynthesis of fatty acid methyl ester (FAME) in Synechocystis sp. PCC 6803. Metab Eng 57:217–227. https://doi.org/10.1016/j.ymben.2019.12.001
55. Grillo MA, Colombatto S (2008) S-adenosylmethionine and its products. Amino Acids 34:187–193. https://doi.org/10.1007/s00726-007-0500-9
56. Lee HJ, Choi J, Lee SM, Um Y, Sim SJ, Kim Y et al (2017) Photosynthetic CO2 conversion to fatty acid ethyl esters (FAEEs) using engineered cyanobacteria. J Agric Food Chem 65:1087–1092. https://doi.org/10.1021/acs.jafc.7b00002
57. Sattayawat P, Sofian Yunus I, Jones PR (2020) Bioderivatization as a concept for renewable production of chemicals that are toxic or poorly soluble in the liquid phase. Proc Natl Acad Sci U S A 117:1404–1413. https://doi.org/10.1073/pnas.1914069117
58. Inada T, Machida S, Awai K, Suzuki I (2021) Production of hydroxy fatty acids and its effects on photosynthesis in the cyanobacterium Synechocystis sp. PCC 6803. Algal Res 53:102155. https://doi.org/10.1016/j.algal.2020.102155
59. Kato A, Use K, Takatani N, Ikeda K, Matsuura M, Kojima K et al (2016) Modulation of the balance of fatty acid production and secretion is crucial for enhancement of growth and

productivity of the engineered mutant of the cyanobacterium Synechococcus elongatus. Biotechnol Biofuels 9:1–10. https://doi.org/10.1186/s13068-016-0506-1
60. Liu X, Sheng J, Curtiss R (2011) Fatty acid production in genetically modified cyanobacteria. Proc Natl Acad Sci U S A 108:6899–6904. https://doi.org/10.1073/pnas.1103014108
61. Ruffing AM (2014) Improved free fatty acid production in cyanobacteria with Synechococcus sp. PCC 7002 as host. Front Bioeng Biotechnol 2:1–10. https://doi.org/10.3389/fbioe.2014.00017
62. Eungrasamee K, Incharoensakdi A, Lindblad P, Jantaro S (2020) Synechocystis sp. PCC 6803 overexpressing genes involved in CBB cycle and free fatty acid cycling enhances the significant levels of intracellular lipids and secreted free fatty acids. Sci Rep:1–13. https://doi.org/10.1038/s41598-020-61100-4
63. Swarbrick CMD, Nanson JD, Patterson EI, Forwood JK (2020) Structure, function, and regulation of thioesterases. Prog Lipid Res 79:101036. https://doi.org/10.1016/j.plipres.2020.101036
64. Cho H, Cronan JE (1995) Defective export of a periplasmic enzyme disrupts regulation of fatty acid synthesis. J Biol Chem 270:4216–4219. https://doi.org/10.1074/jbc.270.9.4216
65. Ruffing AM, Jones HDT (2012) Physiological effects of free fatty acid production in genetically engineered Synechococcus elongatus PCC 7942. Biotechnol Bioeng 109:2190–2199. https://doi.org/10.1002/bit.24509
66. Work VH, Melnicki MR, Hill EA, Davies FK, Kucek LA, Beliaev AS et al (2015) Lauric acid production in a glycogen-less strain of Synechococcus sp. PCC 7002. Front Bioeng Biotechnol 3:1–12. https://doi.org/10.3389/fbioe.2015.00048
67. Kawahara A, Sato Y, Saito Y, Kaneko Y, Takimura Y, Hagihara H et al (2016) Free fatty acid production in the cyanobacterium Synechocystis sp. PCC 6803 is enhanced by deletion of the cyAbrB2 transcriptional regulator. J Biotechnol 220:1–11. https://doi.org/10.1016/j.jbiotec.2015.12.035
68. Lu X, Vora H, Khosla C (2008) Overproduction of free fatty acids in E. coli: implications for biodiesel production. Metab Eng 10:333–339. https://doi.org/10.1016/j.ymben.2008.08.006
69. Gu H, Jinkerson RE, Davies FK, Sisson LA, Schneider PE, Posewitz MC (2016) Modulation of medium-chain fatty acid synthesis in synechococcus sp. PCC 7002 by replacing FabH with a chaetoceros Ketoacyl-ACP synthase. Front Plant Sci 7:1–13. https://doi.org/10.3389/fpls.2016.00690
70. Wang X, Xiong X, Sa N, Roje S, Chen S (2016) Metabolic engineering of enhanced glycerol-3-phosphate synthesis to increase lipid production in Synechocystis sp. PCC 6803. Appl Microbiol Biotechnol 100:6091–6101. https://doi.org/10.1007/s00253-016-7521-9
71. Towijit U, Songruk N, Lindblad P, Incharoensakdi A, Jantaro S (2018) Co-overexpression of native phospholipid-biosynthetic genes plsX and plsC enhances lipid production in Synechocystis sp. PCC 6803. Sci Rep 8:1–13. https://doi.org/10.1038/s41598-018-31789-5
72. Ruffing AM (2013) RNA-Seq analysis and targeted mutagenesis for improved free fatty acid production in an engineered cyanobacterium. Biotechnol Biofuels 6:1–15. https://doi.org/10.1186/1754-6834-6-113
73. Kato A, Takatani N, Ikeda K, Maeda SI, Omata T (2017) Removal of the product from the culture medium strongly enhances free fatty acid production by genetically engineered Synechococcus elongatus. Biotechnol Biofuels 10:1–8. https://doi.org/10.1186/s13068-017-0831-z
74. Kizawa A, Kawahara A, Takashima K, Takimura Y, Nishiyama Y, Hihara Y (2017) The LexA transcription factor regulates fatty acid biosynthetic genes in the cyanobacterium Synechocystis sp. PCC 6803. Plant J 92:189–198. https://doi.org/10.1111/tpj.13644
75. Kodama Y, Kawahara A, Miyagi A, Ishikawa T, Kawai-Yamada M, Kaneko Y et al (2018) Effects of inactivation of the cyAbrB2 transcription factor together with glycogen synthesis on cellular metabolism and free fatty acid production in the cyanobacterium Synechocystis sp. PCC 6803. Biotechnol Bioeng 115:2974–2985. https://doi.org/10.1002/bit.26842

76. Dunlop MJ, Dossani ZY, Szmidt HL, Chu HC, Lee TS, Keasling JD et al (2011) Engineering microbial biofuel tolerance and export using efflux pumps. Mol Syst Biol 7:1–7. https://doi.org/10.1038/msb.2011.21
77. Kato A, Takatani N, Use K, Uesaka K, Ikeda K, Chang Y et al (2015) Identification of a cyanobacterial RND-type efflux system involved in export of free fatty acids. Plant Cell Physiol 56:2467–2477. https://doi.org/10.1093/pcp/pcv150
78. Bellefleur MPA, Wanda S-Y, III, R. C. (2019) Characterizing active transportation mechanisms for free fatty acids and antibiotics in Synechocystis sp. PCC 6803. BMC Biotechnol 19:1–17. https://doi.org/10.1186/s12896-019-0505-y
79. Liu X, Fallon S, Sheng J, Curtiss R (2011) CO2-limitation-inducible green recovery of fatty acids from cyanobacterial biomass. 15–18. https://doi.org/10.1073/pnas.1103016108
80. Choi Y, Lee JW, Kim JW, Park JM, Park JM (2020) Acetyl-CoA-derived biofuel and biochemical production in cyanobacteria: a mini review. J Appl Phycol:1643–1653
81. Hauf W, Schmid K, Gerhardt ECM, Huergo LF, Forchhammer K (2016) Interaction of the nitrogen regulatory protein GlnB (PII) with biotin carboxyl carrier protein (BCCP) controls acetyl-Coa levels in the cyanobacterium synechocystis sp. PCC 6803. Front Microbiol 7:1–14. https://doi.org/10.3389/fmicb.2016.01700
82. Liu X, Miao R, Lindberg P, Lindblad P (2019) Modular engineering for efficient photosynthetic biosynthesis of 1-butanol from CO2 in cyanobacteria. Energ Environ Sci 12:2765–2777. https://doi.org/10.1039/c9ee01214a
83. Salis HM, Mirsky EA, Voigt CA (2009) Automated design of synthetic ribosome binding sites to control protein expression. Nat Biotechnol 27. https://doi.org/10.1038/nbt.1568
84. Rodrigues GS, Martins CR, de Barros I (2018) Sustainability assessment of ecological intensification practices in coconut production. Agr Syst 165:71–84. https://doi.org/10.1016/j.agsy.2018.06.001
85. Asveld L, Stemerding D (2016) Algae oil on trial: conflicting views of technology and nature. 1–33

Adv Biochem Eng Biotechnol (2023) 183: 171–252
https://doi.org/10.1007/10_2022_211

Published online: 27 December 2022

Sustainable Production of Pigments from Cyanobacteria

Charu Deepika, Juliane Wolf, John Roles, Ian Ross, and Ben Hankamer

Contents

C. Deepika, J. Wolf, J. Roles, I. Ross, and B. Hankamer (✉)
Institute of Molecular Bioscience, The University of Queensland, Brisbane, QLD, Australia
e-mail: b.hankamer@uq.edu.au

Abstract Pigments are intensely coloured compounds used in many industries to colour other materials. The demand for naturally synthesised pigments is increasing and their production can be incorporated into circular bioeconomy approaches. Natural pigments are produced by bacteria, cyanobacteria, microalgae, macroalgae, plants and animals. There is a huge unexplored biodiversity of prokaryotic cyanobacteria which are microscopic phototrophic microorganisms that have the ability to capture solar energy and CO_2 and use it to synthesise a diverse range of sugars, lipids, amino acids and biochemicals including pigments. This makes them attractive for the sustainable production of a wide range of high-value products including industrial chemicals, pharmaceuticals, nutraceuticals and animal-feed supplements. The advantages of cyanobacteria production platforms include comparatively high growth rates, their ability to use freshwater, seawater or brackish water and the ability to cultivate them on non-arable land. The pigments derived from cyanobacteria and microalgae include chlorophylls, carotenoids and phycobiliproteins that have useful properties for advanced technical and commercial products. Development and optimisation of strain-specific pigment-based cultivation strategies support the development of economically feasible pigment biorefinery scenarios with enhanced pigment yields, quality and price. Thus, this chapter discusses the origin, properties, strain selection, production techniques and market opportunities of cyanobacterial pigments.

Graphical Abstract

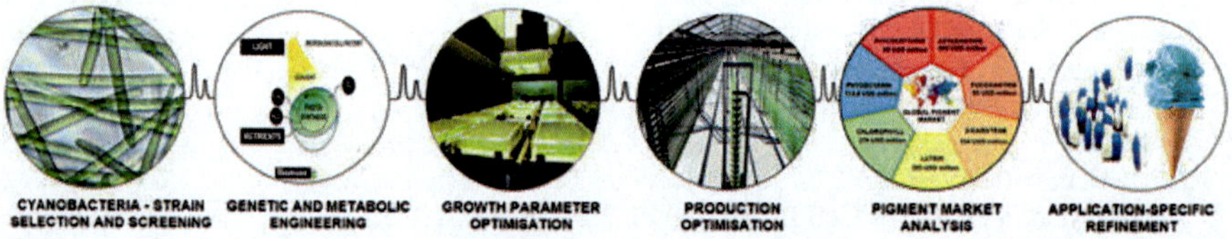

Keywords Astaxanthin, Chlorophyll, Fucoxanthin, Lutein, Phycocyanin, *Spirulina*

Abbreviations

ASE	Accelerated solvent extraction
ATP	Adenosine triphosphate
BDW	Biomass dry weight
CAGR	Compound annual growth rate
Chl	Chlorophyll
Cytb6	Cytochrome b6
EFSA	European Food Safety Authority
ETC	Electron transport chain
Fd	Ferredoxin
FDA	Food and Drug Administration

FNR	Ferredoxin $NADP^+$ reductase
FRP	Fluorescence recovery protein
HPH	High-pressure homogenisation
HRP	High-rate pond
LCA	Life-cycle assessment
L_{CM}	Linker (protein) core membrane
MEP	Methylerythritol phosphate
NADPH	Nicotinamide adenine dinucleotide phosphate
NPQ	Non-photochemical quenching
OCP	Orange carotenoid protein
PAR	Photosynthetically active radiation
PBP	Phycobiliproteins
PBR	Photobioreactor
PC	Phycocyanin
PCB	Phycocyanobilin
PE	Phycoerythrin
PEB	Phycoerythrobilin
PEF	Pulsed-electric field
PLE	Pressurised liquid extraction
PQ	Plastoquinone
PS	Photosystem
PUB	Phycourobilin
PVB	Phycoviolobilin
RC	Reaction centre
$SCCO_2$	Super critical carbon dioxide
TEA	Techno-economic assessment

1 Introduction

Earth formed around 4.6 billion years ago [1] and the Sun remains its largest energy source, delivering 3,020 ZJ $year^{-1}$ to the Earth's surface. The massive scale of this energy supply is highlighted by the fact that every 2 h Earth receives more energy than we need to power our total global economy for an entire year (~0.56 ZJ $year^{-1}$) [2]. Geological records indicate that around 3.4 billion years ago, early anoxygenic photosynthetic organisms evolved [3] using light absorbing pigments, today typified by chlorophylls and carotenoids bound as cofactors to proteins. These organisms were not yet able to catalyse the highly oxidising photosynthetic water splitting reaction of oxygenic photosynthesis. As a result, instead of water, purple bacteria, green sulphur bacteria, acidobacteria and heliobacteria used a range of alternative, available and more energetically accessible substrates as electron donors. These included hydrogen sulphide, dihydrogen, thiosulphate, elemental sulphur and ferrous iron [4]. Of these, early cyanobacteria evolved to use sulphides [5]. About 2.4 billion years ago, a genetic fusion event is thought to have taken place between two bacteria, one with a pheophytin-quinone reaction centre (Type II – an archetypal form of Photosystem II; Q-type) and the other with an iron-sulphur reaction centre (Type I – an archetypal form of Photosystem I; FeS-type) to produce a chimeric photosynthetic organism with two unlinked photosystems [3]. Subsequently, these two archetypal photosystems evolved further and were linked into one operational photosynthetic electron transport chain. Development of the oxygen evolving complex of PSII [6, 7] enabled it to catalyse the most oxidising reaction in biology (water photolysis). This photosynthetic electron transport chain enabled cyanobacteria to use the huge energy resource of the Sun to split water into protons, electrons and oxygen to provide ATP and reducing equivalents such as NADPH [7]. Cyanobacteria remained the principal oxygenic photosynthetic organisms throughout the Proterozoic Eon (2,500 to 541 mya) and are thought to be responsible for the Great Oxidation Event (i.e. the rise of the oxygen concentrations in the atmosphere and oceans [8]). Later, capture of cyanobacteria by eukaryotes expanded oxygenic photosynthesis into a range of other organisms, including red algae, glaucophyta, green algae and higher plants, capable of producing and coordinating a range of pigments involved in photosynthesis to provide the food, fuel, biomaterials and atmospheric oxygen that support aerobic life on Earth [8]. This chapter elaborates on the many pigments coordinated within these intricate cyanobacterial cells and particularly their role in photosynthesis and the economic opportunities that these provide for commercial scale sustainable production platforms across the food, pharmaceutical, biomaterials and primary production (aquaculture and livestock feed) sectors.

Cyanobacteria are commonly referred to as blue-green algae but are strictly speaking microscopic prokaryotic photosynthetic bacteria. They exist as single cells, filaments, sheets or spherical clusters of cells and are found in diverse habitats including fresh, brackish and salt water. Under favourable environmental conditions, cyanobacteria can exhibit high growth rates but can also resist harsh environments through dormancy [9]. Cyanobacteria contain a range of pigments including chlorophylls (green), carotenoids (red, orange and yellow), phycobiliproteins (red and blue) and scytonemin (yellow-brown). These pigments function largely in photosynthesis and photoprotection and have useful properties that can be translated into advanced technical and commercial products [10, 11] and in certain cases (e.g. phycocyanin which has been explored to treat autoimmune encephalomyelitis [12]) are potentially beneficial to human health [13–15] and the environment (through biodegradability) [16].

Pigments are intensely coloured compounds that are used in a broad range of industries to colour other materials. They are extensively used to enhance the attractiveness of industrial products and are usually termed *'pigments'* in the pharmaceutical, ink and cosmetic industries and *'dyes'* in the food and textile industries [17]. They are broadly classified into organic vs. inorganic as well as natural vs. synthetic categories [17]. *Organic pigments* are carbon-based compounds with conjugated chains and rings, either synthetic or natural. *Inorganic pigments* are usually metals and metallic salts that are typically insoluble, heat stable opaque oxides such as Prussian blue (Iron (III) ferrocyanide, produced by the oxidation of ferrous ferrocyanide salts), cobalt blue, cadmium yellow, lead oxide and titanium yellow. *Natural pigments* are mainly organic and include chlorophyll, lutein, β-carotene, astaxanthin, indole based dyes and anthocyanins and are widely used as food colourants (e.g. chlorophyll derivatives) and nutraceuticals (e.g. lutein from marigold flowers used in functional foods) for human consumption [18]. *Synthetic pigments* are usually carbon-based molecules chemically derived from petrochemical products, acids and other chemicals. Even when synthetic pigments are copies of natural products, their activity may not be the same. This is because natural products are often chiral in nature while their synthetic counterparts may be racemic. For example, synthetic astaxanthin produced from petrochemical products (e.g. the Wittig reaction) is reported to provide less antioxidative activity than natural astaxanthin (55x less singlet oxygen quenching capacity and 20x less free radical elimination [19]). Some synthetic pigments (e.g. citrus red II, metanil yellow and rhodamine B) are reported to have various toxicological effects, including carcinogenesis, oestrogenic activity and neurotoxicity [20] which has increased the desirability of natural pigments. Pigments in the food sector are strictly regulated due to health and safety concerns [21, 22]. Synthetic pigments are inexpensive and typically stable, but increasing health and environmental awareness has led to market-driven expansion of the naturally derived pigment sector as part of an expanding circular bioeconomy [23, 24]. In terms of industrial-scale pigment production it is important to note that pigments can be produced as isolated *coloured chromophores* such as chlorophylls, carotenoids and pheophytin (Fig. 1b), phycoerythrobilin (PEB) and phycocyanobilin (PCB; Fig. 1c), or as the *coloured proteins* that coordinate them (e.g. phycoerythrin, phycocyanin and allophycocyanin). To avoid confusion, isolated chromophores are here referred to as *chromophores* and chromophore binding proteins as *coloured proteins*. Collectively, along with other coloured molecules, they are referred to as *pigments*.

The *global pigment market* including both natural and synthetic pigments was estimated to be USD $36.4 billion in 2020 and based on a 5.1% Compound Annual Growth Rate (CAGR) between 2021–2028 is forecast to expand to USD $51.7 billion in 2028 [25]. Different market sectors comprising textiles (62%), leather (10%), printing inks (10%) and others (food, nutraceuticals, pharmaceuticals and cosmetics, 18%) provide significant opportunities for high quality natural pigments. Compared to plant and animal sources, microbial pigment production is more sustainable [26], providing opportunities for the production of biodegradable colourants (e.g. phycocyanin from *Arthrospira platensis* (*Spirulina*)). For large-scale production, cyanobacteria offer specific advantages for pigments unique to cyanobacteria (e.g. phycocyanin and scytonemin) or that they can deliver higher yields (e.g. lutein yields are reported to be three- to sixfold higher than in marigold). Other potential benefits of cyanobacterial systems include lower cultivation time (compared to plants; days/weeks vs season), lower cultivation cost [27], less arable land (ability to use non-arable land and floating systems), low freshwater demand (ability to grow in closed systems using recycled freshwater/seawater/brackish water) and labour requirements [28–30]. Furthermore, cyanobacteria are amenable to genetic engineering to support further improvement.

This chapter focusses specifically on natural pigment production from cyanobacteria – their properties, applications, current extraction technologies and market trends.

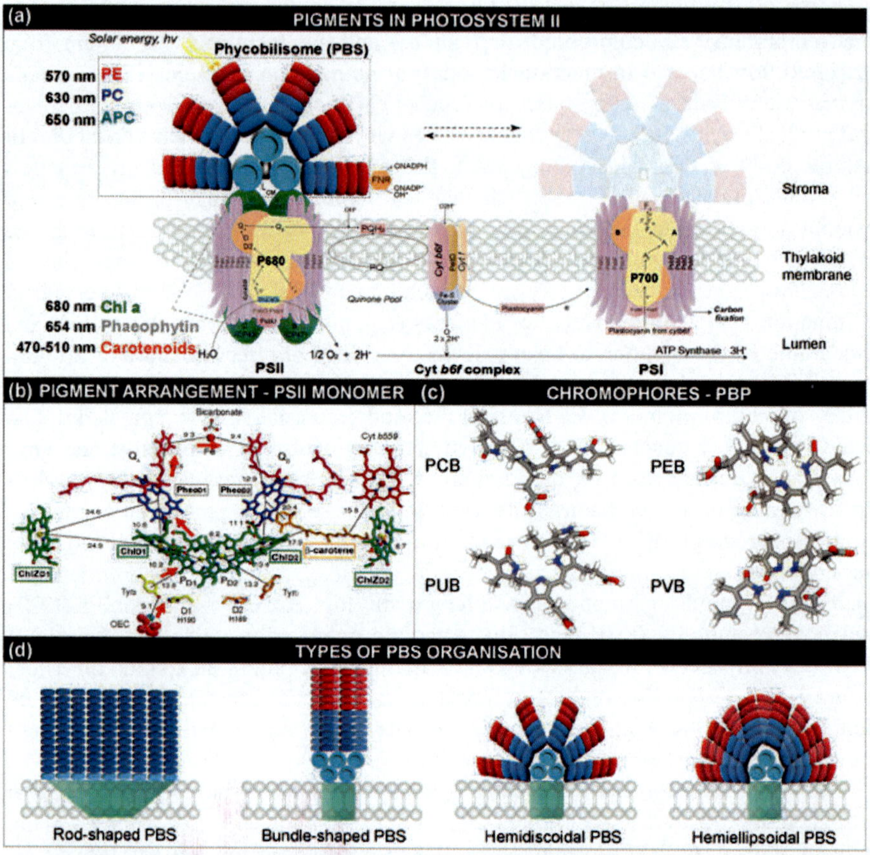

Fig. 1 Cyanobacterial light harvesting antenna and pigment organisation. (**a**) Cyanobacterial photosynthetic electron transport chain including the dynamic extrinsic antenna system consisting of phycoerythrin (PE), phycocyanin (PC), allophycocyanin (APC) is connected to the stromal surface of the PSI and PSII core complexes via the Core-Membrane Linker (L_{CM}). (**b**) Example of pigment coordination within the PSII monomer. (**c**) Four major chromophores in cyanobacteria. The chromophores Phycocyanobilin (PCB; $C_{33}H_{40}N_4O_6$), Phycoerythrobilin (PEB; $C_{33}H_{38}N_4O_6$), Phycourobilin (PUB; $C_{33}H_{42}N_4O_6$) and Phycoviolobilin (PVB; $C_{33}H_{34}N_4O_6$). (**d**) Typical phycobilisome (PBS) organisation: rod-shaped, bundle-shaped, hemi-discoidal and hemi-ellipsoidal. In most cyanobacteria the hemi-discoidal organisation occurs but the pigment composition within these rods is species-specific

2 Cyanobacterial Pigments

The first step of photosynthesis is light capture, which is mediated by the light harvesting antenna proteins of photosystems I (PSI) and II (PSII). These light harvesting antenna systems are designed to capture Photosynthetically Active Radiation (PAR) in the visible spectrum (400–700 nm). In cyanobacteria, these antenna systems consist of pigment-protein complexes located on and in the thylakoid membranes, which lie under the cell membrane (see Fig. 1), typically in a dense multilayered wrapping (Fig. 6, Sect. 5.2). The extrinsic and intrinsic antenna proteins have evolved to provide a dynamic scaffold that coordinates an intricate and excitonically coupled network of chromophores including phycoerythrobilin (PEB; Fig. 1c), phycocyanobilin (PCB; Fig. 1c), phycourobilin (PUB; Fig. 1c), phycoviolobilin (PVB; Fig. 1c), chlorophylls, pheophytins and carotenoids that collectively support the dual function of PSI and PSII light-driven charge separation and photoprotection. The extrinsic antenna systems include the light harvesting protein complexes (phycoerythrin, phycocyanin and allophycocyanin) which usually coordinate the chromophores phycoerythrobilin and phycocyanobilin within them and connect them into the excitonically coupled chromophore network coordinated by the PSI and PSII core complexes [31].

The cyanobacterial PSII core complex is composed of around 20 subunits (Fig. 1a). In 2001 a 3.8 Å resolution PSII core complex structure from *Synechococcus elongatus* was described [32]. Each 350 kDa PSII monomer (Fig. 1b) is reported to contain 17 membrane spanning protein subunits as well, three extrinsic proteins, 99 cofactors, 35 chlorophyll a, 12 β-carotene, 2 pheophytin, 2 plastoquinone and 2 heme molecules, the water splitting Mn_4CaO_5 cluster and one non-heme Fe^{2+} [33]. The electrons extracted from water by PSII are passed, via the cytochrome b_6f complex (a dimer which includes one chlorophyll and one carotenoid per monomer) to PSI, contributing to the generation of an electrochemical gradient across the membrane that drives ATP production [34]. At PSI, photons harvested by its phycoerythrin, phycocyanin and allophycocyanin antenna system are passed on to the PSI core complex to drive charge separation and raise the redox potential of the donated electrons [35]. Specifically, PSI catalyses the light-induced electron transfer from plastocyanin or cytochrome *c6* to ferredoxin or flavodoxin via its chain of electron carriers [36, 37]. The first crystal structure (2.5 Å resolution) of the cyanobacterial *Synechococcus elongatus* PSI complex was also reported in 2001 [38]. Cyanobacterial PSI core complexes are typically trimeric with each monomer core consisting of 12 subunits and 127 cofactors which include 96 chlorophylls, 22 carotenoids, two phylloquinones and three iron-sulphur (4Fe4S) clusters [36, 37]. The subunits collectively stabilise the core-antenna system and help them interconnect with peripheral antenna systems. Within the PSI core is the redox active PSI reaction complex which consists of PsaA and PsaB which coordinate the key intrinsic redox active cofactors in the membrane [37]. Plastocyanin/cytochrome *c6* are soluble electron carrier proteins that donate electrons at the luminal surface of PSI. Cytochrome c6 is likely the evolutionary older electron donor as it can be found in most cyanobacteria [39, 40]. Excitation energy transfer from the antenna chlorophylls leads to excitation of P_{700} to the excited state P_{700}*, which catalyses the primary charge separation [41]. Upon illumination, electrons are transferred from plastocyanin/cytochrome c6 at the luminal surface of the PSI reaction centre to ferredoxin/flavodoxin at the PSI stromal surface.

2.1 Phycobiliproteins

Definition: Cyanobacterial phycobilisomes (PBS) (Fig. 1a) are large organised complexes of water-soluble phycobiliproteins (PBPs), phycoerythrin (PE), phycocyanin (PC), allophycocyanin (APC) and their chromophores [42, 43]. Their chromophores (phycocyanobilin and phycoerythrobilin) are synthesised from glutamic acid, which is converted to aminolevulinic acid (ALA), two molecules of which form porphobilinogen and ultimately protoporphyrin IX by the action of three enzymes (Fig. 2a). The enzyme Fe-chelatase catalyses the formation of protoheme from protoporphyrin IX. Subsequently, this protoheme is converted to biliverdin IX, from which phycocyanobilin and phycoerythrobilin are produced.

Classes: The 3 major PBPs (PE, PC and APC) [35] have been further classified into six groups based on their light absorption and fluorescence properties: phycoerythrocyanin, C-phycoerythrin (C-PE) and R-phycoerythrin (R-PE), C-phycocyanin (C-PC), allophycocyanin (APC) and allophycocyanin-B (AP-B) [35] (Table 1).

Sources: Phycobilisomes (PBS) are unique to cyanobacteria and some red macroalgae [45]. In green microalgae and higher plants they were replaced by transmembrane chlorophyll a/b binding proteins [46]. In cyanobacteria, phycobiliproteins make up a large proportion of soluble proteins; e.g. *Nostoc commune* (54%), *Scytonema* sp. (37%), *Lyngbya* sp. (32%) and *Anabaena* sp. (8%) [47].

Structures & Properties: The PBS consist of water-soluble phycobiliproteins (PBPs) and hydrophobic linker peptides and are classified into 4 structural types which are both species and light-dependent: rod-shaped, hemi-ellipsoidal, hemi-discoidal and bundle-shaped (Fig. 1b). The most common and stable type of PBS organisation is reported to be the hemi-discoidal form (4.5–15 MDa) [48]. It is thought to accommodate a maximum of 800 chromophores per PSII dimer [49]. The bundle-shaped PBS was found in *Gloeobacter violaceus* and reported to support among the fastest energy transfer rates [49]. The rod-shaped PBS was found in *Acaryochloris marina* and the excitation energy transfer is reported to be unidirectional and faster in PS II (compared to hemi-discoidal form) because of its differential organisation of APC and PC [50].

PC is ubiquitous in cyanobacteria and present at high intracellular levels. It consists of two subunits: α-PC (15 kDa) and β-PC (19 kDa). These subunits coordinate three PCBs via thioether bonds within each αβ PC monomer [51]. These αβ PC monomers can in turn form PC trimers $(\alpha\beta)_3$ and hexamers $(\alpha\beta)_6$. The fluorescence of PC has been attributed to the covalent linkage of phycocyanobilin to cysteine-84 of α-subunits as well as cysteine-82 and cysteine-153 residues of β-subunits [51]. These coordinated phycocyanobilins collectively contribute to the high Stokes shift of PC (i.e. the difference between the band maxima of the absorption and emission spectra [51]) and its high quantum yield, with maximum fluorescence emission at ~640 nm, and the molar extinction coefficient at ε_{620} is $1.54 \times 10^6\ M^{-1}\ cm^{-1}$ for a 242 kDa C-PC hexamer [52].

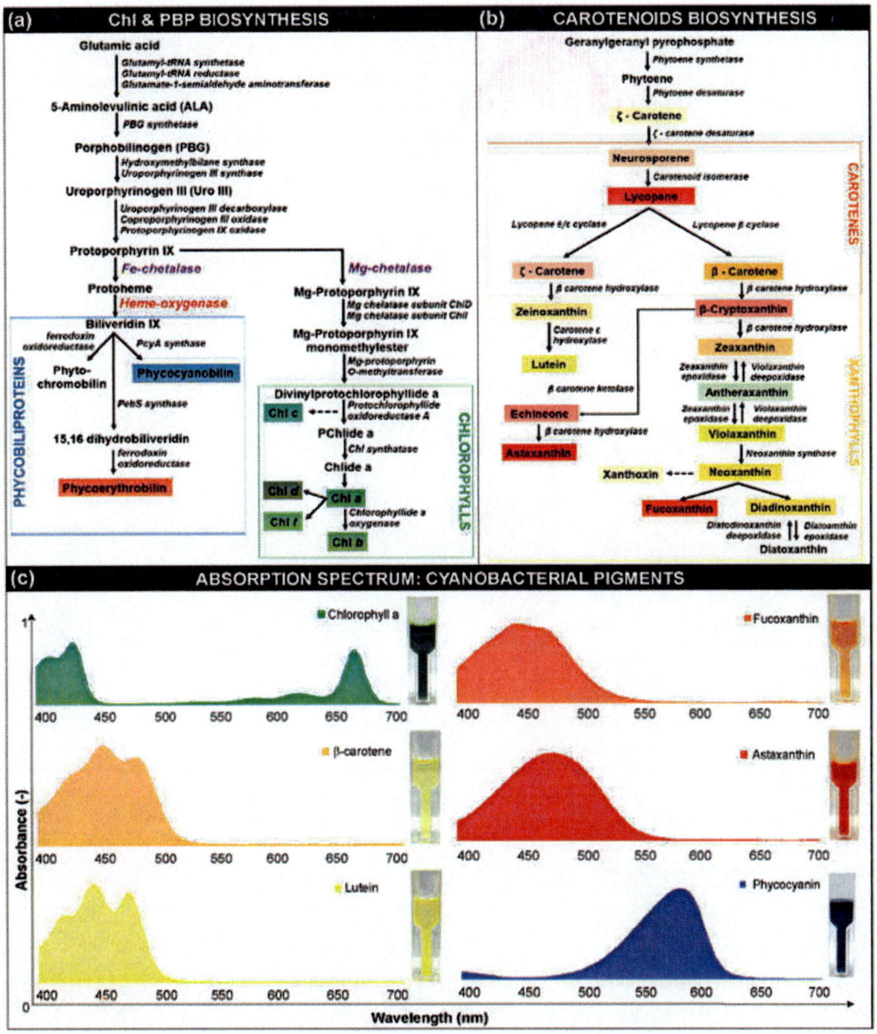

Fig. 2 Cyanobacterial pigments – biosynthesis and absorption spectra. (**a**) Phycobiliprotein and Chlorophyll biosynthesis. The enzymes Fe-chelatase, Mg-chelatase and Heme oxygenase play important regulatory roles in chlorophyll and bilin synthesis. The enzymes PebS synthase and PcyA synthase catalyse key steps in phycoerythrobilin and phycocyanobilin synthesis, respectively, and are either NAD(P)H- or ferredoxin-dependent bilin reductases. During chlorophyll biosynthesis, Mg-chelatase catalyses the insertion of Mg^{2+} into protoporphyrin IX at the branch point between bilin synthesis and chlorophyll biosynthesis [35]. (**b**) Carotenoid biosynthetic pathway via the Methyl-Erythritol 4-Phosphate (MEP) pathway [44]. Phytoene synthase and phytoene desaturase (red dotted boxes) are both important enzymes in carotenoid biosynthesis. The carotenes and xanthophyll pathways are highlighted by the orange and yellow boxes, respectively. (**c**) Absorption spectra of major cyanobacterial pigments of commercial interest – Chlorophyll (Chlorophyll a), Carotenoids (β-carotene, lutein, fucoxanthin, astaxanthin) and Phycobiliproteins (phycocyanin)

Table 1 Phycobiliproteins structure (PDB; scale bar 10 nm) and spectral properties (λ_{exc} – excitation wavelength)

PBP pigments	Structure	Colour	Absorption maxima (nm)	Fluorescence emission maxima (nm)
Allophycocyanin (4RMP)		Bright blue	652	657 ($\lambda_{exc} = 633$)
C-phycocyanin (1HA7)		Dark blue	621	642 ($\lambda_{exc} = 620$)
R-phycocyanin (1F99)		Blue	533,544	636 ($\lambda_{exc} = 580$)
C-phycoerythrin (5FVB)	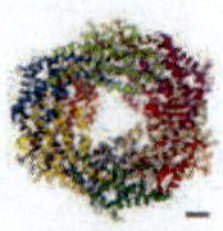	Reddish pink	565	573 ($\lambda_{exc} = 560$)
R-phycoerythrin (1B8D)		Red	566	578 ($\lambda_{exc} = 561$)
B-phycoerythrin (3 V58)		Orange	545	572 ($\lambda_{exc} = 545$)

APC consists of the two subunits α-APC (15 kDa) and β-APC (17 kDa). They coordinate 2 PCB per αβ-APC monomer via thioether bonds [42, 53]. These αβ PC monomers usually form trimeric APC ($(\alpha\beta)_3$). As for PC, the fluorescence of APC has been attributed to the covalent linkage of phycocyanobilin to cysteine-84 of the α-subunit as well as to cysteine-84 and cysteine-155 residues of β-subunit. The APC core (Fig. 1a) is formed by four APC trimers in *Synechocystis* sp. PCC6803 [54] and has a maximum fluorescence emission at ~660 nm, and the molar extinction coefficient at ε_{650} is 0.7×10^6 M^{-1} cm^{-1} for the 104 kDa APC trimer [55].

The two subunits of *PE* named α-PE (20 kDa) and β-PE (22 kDa) are reported to coordinate from 2–6 chromophores via thioether bonds (i.e. 2–6. PEB, PUB or PVB or a combination thereof; Fig. 1) per αβ monomer $(\alpha\beta)_1$ [56]. These αβ-PE monomers are generally organised into disc-shaped trimers $(\alpha\beta)_3$ or hexamers $(\alpha\beta)_6$. As an example, PE in *Gloeobacter violaceus* (PDB: 2VJH) is reported to form hexamers coordinating 4 PEB and 1 PUB per αβ monomer. The maximum fluorescence emission occurs at ~578 nm and the molar extinction coefficient at ε_{578} is $2 \times 10^6\ M^{-1}\ cm^{-1}$ for a 240 kDa R-PE hexamer [52].

PBPs emit an intense autofluorescence which results from their strong light absorption and intense fluorescence emission within the visible spectrum when not coupled into the photosystems [57]. Wynam et al. (1985) [57] reported that a proportion of the light energy is absorbed by PE in PBS of *Synechococcus* sp. DC2 when cultivated under excess nitrate. As a result the cells exhibited high autofluorescence as the PE granules accumulated (as a form of stored nitrogen) and were uncoupled from PBS in the photosystems. Efficient excitation energy coupling among the chromophores in the PBP trimers and hexamers in the PBS contributes to high autofluorescence.

Biological functions: PE, PC and APC absorb radiation in regions of the visible spectrum in which Chl has a low absorptivity (Fig. 2, 470–620 nm). Photosynthetic organisms typically have antenna systems that are tuned to their environmental conditions to best capture the light energy that they require. For example at the illuminated surface of a water column (euphotic zone) PAR in the 400–700 nm range is abundant, while below this (disphotic zone) less red, yellow and green light is available, resulting in dim blue illumination [58]. Consequently, organisms have evolved antenna systems best adapted to capture differing wavelengths of light under a range of light intensities to support optimal light to chemical energy conversion [35, 59]. Phycoerythrin is adapted to capture high energy wavelengths ($\lambda_{max} \sim 565$ nm), phycocyanin intermediate energy wavelengths ($\lambda_{max} \sim 620$ nm) and allophycocyanin low energy wavelengths ($\lambda_{max} \sim 650$ nm) [60]. Their major biological function is to increase the energy absorbed from light and its transfer to the redox active reaction centres and the special pair chlorophylls (i.e. P_{680} in PSII and P_{700} in PSI). In cyanobacteria, they also offer protection against photodamage [61].

2.2 Chlorophylls

Definition: Chlorophylls are tetrapyrrole based chromophores that are generally green in colour.

Classes: Chlorophylls are classified as Chl *a*, *b*, c_1, c_2, c_3, *d* and *f* in the order that they were discovered [62] (Table 2).

Sources: Chlorophylls are abundant in the photosynthetic machinery of cyanobacteria, algae and plants where they are coordinated within specific light harvesting antenna proteins and the redox active reaction centres of PSI and PSII. In cyanobacteria, green plants and green microalgae, Chl *a* is the predominant form of chlorophyll with other chlorophylls usually considered to be accessory chlorophylls. Chl *b* is common in land plants and microalgae while Chl *c* has been reported in marine algae including diatoms, brown algae and dinoflagellates [63].

Table 2 Chlorophyll structure (ChemDraw 20.1.0) and spectral properties. (λ_{exc} – excitation wavelength; NA – Not available)

Chlorophyll pigments	Chemical structure	Chemical formula	Colour	Absorption maxima (nm)	Fluorescence emission maxima (nm)
Chl *a*	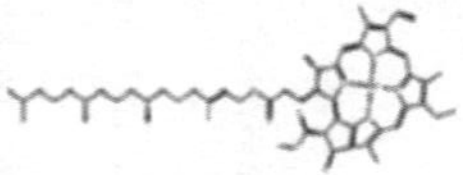	$C_{55}H_{72}O_5N_4Mg$	Blue/green	430,664	668 ($\lambda_{exc} = 430$)
Chl *b*	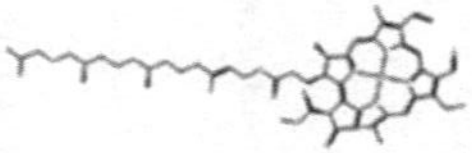	$C_{55}H_{70}MgN_4O_6$	Green/yellow	460,647	652 ($\lambda_{exc} = 453$)
Chl *c1*		$C_{35}H_{28}MgN_4O_5$	Green/ yellow	442,630	633,694 ($\lambda_{exc} = 450$)
Chl *c2*	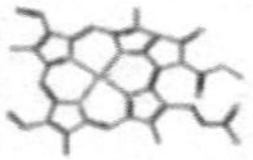	$C_{35}H_{28}MgN_4O_5$	Green/yellow	444,630	635,696 ($\lambda_{exc} = 453$)
Chl *c3*	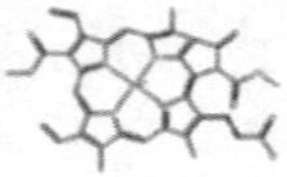	$C_{35}H_{28}MgN_4O_5$	Green/yellow	452,627	635,690 ($\lambda_{exc} = 452$)
Chl *d*		$C_{54}H_{70}MgO_6N_4$	Green	401,696	NA
Chl *f*		$C_{55}H_{70}MgO_6N_4$	Green/yellow	700	720 ($\lambda_{exc} = 425$)

Chl *d* has been reported in certain cyanobacteria, for example in the cyanobacterium *Acaryochloris marina* it makes up 99% of the chlorophyll [64]. Chl *f* was found in extracts from stromatolytes, layered sedimentary formations which are rich in cyanobacteria [65].

Chlorophyll synthesis (Fig. 2a) involves the reduction of protochlorophyllide. Two pathways exist for chlorophyll biosynthesis, one taking place in darkness (using the enzyme dark-operative protochlorophyllide oxidoreductase) and the other requiring continuous light (light-dependent protochlorophyllide oxidoreductase).

Structures & Properties: Chlorophylls *a, b,* c_1, c_2, c_3, *d* and *f* consist of a large aromatic tetrapyrrole macrocycle with a fifth modified cyclopentane, responsible for their light absorption and redox chemistry [66, 67]. A central Mg ion maximises excited state lifetime and the interactions of Chls with their proteins, and in many cases a hydrophobic phytyl tail is present (Chl *a, b, d & f*) although this tail is absent *in* Chl *c1, c2* and *c3* [68]. Chlorophylls differ in their chemical formulae at their C2, C3, C7, C8, C17 positions and in their C17-C18 bonds (Table 2). The only difference between Chl *a* and Chl *b* is that at the C-7 position on the pyrrole ring B, there is a methyl group (–CH_3) in Chl *a*, while in Chl *b* there is a formyl group (–CHO) at the same position. In Chl *d* a formyl group (–CHO) replaces the vinyl group (–$CH{=}CH_2$) at the C-3 position of the pyrrole ring A of Chl *a* (Table 2). In Chl *f* a formyl group (–CHO) instead replaces the methyl group (–CH_3) at the C-2 position of the pyrrole ring A of Chl *a* (Table 2).

Although most chlorophylls absorb in the red (660–665 nm) and blue (~430 nm) regions of the spectrum, these structural differences result in subtle shifts in their respective absorption and fluorescence spectra. Consequently, chlorophylls differ somewhat in their colour: Chl *a* is blue-green (absorbs predominantly violet-blue and orange-red light), Chl *b* is yellow-green, Chl *c*'s are blue-green, *Chl d* is green and absorbs in the far-red region of the spectrum (710 nm, outside of the visible range) as does Chl *f* (yellow-green). The phytyl chains of Chl *a, b, d* and *f* make these chlorophylls oil soluble and give them a wax like consistency as solids [69].

Biological functions: Collectively chlorophylls have four major biological functions including light capture, excitation energy transfer, acting as electron donors, and energy dissipation (Fig. 1a).

Light capture: The first function is to capture light. Different chlorophylls have different absorption spectra. Consequently, by coordinating different combinations of chlorophylls within the antenna systems (e.g. Chl *a* and *b* in the light harvesting systems of microalgae and higher plants) photosynthetic organisms can use chlorophylls to optimise their absorption spectra to capture the light that they require. The broader the absorption spectra and the larger the cross-sectional area of a given antenna, the more light can theoretically be captured [37]. Interestingly in Chl *d* and f the typical red peaks of Chl *a* and *b* are shifted towards the far red (which enables capture of the infra-red portion of the spectrum). Consistent with this it was recently suggested that Chl *f* may function solely as an antenna chromophore [70], but in *Acaryochloris marina*, Chl *d* makes up 99% of the chlorophyll (~80% of total lipid soluble pigment and >2% cell dry weight) suggesting that it also has a role in primary light harvesting in certain organisms [64, 71]. Chl d assists in the capture of far-red light (FRL) and is thus thought to be responsible for remodelling PSI under FRL-induced photoacclimation (FaRLiP) [64].

Excitation energy transfer: The second function of chlorophylls is to support the transfer of excitation energy from the antenna to the redox active Chl *a* dimer (P_{680} and P_{700}) in PSII and PSI reaction centres, respectively. Chlorophylls can support long-lived excited states, making them powerful photosensitisers that play an important role in excitation energy transfer. The safe transduction of this excited state into chemical energy is the basis of photosynthesis. Typically, the absorption spectra shift from blue (shorter/higher energy wavelength) towards the red (longer/lower energy wavelength) towards the reaction centres to facilitate energy transfer.

Electron donor: The third biological function of chlorophylls is to drive P_{680} and P_{700}-mediated redox chemistry. Chlorophylls and chlorophyll derivatives (e.g. pheophytin) can act as primary electron donors and acceptors, transporting electrons within a few picoseconds across half the thylakoid membrane [72]. Here again the ability to support long-lived excited states is important.

Energy dissipation: The fourth function of chlorophylls is photoprotection. Under conditions of excess light, the photosystems and particularly PSII are subject to photodamage due to the formation of reactive oxygen species. To prevent this, certain photosynthetic organisms including higher plants and microalgae have evolved mechanisms to dissipate excess light (up to 85–90%) derived energy from chlorophyll-containing proteins [73].

2.3 Carotenoids

Definition: Carotenoids are lipophilic tetraterpene derivatives which consist of eight isoprene molecules and typically contain 40 carbon atoms [74, 75].

Classes: Approximately 1,100 carotenoids [76] have been reported and these have been categorised into carotenes (hydrocarbons) and xanthophylls, which additionally contain oxygen. The structure and properties of some of the most industrially relevant carotenoids are summarised in Table 3. Of these, the *carotenes* include α-carotene, β-carotene, γ-carotene and lycopene. The *xanthophylls* include lutein, zeaxanthin, neoxanthin, violaxanthin, canthaxanthin, fucoxanthin, antheraxanthin, myxoxanthophyll, β-cryptoxanthin and echinenone.

Sources: Carotenoids are produced by bacteria, fungi, cyanobacteria, algae, plants and animals, where they fulfil a plethora of different roles, but they are most abundant in photosynthetic organisms. Of these 1,100 carotenoids about 30 are reported to have a function in photosynthesis [77]. Consequently, in photosynthetic organisms, these hydrophobic molecules are often enriched in the thylakoid membrane [74]. In higher plants certain xanthophylls (i.e. zeaxanthin, antheraxanthin and violaxanthin) that are involved in the photoprotective xanthophyll cycle and so are located in the light harvesting complexes in the thylakoid membranes. In cyanobacteria, xanthophylls have been reported to be located in the

Table 3 Major carotenoid structures (ChemDraw 20.1.0) and spectral properties

Carotenoid pigments	Chemical structure	Chemical formula	Colour	Absorption maxima (nm)
Carotenes				
α-Carotene		$C_{40}H_{56}$	Light-yellow	378, 400 and 425
β-Carotene	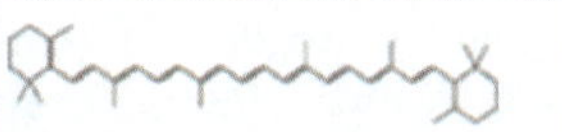	$C_{40}H_{56}$	Orange	425, 450 and 480
γ-Carotene	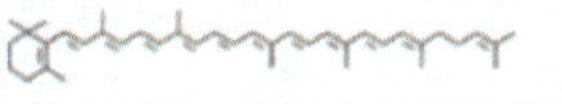	$C_{40}H_{56}$	Yellowish-orange	437, 462 and 492
Lycopene	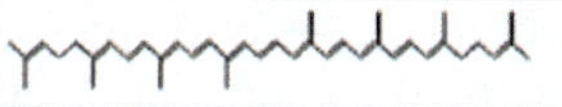	$C_{40}H_{56}$	Red	443, 471 and 502
Xanthophylls				
Astaxanthin	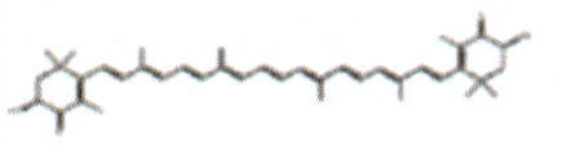	$C_{40}H_{52}O_{24}$	Red	482
Lutein	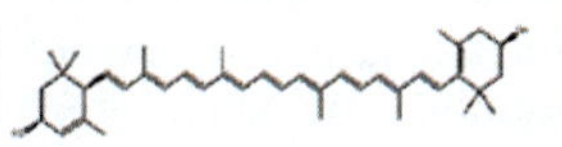	$C_{40}H_{56}O_2$	Yellowish-red	425, 448 and 476
Zeaxanthin	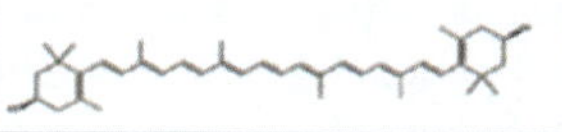	$C_{40}H_{56}O_2$	Yellow	428, 454 and 481
Neoxanthin		$C_{40}H_{56}O_5$	Yellow	486,495
Violaxanthin	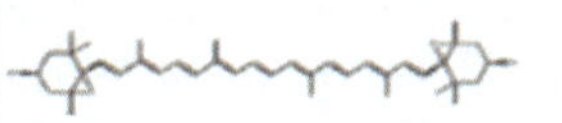	$C_{40}H_{56}O_5$	Orange	417, 440 and 470
Canthaxanthin	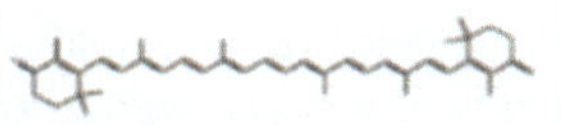	$C_{40}H_{56}O_2$	Yellowish-orange	450, 475 and 506
Fucoxanthin	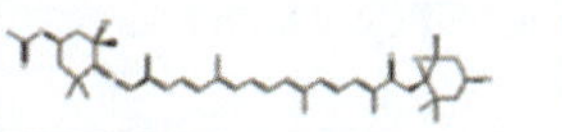	$C_{40}H_{56}O_6$	Orange	423 and 445
Myxoxanthophyll	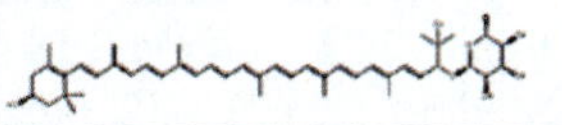	$C_{46}H_{66}O_8$	Bright red	450, 475 and 506
β-Cryptoxanthin	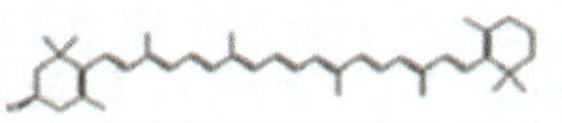	$C_{40}H_{56}O$	Yellowish-orange	425, 449 and 476
Echinenone		$C_{40}H_{54}O$	Brownish-red	452

hydrophobic part of the cytoplasmic membranes [78] but they may also be present in the thylakoids [79].

The carotenoids are typically synthesised from isopentenyl pyrophosphate (IPP) via the methylerythritol-4-phosphate (MEP) pathway in cyanobacteria and in chloroplasts of microalgae and higher plants (Fig. 2a) and via the mevalonic acid (MVA) pathway in the cytosol of bacteria and fungi [77]. Two important enzymes which regulate the first committed steps towards carotene biosynthesis are phytoene synthase and phytoene desaturase. Silencing the genes encoding these enzymes is reported to completely eliminate carotenoid production [80, 81].

Structures & Properties: Carotenoids are unsaturated hydrocarbons with extended conjugated double bond networks that are an essential component of their light absorbing (chromophore) [82] and antioxidant properties [77]. Carotenoids generally absorb light in the violet to green (400–550 nm) region of the spectrum and so tend to be yellow, orange and red in colour [83]. Carotenoids which capture light from shorter wavelengths (e.g. 400 nm) are redder. Their individual colours depend on the length of the polyene component (3–13 conjugate double bond systems) which influences the delocalisation of electrons along the entire length of the polyene chain [72, 77]. The longer the conjugated bond system, the more delocalised the electrons within and the lower the energy required to change state. The range of the light energy captured reduces as the length of the conjugated bond system increases [72, 77]. Xanthophylls, which additionally contain oxygen, may possess hydroxyl groups (e.g. hydroxycarotenoids such as zeaxanthin and lutein), keto groups (canthaxanthin and echinenone) and epoxy groups (violaxanthin and diadinoxanthin) [77]. The structures of some xanthophylls are even more complex, combining several functional groups, for example astaxanthin (keto-hydroxy groups), dinoxanthin and fucoxanthin (epoxy-acetylated groups and allene linkages) and monadoxanthin (acetylene linkages) [21].

Biological functions: Carotenoids are indispensable components of chlorophyll/carotenoid binding photosystems (Fig. 2a) of photoautotrophs (e.g. cyanobacteria, eukaryotic algae and plants) but also have other roles including the protection of membranes from oxidation [79, 84]. In photosynthesis carotenoids have three key roles: *Structural stabilisation* of the photosystems [85], regulation of *light capture* [86] and supporting *energy dissipation* and *photoprotection*, for example through the process of Non-Photochemical Quenching (NPQ) which dissipates excess energy as heat [86].

Structural stabilisation: β-carotene is the only carotenoid reported in the atomic resolution structure of the cyanobacterial PSII complex [84]. For example, *Synechococcus* sp. PCC7335 was reported to have 11–12 β-carotene molecules [87, 88] in PSI (19 β-carotene molecules per monomer of the PSI trimer) when cultivated under far-red light [89]. Carotenoids are reported to assist in maintaining the stability of the PSII structure [90]. For example, the *Synechocystis* sp. PCC 6803, the Δ*crtB* mutant (deletion of the *crtB* gene coding for phytoene synthase) exhibited limited carotenoid biosynthesis and the absence of xanthophylls. Yet although cyanobacterial phycobilisomes, PSII and PSI reportedly lack xanthophyll, these mutants produced intact phycobilisomes while displaying reduced PSI and PSII oligomerisation. Interestingly, xanthophylls reportedly rigidify the fluid phase of the membranes and limit oxygen penetration to the hydrophobic membrane core (susceptible to oxidative degradation) [78]. This is due to the presence of lipid acyl chains in xanthophyll molecules that are responsible for van-der-Waals interactions [78]. In thylakoids, therefore, this may be important for the correct assembly of PSI,

PSII and their antenna systems [79]. It may also be important for the protection of other membranes against oxidative damage.

Light capture: Carotenoids can capture violet-green light. Excited β-carotene molecules that are excitonically coupled to chlorophylls within a light harvesting antenna system can transfer the derived excitation energy to a neighbouring chlorophyll molecule (usually Chl *a*), thereby broadening the absorption spectrum or antenna size of the photosystem [75]. Carotenoids can account for ~20–30% of all light harvested [4, 91].

Energy dissipation and photoprotection: In cyanobacteria, the water-soluble Orange Carotenoid Proteins (OCP) which bind a single carotenoid (3′-hydroxy-echinenone; chromophore) can act as photosensors that can trigger light-activation [92, 93] and quenching of excess light energy in the PBS through the release of excess heat. This can prevent oxidative damage to proteins, DNA and lipids [94]. Absorption of blue-green light induces structural changes in both the protein and carotenoid, which triggers NPQ induction, although the NPQ mechanism is still under active investigation [93]. Under low light or in darkness, OCP converts back to the inactive state. This process has been shown to be mediated by another protein called the Fluorescence Recovery Protein (FRP) that interacts with the active form of OCP and accelerates the reconversion of active OCP to the inactive form [95]. Carotenoids also serve as sacrificial molecules to neutralise reactive species (e.g. oxygen free radicals) [4, 96, 97]. Here, β-carotene helps to quench excess light in the chlorophyll triplet state by releasing it as heat [77]. It is the only carotenoid bound to the core reaction centre complex of photosystem II and offers protection against UV radiation [4, 98]. Zeaxanthin and echinenone are reported to protect the repair stage of the PSII recovery cycle from photoinhibition in cyanobacteria by decreasing the level of singlet oxygen that inhibits protein synthesis [99].

2.4 Scytonemin

Definition: Scytonemin is an aromatic indole alkaloid (Table 4).

Sources: Scytonemin has been reported to accumulate in the extracellular matrix of a broad range of cyanobacteria [100] including species of the genera *Scytonema*, *Aulosira* (*A. fertilissima*), *Nostoc* (*N. linckia*, *N. spongiaeforme*, *N. punctiforme*), *Schizothrix* (*S. coriacea*), *Lyngbya* (*L. majuscule*, *L. aestuarii*), *Leptolyngbya* (*L. boryana*), *Laspinema* (*L. thermale*) and *Chlorogloeopsis* (*C. fritschii*). It has been reported that an 18-gene cluster responsible for scytonemin synthesis in *N. punctiforme* is upregulated upon exposure to UV-A radiation and co-transcribed as a single operon [101].

Structures & Properties: Scytonemin is a secondary metabolite that absorbs UV-C (100–280 nm), UV-B (280–315 nm) and UV-A (315–400 nm) radiation but has a low absorbance in the PAR (400–700 nm) range. It is generally insoluble in water and moderately soluble in organic solvents. Derivatives of scytonemin include scytonine, dimethoxy-scytonemin, tetramethoxy-scytonemin and scytonemin-imine (Table 4) [101, 102].

Table 4 Scytonemin-derivatives structure (ChemDraw 20.1.0) and spectral properties

Scytonemin derivatives	Chemical structure	Chemical formula	Colour	Absorption maxima (nm)
Scytonemin	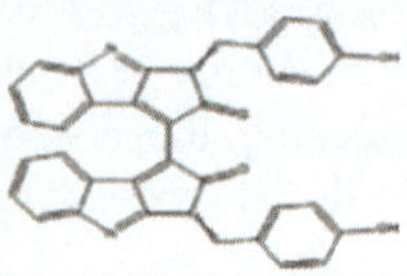	$C_{36}H_{20}N_2O_4$	Yellowish brown	252,278,300,386
Reduced scytonemin	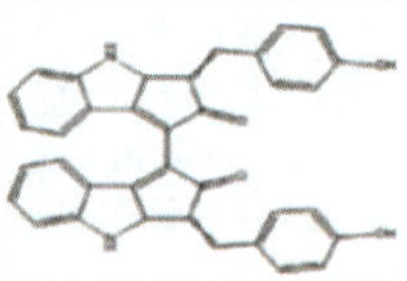	$C_{36}H_{24}N_2O_4$	Bright red	246,276,314,378
Dimethoxy scytonemin	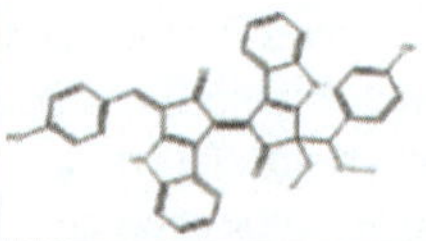	$C_{38}H_{28}N_2O_6$	Red	215,316,422
Tetramethoxy scytonemin	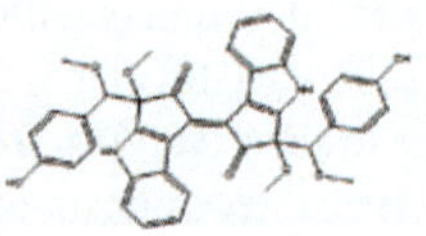	$C_{40}H_{36}N_2O_8$	Purple	212,562
Scytonine	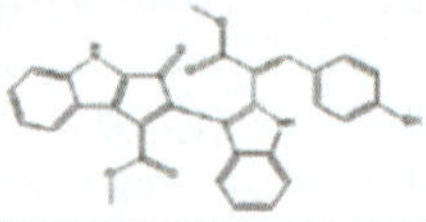	$C_{31}H_{22}N_2O_6$	Reddish pink	207,224,270
Scytonemin-3a-imine	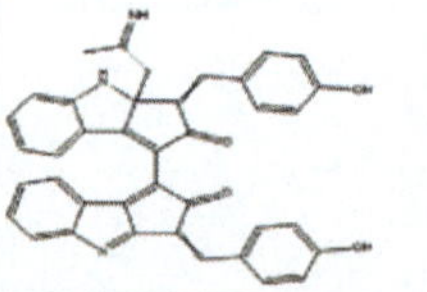	$C_{38}H_{25}N_3O_4$	Reddish brown	237, 366, 437, 564

Biological functions: The location of scytonemin in the extracellular matrix and its UV absorbing and PAR light transmitting properties likely provide cyanobacterial cells with UV protection while allowing PAR light (400–700 nm) into the cell to drive photosynthesis. The energy captured in the UV range is thought to be released as heat [103]. Scytonemin synthesis is induced by high irradiance and most effectively by UV-A and UV-B radiation (~85%) [104]. Cells surrounded by a scytonemin containing sheath [105] exhibited resistance to UV-A induced photobleaching of Chl *a*. In *Chlorogloeopsis* sp., photosynthesis was inhibited and growth delayed until substantial amounts of scytonemin had been deposited in the sheaths [105].

3 Applications

This diverse array of pigments derived from cyanobacteria, i.e. phycobiliproteins (blue and red, Table 1), chlorophylls (green, Table 2), carotenoids (red, orange and yellow, Table 3) and scytonemin (Table 4), can be translated into advanced technical and commercial products [9, 10]. Indeed, cyanobacterial pigments already have a wide range of industrial applications (Fig. 3) especially in the food, cosmetics, nutraceutical and pharmaceutical sectors [17, 106]. Besides their use as colourants and dyes, they are used as food additives, nutraceuticals, putative pharmaceuticals, cosmetics, molecular assays, aquaculture feeds and textiles. One of the first potential

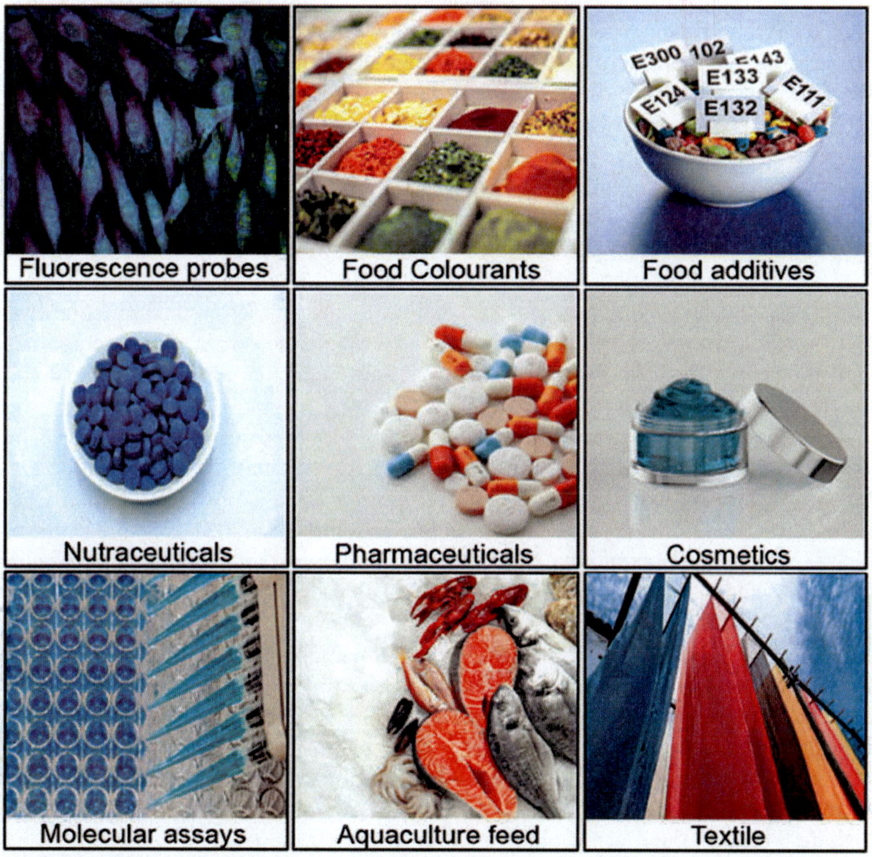

Fig. 3 Applications of cyanobacterial pigments. Cyanobacterial pigments have been reportedly used as fluorescence probes (Single-Cell Imaging – e.g. Supernova 428 dye), food colourants, food additives, nutraceuticals, putative pharmaceuticals, cosmetics, molecular assays, aquaculture feed and textiles

industrial uses for chlorophyll was during experiments in early colour photography by Becquerel (1874) [107] by employing chlorophyll as a photosensitiser of collodion (a flammable, viscous solution of nitrocellulose in ether and alcohol) and silver bromide. Chlorophylls were also used in surgical dressings and as chelators (carriers of micronutrients like cobalt, zinc, manganese, iron and molybdenum) in hydroponics [11, 16, 21].

3.1 Food and Nutraceuticals

Commercially, phycobiliproteins (PBP) are broadly classified into two categories – phycocyanin and phycoerythrin, based on their colour. Phycocyanin has a bright blue colour and is considered versatile, although it is heat and light sensitive. Phycoerythrin is a bright red water-soluble pigment used as a natural food colourant. Both are non-toxic and have been reported to provide antioxidant [108], anti-cancer [109], anti-inflammatory [110], anti-obesity [111], anti-angiogenic [112], neuroprotective [113] and anti-ageing properties [51, 114], though in many cases this may require further study to verify these claims. Phycocyanin is widely used as a natural colourant in ice cream, soft drinks, candies, chewing gum, desserts, cake decorations, icings and frostings, milk shakes as well as lipsticks and eyeliners [51]. Although PBP-rich *Spirulina* extracts are FDA approved (2013) food colourants and additives, they are susceptible to heavy metal contamination and therefore, human use is tightly regulated [115]. Stable isotope labelled metabolites with phycoerythrin have gained attention as fluorescent probes for cytometry and immunodiagnostics [116, 117].

Cyanobacteria can be produced to contain high levels of carotenoids [118]. The global carotenoid market in 2016 was valued at approximately USD 1.24 billion and forecast to increase to USD 1.74 billion by 2025 at a 4.3% CAGR [119]. The market share of the major carotenoids in this sector, anticipated in 2021 is in the order of β-carotene (26%), astaxanthin (25%), lutein (18%), fucoxanthin (15%), canthaxanthin (10%) and lycopene (6%) [120]. The global chlorophyll market was valued to be USD 279.5 million in 2018 and is anticipated to reach USD 463.7 million by 2025 with a 7.5% CAGR from 2018 to 2025 [121]. In Europe, both carotenoids (yellow, orange and red colour) and chlorophyllins (90% of green colour in food) are widely used as food-colouring agents (approved as Group II food additives; authorised by the European Commission).

Carotenoids play an important role in the global food industry as food additives. Of the many known carotenoids, only ~40 are produced commercially. These include β-carotene and astaxanthin, and, to a lesser extent, lutein, zeaxanthin and lycopene. The major carotenoids produced commercially today are β-carotene and astaxanthin, which are currently produced from the commercial strains *Dunaliella salina* (14% β-carotene of dry weight) [122] and *Haematococcus pluvialis* (3% astaxanthin of dry weight), respectively [123]. The largest astaxanthin consumer is the salmon feed industry (FDA approved in 1987) [124]. Astaxanthin is widely used in aquaculture feeds [106] as a colourant for fish and shrimp; the reddish pink pigmentation of salmon is considered an important consumer criterion of quality [125]. The annual aquaculture market of this pigment is estimated at USD 200 million, with an average price of USD 2,500 kg^{-1} [123]. Astaxanthin is also known as 'super vitamin E' as it exhibits the highest antioxidant property (500× more potent than α-tocopherol). Natural carotenoids from cyanobacteria have potential to replace commonly used synthetic colourants such as Erythrosine (pinkish red; E127), Sunset Yellow FCF (yellowish orange; E110), Tartrazine (lemon yellow; E102) and Allura red (red; E129). β-Carotene is used as a food-colouring agent with the E number E160. Lutein (bright yellow) cannot be synthesised by humans and has a protective role against macular degeneration of the eye. It is therefore an important dietary supplement (E161b in the European Union) [126, 127]. Hammond et al. (2014) studied the effect of daily uptake of lutein (10 mg) and zeaxanthin (2 mg) supplement in 100 healthy adults over a period of 1 year and regularly recorded their contrast sensitivity and glare tolerance. The study concluded good improvement in both the parameters and thus suggested lutein and zeaxanthin good for ocular health. Carotenoids are also used in nutraceuticals (e.g. astaxanthin approved by FDA as a human nutraceutical ingredient in 2004 [128]). Carotenoids extracted from *Spirulina* sp. are used to treat vitamin A deficiency, β-carotene and cryptoxanthin being precursors of vitamin A [30, 129].

3.2 Cosmetics

The global pigment-based cosmetic market was valued at USD $10 billion in 2020 and is anticipated to increase to USD $17 billion by 2028 at a ~7% CAGR [130]. The demand for natural pigments in the cosmetic industry has significant traction due to the increasing safety concerns associated with synthetic sunscreen compounds that exhibit cytotoxicity [20, 131]. The interest in cyanobacterial pigments in cosmetics (e.g. sunscreens, creams, lotions) is mainly due to their reported photoprotective property (see biological functions in Sect. 2.4) that prevents skin cancer and suppresses ageing-related skin issues (demonstrated through increased cell viability in keratinocyte cell line HaCat, fibroblast cell line 3T3L1 and endothelial cell line hCMEC/D3 exposed to 10 μg mL^{-1} aqueous cyanobacterial extract containing high levels of phycocyanin) [132]. Scytonemin is a yellow to brown lipophilic pigment that is exclusively found in cyanobacteria and is employed in sunscreens due to their promising effect on protection from UV radiation [104, 105]. Scytonemin is extracted from the cell wall of cyanobacteria cultivated under harsh conditions (e.g. exposure to high solar radiation; desiccation). The UV radiation trigger for natural scytonemin production prevented ~92% of radiation from entering the cell, making it a promising ingredient for cosmetics [110, 133]. Further, the cyanobacterial carotenoids, including β-carotene, fucoxanthin, zeaxanthin, lutein, echinenone, astaxanthin and canthaxanthin also exhibit strong antioxidative properties which help in the reduction of UV-induced oxidative damage [123, 134]. Darvin et al. [135] performed in-vivo carotenoid assays on human skin from healthy normal skin volunteers (20–70 years old) at multiple points over a year and also studied differences in absorption capacity based on the application. They concluded that carotenoids are crucial components of the antioxidative protective system of the human skin and ideally supplied as a topical application. Scarmo et al. [136] demonstrated the effect of carotenoids on skin health by performing dermal biopsies and analysing blood samples to generate a correlation of individual and total carotenoid content in human skin. Carotenoids absorbed in the gut are transported to the epidermis and the two abundant carotenoids found in skin were beta-carotene and lycopene which suggested their role in photoprotection. Lutein and zeaxanthin are marketed as nutraceutical tablets to be ingested and then deposited in lipophilic tissues in humans. Phycobiliproteins have an already established market in the cosmetic sector and are mainly derived from *Arthrospira platensis* (commonly known as *Spirulina platensis*) [51, 137]. Similarly, phycocyanin and phycoerythrin are widely incorporated into hair conditioners, anti-ageing, skin-whitening and anti-wrinkle skin creams and moisturisers, colourant in eye shadow, eye liners, soaps, nail polish and lipsticks [138]. Given the potential of scytonemin in UV screening and free radical scavenging, together with its non-toxic properties [139], this highly stable pigment [133] offers biotechnological opportunities for exploitation by the cosmetics industry [104]. Examples of companies that use cyanobacterial pigments in their cosmetic products today include Lush Cosmetics Pty. Ltd., L'Oreal Pty. Ltd. and Aubrey Organics Inc.

3.3 Pharmaceuticals and Diagnostics

PC is commonly used in immunoassays such as flow cytometry and high-throughput screening [35, 51, 59]. PE is considered one of the world's brightest fluorophores and is widely employed in Time Resolved Laser Induced Fluorescence (TR-LIF), flow cytometry and immunofluorescent staining [140]. Similarly, fluorescent phycobiliproteins are used in fluorescent microscopy, flow cytometry, fluorescence-activated cell sorting, diagnostics, immunolabelling, Fluorescence Resonance Energy Transfer (FRET) assays and immunohistochemistry [59, 60, 137]. Phycobiliproteins are also reported to possess therapeutic properties such as anti-inflammatory and anti-tumour activities [138, 141]. Czerwonka et al. 2018 [142] demonstrated anti-tumour activity of phycocyanin extracts from *Spirulina* sp. Using A549 lung adenocarcinoma cells, and recording cell viability, proliferation and morphology, the cell viability and proliferation of A549 tumour cells were found to be significantly reduced (cell cycle inhibited in G1 phase). The tumour cells were also much more sensitive to PC than the normal skin fibroblasts. Lopes et al. [118] reported the effective treatment of psoriasis using carotenoid extracts from five different cyanobacterial strains from the genera *Alkalinema*, *Cyanobium*, *Nodosilinea*, *Cuspidothrix* and *Leptolyngbya*. HPLC analysis of acetone carotenoid extracts showed high levels of β-carotene, zeaxanthin, echinenone and lutein.

Lutein also has applications in maintaining ocular health, reportedly acting as a photoprotective agent for macular cells [126]. Reynoso-Camacho et al. [15] demonstrated the efficacy of lutein to treat colon cancer in rat models, by investigating the protein expression levels of K-ras (coded by Kirsten rat sarcoma virus gene, responsible for delivering signals to the cell's nucleus), PKB (Protein Kinase-B, regulates cell survival and apoptosis), and β-catenin (regulates cell–cell adhesion and signal transduction) in rats. Lutein treatment reduced these levels by 25%, 32% and 28% in the prevention phase and by 39%, 26% and 26% in the treatment phase. In another study, FloraGLO® Lutein was found to increase the sensitivity/response of transformed and tumour cells to chemotherapy agents, inducing apoptosis in MCF-7 tumour cells [143]. Scytonemin has antioxidant activity and functions as a radical scavenger to prevent cellular damage resulting from reactive oxygen species produced upon UV radiation exposure and thus has potential applications in biomedical products [104]. Scytonemin is reported to repress proliferation of T-cell leukaemia Jurkat cells ($IC_{50} = 7.8\ \mu M$) in humans [61] and to act as an inhibitor of human polo-like kinase 1 (PLK1), the enzyme involved in regulating the G2/M transition in the cell cycle. Zhang et al. (2013) [144] demonstrated the antiproliferative activity of scytonemin (3–4 μmol/l) against multiple myeloma (anti-tumour activity) targeting PLK1 on three different myeloma cell lines (U266, RPMI8226 and NCI-H929). The study concluded that scytonemin significantly decreased cell proliferation. Thus scytonemin could be used as a therapeutic agent for the management of chronic disorders involving inflammation and proliferation (such as Alzheimer's, arthritis and cystic fibrosis) [145]. Consequently, cyanobacterial pigments offer a broad array of opportunities for further evaluation and industrial scale-up to supply existing markets and realise new opportunities.

4 Pigment Production in Cyanobacteria

Cyanobacteria can be used as renewable microbial cell factories [146]. Their optimisation for pigment production requires augmentation of both biomass productivity and pigment yield [11, 17, 147]. The interdependence of these two variables depends on pigment type, and whether the pigments are primary or secondary metabolites. Understanding pigment synthesis pathways and the growth characteristics of production strains are therefore both important.

Cyanobacterial biomass and pigment yields rely on strain-specific characteristics and their alignment with cultivation parameters, such as light intensity and spectral quality [34], the availability of macro and micronutrients [148–150], CO_2 supply [150, 151], temperature [152, 153] and mixing rates [151, 154].

4.1 *Cultivation Parameters and Their Impact on Biomass and Pigment Yields*

4.1.1 Carbon and Energy Supply

The industrial production modes for microbes differ in their supply strategy for carbon (e.g. hetero- and mixotrophic) and energy (e.g. photo-, chemotrophic). *Chemo-heterotrophic organisms* have a metabolic strategy that derives both energy and carbon from organic compounds (chemosynthesis) to enable growth. Thus, the production processes applying chemo-heterotrophs are essentially depending on the organic carbon source, typically sugars, which can add cost (both media costs and the cost of maintaining sterile cultures) and limit viable options for specific-applications. That said *photo-autotrophic* cultures have added costs due to the need for light and CO_2 delivery. Economic and environmental feasibility is thus product-, process and location-specific and can be assessed using techno-economic and life-cycle analysis tools [172].

However, many cyanobacteria are neither completely photo-autotrophic nor completely chemo-heterotrophic; they can perform both photosynthesis and chemosynthesis in a mixed mode of growth called mixotrophy, which has advantages for commercial production. *Photo-heterotrophic* growth is a specific type of mixotrophy, where light is an essential energy source for the cells but can be supplemented with energy derived from the metabolisation of organic carbon compounds, e.g. when growing under light limiting conditions. Under *facultative mixotrophic* growth light is not essential anymore and the organisms can be grown either heterotrophically or autotrophically, and modes can be changed throughout the production process [173]. Under *obligate mixotrophic* growth, the organism utilises both, organic and inorganic carbon (CO_2), simultaneously to support growth and maintenance.

Several studies found that mixotrophic and particularly photo-heterotrophic cultivation modes resulted in higher biomass yields compared to chemo-heterotrophic cultivation [174–178] (Table 5). Schwarz et al. (2020) [179] studied the influence of different growth modes (using different carbon sources; mixotrophic and heterotrophic) on two xenic cyanobacterial strains – *Trichocoleus sociatus* and *Nostoc muscorum*. Mixotrophic cultivation at a light intensity of 100 μmol photons m^{-2} s^{-1} led to the highest biomass concentrations. Glucose was identified as the best organic carbon source for *N. muscorum* (2.46 g L^{-1}) while raffinose was best for *T. sociatus* (3.77 g L^{-1}) [179]. The uptake of complex sugars such as raffinose in cyanobacteria is believed to be mediated through sugar transporters such as the GlcP transporter (fructose/glucose transport system) which was identified in the model organism *Synechocystis* sp. PCC6803 [180] and the ABC fructose transporter which was identified in *Nostoc punctiforme* [181]. *Synechococcus elongatus* PCC7942 was identified to have three different sugar transporters, including galP (glucose), cscB

Table 5 Reported biomass and pigment yields achieved in cyanobacteria

Cyanobacteria strain	Biomass productivity ($g\ L^{-1}\ day^{-1}$)	Pigment productivity ($mg\ L^{-1}\ day^{-1}$)	Reactor type/ scale	Growth condition if different from BG11 (photo-autotroph)	Illumination intensity ($\mu\ mol\ m^{-2}\ s^{-1}$)	Reference
Phycocyanin (PC)						
Spirulina platensis M2	0.18	15.00	300 L Raceway Pond	Zarrouk	Sunlight (Italy)	[155]
Spirulina platensis	0.32	24.00	282 L Raceway Pond	Zarrouk	Sunlight (Italy)	[155]
Spirulina platensis	0.05	3.00	135000 L Raceway Pond	Zarrouk	Sunlight (Spain)	[26]
Anabaena sp. ATCC 33047	0.24	13.00	Raceway Pond	Custom	200	[156]
Spirulina platensis M2	1.32	92.00	11 L Tubular PBR	Zarrouk	Sunlight (Italy)	[157]
Spirulina platensis	0.12	14.00	500 mL flask	Zarrouk	140	[158]
Spirulina platensis	0.06	10.00	Open tank	Zarrouk	30	[159]
Synechocystis sp.	–	12.00	100 mL flask	–	75 (16 h light)	[160]
Spirulina platensis	0.33	50.00	Tubular PBR	Zarrouk	200 (14 h light)	[161]
Spirulina platensis TISTR 8172	0.03	1.3[a]	9 L tank	Zarrouk	Sunlight; white filter	[162]
Spirulina platensis TISTR 8172	0.038	4.3[a]	9 L tank	Zarrouk +16.8 $g\ L^{-1}$ $NaHCO_3$	Sunlight; white filter	

(continued)

Table 5 (continued)

Cyanobacteria strain	Biomass productivity ($g\ L^{-1}\ day^{-1}$)	Pigment productivity ($mg\ L^{-1}\ day^{-1}$)	Reactor type/ scale	Growth condition if different from BG11 (photo-autotroph)	Illumination intensity ($\mu\ mol\ m^{-2}\ s^{-1}$)	Reference
Spirulina platensis TISTR 8172	0.034	6.4[a]	9 L tank		Sunlight; yellow filter	
Spirulina platensis	0.74	13.00	1 L Flat panel PBR	Spirul	50	[163]
Spirulina platensis WH879	0.436	94.00	1 L Flat panel PBR	Zarrouk	450	[164]
Anabaena oryzae SOS13	N/A	0.12	250 mL flask	BG11-N_0	30	[165]
Nostoc sp. LAUN0015	0.057	0.01	500 mL flask	–	156 (12 h light)	[166]
Nostoc sp. UAM206	0.064	0.01				
Anabaena sp. 1	0.141	0.08				
Anabaena sp. 2	0.115	0.06				
Nostoc sp. NK	0.32	57.00	1 L Column PBR	BG11-N_0	100, Red light	[167]
Spirulina sp. S1	0.108	0.07	300 mL flask	–	100	[168]
Spirulina sp. S2	0.057	0.03		BG11 + 0.3% glucose		
Anabaena sp. C2	0.068	0.02		BG11 + 0.15% glycerol		
Anabaena sp. C5	0.059	0.09		BG11 + 0.3% glucose		
Nostoc sp. 2S7B	0.071	0.01		BG11 + 0.3% glycerol		
Nostoc sp. 2S9B	0.024	0.03		BG11 + 0.3% glycerol		

Synechocystis sp. PCC 7338	0.07	0.0006	250 mL flask	ASN-III + 1.2 M NaCl	30	[169]
Nostoc sp. NK	0.32	0.057	1 L column PBR	BG11-N_0	100	[167]
Synechocystis salina LEGE 06,155	N/A	7[a]	5 L flask	Z8 + 25 g L^{-1} NaCl	100 (16 h light)	[170]
Phycoerythrin (PE)						
Anabaena oryzae SOS15	N/A	0.49	250 mL flask	BG11-N_0	30	[165]
Nostoc sp. LAUN0015	0.057	0.0005	500 mL flask	–	156 (12 h light)	[166]
Nostoc sp. UAM206	0.064	0.0003				
Anabaena sp. 1	0.141	0.10				
Anabaena sp. 2	0.115	0.08				
Synechocystis sp. PCC 7338	0.07	0.10	250 mL flask	ASN-III + 1.2 M NaCl	30	[169]
Synechocystis salina LEGE 06,155	N/A	4.3[a]	5 L flask	Z8 + 25 g L^{-1} NaCl	100 (16 h light)	[170]
Allophycocyanin (APC)						
Anabaena oryzae SOS14	N/A	0.28	250 mL flask	BG11-N_0	30	[165]
Spirulina sp. S1	0.108	0.01	300 mL flask	–	100	[168]
Spirulina sp. S2	0.057	0.0004		BG11 + 0.3% glucose		
Anabaena sp. C2	0.068	0.0009		BG11 + 0.15% glycerol		
Anabaena sp. C5	0.059	0.05		BG11 + 0.3% glucose		
Nostoc sp. 2S7B	0.071	0.03		BG11 + 0.3% glycerol		

(continued)

Table 5 (continued)

Cyanobacteria strain	Biomass productivity ($g\ L^{-1}\ day^{-1}$)	Pigment productivity ($mg\ L^{-1}\ day^{-1}$)	Reactor type/ scale	Growth condition if different from BG11 (photo-autotroph)	Illumination intensity ($\mu\ mol\ m^{-2}\ s^{-1}$)	Reference
Nostoc sp. 2S9B	0.024	0.02		BG11 + 0.3% glycerol		
Synechocystis sp. PCC 7338	0.07	0.30	250 mL flask	ASN-III + 1.2 M NaCl	30	[169]
Synechocystis salina LEGE 06,155	N/A	8.7[a]	5 L flask	Z8 + 25 g L^{-1} NaCl	100 (16 h light)	[170]
β-Carotene						
Synechococcus elongatus PCC 7942	0.13	0.70	250 mL flask	–	120	[171]
Synechococcus elongatus R48	0.12	0.60		–		
Synechococcus elongatus RG48	0.91	0.50		–		
Synechocystis salina LEGE 06,155	N/A	0.11[a]	5 L flask	Z8 + 25 g L^{-1} NaCl	100 (16 h light)	[170]
Zeaxanthin						
Synechococcus elongatus PCC 7942	0.13	0.50	250 mL flask	–	120	[171]
Synechococcus elongatus R48	0.12	0.80	250 mL flask	–		
Synechococcus elongatus RG48	0.91	1.10	250 mL flask	–		

Synechocystis salina LEGE 06,155	N/A	0.05[a]	5 L flask	Z8 + 25 g L^{-1} NaCl	100 (16 h light)	[170]
Lutein						
Synechocystis salina LEGE 06,155	N/A	0.14[a]	5 L flask	Z8 + 25 g L^{-1} NaCl	100 (16 h light)	[170]
Echinenone						
Synechocystis salina LEGE 06,155	N/A	0.48[a]	5 L flask	Z8 + 25 g L^{-1} NaCl	100 (16 h light)	[170]
Chlorophyll a						
Spirulina platensis TISTR 8172	0.031	0.41[a]	9 L tank	Zarrouk	Sunlight; white filter	[162]
	0.017	0.44[a]			Sunlight; blue filter	
	0.027	0.48[a]			Sunlight; red filter	
	0.038	0.47[a]		Zarrouk +16.8 g L^{-1} $NaHCO_3$	Sunlight; white filter	
	0.024	0.49[a]			Sunlight; blue filter	
	0.031	0.52[a]			Sunlight; red filter	
Spirulina platensis	0.12	0.24	500 mL flask	Zarrouk	140	[158]
Synechocystis sp. PCC 7338	0.07	0.30	250 mL flask	ASN-III + 1.2 M NaCl	30	[169]

[a] Denotes pigment yields in mg g_{DW}^{-1} day^{-1}

(sucrose) and xylEAB (xylose) [182]. The variability in the carbohydrate uptake rates between strains were attributed to their metabolic activity and the varying membrane permeability to different organic substrates [183]. The mixotrophic cultivation of *Spirulina platensis* using glucose as a carbon source under continuous light yielded the highest biomass (2x that obtained in phototrophic and heterotrophic cultures). This led to the suggestion that photo-driven and oxidative glucose metabolism function efficiently and independently. The photosynthetic pigment content was also found to be 1.5–2× higher in mixotrophic cultures [162, 184, 185].

4.1.2 Key Macro- and Micronutrients Optimisation

Given the diversity of cyanobacteria and their ability to thrive in diverse habitats, it is not surprising that high-efficiency cyanobacterial production requires the optimisation of all species-specific production parameters. In addition to light, CO_2 and water, cyanobacteria also need other macro- and microelements, to enable growth. Strain-specific optimisation of chemical media composition for commercial production is therefore one of the most important processes to increase not only biomass yields and product quality but also economic viability. This in turn reduces the cost and complexity of downstream processing and increases the economic sustainability of the cultivation system.

Collectively, there are 21 elements (C, O, H, N, P, Ca, Mg, K, Cu, Mn, Zn, Fe, Co, Mo, Se, Ni, V, B, Na, Cl and S) and several vitamins broadly needed for cyanobacterial growth [186]. However, bioavailability of each element depends significantly on various factors such as solubility, chemical speciation, pH, temperature, ionic strength, inorganic anions, chelates or interaction with other elements. The biological significance of each nutrient and examples of cultivation impacts on pigment synthesis are given in Table 6.

The *elemental stoichiometry* of phytoplankton (with cyanobacteria being a major constituent) has been reported to be 106C: 16N: 1P (molar ratio) [235], the so-called Redfield ratio. Subsequent studies [236, 237] expanded this ratio and have included trace elements to C(124): N(16): P(1): S(1.3): K(1.7): Mg(0.56): Ca(0.5): Fe (0.0075): Zn(0.0008): Cu(0.0038): Cd(0.00021): Co(0.00019). Many cyanobacterial media formulations (e.g. BG11, Zarrouk) are based on this Redfield ratio [238] assuming that this reflects the essential nutrient requirements of the organism. Such media are most successful in enabling the survival for a vast diversity of cyanobacteria strains, however, for a given species or a specific product target, such media are not necessarily perfectly optimal. Fine-tuning of cultivation medium composition for commercial production can significantly influence product concentration, yield, volumetric productivity as well as overall process economics. Nutrient optimisation is often a laborious, expensive, open-ended and time-consuming process that involves many steps and iterations.

The *selection of culture media component and growth conditions* involve target literature reviews on the selected strain and growth medium to optimise the yield of the final pigment product. Either simple or complex salts may be used. For example, the triple superphosphate ($Ca(H_2PO_4)2H_2O$; ingredient in *Spirulina* sp. growth

Table 6 Elements important in cyanobacterial cultivation and pigment pathways

Nutrient (~abundance in biomass, %w/w)	Biological role	Impact on pigment synthesis	Reference
Macronutrients			
Carbon (20–65%)	Basic component of biomass	Increasing ambient CO_2 supply accelerates growth. Supplementation with organic carbon sources can improve pigment yields High concentration of glucose and glycerol exhibits increase in the production of PBPs in *Anabaena* and *Spirulina* strains	[187–189]
Nitrogen (1–14%)	Required for nucleic acid and protein synthesis	Ammonium toxicity reduces growth rates by disturbing the high inter-thylakoid pH and uncouples photosynthesis *Fischerella* sp. produced more phycobiliproteins under high nitrogen (nitrate or ammonium) conditions	[46, 189]
Phosphorus (0.05–3.3%)	Significance in the production of phospholipids, and nucleic acids Involved in regulatory phosphorylation events, critical for the synthesis of ATP and NADPH Accumulates as polyphosphate granules (used in P-starvation)	Higher concentrations lead to precipitation Phosphate optimisation in *Phormidium ceylanicum* cultures resulted in 2.3-fold increase in PC production	[190–192]
Calcium (0.2–8%)	Integral part of the water splitting manganese cluster in PSII Involved in intracellular signalling and CO_2 fixation Stabilises lipid bilayers Critical to the abiotic and biotic stress related signalling cascades (blooms)	Calcium optimised cultures of *Anabaena fertilissima* PUPCCC 410.5 were reported to have 1.6-fold increase in phycocyanin and 4.5-fold increase in phycoerythrin Calcium was reported to prevent the significant degradation of pigments during high cadmium uptake in *N. muscorum*	[193–197]
Magnesium (0.35–7.5%)	Central atom of all chlorophylls Cofactor for the enzymes involved in Chl synthesis pathway (e.g. Mg-chelatase)	Magnesium starvation was reported to lead to chlorosis in *Synechocystis* sp. PCC6803	[198–202]
Micronutrients			
Iron	Involved in DNA and RNA synthesis, N assimilation and Chl synthesis Component of non-heme and	Increased C-PC (45 mg g^{-1}) was reported in *Euhalothece* sp. KZN with iron optimisation	[203–205]

(continued)

Table 6 (continued)

Nutrient (~abundance in biomass, %w/w)	Biological role	Impact on pigment synthesis	Reference
	heme-containing proteins Crucial part of iron-sulphur proteins (e.g. ferredoxin); necessary for cyclic and non-cyclic photophosphorylation events		
Manganese	Assists proper functioning of malic dehydrogenases, superoxide dismutase and oxalosuccinate decarboxylases It is a key component for water splitting (Mn-cluster of PS II)	Manganese is reported to support the growth of *Anabaena* sp. PCC 7120 under iron-starved conditions (oxidative stress) and showed increased Chl *a* and phycocyanin yields	[206–208]
Copper	Cofactor for enzymes involved in the elimination of superoxide radicals such as ammonia monooxidase, lysyl oxidase and amine oxidases Copper limitation leads to a copper-sparing reorganisation of metabolism and photosynthetic complexes	Increased Cu concentrations reduced the pigment content in *Nostoc muscorum*	[209–214]
Zinc	Maintains membrane integrity Offers protection to the phospholipid membrane bilayer from photodamage Cofactor for a multitude of enzymes including RNA polymerase, carbonic anhydrase and proteases Aids formation of carbohydrates and catalyses the oxidation processes	Zinc stress limited growth rates but increased phycocyanin content in *Spirulina platensis* Higher pigment content was reported in zinc-adapted cells of *Synechococcus* sp. PCC 6803	[209, 215–217]
Boron	Absence inhibits nitrogenase activity in *Nodularia sp.*, *Chlorogloeopsis sp.* and *Nostoc sp.* cultures Stimulates growth rates in the absence of combined nitrogen in *Nostoc muscorum* and *Anabaena cylindrica* Boron deficiency in *Nostoc sp.* leads to chlorosis	Phycocyanin content increased in *Spirulina* sp. under boron-limitation	[218–221]
Cobalt	Cobalt is an integral part of cobalamin (vitamin B_{12}) and helps to convert ribonucleotides to deoxyribonucleotides required for RNA synthesis	*Spirulina sp.* grown in the presence of cobalt ($CoCl_2$) exhibited higher levels of phycocyanin and carotenoids, while showed a decrease in the content of chlorophylls	[222, 223]

(continued)

Table 6 (continued)

Nutrient (~abundance in biomass, %w/w)	Biological role	Impact on pigment synthesis	Reference
Vanadium	Influences chlorophyll synthesis Integral part of V-haloperoxidases	Presence of vanadium stimulated heterocyst formation and resulted in lower pigment content in *Anabaena cylindrica*	[224]
Molybdenum	Essential for nitrate assimilation and nitrate reduction Cofactor for enzymes such as nitrate reductase, molybdopterin adenylyl transferase and xanthine oxidase	Pigment content and nitrogen-fixing activity were higher in cultures containing molybdenum in *Anabaena cylindrica* cultures	[207, 224, 225]
Selenium	Role of a cofactor in enzymes regulating the metabolic pathways Essential for the formation of selenoproteins (oxidoreductases)	High-selenium concentration (450 mg L^{-1}) resulted in both high biomass and high pigment accumulation in *Spirulina platensis* Formation of Se-PC (selenium bound phycocyanin) has higher superoxide and hydrogen peroxide radical scavenging activities than PC	[175, 226, 227]
Counter ions			
Potassium (1.2–7.5%)	Balances the charge in the cytoplasm; controlling the turgor pressure Dominant counter ion (K^+) for the large excess of negative charge on proteins, nucleic acids and lipids	*Microcystis aeruginosa* buoyancy weakened with the increase in the K^+ concentration leading to cell death High K concentrations also led to gas vacuole formation reducing pigment content	[200, 228, 229]
Sodium	Impacts salinity, osmotic stress and membrane transport Essential for the translocation of pyruvate and promotes the biomass growth under K-limited conditions	Sodium glutamate stress in *Spirulina platensis* FACHB-314 resulted in phycocyanin hyperaccumulation	[200, 230, 231]
Chloride	Key role in osmoregulation Balances electrical neutrality in the cells and aids in the uptake macronutrients (N and P)	Increased salinity was reported to increase the carotenoid and allophycocyanin content but decrease the phycocyanin and phycoerythrin content in *Spirulina platensis*	[232–234]

recipe is a mixture of 20% total P (44–48% P_2O_5), 13–15% calcium (Ca) and about 4% residual phosphoric acid (H_3PO_4). The availability of certain elements is frequently hindered by precipitation (e.g. of magnesium salts, forming insoluble $Mg_3(PO_4)_2$) and further complicated by nutrient carryover (e.g. intracellular granules stored in vesicles or from the material of the reactor walls). Thus, understanding the effect of different elemental interactions is essential to determine their availability and perform nutrient optimisation. Additionally, the selection of nutrient components for commercial scale production also involves cost consideration. Commonly used N-sources include nitrate, ammonia and/or urea. To reduce cost, waste streams (e.g. non-toxic or non-pathogenic industrial waste) are sometimes employed to supply nutrients in large scale (depending on the reactor type and final product) [239, 240].

Both *media design and the optimisation strategy* (based on a suitable mathematical model) are pre-requisites to conduct media optimisation experiments. Strategies for media optimisation include component exchange (different sources for the same element), bioavailability controls and culture parameter modifications (e.g. temperature, pH). Media optimisation methods have significantly evolved in the past two decades, from using biomass elemental composition to the use of complete and incomplete factorial statistical approaches (e.g. using approaches such as Plackett-Burman or Box-Behnken designs) [149, 241]. The data analysis for a large dataset with many variables is usually performed using Response Surface Methodology (RSM) to select the best condition and Analysis Of Variance (ANOVA) to establish statistical significance [149, 241].

4.2 Mass Cultivation Systems and Process Management

Mass cultivation of cyanobacteria can be performed in open systems (mixed ponds), closed systems (photobioreactors), or hybrids thereof. Biomass (dry weight) productivities are reported to range from 35 to 70 T ha^{-1} $year^{-1}$ in commercial systems [242–244]. In comparison, soybeans typically yield a harvest of up to ~3.5 T ha^{-1} $year^{-1}$, corn ~10 T ha^{-1} $year^{-1}$ and sugarcane ~70 T ha^{-1} $year^{-1}$ [245].

4.2.1 Open Systems

Open cultivation systems are typically circular raceway ponds and offer simplicity of design, low capital cost and a relatively easy scalability. In commercial production, raceway systems are most common and consist of a circuit of parallel channels in which the microalgae culture is circulated (e.g. by paddle wheels or pumps) [246]. Disadvantages include higher evaporation rates, poor light distribution, dilute cultures which increase the cost of harvesting, nutrient and biomass dilution with rainfall and higher susceptibility to contamination. Advanced pond systems are often called High-Rate Ponds (HRP) and are relatively shallow, mixed by paddle wheels (or equivalent) and the cultivation solution circulates in a circuit leading to reduced energy consumption and water usage, optimised water depths and increased algae biomass yields.

4.2.2 Closed Systems (Photobioreactors)

Closed cultivation systems were mainly designed to overcome the challenges associated with contamination, illumination, harvest efficiency and evaporative water loss in open ponds. Photobioreactors (PBRs) provide a closed (but rarely axenic) environment, which allows better control of culture parameters compared to High-Rate Ponds (HRPs). Different types of PBR (Fig. 4) have been employed to increase the biomass and bioproduct productivity. Closed systems include both indoor (artificial light) and outdoor cultivation (sunlight). Most importantly PBRs are selected based on the target product and the associated need for high quality control to attain regulatory approvals.

Many photobioreactors that differ in design and size have been evaluated at lab, pilot, or commercial scale. Examples include flat panel PBR (used at, e.g., Subitec GmbH Germany; Arizona State University, USA), tubular PBRs (used at, e.g., Roquette GmbH, Kloetze, Germany; University of Almeria, Spain; Microphyt, France) and submerged flat panel systems (used at, e.g., Proviron Inc., USA). PBRs can be further classified into horizontal, inclined, vertical or spiral designs based on the shape and inclination of the PBR. Biofilms or hybrid systems combine features of HRP and PBRs such as floating PBRs (used at, e.g., AlgaeStream SA, France). Each PBR design has its own characteristics, and each differs in mixing and fluid dynamics, light dilution properties, surface area to volume ratio, illumination per footprint area, gas exchange and mass transfer. The main drawbacks for most closed PBR designs compared to open cultivation systems are their high capital cost, high operating costs and scalability challenges. The major advantage of PBR systems is that they achieve higher product yields per unit volume due to the improved supply of light, whether the product is biomass, a secondary metabolite, or an overexpressed protein of interest (e.g. phycocyanin, phycoerythrin). Other advantages include higher culture density, light dilution (allows light to reach deeper areas of a culture via a larger surface area to volume ratio), reduced evaporation, lower contamination, the ability to filter out IR heat load and minimisation of stress which can reduce aggregation and increase product quality. Light dilution and larger surface area to volume ratios through vertical systems minimises photoinhibition (e.g. NPQ) and hence increases photosynthetic conversion efficiencies (PCE) (further discussion in Sect. 4.2.4). PBRs offer the advantage of reproducible cultivation, controlled illumination and spectral quality. Material properties (e.g. durability, spectral quality, UV and thermal resistance, sterilisation efficiency, brittleness) play an important role in production costs and require case-specific analysis.

Generally, to attract investment for cyanobacteria cultivation systems, they should be proven economically viable under operational field conditions, scalable and ideally have a low capital expenditure (CAPEX) and operational expenditure (OPEX).

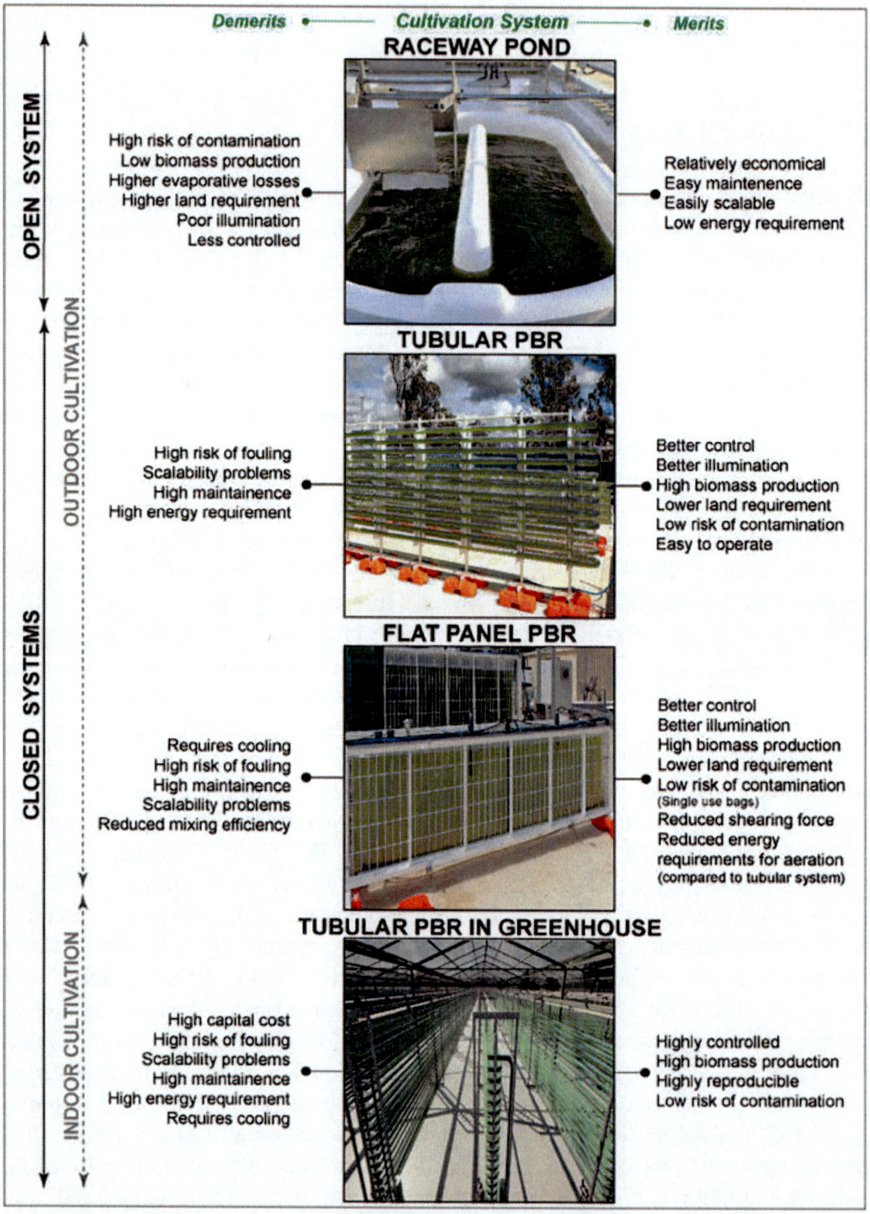

Fig. 4 Cyanobacterial cultivation systems. The different types of cultivation system components are broadly classified into two categories – open/closed production systems and indoor and outdoor cultivation facilities. Open production systems include raceway ponds while closed systems include a range of photobioreactors (PBR) such as tubular and flat panel PBRs. More expensive production systems (e.g. tubular bioreactors) are used to provide higher yields and control, while cheaper systems (e.g. open ponds) tend to be used more for commodity products. Production systems can be used both in indoor and outdoor cultivation facilities depending on the final product requirements. The indoor or closed greenhouse facility installed with tubular PBRs offers a highly controlled

In parallel with economic assessment (Techno-Economic Analysis; TEA), environmental sustainability can be evaluated through comprehensive Life-Cycle Assessment (LCA) by accounting for all energy and material inputs and outputs associated with a particular product or process over all stages of its life cycle: extraction of raw materials, manufacturing, transport, use and recycling or disposition [247]. Life-Cycle Costing (LCC) assesses economic sustainability through similarly comprehensive financial accounting [248].

Photoautotrophic *Spirulina* cultivation in different PBR designs have achieved productivities of 0.40 g L^{-1} day^{-1} (bench-top helical tubular PBR [249]), 0.46 g L^{-1} day^{-1} (tubular PBR [250]), 0.021 g L^{-1} day^{-1} (air-lift PBR [251]) and 0.018 g L^{-1} day^{-1} (bubble-column PBR [251]) and 0.15 ± 0.005 g L^{-1} day^{-1} (low-cost a floating horizontal PBR without mixing [151]). Under photosynthetic conditions both the growth and product accumulation in cyanobacteria are highly light-dependent. Most commercial strains of cyanobacteria are filamentous strains which are often both shear sensitive and extremely adhesive due to their outer mucilaginous sheath, which can cause biofouling and increase the cleaning and sterilisation requirements particularly in tubular PBRs. For example, Zhang et al. (2021) [252] developed a miniature bubble-column PBR (50 L, 60 cm × 60 mm × 137 mm) for *Spirulina* sp. cultivation and achieved a biomass yield of 0.34 g L^{-1} day^{-1} during a 25-day cultivation. Even though globally cyanobacteria cultivation is currently largely conducted in open ponds, higher biomass productivities are achieved in PBRs. In Europe, a 2021 study on commercial microalgae production systems showed that 71% are produced in PBRs, 19% in open ponds and 10% in fermenters [253]. Further biomass and pigment yields in different closed bioreactors are summarised in Table 5.

4.2.3 Performance Comparison, Transfer of Scale and Process Control

Photosynthetic performance of cyanobacteria can be measured in terms of *energy conversion efficiency* (PCE) or *energy conversion rate* (productivity), both of which can be used to compare the performance of different cultivation system designs.

Cyanobacteria culture performance is often defined in terms of growth rate μ (h^{-1} or day^{-1}) which measures the increase in biomass fraction per unit time. However, a high growth rate is not necessarily equivalent to a high productivity P (g m^{-2} day^{-1}). Productivity is the product of specific growth rate and the total biomass (typically expressed as biomass concentration Y, g L^{-1}). The productivity can be expressed as *volumetric biomass productivity* P_{vol} (g L^{-1} day^{-1}; biomass increase per unit reactor

Fig. 4 (continued) environment. However, low-cost open pond systems can be operated in closed environments to enhance control. The advantages and disadvantages of each cultivation system are summarised. (Photographs were obtained from the Centre for Solar Biotechnology, University of Queensland Australia). The rendered image (bottom) provided courtesy of Dr. Fred Fialho Leandro Alves Teixeira (University of Queensland Australia) volume), or as *areal biomass productivity,* either P_{areal} (g m^{-2} day^{-1}; biomass increase per unit reactor footprint) or P_{SA} (g m^{-2} day^{-1}; biomass per unit illuminated surface of reactor, based on surface area to volume ratio). The photosynthetic performance varies during the cultivation process of a batch regime due to self-shading of the cells or aggregated filaments experienced with high biomass density.

Transfer of Scale: Smaller-scale analyses in flasks or microwell plates help to determine the criteria for optimal productivity conditions while large-scale studies provide context and constraints for analyses at smaller scale systems and help to define criteria for the optimisation for high-efficiency systems. At larger scales, engineering parameters become more important and focussed on providing technical solutions for a more economically viable process. Traditionally, system designs and inoculum preparation are often scaled up stepwise in approximately 10-fold volume increases for cyanobacteria. Monitoring culture parameters (light, temperature, pH, CO_2) on a regular basis and logging them using suitable software offers significant benefits to achieve a target culture condition.

Process control aims to maintain the culture at optimal growth conditions to maximise productivity for a given bioreactor design. Growth rates and maximum biomass yields vary for different system designs due to differences in factors such as SA:V ratio and light supply. Successful process control requires suitable dimensioning and drivers of dosing equipment (e.g. nutrients, water, CO_2, base or acid, crop protection agents, anti-foam agent) to balance and maintain process parameters at adequately fast time scales and to attain high energy efficiency. The development of reactor-specific computer simulations may enhance process control reducing material wastage and time. Ideally, growth and production models and machine learning approaches can help to identify which of the 'easy-to-measure' parameters can be used and how they can be implemented to predict culture behaviour and hence optimise process control to reduce costs and increase cultivation robustness.

Process regime: In biotechnological processes, it is possible to maintain a culture at a target growth phase using a continuous cultivation regime (exponential/stationary phase to increase pigment accumulation). In laboratories this is achieved by simultaneously feeding fresh media (feed flow rate F) and harvesting (effluent) the culture at the same rate (inflow = outflow) to keep the culture volume (V) constant. The resulting dilution rate (D) equals the specific growth rate (μ) and is defined by the quotient of the feed flow rate (F) to working volume (V). For a batch regime cultivation, the dilution rate (D) equals zero. Cell aggregation (common in filamentous strains) and product accumulation in the cultivation media can disturb the accuracy of process control. For example, if optical density is used for monitoring culture density cell aggregates may interfere with accuracy. The closer the dilution rate of a steady state is kept to the maximum specific growth rate (μ_{max}), the more difficult it is to maintain a robust cultivation.

In cyanobacteria cultivation platforms, the energy source (solar energy or artificial light) and the carbon source (CO_2) are interdependent, and their supply must be matched to one another. Light serves as the main energy source, being supplied depending on weather conditions, while CO_2 as the main C-source is supplied with the air flow rate ideally in response to available light. Nutrients such as N and P are supplied via the media feed flow (*F*) (Dilution rate, $D = \mu = F/V$) with the aim of maintaining sufficiency. The energy supply is indirectly controlled by the degree of light dilution depending on biomass concentration which makes the process control more difficult compared to heterotrophic cultivation regimes. The biomass concentration varies between different cultivation system designs as the optical properties and hence light energy received by the culture are also influenced by cultivation system optical path length (PBR thickness) or light dilution effects due to the spacing between vertical PBR modules. As a result of periodic fluctuations in the irradiance in outdoor systems (day/night), light availability is often synchronised with the cell division time (circadian rhythm), which makes the prediction models less accurate. Growth models dealing with light and nutrient limitation [254] assist with the development of new concepts to maintain high productivity levels and robust process control during dynamically changing weather conditions. Real-time experimental data can provide feedback to specifically developed models for cyanobacterial pigment production platforms with a selected strain and reactor at a selected geographical location.

4.2.4 Light Supply and Optimisation

In dense cultures, light intensity decreases dramatically with the distance from the illuminated surface, due to self-shading of the cells and light absorption by intracellular pigments. In a well-mixed culture this creates cycles of light and dark phases for each cell, which can be observed in an air-lift reactor, in which the light seems to form a gradient as it penetrates the reactor [255]. Antenna engineering in cyanobacteria, for example through the reduction of the light harvesting antenna size, has the potential to increase the productivity of cyanobacteria cultivation systems at a commercial scale [256].

The illumination intensity determines the amount of light energy available for photosynthesis and thus directly affects the rate of pigment production [148]. As photosynthetic pigments are directly related to and influenced by the composition of the light provided to the culture, optimisation of light intensity and quality is critical for higher pigment yields [257–259]. Light harvesting in cyanobacteria is carried out primarily by phycobilisomes (PBS). The functioning of PBS is continuously modulated to enable adaptation to variations in light (intensity and spectral quality). During high light stress, PBS rapidly saturate the photosynthetic electron transport chain (ETC), which leads to the accumulation of over-excited Chl molecules within the RC, which in turn increases the generation of Reactive Oxygen Species (ROS) which damage the photosynthetic apparatus.

Strategies employed by cyanobacteria under high light stress include:

- Orange Carotenoid Protein (OCP)-dependent NPQ: NPQ of PBS fluorescence occurs in a process mediated by the OCP, which is induced by blue light [260–262].

- State transitions: These regulate the distribution of excitation energy between PSII and PSI [263, 264].
- Quenching of PSI chlorophylls by P700 cation radical or triplet state (based on P700 redox state) [265–267].
- Excitonic delocalisation of the antenna complexes from the RC [268].

Tamary et al. (2012) [269] studied the structural and functional alterations (energetic coupling, stability and membrane association) of PBS induced by high light stress in *Synechocystis* sp. PCC 6803. They identified that high light intensity with white light leads to electronic decoupling of the PBSs due to over-excitation of PBP-chromophores and Chl molecules.

It has been shown that both light intensity and spectral quality affect the phycocyanin content in cyanobacteria [159, 270]. Interestingly, *Spirulina platensis* possesses a very low energy Chl *a* in PSI and only PC in their PBS for energy capture, so PE cannot be produced using this species [271]. High light conditions were found to favour PC accumulation in *Spirulina platensis* [159] (Table 5). Chaiklahan et al. (2022) [272] reported that light optimisation as a cultivation management strategy of a 10 L PBR increased the biomass concentration of *Spirulina* sp. from 0.67 to 1.23 g L^{-1} and the PC content from 16% to 24% by increasing the illumination intensity from 140 to 2,300 μmol m^{-2} s^{-1} demonstrating that cyanobacterial pigment production is highly dependent on the illumination intensity and exposure time (12:12 light:dark cycle).

4.2.5 Salinity and pH

The availability of saline, brackish or wastewater streams at a cultivation site can significantly reduce the 'freshwater' consumption of a cyanobacterial system and improve its competitiveness. In large-scale continuous production systems salinity levels must be maintained within prescribed limits, therefore blowdown of water is required to remove excess salts. The vast amount of counter ions (e.g. Na, Cl) from supplied nutrients (if applied as salts) remain in the water as the nutrients are taken up by the microbes (e.g. N, P, Mg, Ca). Their concentration is further increased by evaporative water losses. The use of closed bioreactor systems offers the potential to increase efficiency, minimise evaporation and enable water and nutrient recycling. The challenge is to do so cost effectively.

Salinity levels play a significant role both in biomass and pigment productivity in cyanobacteria [231, 233, 273]. Strain-specific optimisation of salinity is crucial for proper cell function, filament elongation, metabolic activity, ion regulation (membrane potential) and osmotic balance (turgor pressure in gas vacuoles) [274]. Increases in salinity have been reported to have adverse effects on non-tolerant cyanobacteria and are indicated to cause inhibition of electron transport [233]. For example, it is thought that high levels of salinity lead to a higher influx of Na^+ ions which in turn induce PBS detachment from the PSI/PSII in the thylakoid membrane, reducing photosynthetic activity and thus lowering growth rates [233].

Strategies employed by cyanobacteria to survive salt stress include:

1. Na^+/H^+ antiport – Reduces the uptake of Na^+ ions and promotes an active efflux [275].
2. Enhanced antioxidative defence system – Triggers the expression of salt-induced and osmotic-induced proteins to tolerate salt stress [276].
3. Active extrusion of toxic inorganic ions and the accumulation of compatible solutes (to compensate the difference in water potential) [277] which are low-molecular mass organic compounds (e.g. sucrose, trehalose and glycine betaine), that do not have a net charge and can be accumulated in high (molar) amounts without negatively interfering with cellular metabolism [278].

Salt stress modulates the composition of phycobilisomes (PBS; PE:PC ratio). *Anabaena* sp. NCCU-9 cultivated under low salinity levels (~10 mM) was reported to have increased PBP content [279]. Abd El-Baky et al. [280] reported that C-PC productivity and the antioxidant capacity were higher in *Spirulina maxima* cultures cultivated under high salinity levels (Zarrouk medium supplemented with 0.1 M NaCl). Lee et al. [169, 202] studied the effect of salt stress on *Synechocystis* sp. PCC 7338 cultivated in ASN-III medium supplemented with 1.2 M NaCl (high salinity) and achieved an increased yield of *Chl a* (4.18 mg L^{-1}), PE (1.70 mg L^{-1}) and APC (4.08 mg L^{-1}).

Similar to salinity, the pH of a culture medium affects cyanobacteria growth and is altered during the cultivation process by the supply and uptake of CO_2 and nutrients. Many studies have reported the effect of pH on the growth of cyanobacteria and identified that the optimum pH for mostly used strains to date generally ranged between 7.4 and 9 [153, 281, 282]. However, some cyanobacteria are extremophiles that prefer highly alkaline or more acidic conditions, which can be used as a competitive advantage in the cultivation regime for contamination control.

4.2.6 Temperature

Cyanobacteria, with the ability to perform adaptive cell differentiation, are known to survive in a diverse range of temperatures (−20–70°C). These temperature-tolerant cyanobacteria are classified into 4 groups – psychrophilic (−20–10°C), psychrotrophic (>20°C), mesophilic (50°C) and thermophilic (>80°C). The fatty acid composition, fluidity and integrity of the membrane changes, based on the temperature. High temperature stress inhibits photosynthetic machinery and results in uncoupling of PBS [283]. The heat shock proteins (Hsps; Hsp_{100} in unicellular cyanobacteria, e.g. *Synechocystis* sp. and Hsp_{60} in filamentous cyanobacteria, e.g. *Anabaena* sp.) function as chaperones and assist in protein refolding required for high-temperature tolerance [284]. The HtpG protein from the Hsp_{90} family protects the photosynthetic apparatus by interacting with PBS, preventing PBP aggregation [284]. At low temperatures, cyanobacteria were observed to desaturate membrane fatty acids and induce enzymes that improve transcription and translational efficiency. Tiwari et al. (2016) [285] reported that heat stress (45°C) reduced the pigment content in *Anabaena* sp. PCC7120, but this effect was countered by the addition of calcium to the cultures (0.25 mM Ca supplementation in BG11; increased PC, Chl *a* and carotenoid levels).

4.2.7 Mixing and Shear Sensitivity

In most PBR systems, mixing is coupled to aeration and degassing to balance aerobic conditions and inhibiting oxygen concentrations in the culture. Mixing is also needed for the optimal nutrient distribution and, in contrast to heterotrophic cultivations, for optimal light penetration as it avoids sedimentation and self-shading of cells [286]. The sensitivity to mixing is highly strain-specific for cyanobacteria due to their range of morphologies (unicellular, colonial and filamentous). Ravelonardo et al. [154] examined the effect of agitation on biomass growth of *Spirulina platensis*, comparing air-lift systems, pumping and mechanical stirring methods for mixing. They conclude that filamentous cells were highly fragile and achieved the highest biomass productivity (1.8 g L^{-1}) in the mixing regime with lowest shearing force, a bubble-column reactor without additional mixing. Xiao et al. [287] reported that both unicellular (*Microcystis flos-aquae*) and filamentous (*Anabaena flos-aquae*) cyanobacteria can modulate their growth rates in response to the mixing rates via asynchronous cellular stoichiometry of C, N and P, for better nutrient uptake. Further research in association with shear regime and growth rate-dependent sensitivity to turbulence would improve the understanding and optimisation of mixing in commercial-scale ponds and PBRs.

5 Downstream Processing

Pigment extraction requires biomass dewatering to harvest cells, cell disruption to release the pigment followed by pigment extraction and purification. These steps are further elucidated below.

5.1 Biomass Harvesting

The first step of biomass harvesting (dewatering) describes the separation of solids (cells) which are mixed in a dilute suspension, from the liquid phase (media). Dewatering efficiency depends on several factors including viscosity, particle size and density, specific gravity of the particle compared to the medium. The choice of technique depends on the properties of the cyanobacterial species and the final product requirements. The dewatering strategy of an industrial-scale process impacts both economic viability and product quality, while it must be aligned with the other processing steps, such as lysis, extraction and refinement.

Cyanobacterial cells in culture can generally be considered to be particles whose stability is due to surface charge (electronegative; *p*H of 2.5–11.5 [273, 288]), steric effects (due to water molecules bound to the microalgal surface) and adsorbed macromolecules or extracellular organic matter. When compared to other particles in suspension, cyanobacteria species differ in characteristics such as size, shape and motility that each influences their harvesting behaviour. The main techniques currently employed in microalgae harvesting include flocculation, gravity sedimentation, flotation, electrophoresis techniques, filtration and screening as well as centrifugation. The performance of each dewatering technique can be quantitatively evaluated by the rate of water removal, the solid content of the recovered cyanobacteria-water slurry and the efficiency/yield of the dewatering technique.

Sedimentation can be applied as the first step of dewatering (Fig. 5a). During sedimentation different materials are separated from one another based on their density and/or particle size [289]. Gravity sedimentation naturally separates a feed suspension into a concentrated slurry and clear liquid. Harvesting by sedimentation at natural gravity can be accomplished via lamella separators (plates installed to increase settling area) and sedimentation tanks. In these systems the highest energy demand is related to pumping the slurry. Typically higher biomass concentrations result in improved sedimentation rates and 95% biomass recovery has been reported after 24 h of settling for *Spirulina platensis* [290]. However, the settling rate is low compared to other dewatering techniques, due to the small difference in density between water (freshwater = 1,000 kg m^{-3} or saltwater = 1,025 kg m^{-3}) and cyanobacteria (1,040–1,140 kg m^{-3}) [291]. Collectively these properties make sedimentation a low-cost but time-consuming process.

Flocculation is used to increase the efficiency of sedimentation or flotation-based dewatering (Fig. 5b). Here, a particle in a solution forms an aggregate with other particles to form flocs [292–294]. Flocculation occurs when the solute particles interact and adhere to each other. Chemical flocculation can be induced by inorganic flocculants (e.g. alum, ferric sulphate, lime) [294] or organic polymer and polyelectrolyte flocculants (e.g. Purifloc, Zetac 51, Dow 21M, Dow C-31, Chitosan [295]) which are usually positively charged [293]. The stability of the flocs is dependent on the forces that interact between the particles themselves and the particles and water. Electroflocculation is induced by the passage of electric current passed between the two electrodes (anode and cathode) immersed in the culture. The negatively charged cells tend to move towards the positive electrode (anode) leading to neutralisation and formation of cell flocs/aggregates [295]. Certain cyanobacterial species have the ability to self-flocculate in response to a change in their environment or stress. This phenomenon is known as auto-flocculation [289]. Flocculation can also be induced by adjusting CO_2 supply in the cultivation system [296]. Typically, while flocculation increases the efficiency of flotation or sedimentation, the dewatered biomass likely contains the flocculant, which may lead to the requirement of further refinement processes and increases cost.

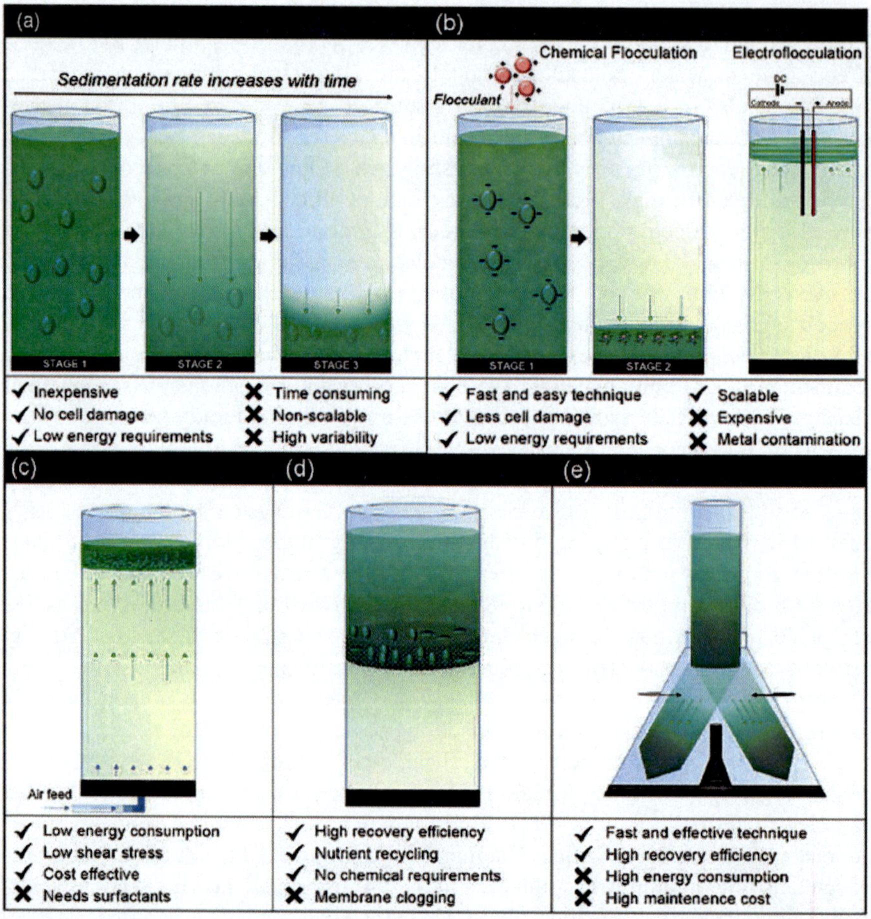

Fig. 5 Cyanobacterial biomass harvesting techniques. **(a)** Sedimentation, **(b)** Flocculation, **(c)** Filtration, **(d)** Froth flotation and **(e)** Centrifugation. The advantages (✓) and disadvantages (**X**) of each techniques mentioned

Froth flotation is a physiochemical gravity separation technique based on density differences between the cells and the aqueous phase [297–299]. Air is pumped into the flotation unit with or without an additional organic/inorganic chemical, and the resultant bubbling causes biomass accumulation along with the froth of bubbles at the top phase (Fig. 5c) [300]. This froth layer is separated and treated to harvest the biomass [301]. The flotation process can be subdivided according to the methods used for the bubble formation (e.g. dispersed air flotation, dissolved air flotation, microbubble generation and electrolytic flotation [300]). Flotation can also be combined with flocculation technique to separate a floating floc layer [300]. The advantages and disadvantages of froth flotation are summarised in Fig. 5c.

Filtration utilises a permeable size-exclusion based material through which a suspension is passed to separate smaller (e.g. aqueous phase) from larger molecules or particles (e.g. cells). Membrane filtration (tangential flow/cross-flow filtration) is the most commonly used harvest technique in *Spirulina* sp. farms [302, 303] (Fig. 5d). Filtration requires a pressure difference across the filter which can be driven by gravity, applied pressure or the use of a vacuum [303, 304]. Membrane filters are classified based on the pore size into macro- (greater than 10 μm), micro- (0.1–10 μm) and ultrafiltration (0.02–0.20 μm) as well as reverse osmosis (<0.001 μm) [305]. Filtration is widely used as a secondary dewatering step because it is reasonably cheap, fast and effective. The main drawbacks are increased energy requirements and the cost of downtime associated with washing membranes or replacing membranes due to fouling [306].

Centrifugation applies centrifugal force to enhance the dewatering efficiency (Fig. 5e) [307, 308]. Spinning the cell suspension creates the differential pressure necessary for a particle to separate from the liquid. The efficiency of the recovery process is dependent on the centrifugal force, particle size and density [308]. The two common types of centrifuges used for microalgae are disc stack and decanter centrifuges [307]. The major drawback of this technique is its high energy requirement; decanter centrifuges have been estimated to consume 3,000 kWh per ton of dry biomass [309]. Centrifugation for *Spirulina* sp. biomass harvest has been reported by many studies over the years from 1980 to 2001 and was later mostly replaced by membrane filtration or flocculation-flotation techniques due to more cost-effective scale-up. Centrifugation is often employed for small-scale laboratory harvests or very high-value products. De Souza Sossella et al. (2020) [310] compared the effect of harvesting techniques (centrifugation, chemical flocculation and froth flotation) on enzyme hydrolysis for *Spirulina platensis* biomass, where chemical flocculation yielded the highest harvesting and hydrolysis efficiency [310].

5.2 *Product Release via Cell Disruption or Pre-Treatment*

To extract intracellular products such as pigments, a suitable cell disruption step is a crucial part of downstream processing. Typically, cell disruption is performed after the first dewatering step to reduce treatment volume and save costs. An ideal cell disruption method would be one that lyses the cells and selectively releases the target product without damaging it, while using the least possible energy. The major challenge is to obtain high disruption efficiency while maintaining the functionality of the cyanobacteria pigments during cell disruption [311]. The major barrier to cell disruption is the sturdy cell wall and the thylakoid membrane, which encloses the pigments [311].

Typically, cyanobacterial cell walls consist of six layers. A rigid peptidoglycan layer overlays the inner cell membrane, and this is tightly connected to the outer membrane of the wall and contains muramic acid on the upper surface. The microfibrillar framework and an amorphous outer mucilaginous sheath are composed of polysaccharides, lipids and proteins (Fig. 6). Cell disruption methods can be classified into mechanical (e.g. bead milling, homogenisation, ultrasonication), physical (e.g. drying, pulsed electric field) or chemical/biological (e.g. acid, base,

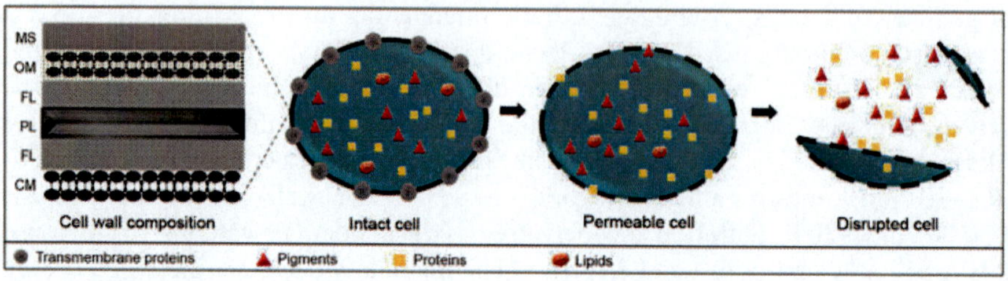

Fig. 6 Different stages of cell disruption. Schematic view of cell wall composition of commercial pigment producing cyanobacteria (*Spirulina* sp.) – MS – Mucilaginous Sheath, OM – Outer Membrane, FL – Fibrillar Layer, PL – Peptidoglycan Layer, CM – Cell membrane. Pigment extraction proceeds from left to right, as the cells are lysed. The cyanobacterial filaments typically quickly break into single cells. Intact cells are usually permeabilised more slowly and once broken typically release their intracellular contents quickly

enzymes) and are optimised depending on strain-specific parameters, including cell wall structure, cell size, product location, product solubility and energy requirements. The merits and demerits of some commonly used cell disruption methods are summarised in Table 7.

5.2.1 Mechanical Pre-Treatment Methods

The most commonly used mechanical methods for microalgal cell disruption are bead milling, high-pressure homogenisation and ultrasonication. To date, for soluble proteins such as PC and hydrophobic pigments such as carotenoids the most used cell disruption technique in large scale is bead milling.

Bead milling is a high-intensity cell lysis method which uses kinetic energy to force small beads (glass, ceramic, plastic, or steel) to collide with each other and the cells [320, 322]. The factors that affect the efficiency of the bead milling process include cell size, bead size and type of material, cell density, feed flow rate and chamber volume. For cyanobacterial species such as *Nodularia* sp., *Anabaena* sp., *Nostoc* sp., and *Spirulina* sp., smaller bead sizes of about ~0.1–0.3 mm are reported to be optimal [167, 326, 327]. The bead milling method has been reported to be more effective than ultrasonication or pulsed electric field treatment, achieving extraction efficiencies of up to 95% for total proteins at low to moderate energy consumption rates [304].

High-Pressure Homogenisation (HPH) is a mechanical cell lysis method during which the cells are subjected to high pressure that forces them to pass through a narrow opening, causing a rapid pressure release that breaks open the cell wall. The combination of intense shear force, cavitation and turbulent flow induces rapid cell disruption (that leads to emulsion formation) and is suitable even for strains with highly stable cell wall structures. One of the highest yields of C-phycocyanin extraction from *Cyanobacterium aponinum* PCC10605 [328] was achieved by applying three to six passes of HPH (1,000–1,500 bar) to a 1–2% biomass suspensions resulting in a total protein release of 70–90% [329]. HPH can achieve a high

Table 7 Comparison of different cyanobacterial cell disruption methods

Cell disruption technique	Merits	Demerits	Reference
Mechanical methods			
Freeze-drying	• Gentle extraction of fragile compounds	• High energy requirement and time-consuming • High maintenance cost • Difficult to scale up	[312]
Bead milling	• Simple equipment • Rapid process • High disruption efficiency • Easy scale-up • High degree of automation	• High energy requirements • High cooling requirements for thermolabile compounds	[313]
High-pressure homogenisation	• High disruption efficiency • No biomass drying required • Easy scale-up	• High energy requirement • Temperature rise may lead to degradation of thermolabile compounds • Rigid cell wall may hinder product release	[314, 315]
Ultrasonication	• Simple • Rapid process • High reproducibility • High disruption efficiency	• Moderate energetic costs • Temperature rise • Rigid cell wall hinders product release • Risk of reactive hydroxyl radicals • Difficult scale-up	[315]
Physical methods			
Pulsed electric field (PEF) treatment	• Simple operation • High energy efficiency • Rapid process • Easy scale-up	• High maintenance costs • Temperature rise affects product stability • Dependence on medium composition • Degradation of fragile compounds	[314, 316–318]
Microwave-assisted cell disruption	• Simple operation • Rapid process • High disruption efficiency • Easy scale-up • Eliminates the requirement or dewatering of algal biomass	• High energy requirements • High maintenance costs • Requires extensive cooling for thermolabile compounds • Lipid degradation and protein aggregation, denaturation, and formation of free radicals	[319]
Chemical methods			
Acid/alkali treatment	• No special equipment required • Rapid process • High disruption	• Hazardous • Possibility of damaging desired product • Difficulties in purification	[320–322]

(continued)

Table 7 (continued)

Cell disruption technique	Merits	Demerits	Reference
	efficiency • Low product contamination risk	• Extreme pH changes can cause protein denaturation	
Osmotic lysis	• Lower energy requirement • Easy scale-up • Low cost	• Time consuming • Generates high salinity of wastewater • Low efficiency	[320, 322]
Enzymatic treatment	• Highly species-specific • No special equipment required	• High cost at scale-up • Enzyme stability and incubation time is critical	[313, 323–325]

disruption efficiency and protein extraction yield, is relatively energy efficient [329], but has drawbacks in potential pigment degradation due to temperature elevation up to 84°C [330].

Ultrasonication is a mechanical treatment based on the use of ultrasound to generate cavitation bubbles (i.e. vapour bubbles formed from a flowing liquid in a region where the pressure of the liquid falls below its vapour pressure). These promote a non-specific cell-surface barrier disruption as bursting of the cavitation bubbles exerts a pressure that bursts the cyanobacterial cells. Sonication permeabilises both the cell wall and the membrane, a key difference from pulsed electric field (PEF) treatment (see below), which permeabilises only cell membranes [331]. In certain cases, sonication can be enhanced by the application of a secondary cell disruption method, such as high-shear mixing, enzymatic cell wall hydrolysis or chemical treatment. Such dual approaches can increase the release of soluble proteins (e.g. phycobiliproteins) from *Spirulina platensis* but in turn increases the cost of cell disruption [332–334].

5.2.2 Physical Pre-Treatment Methods

Pulsed Electric Fields (PEF) can provide a non-thermal approach to disrupt lipid cell membranes and allow low-molecular-weight molecules to enter into and diffuse out of the cells. The application of a high voltage PEF treatment for 10^{-4}–10^{-8} s typically disrupts cyanobacteria cells through electromechanical stress that causes irreversible permeabilisation [335]. As the phycobiliproteins are soluble rather than in the membrane fraction, PEF treatment is highly efficient at releasing them into the suspension. Jaeschke et al. (2019) [336] compared bead milling and PEF treatment on *Spirulina platensis* for C-PC extraction and recorded the highest C-PC yield of 85.2 ± 5.7 mg g^{-1} biomass using a PEF at 122 J mL^{-1} specific energy input. It was also noted that the antioxidant capacity of C-PC extracted with PEF was higher than that extracted with bead milling. Thus, PEF treatment may offer both an economic and functional benefit for pigment extraction. Furthermore, Akaberia et al. [337]

analysed the interaction of PEF (1 μs pulses; 40 kV cm^{-1} field strength; energy input, 56–114 kJ kg^{-1}) and pH on C-phycocyanin extraction and found extraction yield and PC stability to be highest at pH 8 [337].

Microwave-assisted cell disruption is a rapid and efficient physical pre-treatment technique. The transmission of microwave energy into a cell suspension results in rapid vibration of the water molecules and rapid heating of the cell matrix [338]. The microwaves heat the solvent and increase the partition between the cell matrix and the solvent [319, 338]. The increase in intracellular kinetic energy exerts pressure on the cell wall and leads to cell rupture [120]. Similar to high-pressure homogenisation, the use of microwave-assisted pre-treatment leads to significant emulsion formation that impedes solvent recovery [329].

5.2.3 Chemical Pre-Treatment Methods

Chemical disruption is achieved through the addition of a chemical additive such as a detergent, solvent, acid, chaotrope or a chelator to the biomass to degrade the cell membrane.

Acid or alkali treatment is a highly effective chemical pre-treatment method in which the cell wall of cyanobacteria is hydrolysed [339]. Chemical treatments work by attacking certain linkages in the cell wall, hydrolysing the phosphodiester bonds and ultimately dissolving the entire cell wall or permeabilising the cell wall layers [340]. Pigments extracted using ethanol require stronger acid concentrations than those extracted using methanol or acetone, to lower pH to an acceptable range for pigment determination (pH causes spectral shift). It was reported that the spectral shifts in Chl *a* absorbance persisted for about 30 min after acidification (0.005 mol HCl L^{-1}) [341]. The major advantage of chemical over mechanical pre-treatment is the lower energy requirement and elimination of emulsion formation (as long chain polysaccharides, proteins and phospholipids are hydrolysed into their constituent units) [342, 343]. However the pigments can be damaged, costs can be higher and neutralisation is required.

Osmotic lysis (osmotic shock) involves exposing cyanobacterial cells to a low-salt or hypotonic extracellular environment that results in a net flow of water into the cells, which can ultimately result in them bursting [322]. However, due to its low efficiency, it is one of the less commonly used methods of cell disruption [344].

Enzymatic treatment is generally performed under mild conditions and is a relatively eco-friendly, non-hazardous, and a low energy alternative to mechanical and chemical techniques. In some instances, enzymatic disruption can result in more efficient protein extraction than mechanical and chemical cell disruption [345]. Cyanobacteria have rigid cell walls to protect the cell from the environment and thus increase their survival capacity. The cyanobacterial cell wall comprises tri-layered structures of cellulose and proteins with other components such as mannose, xylan, algenan and glycoproteins, with minerals (Fig. 6) [321]. Thus, the selection of enzymes is typically biomass-specific and based on composition and cell wall structure for the target species. Cyanobacterial cell disruption employing lytic enzymes is quite popular in product and application-based scenarios (e.g. phycocyanin-based pharmaceutical production), owing to their ability to avoid the harsh conditions that phycobiliproteins are subjected to in other techniques. Several lytic enzymes can be used, such as cellulase which can effectively hydrolyse the cellulosic structure of the cell walls, and lysozyme which can hydrolyse the linkage between peptidoglycan residues [345]. Enzymatic treatment is most commonly used to improve extraction yields rather than for cell disruption alone and normally requires a preceding cell permeabilisation or disruption step [338]. However, the use of lytic enzymes typically increases extraction costs. The exploitation of biological mechanisms for in situ production of a lytic enzyme that can activate to cleave the cell wall is a promising solution to overcome high purified enzyme prices [36, 346].

5.3 *Product Recovery via Pigment Extraction*

After cell rupture, released intracellular pigments are recovered from the ruptured cell sample using an extraction solvent. For water-soluble pigments (such as phycobiliproteins), the extraction solvent can be water, the culture medium itself or an aqueous buffer. For water-insoluble pigments such as chlorophylls and carotenoids, one of the following extraction systems is typically used: organic solvent extraction, pressurised solvent extraction, microwave-assisted extraction, ultrasound-assisted extraction, pulsed-electric field extraction, ionic liquid extraction and supercritical carbon dioxide extraction [129]. Commercially, extraction processes with lower solvent consumptions, shorter extraction times and higher environment sustainability are preferred, as over the lifetime of a production plant the associated operational costs can exceed initial capital expenditure on infrastructure. The cell disruption and pigment extraction methods used in previous studies investigating pigment recovery from commercial pigment production strains are summarised in Table 8.

5.3.1 Conventional Organic Solvent Extraction

Water-soluble pigment binding proteins such as phycocyanin need to be extracted in an aqueous medium (using solvents such as water, sodium phosphate buffer or phosphate buffer saline) following cell lysis. In contrast, chlorophylls and carotenoids with high partition coefficients in organic solvents migrate out of the biomass into the solvent phase during organic solvent extraction, with a rate and extent of extraction limited primarily by the cell wall and the solubility of the target pigment. Solvents used for the extraction of chlorophylls and carotenoids include acetone, chloroform/methanol mixtures, DMSO, dodecane, ethanol, ethyl acetate, hexane, methanol, methylene chloride, vegetable oil and a mixture of one or more of the above solvents [354]. As global bio-economies expand and the importance of

Table 8 Summary of cell disruption methods and their pigment yields

Pigment – production species	Biomass state	Cell disruption	Pigment extraction	Max. pigment recovery	Key findings	Reference
Phycocyanin *S. platensis*	Oven dried (60° C)	Ultrasound-assisted protic ionic liquid extraction – 25 kHz, 25°C	Solvent extraction: 2-HEAA or 2-HEAF and 0.1 M sodium phosphate buffer	6.34 mg APC g_{BDW}^{-1} 5.95 mg PC g_{BDW}^{-1} 2.62 mg PE g_{BDW}^{-1}	Highly effective in extracting all PBP due to their greater diffusional power and interaction with the pigments	[347]
	Freeze dried biomass	Ultrasonication	Suspended in 0.1 M phosphate buffer, sonicated at 80 W, 20 kHz at 10°C	60 mg PC g_{BDW}^{-1} dry biomass), purity of 80%	Other biological compounds such as chlorophyll were also released during the ultrasonic disruption	[348]
B-Phycoerythrin *Porphyridium cruentum*	Direct wet biomass	Low pressure homogenisation (~90 MPa)	Resuspended in distilled water and HPH (~270 MPa)	3.6 mg B-PE g_{BDW}^{-1}, 0.79 purity ratio	Intracellular proteins are extracted more effectively at low HPH pressure than B-PE	[349]
Astaxanthin H. pluvialis	Freeze dried	–	Pressurised solvent extraction, methylene chloride: methanol (1:3 v/v) at 1500 psi, 40°C	10.9 mg astaxanthin g_{BDW}^{-1}	Increasing temperature degraded pigments.	[350]
	Powder	–	Supercritical carbon dioxide extraction with ethanol as co-solvent	77.9 wt% astaxanthin at 55 MPa, 343 K and 3 mL min^{-1} CO_2	Astaxanthin content increased with increasing temperature and pressure	[351]
β-Carotene *D. salina*	Direct culture from cultivation	–	Centrifugal partition chromatography with ethyl oleate and dichloromethane	65% β-carotene recovery	Solvent biocompatibility is critical	[352]
	Spray dried	Bead milling (glass beads)	Supercritical carbon dioxide extraction	115.43 μg β-carotene/g biomass at 400 bar, 55°C	At optimum condition, the carotenoid recovery of $SCCO_2$ extraction was half that of solvent extraction	[353]

2-HEAA 2-hydroxy ethylammonium acetate, *2-HEAF* 2-hydroxy ethylammonium formate process sustainability increases, use of bio-based solvents (e.g. acetone and ethyl lactate) is increasing. The solvent selected for industrial-scale pigment extraction should also be relatively volatile to support recovery [355, 356]. If the extracted pigments will be used for nutraceutical or food applications, it is also preferable to use non-toxic solvents for the extraction (e.g. acetone and ethanol instead of chloroform and methanol) [354].

5.3.2 Accelerated Organic Solvent Extraction

Accelerated Solvent Extraction (ASE) or Pressurised Liquid Extraction (PLE) is an organic solvent extraction performed at elevated temperature or pressure. The solvent is maintained in its liquid state throughout the extraction process [329]. Increased temperature and pressure aids in cyanobacterial cell wall disintegration, accelerating the mass transfer kinetics. Thus, PLE uses less solvent and can complete the extraction process in a shorter timeframe compared to a conventional organic solvent extraction but has higher energy requirements. High temperature (~200°C) introduces the risk of degradation of temperature sensitive pigments. PLE used during astaxanthin production from *H. pluvialis* improved apparent extraction yield but reduced the actual yield of intact astaxanthin, as assessed by the antioxidant activity of the extract [350] (Table 8).

5.3.3 Ionic Liquid Extraction

Ionic liquids are salts of weakly held anions and cations that remain in a liquid state over a wide range of temperatures. For pigment extraction, ionic liquids with a melting point less than 100°C are typically selected to eliminate the need for additional energy inputs to bring them into the liquid phase. Ionic liquids are broadly classified into two categories – protic and aprotic (based on their ability to transfer protons from acid to the base). Protic ionic liquids are generally molten salts (synthesised by transferring protons from a Bronsted-Lowry acid to a Bronsted-Lowry base). Many of their physical properties, such as polarity, hydrophobicity and viscosity, are adjustable and can be controlled by the exchange or combination of ions [129, 357] so that their solvating power can be specifically tailored to the target compound to enhance solvent–solute interaction and increase extraction efficiency [358]. Their high extraction efficiency is attributed to the liquid's high diffusivity and strong interaction with the pigment molecules. Indeed ionic extraction has been efficiently used for lipid extraction [359]. Chang et al. [360] reported the use of an Aqueous Two-Phase System (ATPS) with ionic liquids and achieved an extraction efficiency of 99% with the separation factor being 5.8 [360]. Sanchez-Laso et al. [361] reported the use of 1-ethyl-3-methylimidazolium ethyl sulphate for the extraction of phycobiliproteins from *Spirulina platensis* and achieved a recovery of 67.2 mg g_{BDW}^{-1} PC, 20.9 mg g_{BDW}^{-1} APC and 5.3 mg g_{BDW}^{-1} PE, the overall extraction efficiency was ~80% which was slightly lower than the ATPS approach reported previously (99%) [360].

5.3.4 Supercritical Carbon Dioxide Extraction

When carbon dioxide gas is subjected to temperature and pressure beyond its critical values (Critical Temperature – T_c at 31.1°C and Critical Pressure – P_c at 73.86 bar), it is transformed into a supercritical fluid that exhibits physical properties intermediate between a liquid and a gas. Supercritical carbon dioxide ($SCCO_2$) is a highly effective extraction solvent because of its transitional properties, such as high diffusivity and adjustable solvating power, that can be varied by pressure. The high diffusivity of $SCCO_2$ enables it to rapidly penetrate the cellular matrix and complete extraction within a shorter timeframe than less diffuse solvents [362, 363]. The fluid's solvating power is a direct function of its density and can thus be tailored to the targeted pigment through pressure and temperature adjustments. In addition to being a powerful extraction solvent, $SCCO_2$ is non-toxic, produces a solvent-free pigment extract (no additional solvent recovery step) and operates at a moderate temperature range which minimises degradation of thermally sensitive pigments [364, 365]. A major disadvantage of $SCCO_2$ extraction is the higher capital cost of the extraction apparatus but this can potentially be mitigated by lower longer term operating costs.

5.4 Pigment Purification

Crude cyanobacterial pigment extracts often contain other impurities that interfere with their stability and function. These impurities include chlorophylls and lipids in carotenoid extracts and proteins in phycobiliprotein extracts and can be separated from the desired pigments using chemical and/or chromatographic techniques. Selection of a purification method is highly affected by the final product application as PC with a purity (A_{620}/A_{280}) ≥0.7 is considered as food grade, >3.9 as reagent grade and >4.0 as analytical grade. Thin layer chromatography (Chl and carotenoids), liquid chromatography and spectrophotometric analysis are widely used for the analysis of the purified pigments [325, 366]. Calcium hydroxide precipitation, acid precipitation and column chromatography have previously been used to remove chlorophylls from astaxanthin and β-carotene extracts [367–369]. Phycobiliprotein purification generally involves an initial lysis step (e.g. freeze-thaw, sonication) followed by subjecting the lysate supernatant to one or more of the following steps: ammonium sulphate precipitation, activated carbon and chitosan precipitation, aqueous two-phase purification with polyethylene glycol, gel permeation chromatography, for example, with a Sephadex G-150 column (Fig. 7a) and anionic chromatography with diethylaminoethyl cellulose (DEAE) [370, 371], anion exchange chromatography with a Q-Sepharose column (Fig. 7b) and concentration by ultrafiltration (Fig. 7c) or tangential flow ultrafiltration (30–50 kDa). Different stages of PC extraction from cyanobacterial biomass are shown in Fig. 7d. Halim et al. [30] described the extraction of PC from *Galderis sulphuraria* in which

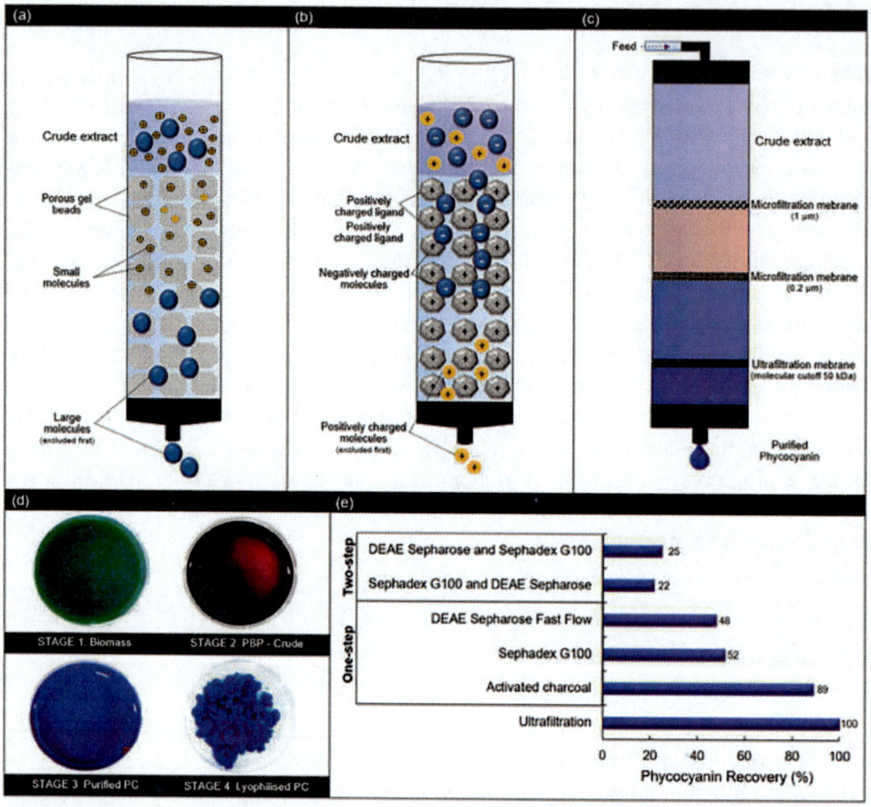

Fig. 7 Cyanobacterial pigment purification. Schematic of the most commonly employed PC purification techniques – (**a**) Gel filtration/permeation chromatography, (**b**) Anion exchange chromatography and (**c**) Ultrafiltration. (**d**) Different stages in PC production – harvested cyanobacterial biomass, PBP aqueous crude extract (contains PE, APC and other soluble proteins), purified PC and lyophilised PC powder. (**e**) Comparison of different PC purification techniques based on PC recovery (%). Ultrafiltration method using microfiltration membranes (1 μm, 0.2 μm) and ultrafiltration membrane with molecular cut-off of 50 kDa has recorded among the highest recovery rates but achieved comparatively low purity [372]

ammonium sulphate precipitation with aqueous two-phase extraction and ultrafiltration resulted in both the highest PC yield (42 wt% of PC in the crude extract) and the highest product purity ($A_{620}/A_{280} = 4.5$). Chaiklan et al. [372] investigated stepwise extraction of PC and economic feasibility analyses by comparing different PC purification techniques from *Spirulina* sp. which included ultrafiltration, one-step and two-step chromatography techniques using three different matrixes: activated charcoal, Sephadex G100 and DEAE Sepharose Fast Flow (Fig. 7e). The highest PC recovery rate was recorded using ultrafiltration (Yield: 6.43 mg/mL) but the purity achieved was comparatively low (A_{260}/A_{280}- = 1.22;Fig. 7e) [372].

6 Pigment Bioprocessing Challenges

The development of more cost-effective cyanobacterial pigment production processes requires improved production (Sect. 4.2), disruption (Sect. 5.2) and extraction techniques (Sect. 5.3) to drive down the costs and enhance quality and value. The main challenges of natural pigments production include optimising species selection, cost of production as well as the product, quality and stability.

Increasing pigment yield: As cyanobacteria can be relatively slow growing, biomass and pigment yields can be low compared to other microbes (e.g. microalgae; growth rate of *Chlorella* sp. ~0.047 day^{-1} [241] while *Spirulina* sp. is only 0.0027 day^{-1} [162]). This explains why the first pigments commercially produced (e.g. phycocyanin) were unique to cyanobacteria, of high value and expressed at high levels. Recent technological advances in photobioreactor development and process optimisation parameters are overcoming scale-up associated challenges [373–375]. Bio-process optimisation and genetic engineering of the strain are two-key ways to increase biomass and pigment accumulation.

Disruption and extraction techniques: Cost and efficiency require optimisation for each target product. For example, microwave-assisted cell disruption is an efficient method to disrupt biomass, but the use of high temperatures can also result in pigment degradation. During traditional solvent-extraction of chlorophylls and carotenoids, the choice of solvent and biomass-solvent ratio is critical to achieve high final pigment yields. The choice of solvent is often also influenced by regulatory policies. For example, although hexane is an excellent solvent for carotenoid extraction, it must be completely removed to comply with regulations for human consumption. This hurdle can technically be overcome by replacing hexane with green solvents such as ethanol, ethyl acetate or critical CO_2 extraction, but this can compromise pigment yields. To date, lead disruption processes for pigments are based on bead milling for both phycobiliproteins and carotenoids.

Enhancing product stability: Natural pigments such as carotenoids and chlorophylls are generally sensitive to light, pH, UV, temperature and oxygen as oxidation of their conjugated bond systems results in fading (e.g. in β-carotene and astaxanthin) and a reduced shelf life. Other natural pigments such as phycobiliproteins and chlorophylls are sensitive to other ambient conditions like metal ion concentrations, heat or organic solvents that can denature proteins. C-phycocyanin (C-PC) has been approved as a food additive and blue colourant and it is typically used in the αβ-monomeric and trimeric forms which coordinate the Phycocyanobilin (PCB) chromophore. The hexamer may, however, offer improved stability and colour properties [337, 376, 377]. C-PC has been reported to retain its hexameric form (Fig. 8) in the pH 5–7 range and to be more stable below 46°C [377]. Therefore, PC application in the food sector is mainly limited due to its sensitivity to external factors. The use of effective encapsulation techniques or stabilising agents such as glucose, alginate, pectin, whey protein and carrageenan would help overcome this challenge.

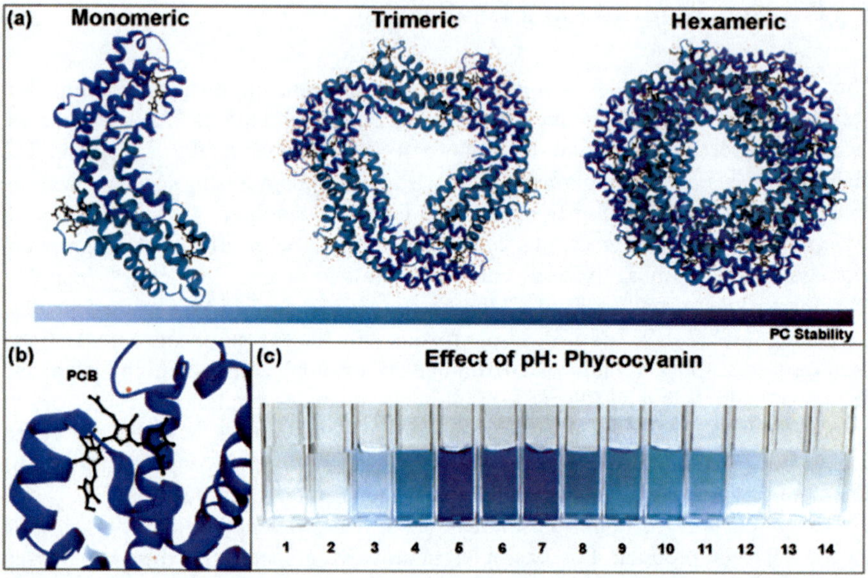

Fig. 8 Stability of C-Phycocyanin. (**a**) Crystal structure of monomeric, trimeric and hexameric forms of C-phycocyanin (from *Thermosynechococcus vulcanus*; acquired from PDB) – monomeric (least stable; 1ON7), trimeric (3O2C) and hexameric (most stable; 1I7Y). (**b**) Phycocyanobilin (PCB), the chromophore responsible for the blue colour of PC. (**c**) Effect of pH on PC. The PC extracts were derived from *Spirulina platensis* wet biomass using the freeze-thaw method with water as solvent. The pH of the extracts was adjusted using 0.1 N HCl/NaOH

To be economically and environmentally beneficial, pigment production (as a single product or as co-product) in biorefineries requires strong process intensification strategies. The final pigment product should be stable under environmental factors such as light, pH, temperature, UV and food matrices. Development of novel encapsulation techniques based on the market value of pigments will thus assist in the production of more stable natural pigments with a higher shelf life (expanding their applications). Understanding the biosynthetic pathways of cyanobacterial pigments is an important starting point, followed by identifying genes and the gene cascades responsible for pigment production, which supports metabolic engineering approaches for pigment accumulation.

7 Commercial Pigment Production Technologies

Currently commercial production of cyanobacteria strains is confined to phycocyanin production but has the opportunity to be expanded for the production of other pigments including chlorophyll and carotenoids. Cyanobacteria produce most of the major carotenoids present in microalgae. With expansion of strain phytoprospecting and cultivation optimisation, they are promising candidates for industrial production for many pigments. Cyanobacteria strains reported to produce different pigments of commercial interest (serve as alternatives) and their corresponding production strains are summarised in Fig. 9.

7.1 Patents and Technology Transfer

Patents are public documents and effectively part of the open access literature that document recent technical developments that have commercial potential [381]. A patent search on Patent Lens (Lens.org), a patent database with an integrated framework that serves nearly all the patent documents in the world, for the pigment 'Phycocyanin' showcases an example for current cyanobacterial pigments in the market and is represented in Fig. 10.

Technological developments and transfer can help to address existing scalability challenges and increase the economic feasibility of production platforms [382, 383]. The selection of production technology and process optimisation is highly application-specific in the case of pigments. For example, phycocyanin marketed as a food colourant (blue Spirulina powder with 2–6% PC – selling price USD \$160 kg^{-1}) is produced in open ponds with a low number of extraction steps while the pure phycocyanin marketed for flow cytometry applications (~98% pure; selling price USD \$217,000 g^{-1}) is produced under highly controlled environments with a series of purification (chromatography) steps. Examples of some recent patents that focus on cyanobacterial pigment-based technological innovations include:

- Method for separating and purifying high-purity phycobiliprotein from nitrogen-fixing cyanobacteria (080-530-697-056-493; August 2021; Pending).
- Phycocyanin-casein/porous starch microgel as well as preparation method and application thereof (091-869-437-651-829; June 2021; Pending).
- Supercritical cracking process of phycocyanin (002-379-984-590-359; April 2021; Pending).
- Method for extracting phycocyanin from *Spirulina* sp. through low-salt flocculation method (051-541-487-645-566; Jan 2021; Pending).
- Mixing temperature tank for phycocyanin (055-912-539-763-114; Nov 2020; Active).
- Spray drying device applied to phycocyanin production (067-605-942-811-388; Nov 2020; Active).

The increasing number of natural pigment-based patents (related to cyanobacteria and microalgae) is considered as evidence to consolidate the growth of cyanobacterial pigments market, which is expected to grow further in the upcoming years (increasing the likelihood of replacing synthetic pigments). Most of the published patents are reported to be technological patents in association with novel cultivation and extraction techniques [384].

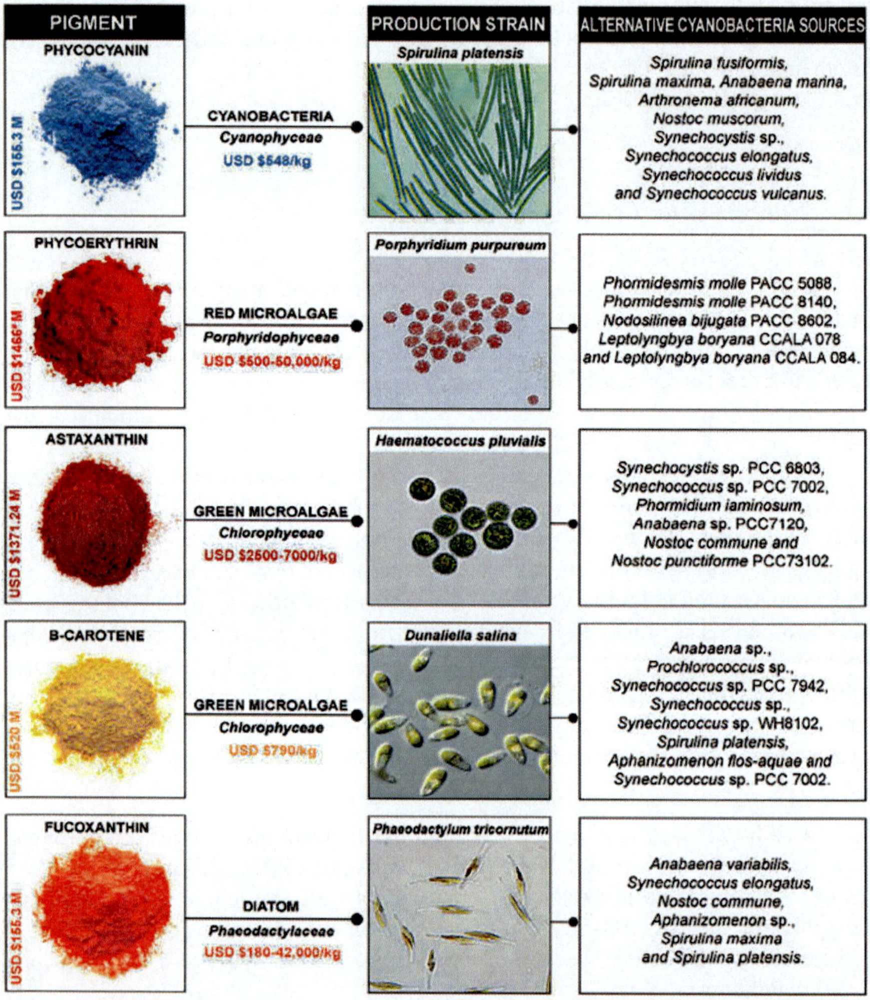

Fig. 9 Current commercial algae and cyanobacterial-based pigments. The pigments (left), the commercial strain (source, middle) and potential cyanobacterial strains with high pigment content. The micrographs of the commercial strains were obtained from the 'Microalgae Strain Catalogue' [378]. The reported strains from the literature are listed as potential candidates to replace or supplement the current production strains. The market size of the pigments in 2020 (USD millions) are denoted for each of the pigments (according to BCC research – https://www.bccresearch.com/). The selling price for each pigment (per kg) is also provided [379, 380]. #indicates the market size in 2019. (Lutein and chlorophyll are not listed as they are commercially produced only from plant sources, marigold flowers and alfalfa, respectively)

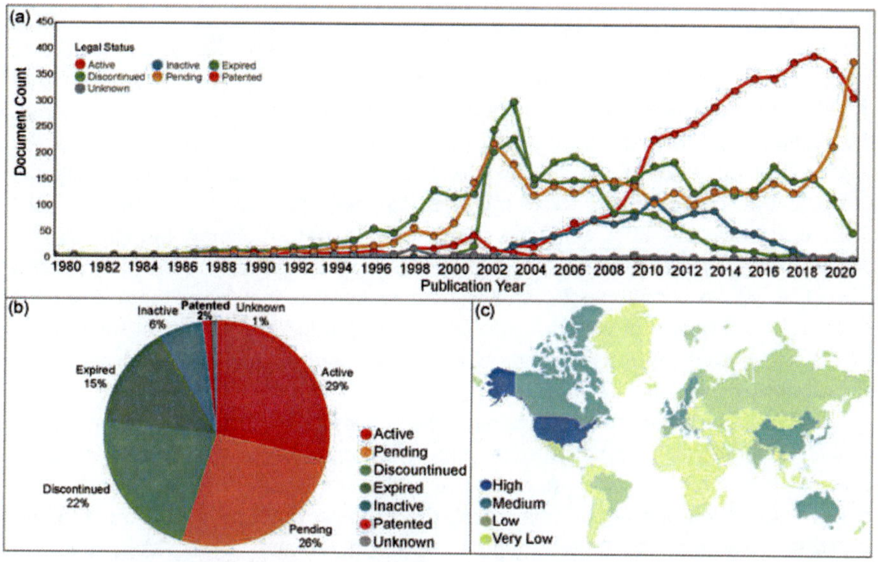

Fig. 10 Phycocyanin patent analysis. (**a**) Phycocyanin-based patent document count vs publication year (with legal status). The number of active patents significantly increased after 2004 but saw a general drop after about 2018 (**b**) Legal status of the patents vs document count. There are many patent applications pending (latency period) and discontinued categories which still provide useful literature for competitor analysis. (**c**) Patent performance by jurisdiction (country). Currently, USA holds the highest number of PC patents ($n = 9{,}875$)

7.2 Techno-Economic Analysis and Life-Cycle Analysis: CAPEX/OPEX and Price Points

Cyanobacteria provide the basis for a range of light-driven biotechnologies and exhibit promising characteristics such as high biomass yields (30–33 T dry weight ha^{-1} $year^{-1}$ [26, 385]), utilisation of non-arable land and ocean water, and integrate CO_2 utilisation and capture opportunities [386].

The global production of *Spirulina* sp. comprises about 10,000 tons of dry biomass per annum [387]. The focused attention on the improvement of production and processing steps for microalgae is used to derive both low volume, high-value products and high volume, low value commodities [388, 389]. Techno-Economic Assessments (TEA) and Life-Cycle Assessments (LCA) are important foundational tools to evaluate the economic, social and environmental benefits of specific cyanobacteria processes. TEA is used to analyse and optimise the economics of the process (e.g. production systems, dewatering, cell disruption, purification) by calculating, comparing and simulating the Capital expenditure (CAPEX), operational expenditure (OPEX) and product sales which provide the income stream. TEA analysis has been widely used to evaluate and optimise the efficiency and economic performance of various production processes [172, 390]. TEA includes analysis of cost parameters such as energy inputs and outputs which accounts for delivered energy and energy losses associated with the production. It enables the calculation of Energy Return on Energy Invested (EROEI) based on operating conditions, total capital investment, production cost and payback period [9]. LCA, on the other hand, is a method to perform environmental analysis of the complete production process cycle and includes parameters such as GHG emissions, cumulative energy demands, eutrophication potential and waste management. Individually, TEA evaluates economic efficiency and LCA evaluates the environmental efficiency and can also be used to assess social benefit (e.g. jobs and eco-system services) [172]. Integrated TEA/LCA allows simultaneous analysis of economic, social and environmental factors and is a powerful tool that enables model guided design to fast-track triple bottom line system optimisation, de-risk scale-up and enable the development of robust business models [172].

TEA/LCA has been used to evaluate a wide variety of cultivation technologies (which include open pond systems and different types of photobioreactors [172, 390]) to evaluate their product yield and quality and ultimately commercial viability. The open pond system is among the simplest in terms of construction and operation, leading to lower capital and operational costs compared to photobioreactors (PBRs) [255, 391]. However, PBRs have advantages in terms of maintaining strain purity, biomass productivity, optimising light delivery, CO_2 supply and use efficiency, and controllability. TEA/LCA is also used to simulate different downstream processes (e.g. cell disruption, product recovery/extraction, purification, formulation) and to compare, evaluate, integrate and optimise different process components as well as the complete process [305, 392, 393].

Biorefinery strategies designed to produce multiple products can offer economic benefits, but this is not always the case. Chaiklan et al. (2018) [372] performed an economic feasibility study on extracting multiple products (phycocyanin produced with lipids and polysaccharides) from *Spirulina platensis*. They concluded that single-product production of phycocyanin was economically feasible, but the multiple-product approach (coproduced with lipids and polysaccharides) was not feasible. The estimated production cost of phycocyanin was USD $250 kg^{-1} which is an encouraging figure for large-scale production.

In summary, the use of TEA, LCA or integrated TEA/LCA (TELCA) is very important to fast-track systems optimisation, de-risk scale-up and establish robust business models [172, 390]. In particular, our international community is faced with the urgent challenge of reducing CO_2 emissions by almost 100% by 2050. This will require an investment of about USD $40 Trillion, and so robust system optimisation is critical as the scale-up cost is equivalent to approximately 31% of the Worlds ~$127 Trillion 2019 Global GDP [394].

At the current cost of USD $3 – 9 kg^{-1} (biomass dry weight), cyanobacteria are already accessible for the production of a range of high-value products in industries. Rapid advancements in high-throughput production strain selection [241, 395], photosynthetic machinery (antenna engineering), product biosynthesis, process optimisation (light, macro and micronutrients, CO_2, pH, temperature), reactor design and scale-up [255], harvesting and purification techniques [396], location selection (climate, land costs, regional jobs), automation (to reduce operational cost), biorefinery (multi-product approach), cryopreservation [397, 398], scale-up (laboratory, pilot scale, and industrial) [255], as well as TEA/LCA [172, 389] and policy adaptations [172] are collectively contributing to improved production systems which in turn are the areas for future development in the cyanobacteria-based industries [399]. It is anticipated that biomass prices can be reduced towards USD ~$1 kg^{-1} allowing the industry to expand from high-value products down to commodity products [172].

8 Global Pigment Market Analysis: Opportunities and Challenges

Opportunities: Growing awareness about the health benefits of natural pigments is supporting the growth in demand. The World Health Organisation (WHO) developed a global action plan for prevention and control of chronic diseases, encouraging a diet with essential nutrients, enriched with bioactive components (e.g. Ω-3 PUFAs (Poly-Unsaturated Fatty Acids) and Polyphenols) [400] and thus further increasing the overall demand.

Cyanobacterial and microalgal biomass are already in the market and have recently gained attention as alternatives to produce nutrient-rich foods. They are known to have a high nutritional value being rich in phycocyanin, chlorophylls, essential fatty acids (e.g. gamma linoleic acid), carbohydrates and trace minerals supporting consumer acceptance and marketing of natural pigments from microalgae and cyanobacteria. The colour and bioactive properties of cyanobacteria pigments are a dual benefit for multiple industrial sectors (e.g. Phycocyanin – blue protein pigment from *Spirulina* sp., termed a 'Diamond Food' in the food sector and also used widely in cosmetics and pharmaceuticals) [139, 401, 402].

In the past few decades there has been a transition to the development and use of natural food products and additives to replace chemically produced additives. The global carotenoid market was estimated to be USD $0.76 billion in 2007 (β-carotene held the largest share). In 10 years, the carotenoid market doubled to USD $1.5 billion (astaxanthin held the largest share) and is anticipated to rise further to USD $2.0 billion by 2022 with a CAGR of 5.7% [403]. This shift from β-carotene to astaxanthin was mainly due to the use of astaxanthin in animal and aquaculture feed (USD $300 million) and in nutraceuticals (as an antioxidant agent; USD $30 million) in 2009. Astaxanthin is still known as the most powerful antioxidant (6,000× stronger than Vitamin C [404]). The astaxanthin market demand is expected to increase to $800 million and $300 million by 2020 for animal feed and for nutraceuticals, respectively [405]. Carotenoid pigments such as astaxanthin, β-carotene, fucoxanthin and lutein from microalgae are attracting attention as yields are much higher compared to their conventional sources (e.g. lutein yields from microalgae is 6x higher than from marigold flowers; yield of astaxanthin from microalgae is ~300× higher than from salmon or krill). Additionally, natural pigment production from cyanobacteria and microalgae is much faster with lower cultivation costs (compared to plants) and can be produced throughout the year around the world.

Market and Competitive landscape: The market value of astaxanthin produced from microalgae is reported to be USD $2,500 kg^{-1} with the production cost of microalgae feedstock of USD $5 – USD $20 kg^{-1} dry weight [406]. Commercially, *Haematococcus pluvialis* and *Dunaliella salina* are widely used production strains for astaxanthin and β-carotene production, respectively. The production of *H. pluvialis* is about 300 tons per year primarily from the USA, Israel, and India [10, 123, 407]. AstaReal, Inc. is the pioneer company that commercialised astaxanthin (1994). They marketed natural astaxanthin in 4 forms AstaReal® L10 oleoresin (10% extract), AstaReal® EL25 (2.5% powder), AstaReal® A1010 (astaxanthin-rich dry algae biomass) and Novasta (animal nutrition). Based on the global carotenoid market analysis, Europe has a strong and potential market due to the increasing demand for animal feed, health supplements and cosmetics. Involvement of leading cosmetic industries such as Unilever, L'Oreal, Henkel and Beiersdorf is expected to underpin the growth of the carotenoid market value in the European market. A number of key vendors are playing a major role in producing carotenoid pigments across the globe such as Lycored, Divis Laboratories, Naturex SA, BASF Corporation, FMC Corporation, and ExcelVite SDN BHD. Some of the top companies for cyanobacteria and microalgae-based pigments (already in the market) are listed in Table 9.

Challenges: There is considerable research and commercial interest to develop reliable natural colourants and to improve their stability. Most pigment-based patents are technological patents that claim efficient and gentle extraction techniques that offer final pigment stability (Sect. 7.1). Meeting the current challenges in the natural pigment market would further help their use and commercialisation.

- Synthetic colourants have already been in use for the past few decades and offer strong pigmentation, stability, easier processing, lower cost, and availability in unlimited quantities.
- The pigments produced from other microbial sources such as fungi, bacteria and yeast (by genetic engineering approaches) are exploited for different commercial applications [16, 17, 21, 147] and can increase market competition.
- Some of the major challenges reported when employing natural pigments in food industries include higher cost of production (e.g. carotenoids require solvent extraction), limited application (non-compatible with some foods), complexity of the process (thermal sensitivity) and inconsistent quality (degradation/fading).

Table 9 Examples of cyanobacteria and microalgae-based pigment production companies

Pigment	Current production strain	Companies	Location
Phycocyanin	*Spirulina* sp.	Earthrise Nutritionals, LLC	USA
		Cyanotech Corporation	USA
		Qingdao ZolanBio Co., Ltd.	China
		Yunnan Green A Biological Project Co., Ltd.	China
		Parry Nutraceuticals	India
		Tianjin Norland Biotech Co., Ltd.	China
		Zhejiang Binmei Biotechnology Co., Ltd.	China
		Fuqing King Dnarmsa Spirulina Co. Ltd.	China
		Japan Algae Co., Ltd.	Japan
		Bluetec Naturals Co., Ltd.	China
		Dongtai City Spirulina Bio-engineering Co., Ltd.	China
		BlueBioTech Int. GmbH	Germany
		AlgoSource Pvt Ltd.	France
		D.D. Williamson & Co., Inc.	USA
		Chr Hansen Holding A/S	Denmark
		Sensient Technologies Corporation	USA
		Naturex Inc.	France
		GNT Group B.V.	Netherlands
		Phyco-Biotech Laboratories	France
		Sigma-Aldrich Corporation	USA
Astaxanthin	*Haematococcus pluvialis*	Cyanotech Corporation	USA
		Parry Nutraceuticals	India
		BlueBioTech International GmbH	Germany
		Algatechnologies Ltd.	Israel
		AlgaeCan Biotech Ltd.	Canada
		AstaReal AB	Japan
		Algae Health Sciences – A BGG company	USA
		Algalif Iceland ehf.	Iceland
		Algamo s.r.o.	Chile
		Piveg, Inc.	USA
β-carotene	*Dunaliella salina*	Algalimento SL	Spain
		Seagrass Tech Private Limited	India
		Plankton Australia Pty Ltd	Australia
		Hangzhou OuQi Food co., Ltd.	China
		Shaanxi Rebecca Bio-Tech Co., LTD	China
		Nutragreenlife Biotechnology Co. Ltd.	China
		Israeli Biotechnology Research (IBR) Ltd	Israel
		Xi'an Fengzu Biological Technology Co., Ltd.	China
		Fuqing King Dnarmsa Spirulina Co., Ltd.	China
		Monzón Biotech S.L	Spain

9 Future Perspectives

Cyanobacterial pigments offer significant potential in multiple industrial sectors, including food and pharmaceuticals. The multidisciplinary aspect considered in natural pigment production for the food sector is that the colourants are used both as dyes and additives providing nutritional benefits. Advancements in phytoprospecting and bioprocess engineering have been useful for enhancement of biomass yields by optimising cultivation and extraction strategies (e.g. biomass harvest, solvent selection, extraction, purification and final formulation) and allow higher pigment yields and easy scalability. The combined identification of both biomass productivity and pigment concentration will enable the development of economically feasible pigment production scenarios with enhanced pigment yields and quality. Development of high-throughput screens helps to fast-track the optimisation of production conditions for the chosen target strains and guides the understanding of differences in strain-specific and pigment-specific production scenarios. Further analysis and understanding of the metabolomics will provide significant insights in developing the strategies for in vitro pigment accumulation. A completely different challenge for cyanobacterial pigments is associated with the regulatory bodies. Their approval depends on whether the pigment is a pure extract or dry biomass powder and the pigment concentration (e.g. Spirulina blue powder is marketed as crude/impure PC). Another challenge involves the effect of pigments on taste (consumer acceptance) and their stability, which can be improved through encapsulation or refinement techniques.

Acknowledgements CD, JW, IR, JR and BH thank the Australian Research Council LP180100269 and The University of Queensland, Australia (International Research Scholarship) for financial support.

References

1. McCarthy T (2013) The story of earth & life: a southern African perspective on a 4.6-billion-year journey. Penguin Random House South Africa
2. Ringsmuth AK, Landsberg MJ, Hankamer B (2016) Can photosynthesis enable a global transition from fossil fuels to solar fuels, to mitigate climate change and fuel-supply limitations? Renew Sustain Energy Rev 62:134–163
3. Blankenship RE, Hartman H (1998) The origin and evolution of oxygenic photosynthesis. Trends Biochem Sci 23(3):94–97
4. Grossman AR et al (1995) Light-harvesting complexes in oxygenic photosynthesis: diversity, control, and evolution. Annu Rev Genet 29(1):231–288
5. Wilmotte A (1994) Molecular evolution and taxonomy of the cyanobacteria. In: The molecular biology of cyanobacteria. Springer, pp 1–25
6. Karapetyan N (1974) Evolution of photosystems of photosynthetic organisms. In: Cosmochemical evolution and the origins of life. Springer, pp 253–256
7. Grotjohann I, Jolley C, Fromme P (2004) Evolution of photosynthesis and oxygen evolution: implications from the structural comparison of photosystems I and II. Phys Chem Chem Phys 6(20):4743–4753
8. Hoek C et al (1995) Algae: an introduction to phycology. Cambridge University Press

9. Hariskos I, Posten C (2014) Biorefinery of microalgae–opportunities and constraints for different production scenarios. Biotechnol J 9(6):739–752
10. Mandal MK, Chanu NK, Chaurasia N (2020) Cyanobacterial pigments and their fluorescence characteristics: applications in research and industry. In: Advances in cyanobacterial biology. Elsevier, pp 55–72
11. Bai M-D et al (2011) Microalgal pigments potential as byproducts in lipid production. J Taiwan Inst Chem Eng 42(5):783–786
12. Cervantes-Llanos M et al (2018) Beneficial effects of oral administration of C-phycocyanin and phycocyanobilin in rodent models of experimental autoimmune encephalomyelitis. Life Sci 194:130–138
13. Liu Q, Li W, Qin S (2020) Therapeutic effect of phycocyanin on acute liver oxidative damage caused by X-ray. Biomed Pharmacother 130:110553
14. Ravi M et al (2015) Molecular mechanism of anti-cancer activity of phycocyanin in triple-negative breast cancer cells. BMC Cancer 15(1):1–13
15. Reynoso-Camacho R et al (2011) Dietary supplementation of lutein reduces colon carcinogenesis in DMH-treated rats by modulating K-ras, PKB, and β-catenin proteins. Nutr Cancer 63(1):39–45
16. Manivasagan P et al (2018) Marine natural pigments as potential sources for therapeutic applications. Crit Rev Biotechnol 38(5):745–761
17. Begum H et al (2016) Availability and utilization of pigments from microalgae. Crit Rev Food Sci Nutr 56(13):2209–2222
18. Eastaugh N et al (2007) Pigment compendium: a dictionary of historical pigments. Routledge
19. Capelli B, Bagchi D, Cysewski GR (2013) Synthetic astaxanthin is significantly inferior to algal-based astaxanthin as an antioxidant and may not be suitable as a human nutraceutical supplement. Forum Nutr 12(4):145–152
20. Morocho-Jacome AL et al (2020) (Bio) technological aspects of microalgae pigments for cosmetics. Appl Microbiol Biotechnol:1–10
21. Joshi V et al (2003) Microbial pigments
22. Cortez R et al (2017) Natural pigments: stabilization methods of anthocyanins for food applications. Compr Rev Food Sci Food Saf 16(1):180–198
23. D'Amato D et al (2017) Green, circular, bio economy: a comparative analysis of sustainability avenues. J Clean Prod 168:716–734
24. Giampietro M (2019) On the circular bioeconomy and decoupling: implications for sustainable growth. Ecol Econ 162:143–156
25. GrandViewResearch (2020) Dyes and pigments market size, share & trends analysis report by product (dyes (reactive, vat, acid, direct, disperse), pigment (organic, inorganic)), by application, by region, and segment forecasts, 2020–2027. [Cited 2021 14/01/2021]; Available from: https://www.grandviewresearch.com/industry-analysis/dyes-and-pigments-market#:~:text=The%20global%20dyes%20and%20pigments%20market%20size%20was%20estimated%20at,USD%2034.7%20billion%20in%202020
26. Jimenez C et al (2003) The feasibility of industrial production of Spirulina (Arthrospira) in southern Spain. Aquaculture 217(1-4):179–190
27. Delrue F et al (2017) Optimization of Arthrospira platensis (Spirulina) growth: from laboratory scale to pilot scale. Fermentation 3(4):59
28. Vanthoor-Koopmans M et al (2013) Biorefinery of microalgae for food and fuel. Bioresour Technol 135:142–149
29. Aslam A et al (2020) Biorefinery of microalgae for nonfuel products. In: Microalgae cultivation for biofuels production. Elsevier, pp 197–209
30. Halim R (2020) Industrial extraction of microalgal pigments. In: Pigments from microalgae handbook. Springer, pp 265–308
31. Mirkovic T et al (2017) Light absorption and energy transfer in the antenna complexes of photosynthetic organisms. Chem Rev 117(2):249–293

32. Zouni A et al (2001) Crystal structure of photosystem II from Synechococcus elongatus at 3.8 Å resolution. Nature 409(6821):739–743
33. Minagawa J, Takahashi Y (2004) Structure, function and assembly of photosystem II and its light-harvesting proteins. Photosynth Res 82(3):241–263
34. Fischer WW, Hemp J, Johnson JE (2016) Evolution of oxygenic photosynthesis. Annu Rev Earth Planet Sci 44:647–683
35. Kannaujiya VK et al (2020) Phycobiliproteins in microalgae: occurrence, distribution, and biosynthesis. In: Pigments from microalgae handbook. Springer, pp 43–68
36. Ben-Shem A, Frolow F, Nelson N (2003) Crystal structure of plant photosystem I. Nature 426(6967):630–635
37. Brettel K, Leibl W (2001) Electron transfer in photosystem I. Biochim Biophys Acta (BBA) – Bioenergetics 1507(1–3):100–114
38. Jordan P et al (2001) Three-dimensional structure of cyanobacterial photosystem I at 2.5 Å resolution. Nature 411(6840):909–917
39. Golbeck JH, Bryant DA (1991) Photosystem I. In: Current topics in bioenergetics, vol 16, pp 83–177
40. Chitnis PR (2001) Photosystem I: function and physiology. Annu Rev Plant Biol 52(1): 593–626
41. MacIntyre HL et al (2002) Photoacclimation of photosynthesis irradiance response curves and photosynthetic pigments in microalgae and cyanobacteria 1. J Phycol 38(1):17–38
42. Glazer AN (1988) [31] Phycobiliproteins. In: Methods in enzymology. Elsevier, pp 291–303
43. Yeremenko NG (2004) Functional flexibility of photosystem I in cyanobacteria. Universiteit van Amsterdam
44. Mikami K, Hosokawa M (2013) Biosynthetic pathway and health benefits of fucoxanthin, an algae-specific xanthophyll in brown seaweeds. Int J Mol Sci 14(7):13763–13781
45. Deepika C (2018) Extraction and spectral characterization of R-phycoerythrin from Macroalgae–Kappaphycus alvarezii. Extraction 5(12)
46. Liotenberg S et al (1996) Effect of the nitrogen source on phycobiliprotein synthesis and cell reserves in a chromatically adapting filamentous cyanobacterium. Microbiology 142(3): 611–622
47. Vega J et al (2020) Cyanobacteria and red macroalgae as potential sources of antioxidants and UV radiation-absorbing compounds for cosmeceutical applications. Mar Drugs 18(12):659
48. Glazer AN (1994) Adaptive variations in phycobilisome structure. In: Advances in molecular and cell biology. Elsevier, pp 119–149
49. Shively J et al (2019) Intracellular structures of prokaryotes: inclusions, compartments and assemblages. In: Encyclopedia of microbiology. Elsevier, pp 716–738
50. Hernandez-Prieto MA, Chen M (2021) Light harvesting modulation in photosynthetic organisms. In: Shen J-R, Satoh K, Allakhverdiev SI (eds) Photosynthesis: molecular approaches to solar energy conversion. Springer, Cham, pp 223–246
51. Pandey V, Pandey A, Sharma V (2013) Biotechnological applications of cyanobacterial phycobiliproteins. Int J Curr Microbiol App Sci 2(9):89–97
52. Dumay J et al (2014) Phycoerythrins: valuable proteinic pigments in red seaweeds. Adv Bot Res 71:321–343
53. Glazer A, Cohen-Bazire G (1971) Subunit structure of the phycobiliproteins of blue-green algae. Proc Natl Acad Sci 68(7):1398–1401
54. Kerfeld CA, Kirilovsky D (2013) Structural, mechanistic and genomic insights into OCP-mediated photoprotection. Adv Bot Res 65:1–26
55. Batard P et al (2002) Use of phycoerythrin and allophycocyanin for fluorescence resonance energy transfer analyzed by flow cytometry: advantages and limitations. Cytometry 48(2): 97–105
56. Hamouda RA, El-Naggar NE-A (2021) Cyanobacteria-based microbial cell factories for production of industrial products. In: Microbial cell factories engineering for production of biomolecules. Elsevier, pp 277–302

57. Wyman M, Gregory R, Carr N (1985) Novel role for phycoerythrin in a marine cyanobacterium, Synechococcus strain DC2. Science 230(4727):818–820
58. Manoa (2022) Light in the ocean. Available from: https://manoa.hawaii.edu/exploringourfluidearth/physical/ocean-depths/light-ocean
59. Stryer L, Glazer AN (1985) Phycobiliprotein fluorescent conjugates. Google Patents
60. Li W et al (2019) Phycobiliproteins: molecular structure, production, applications, and prospects. Biotechnol Adv 37(2):340–353
61. Singh SP, Häder D-P, Sinha RP (2010) Cyanobacteria and ultraviolet radiation (UVR) stress: mitigation strategies. Ageing Res Rev 9(2):79–90
62. Fookes CJ, Jeffrey S (1989) The structure of chlorophyll c 3, a novel marine photosynthetic pigment. J Chem Soc Chem Commun 23:1827–1828
63. Strain HH, Manning WM, Hardin G (1943) Chlorophyll c (chlorofucine) of diatoms and dinoflagellates. J Biol Chem 148(3):655–668
64. Miyashita H et al (1996) Chlorophyll d as a major pigment. Nature 383(6599):402–402
65. Trampe E, Kühl M (2016) Chlorophyll f distribution and dynamics in cyanobacterial beachrock biofilms. J Phycol 52(6):990–996
66. Chen M et al (2010) A red-shifted chlorophyll. Science 329(5997):1318–1319
67. Scheer H (2006) An overview of chlorophylls and bacteriochlorophylls: biochemistry, biophysics, functions and applications. Chlorophylls Bacteriochlorophylls:1–26
68. Mullet JE, Burke JJ, Arntzen CJ (1980) Chlorophyll proteins of photosystem I. Plant Physiol 65(5):814–822
69. Blankenship RE (2014) Molecular mechanisms of photosynthesis. Wiley
70. Cherepanov DA et al (2020) Evidence that chlorophyll f functions solely as an antenna pigment in far-red-light photosystem I from Fischerella thermalis PCC 7521. Biochim Biophys Acta (BBA) – Bioenergetics 1861(5–6):148184
71. Loughlin P, Lin Y, Chen M (2013) Chlorophyll d and Acaryochloris marina: current status. Photosynth Res 116(2):277–293
72. Green BR, Durnford DG (1996) The chlorophyll-carotenoid proteins of oxygenic photosynthesis. Annu Rev Plant Biol 47(1):685–714
73. Niyogi KK, Truong TB (2013) Evolution of flexible non-photochemical quenching mechanisms that regulate light harvesting in oxygenic photosynthesis. Curr Opin Plant Biol 16(3): 307–314
74. Henríquez V et al (2016) Carotenoids in microalgae. In: Carotenoids in nature. Springer, pp 219–237
75. Young A, Britton G (2012) Carotenoids in photosynthesis. Springer Science & Business Media
76. Yabuzaki J (2017) Carotenoids database: structures, chemical fingerprints and distribution among organisms. Database
77. Britton G, Liaaen-Jensen S, Pfander H (2012) Carotenoids: handbook. Birkhäuser
78. Gruszecki WI, Strzałka K (2005) Carotenoids as modulators of lipid membrane physical properties. Biochim Biophys Acta 1740(2):108–115
79. Tóth TN et al (2015) Carotenoids are essential for the assembly of cyanobacterial photosynthetic complexes. Biochim Biophys Acta (BBA) – Bioenergetics 1847(10):1153–1165
80. Qin G et al (2007) Disruption of phytoene desaturase gene results in albino and dwarf phenotypes in Arabidopsis by impairing chlorophyll, carotenoid, and gibberellin biosynthesis. Cell Res 17(5):471–482
81. Armstrong G (1999) Carotenoid genetics and biochemistry
82. Paliwal C et al (2016) Microalgal carotenoids: potential nutraceutical compounds with chemotaxonomic importance. Algal Res 15:24–31
83. Amagata T (2010) Natural products structural diversity-II secondary metabolites: sources, structures and chemical biology. Comp Nat Prod II 2:581–621

84. Zakar T et al (2016) Carotenoids assist in cyanobacterial photosystem II assembly and function. Front Plant Sci 7:295
85. Balevicius V et al (2017) Fine control of chlorophyll-carotenoid interactions defines the functionality of light-harvesting proteins in plants. Sci Rep 7(1):1–10
86. Aro E-M, Virgin I, Andersson B (1993) Photoinhibition of photosystem II. Inactivation, protein damage and turnover. Biochim Biophys Acta (BBA) – Bioenergetics 1143(2):113–134
87. Loll B et al (2005) Towards complete cofactor arrangement in the 3.0 Å resolution structure of photosystem II. Nature 438(7070):1040–1044
88. Umena Y et al (2011) Crystal structure of oxygen-evolving photosystem II at a resolution of 19 Å 473(7345):55
89. Gisriel CJ et al (2022) Structure of a photosystem I-ferredoxin complex from a marine cyanobacterium provides insights into far-red light photoacclimation. J Biol Chem 298(1)
90. Sozer O et al (2010) Involvement of carotenoids in the synthesis and assembly of protein subunits of photosynthetic reaction centers of Synechocystis sp. PCC 6803. Plant Cell Physiol 51(5):823–835
91. Butler W, Kitajima M (1975) Energy transfer between photosystem II and photosystem I in chloroplasts. Biochim Biophys Acta (BBA) – Bioenergetics 396(1):72–85
92. Rakhimberdieva MG et al (2004) Carotenoid-induced quenching of the phycobilisome fluorescence in photosystem II-deficient mutant of Synechocystis sp. FEBS Lett 574(1-3):85–88
93. Wilson A et al (2008) A photoactive carotenoid protein acting as light intensity sensor. Proc Natl Acad Sci 105(33):12075–12080
94. Barber J (1998) Photosystem two. Biochim Biophys Acta (BBA) – Bioenergetics 1365(1-2): 269–277
95. Jahns P, Holzwarth AR (2012) The role of the xanthophyll cycle and of lutein in photoprotection of photosystem II. Biochim Biophys Acta (BBA) – Bioenergetics 1817(1): 182–193
96. Negi S et al (2020) Light regulation of light-harvesting antenna size substantially enhances photosynthetic efficiency and biomass yield in green algae. Plant J
97. Stitt M (1996) Metabolic regulation of photosynthesis. In: Photosynthesis and the environment. Springer, pp 151–190
98. Golbeck JH (2007) Photosystem I: the light-driven plastocyanin: ferredoxin oxidoreductase, vol 24. Springer Science & Business Media
99. Kusama Y et al (2015) Zeaxanthin and echinenone protect the repair of photosystem II from inhibition by singlet oxygen in Synechocystis sp. PCC 6803. Plant Cell Physiol 56(5):906–916
100. Sinha RP, Häder D-P (2008) UV-protectants in cyanobacteria. Plant Sci 174(3):278–289
101. Pathak J et al (2019) Cyanobacterial secondary metabolite scytonemin: a potential photoprotective and pharmaceutical compound. Proc Natl Acad Sci India Sect B Biol Sci:1–15
102. Grant CS, Louda J (2013) Scytonemin-imine, a mahogany-colored UV/Vis sunscreen of cyanobacteria exposed to intense solar radiation. Org Geochem 65:29–36
103. Couradeau E et al (2016) Bacteria increase arid-land soil surface temperature through the production of sunscreens. Nat Commun 7(1):1–7
104. Rastogi RP, Sonani RR, Madamwar D (2015) Cyanobacterial sunscreen scytonemin: role in photoprotection and biomedical research. Appl Biochem Biotechnol 176(6):1551–1563
105. Garcia-Pichel F, Sherry ND, Castenholz RW (1992) Evidence for an ultraviolet sunscreen role of the extracellular pigment scytonemin in the terrestrial cyanobacterium Chiorogloeopsis sp. Photochem Photobiol 56(1):17–23
106. Gupta S et al (2007) Use of natural carotenoids for pigmentation in fishes
107. Becquerel E (1874) Action des rayons differemment refrangibles sur l'iodure et le bromure d'argent; influence des matieres colorantes. Compt Rend Acad Sci [Paris] 79:185
108. Liebler DC (1993) Antioxidant reactions of carotenoids a. Ann N Y Acad Sci 691(1):20–31
109. Metibemu DS et al (2020) Carotenoid isolates of Spondias mombin demonstrate anticancer effects in DMBA-induced breast cancer in Wistar rats through X-linked inhibitor of apoptosis protein (XIAP) antagonism and anti-inflammation. J Food Biochem:e13523

110. Kang MR et al (2020) Inhibition of skin inflammation by scytonemin, an ultraviolet sunscreen pigment. Mar Drugs 18(6):300
111. Hussein MM et al (2020) Anti-obesity effects of individual or combination treatment with Spirulina platensis and green coffee bean aqueous extracts in high-fat diet-induced obese rats. All Life 13(1):328–338
112. Zhang L, Wang H (2015) Multiple mechanisms of anti-cancer effects exerted by astaxanthin. Mar Drugs 13(7):4310–4330
113. Fakhri S et al (2019) The neuroprotective effects of astaxanthin: therapeutic targets and clinical perspective. Molecules 24(14):2640
114. Udayan A, Arumugam M, Pandey A (2017) Nutraceuticals from algae and cyanobacteria. In: Algal Green chemistry. Elsevier, pp 65–89
115. FDA (2020) CFR – code of federal regulations title 21 – Sec. 73.530 Spirulina extract. [Cited 2021 20/10/2021]; Available from: https://www.accessdata.fda.gov/scripts/cdrh/cfdocs/cfcfr/CFRSearch.cfm?fr=73.530
116. Ghosh T, Mishra S (2020) A natural cyanobacterial protein C-phycoerythrin as an HS– selective optical probe in aqueous systems. Spectrochim Acta A Mol Biomol Spectrosc 239: 118469
117. Kechkeche D et al (2018) Semiconductor nanoplatelets: a new class of ultrabright fluorescent probes for cytometric and imaging applications. ACS Appl Mater Interfaces 10(29): 24739–24749
118. Lopes G, Clarinha D, Vasconcelos V (2020) Carotenoids from cyanobacteria: a biotechnological approach for the topical treatment of psoriasis. Microorganisms 8(2):302
119. GrandViewResearch (2016) Carotenoids market size worth $1.74 billion by 2025. [Cited 2021 12/10/2021]; Available from: https://www.grandviewresearch.com/press-release/global-carotenoids-market
120. Rammuni M et al (2019) Comparative assessment on the extraction of carotenoids from microalgal sources: astaxanthin from H. pluvialis and β-carotene from D. salina. Food Chem 277:128–134
121. ValueMarketResearch (2020) Global chlorophyll extract market report by type (liquid, tablet and powder), by application (food additive, cosmetics and dietary supplement) and by regions – industry trends, size, share, growth, estimation and Forecast, 2019–2026. [Cited 2020 14/12/2020]; Available from: https://www.valuemarketresearch.com/report/chlorophyll-extract-market#:~:text=The%20latest%20report%20by%20Value,CAGR%20from%202019%20to%202025
122. Pulz O, Gross W (2004) Valuable products from biotechnology of microalgae. Appl Microbiol Biotechnol 65(6):635–648
123. Spolaore P et al (2006) Commercial applications of microalgae. J Biosci Bioeng 101(2):87–96
124. Ambati RR et al (2014) Astaxanthin: sources, extraction, stability, biological activities and its commercial applications – a review. Mar Drugs 12(1):128–152
125. Lim KC et al (2018) Astaxanthin as feed supplement in aquatic animals. Rev Aquac 10(3): 738–773
126. Krinsky NI, Landrum JT, Bone RA (2003) Biologic mechanisms of the protective role of lutein and zeaxanthin in the eye. Annu Rev Nutr 23(1):171–201
127. Stringham JM, Hammond Jr BR (2005) Dietary lutein and zeaxanthin: possible effects on visual function. Nutr Rev 63(2):59–64
128. Fuji L, Co CI (2004) New dietary ingredient notification for astaxanthin extracted from haematococcus algae. US Food Frug Adm 1:1–6
129. Ngamwonglumlert L, Devahastin S, Chiewchan N (2017) Natural colorants: pigment stability and extraction yield enhancement via utilization of appropriate pretreatment and extraction methods. Crit Rev Food Sci Nutr 57(15):3243–3259
130. VerifiedMarketResearch (2021) Global cosmetic pigments market size by elemental composition, by type, by application, by geographic scope and forecast. 25/08/2021]; Available from: https://www.verifiedmarketresearch.com/product/cosmetic-pigments-market/

131. Pangestuti R et al (2020) Cosmetics and cosmeceutical applications of microalgae pigments. In: Pigments from microalgae handbook. Springer, pp 611–633
132. Alfeus A (2016) Cyanobacteria as a source of compounds with cosmetics potential
133. Gao X et al (2021) Biotechnological production of the sunscreen pigment Scytonemin in cyanobacteria: progress and strategy. Mar Drugs 19(3):129
134. Abed RM, Dobretsov S, Sudesh K (2009) Applications of cyanobacteria in biotechnology. J Appl Microbiol 106(1):1–12
135. Darvin ME et al (2011) The role of carotenoids in human skin. Molecules 16(12): 10491–10506
136. Scarmo S et al (2010) Significant correlations of dermal total carotenoids and dermal lycopene with their respective plasma levels in healthy adults. Arch Biochem Biophys 504(1):34–39
137. Richa RR et al (2011) Biotechnological potential of mycosporine-like amino acids and phycobiliproteins of cyanobacterial origin. Biotechnol Bioinformatics Bioeng 1:159–171
138. Sekar S, Chandramohan M (2008) Phycobiliproteins as a commodity: trends in applied research, patents and commercialization. J Appl Phycol 20(2):113–136
139. Nowruzi B, Sarvari G, Blanco S (2020) The cosmetic application of cyanobacterial secondary metabolites. Algal Res 49:101959
140. Galetović A et al (2020) Use of phycobiliproteins from Atacama cyanobacteria as food colorants in a dairy beverage prototype. Foods 9(2):244
141. Manirafasha E et al (2016) Phycobiliprotein: potential microalgae derived pharmaceutical and biological reagent. Biochem Eng J 109:282–296
142. Czerwonka A et al (2018) Anticancer effect of the water extract of a commercial Spirulina (Arthrospira platensis) product on the human lung cancer A549 cell line. Biomed Pharmacother 106:292–302
143. Sumantran VN et al (2000) Differential regulation of apoptosis in normal versus transformed mammary epithelium by lutein and retinoic acid. Cancer Epidemiol Prev Biomark 9(3): 257–263
144. Zhang G, Zhang Z, Liu Z (2013) Scytonemin inhibits cell proliferation and arrests cell cycle through downregulating Plk1 activity in multiple myeloma cells. Tumor Biol 34(4): 2241–2247
145. Stevenson C et al (2002) Scytonemin-a marine natural product inhibitor of kinases key in hyperproliferative inflammatory diseases. Inflamm Res 51(2):112
146. Hitchcock A, Hunter CN, Canniffe DP (2020) Progress and challenges in engineering cyanobacteria as chassis for light-driven biotechnology. J Microbial Biotechnol 13(2): 363–367
147. Ambati RR et al (2019) Industrial potential of carotenoid pigments from microalgae: current trends and future prospects. Crit Rev Food Sci Nutr 59(12):1880–1902
148. Han Y et al (2020) Exploring nutrient and light limitation of algal production in a shallow turbid reservoir. Environ Pollut:116210
149. Radzun KA et al (2015) Automated nutrient screening system enables high-throughput optimisation of microalgae production conditions. Biotechnol Biofuels 8(1):65
150. Sarma T, Ahuja G, Khattar J (2000) Effect of nutrients and aeration on O 2 evolution and photosynthetic pigments of Anabœna torulosa during akinete differentiation. Folia Microbiol 45(5):434–438
151. Zhu C et al (2018) Large-scale cultivation of Spirulina in a floating horizontal photobioreactor without aeration or an agitation device. Appl Microbiol Biotechnol 102(20):8979–8987
152. Béchet Q et al (2011) Universal temperature model for shallow algal ponds provides improved accuracy. Environ Sci Technol 45(8):3702–3709
153. Rai SV, Rajashekhar M (2014) Effect of pH, salinity and temperature on the growth of six species of marine phytoplankton. J Algal Biomass Util 5(4):55–59
154. Ravelonandro PH et al (2011) Improvement of the growth of Arthrospira (Spirulina) platensis from Toliara (Madagascar): effect of agitation, salinity and CO2 addition. Food Bioprod Process 89(3):209–216

155. Pushparaj B et al (1997) As integrated culture system for outdoor production of microalgae and cyanobacteria. J Appl Phycol 9(2):113–119
156. Moreno J et al (2003) Outdoor cultivation of a nitrogen-fixing marine cyanobacterium, Anabaena sp. ATCC 33047. Biomol Eng 20(4-6):191–197
157. Carlozzi P (2003) Dilution of solar radiation through "culture" lamination in photobioreactor rows facing south–north: a way to improve the efficiency of light utilization by cyanobacteria (Arthrospira platensis). Biotechnol Bioeng 81(3):305–315
158. Leema JM et al (2010) High value pigment production from Arthrospira (Spirulina) platensis cultured in seawater. Bioresour Technol 101(23):9221–9227
159. Walter A et al (2011) Study of phycocyanin production from Spirulina platensis under different light spectra. Braz Arch Biol Technol 54(4):675–682
160. Deshmukh DV, Puranik PR (2012) Statistical evaluation of nutritional components impacting phycocyanin production in Synechocystis sp. Braz J Microbiol 43:348–355
161. Zeng X et al (2012) Autotrophic cultivation of Spirulina platensis for CO2 fixation and phycocyanin production. Chem Eng J 183:192–197
162. Chainapong T, Traichaiyaporn S, Deming RL (2012) Effect of light quality on biomass and pigment production in photoautotrophic and mixotrophic cultures of Spirulina platensis. J Agric Technol 2012(8):1593–1604
163. Chen C-Y et al (2013) Engineering strategies for simultaneous enhancement of C-phycocyanin production and CO2 fixation with Spirulina platensis. Bioresour Technol 145:307–312
164. Xie Y et al (2015) Fed-batch strategy for enhancing cell growth and C-phycocyanin production of Arthrospira (Spirulina) platensis under phototrophic cultivation. Bioresour Technol 180:281–287
165. Salama A et al (2015) Maximising phycocyanin extraction from a newly identified Egyptian cyanobacteria strain: Anabaena oryzae SOS13. Int Food Res J 22(2)
166. Rosales Loaiza N et al (2016) Comparative growth and biochemical composition of four strains of Nostoc and Anabaena (cyanobacteria, Nostocales) in relation to sodium nitrate. Acta Biol Colomb 21(2):347–354
167. Lee NK et al (2017) Higher production of C-phycocyanin by nitrogen-free (diazotrophic) cultivation of Nostoc sp. NK and simplified extraction by dark-cold shock. Bioresour Technol 227:164–170
168. Kovac D et al (2017) The production of biomass and phycobiliprotein pigments in filamentous cyanobacteria: the impact of light and carbon sources. Appl Biochem Microbiol 53(5): 539–545
169. Lee SY, Nielsen J, Stephanopoulos G (2021) Cyanobacteria biotechnology. Wiley
170. Assuncao J et al (2021) Synechocystis salina: potential bioactivity and combined extraction of added-value metabolites. J Appl Phycol 33(6):3731–3746
171. Sarnaik A et al (2018) Recombinant Synechococcus elongatus PCC 7942 for improved zeaxanthin production under natural light conditions. Algal Res 36:139–151
172. Roles J et al (2020) Charting a development path to deliver cost competitive microalgae-based fuels. Algal Res 45:101721
173. Schmidt S, Raven JA, Paungfoo-Lonhienne C (2013) The mixotrophic nature of photosynthetic plants. Funct Plant Biol 40(5):425–438
174. Borsari RRJ et al (2007) Mixotrophic growth of Nostoc sp. on glucose, sucrose and sugarcane molasses for phycobiliprotein production. Acta Sci Biol Sci 29(1):9–13
175. Chen T et al (2006) Mixotrophic culture of high selenium-enriched Spirulina platensis on acetate and the enhanced production of photosynthetic pigments. Enzyme Microb Technol 39(1):103–107
176. Guoce Y et al (2011) Growth and physiological features of cyanobacterium Anabaena sp. strain PCC 7120 in a glucose-mixotrophic culture. Chin J Chem Eng 19(1):108–115
177. Moon M et al (2013) Mixotrophic growth with acetate or volatile fatty acids maximizes growth and lipid production in Chlamydomonas reinhardtii. Algal Res 2(4):352–357

178. Rajendran L, Nagarajan NG, Karuppan M (2020) Enhanced biomass and lutein production by mixotrophic cultivation of Scenedesmus sp. using crude glycerol in an airlift photobioreactor. Biochem Eng J 161:107684
179. Schwarz A et al (2020) Influence of heterotrophic and mixotrophic cultivation on growth behaviour of terrestrial cyanobacteria. Algal Res 52:102125
180. Zhang CC et al (1989) Molecular and genetical analysis of the fructose-glucose transport system in the cyanobacterium Synechocystis PCC6803. Mol Microbiol 3(9):1221–1229
181. Ekman M et al (2013) A Nostoc punctiforme sugar transporter necessary to establish a cyanobacterium-plant symbiosis. Plant Physiol 161(4):1984–1992
182. McEwen JT et al (2013) Engineering Synechococcus elongatus PCC 7942 for continuous growth under diurnal conditions. Appl Environ Microbiol 79(5):1668–1675
183. Wolk CP, Shaffer PW (1976) Heterotrophic micro-and macrocultures of a nitrogen-fixing cyanobacterium. Arch Microbiol 110(2):145–147
184. Marquez FJ et al (1995) Enhancement of biomass and pigment production during growth of Spirulina platensis in mixotrophic culture. J Chem Technol Biotechnol 62(2):159–164
185. Vonshak A, Cheung SM, Chen F (2000) Mixotrophic growth modifies the response of Spirulina (Arthrospira) platensis (cyanobacteria) cells to light. J Phycol 36(4):675–679
186. Markou G, Vandamme D, Muylaert KJWR (2014) Microalgal and cyanobacterial cultivation: the supply of nutrients. 65:186–202
187. Chi Z et al (2013) Bicarbonate-based integrated carbon capture and algae production system with alkalihalophilic cyanobacterium. Bioresour Technol 133:513–521
188. Rubin E, De Coninck HJUCUPTCCFCS (2005) Part, IPCC special report on carbon dioxide capture and storage. 2:14
189. Isleten-Hosoglu M, Gultepe I, Elibol MJBEJ (2012) Optimization of carbon and nitrogen sources for biomass and lipid production by Chlorella saccharophila under heterotrophic conditions and development of Nile red fluorescence based method for quantification of its neutral lipid content. 61:11–19
190. Vaccari DA, Strigul NJC (2011) Extrapolating phosphorus production to estimate resource reserves. 84(6):792–797
191. Powell N et al (2008) Factors influencing luxury uptake of phosphorus by microalgae in waste stabilization ponds. 42(16):5958–5962
192. Singh NK, Parmar A, Madamwar D (2009) Optimization of medium components for increased production of C-phycocyanin from Phormidium ceylanicum and its purification by single step process. Bioresour Technol 100(4):1663–1669
193. Chigri F, Soll J, Vothknecht UCJTPJ (2005) Calcium regulation of chloroplast protein import. 42(6):821–831
194. Roh MH et al (1998) Direct measurement of calcium transport across chloroplast inner-envelope vesicles. 118(4):1447–1454
195. Brand JJ, Becker DWJJOB (1984) Evidence for direct roles of calcium in photosynthesis. J Bioenerg Biomembr 16(4):239–249
196. Khattar J et al (2015) Hyperproduction of phycobiliproteins by the cyanobacterium Anabaena fertilissima PUPCCC 410.5 under optimized culture conditions. Algal Res 12:463–469
197. Ahad RIA, Syiem MB (2019) Influence of calcium on cadmium uptake and toxicity to the cyanobacterium Nostoc muscorum Meg 1. Biotechnol Res Innov 3(2):231–241
198. Kieke M et al (2016) Degradation rates and products of pure magnesium exposed to different aqueous media under physiological conditions. 17(3-4):131–143
199. Finkle BJ, Appleman DJPP (1953) The effect of magnesium concentration on growth of Chlorella. Plant Physiol 28(4):664
200. Huber SC, Maury WJPP (1980) Effects of magnesium on intact chloroplasts: I Evidence for activation of (sodium) potassium/proton exchange across the chloroplast envelope. Plant Physiol 65(2):350–354
201. Shaul OJB (2002) Magnesium transport and function in plants: the tip of the iceberg. Biometals 15(3):307–321

202. Lee J et al (2021) Rapid phosphate uptake via an ABC transporter induced by sulfate deficiency in Synechocystis sp. PCC 6803. Algal Res 60:102530
203. Sutak R et al (2012) A comparative study of iron uptake mechanisms in marine microalgae: iron binding at the cell surface is a critical Step1 [W][OA]
204. Greene RM et al (1992) Iron-induced changes in light harvesting and photochemical energy conversion processes in eukaryotic marine algae. Plant Physiol 100(2):565–575
205. Mogany T et al (2018) Elucidating the role of nutrients in C-phycocyanin production by the halophilic cyanobacterium Euhalothece sp. J Appl Phycol 30(4):2259–2271
206. Cerhan JR et al (2003) Antioxidant micronutrients and risk of rheumatoid arthritis in a cohort of older women. Am J Epidemiol 157(4):345–354
207. Carvalho AP et al (2006) Metabolic relationships between macro-and micronutrients, and the eicosapentaenoic acid and docosahexaenoic acid contents of Pavlova lutheri. 38(3–4): 358–366
208. Kaushik MS et al (2015) Role of manganese in protection against oxidative stress under iron starvation in cyanobacterium Anabaena 7120. J Basic Microbiol 55(6):729–740
209. Johnson HL et al (2007) Copper and zinc tolerance of two tropical microalgae after copper acclimation. Environ Toxicol 22(3):234–244
210. Chay TC, Surif S, Heng LY (2005) A copper toxicity biosensor using immobilized cyanobacteria, *Anabaena torulosa*. Sens Lett 3(1–2):49–54
211. Jardim WF, Pearson H (1985) Copper toxicity to cyanobacteria and its dependence on extracellular ligand concentration and degradation. Microb Ecol 11(2):139–148
212. Khairy HJWASJ (2009) Toxicity and accumulation of copper in Nannochloropsis oculata (Eustigmatophyceae, Heterokonta). 6(3):378–384
213. Levy JL et al (2008) Uptake and internalisation of copper by three marine microalgae: comparison of copper-sensitive and copper-tolerant species. Aquat Toxicol 89(2):82–93
214. Devi YM, Mehta S (2014) Changes in antioxidative enzymes of cyanobacterium Nostoc muscorum under copper (Cu2+) stress. Sci Vision 14:207–214
215. Zhou G-J et al (2012) Biosorption of zinc and copper from aqueous solutions by two freshwater green microalgae Chlorella pyrenoidosa and Scenedesmus obliquus. 19(7): 2918–2929
216. Zhou T et al (2018) Characterization of additional zinc ions on the growth, biochemical composition and photosynthetic performance from Spirulina platensis. Bioresour Technol 269:285–291
217. Mohanty P (1989) Effect of elevated levels of zinc on growth of Synechococcus 6301. Zentralblatt fuer Mikrobiologie 144(7):531–536
218. Carrano CJ et al (2009) Boron and marine life: a new look at an enigmatic bioelement. Mar Biotechnol (NY) 11(4):431
219. Amin SA et al (2007) Boron binding by a siderophore isolated from marine bacteria associated with the toxic dinoflagellate Gymnodinium catenatum. J Am Chem Soc 129(3):478–479
220. Rahman IY et al (2009) Removal of boron from produced water by Co-precipitation/ adsorption for reverse osmosis concentrate. p 156
221. Tarko T, Duda-Chodak A, Kobus M (2012) Influence of growth medium composition on synthesis of bioactive compounds and antioxidant properties of selected strains of Arthrospira cyanobacteria. Czech J Food Sci 30(3):258–267
222. Holm-Hansen O, Gerloff GC, Skoog FJPP (1954) Cobalt as an essential element for blue-green algae. 7(4):665–675
223. Babu TS, Sabat S, Mohanty P (1992) Alterations in photosystem II organization by cobalt treatment in the cyanobacterium Spirulina piatensis. J Plant Biochem Biotechnol 1(1):61–63
224. Fay P, de Vasconcelos L (1974) Nitrogen metabolism and ultrastructure in Anabaena cylindrica. Arch Microbiol 99(1):221–230
225. Glass JB (2011) Molybdenum biogeochemistry in an evolutionary context: nitrogen assimilation, microbial storage and environmental budgets. Arizona State University

226. Pilon-Smits EA, Quinn CF (2010) Selenium metabolism in plants. In: Cell biology of metals and nutrients. Springer, pp 225–241
227. Huang Z et al (2007) Characterization and antioxidant activity of selenium-containing phycocyanin isolated from Spirulina platensis. Food Chem 100(3):1137–1143
228. Malhotra B, Glass ADJPP (1995) Potassium fluxes in Chlamydomonas reinhardtii (I. Kinetics and electrical potentials). Plant Physiol 108(4):1527–1536
229. Shabala S, Cuin TAJPP (2008) Potassium transport and plant salt tolerance. Physiol Plant 133(4):651–669
230. Seale D, Boraas M, Warren GJWR (1987) Effects of sodium and phosphate on growth of cyanobacteria. 21(6):625–631
231. Kebede EJJOAP (1997) Response of Spirulina platensis (= Arthrospira fusiformis) from Lake Chitu, Ethiopia, to salinity stress from sodium salts. 9(6):551–558
232. Rao AR et al (2007) Effect of salinity on growth of green alga Botryococcus braunii and its constituents. Bioresour Technol 98(3):560–564
233. Jepsen PM et al (2019) Effects of salinity, commercial salts, and water type on cultivation of the cryptophyte microalgae Rhodomonas salina and the calanoid copepod Acartia tonsa. 50(1): 104–118
234. Sharma G et al (2014) Effect of carbon content, salinity and pH on Spirulina platensis for phycocyanin, allophycocyanin and phycoerythrin accumulation. J Microb Biochem Technol 6:202–206
235. Ptacnik R, Andersen T, Tamminen T (2010) Performance of the Redfield ratio and a family of nutrient limitation indicators as thresholds for phytoplankton N vs. P limitation. Ecosystems 13(8):1201–1214
236. Ho TY et al (2003) The elemental composition of some marine phytoplankton 1. J Phycol 39(6):1145–1159
237. Quigg A et al (2003) The evolutionary inheritance of elemental stoichiometry in marine phytoplankton. Nature 425(6955):291–294
238. Tett P, Droop M, Heaney S (1985) The Redfield ratio and phytoplankton growth rate. J Mar Biol Assoc U K 65(2):487–504
239. Chaiklahan R et al (2010) Cultivation of Spirulina platensis using pig wastewater in a semi-continuous process. J Microbiol Biotechnol 20(3):609–614
240. Lim HR et al (2021) Perspective of Spirulina culture with wastewater into a sustainable circular bioeconomy. Environ Pollut 284:117492
241. Wolf J et al (2015) High-throughput screen for high performance microalgae strain selection and integrated media design. Algal Res 11:313–325
242. Afroz S, Singh R. Cultivation of super food–Spirulina (Blue-green Algae): an agribusiness outlook
243. Olguin E, Sánchez-Galván G (2011) Phycoremediation: current challenges and applications
244. Schenk P (2016) On-farm algal ponds to provide protein for northern cattle. Meat and Lifestock Australia, North Sydney NSW
245. Wolf J (2015) Effective scale up of microalgal systems for the production of biomass and biofuels
246. Kumar K et al (2015) Recent trends in the mass cultivation of algae in raceway ponds. Renew Sustain Energy Rev 51:875–885
247. Colosi LM et al (2012) Will algae produce the green? Using published life cycle assessments as a starting point for economic evaluation of future algae-to-energy systems. Biofuels 3(2): 129–142
248. Beal CM et al (2012) Energy return on Investment for Algal biofuel production coupled with wastewater treatment. Water Environ Res 84(9):692–710
249. Travieso L et al (2001) A helical tubular photobioreactor producing Spirulina in a semicontinuous mode. Int Biodeter Biodegr 47(3):151–155
250. Ferreira L et al (2012) Arthrospira (spirulina) platensis cultivation in tubular photobioreactor: use of no-cost CO2 from ethanol fermentation. Appl Energy 92:379–385

251. Oncel S, Sukan FV (2008) Comparison of two different pneumatically mixed column photobioreactors for the cultivation of Artrospira platensis (Spirulina platensis). Bioresour Technol 99(11):4755–4760
252. Zhang S et al (2021) Observation of Spirulina platensis cultivation in a prototype household bubble column photobioreactor during 107 days. Biotechnol Biotechnol Equip 35(1): 1669–1679
253. Araújo R et al (2021) Current status of the algae production industry in Europe: an emerging sector of the blue bioeconomy. Front Mar Sci 7:1247
254. Droop MR (1973) Some thoughts on nutrient limitation in algae 1. J Phycol 9(3):264–272
255. Barbosa M (2003) Microalgal photobioreactors: scale-up and optimisation
256. Gaylarde C (2020) Influence of environment on microbial colonization of historic stone buildings with emphasis on cyanobacteria. Heritage 3(4):1469–1482
257. Jeon Y-C, Cho C-W, Yun Y-S (2006) Combined effects of light intensity and acetate concentration on the growth of unicellular microalga Haematococcus pluvialis. Enzyme Microb Technol 39(3):490–495
258. Wang W et al (2019) Structural basis for blue-green light harvesting and energy dissipation in diatoms. Science 363(6427)
259. Pereira S, Otero A (2020) Haematococcus pluvialis bioprocess optimization: effect of light quality, temperature and irradiance on growth, pigment content and photosynthetic response. Algal Res 51:102027
260. Kirilovsky D, Kerfeld CA (2013) The orange carotenoid protein: a blue-green light photoactive protein. Photochem Photobiol Sci 12(7):1135–1143
261. Kirilovsky D, Kerfeld CA (2016) Cyanobacterial photoprotection by the orange carotenoid protein. Nat Plants 2(12):1–7
262. Bondanza M et al (2020) The multiple roles of the protein in the photoactivation of orange carotenoid protein. Chem 6(1):187–203
263. McConnell MD et al (2002) Regulation of the distribution of chlorophyll and phycobilin-absorbed excitation energy in cyanobacteria. A structure-based model for the light state transition. Plant Physiol 130(3):1201–1212
264. Jajoo A et al (2014) Low pH-induced regulation of excitation energy between the two photosystems. FEBS Lett 588(6):970–974
265. Karapetyan N et al (2014) Long-wavelength chlorophylls in photosystem I of cyanobacteria: origin, localization, and functions. Biochemistry (Mosc) 79(3):213–220
266. Shubin VV et al (2008) Quantum yield of P700+ photodestruction in isolated photosystem I complexes of the cyanobacterium Arthrospira platensis. Photochem Photobiol Sci 7(8): 956–962
267. Karapetyan N (2007) Non-photochemical quenching of fluorescence in cyanobacteria. Biochemistry (Mosc) 72(10):1127–1135
268. Abramavicius D, Mukamel S (2009) Exciton delocalization and transport in photosystem I of cyanobacteria Synechococcus elongates: simulation study of coherent two-dimensional optical signals. J Phys Chem B 113(17):6097–6108
269. Tamary E et al (2012) Structural and functional alterations of cyanobacterial phycobilisomes induced by high-light stress. Biochim Biophys Acta (BBA) – Bioenergetics 1817(2):319–327
270. Kilimtzidi E et al (2019) Enhanced phycocyanin and protein content of Arthrospira by applying neutral density and red light shading filters: a small-scale pilot experiment. J Chem Technol Biotechnol 94(6):2047–2054
271. Akimoto S, Yokono M (2017) How light-harvesting and energy-transfer processes are modified under different light conditions: studies by time-resolved fluorescence spectroscopy. In: Photosynthesis: structures, mechanisms, and applications. Springer, pp 169–184
272. Chaiklahan R et al (2022) Enhanced biomass and phycocyanin production of Arthrospira (Spirulina) platensis by a cultivation management strategy: light intensity and cell concentration. Bioresour Technol 343:126077

273. Olsson-Francis K et al (2012) The effect of rock composition on cyanobacterial weathering of crystalline basalt and rhyolite. Geobiology 10(5):434–444
274. Abeynayaka HDL, Asaeda T, Kaneko Y (2017) Buoyancy limitation of filamentous cyanobacteria under prolonged pressure due to the gas vesicles collapse. Environ Manag 60(2):293–303
275. Billini M, Stamatakis K, Sophianopoulou V (2008) Two members of a network of putative Na +/H+ antiporters are involved in salt and pH tolerance of the freshwater cyanobacterium Synechococcus elongatus. J Bacteriol 190(19):6318–6329
276. Rezayian M, Niknam V, Ebrahimzadeh H (2019) Stress response in cyanobacteria. Iran J Plant Physiol 9(3):2773–2787
277. Brown A (1976) Microbial water stress. Bacteriol Rev 40(4):803–846
278. Klähn S, Hagemann M (2011) Compatible solute biosynthesis in cyanobacteria. Environ Microbiol 13(3):551–562
279. Kannaujiya VK, Sundaram S, Sinha RP (2017) Stress response of phycobiliproteins. In: Phycobiliproteins: recent developments and future applications. Springer, pp 71–82
280. Abd El-Baky HH, El-Baroty GS (2012) Characterization and bioactivity of phycocyanin isolated from Spirulina maxima grown under salt stress. Food Funct 3(4):381–388
281. Rippka (1988) Isolation and purification of cyanobacteria. Methods Enzymol. 167:3–27
282. Hinga KR (2002) Effects of pH on coastal marine phytoplankton. Mar Ecol Prog Ser 238:281–300
283. Pedersen D, Miller SR (2017) Photosynthetic temperature adaptation during niche diversification of the thermophilic cyanobacterium Synechococcus A/B clade. ISME J 11(4): 1053–1057
284. Rajaram H, Chaurasia AK, Apte SK (2014) Cyanobacterial heat-shock response: role and regulation of molecular chaperones. Microbiology 160(4):647–658
285. Tiwari A, Singh P, Asthana RK (2016) Role of calcium in the mitigation of heat stress in the cyanobacterium Anabaena PCC 7120. J Plant Physiol 199:67–75
286. Wang B, Lan CQJTCJOCE (2011) Optimising the lipid production of the green alga Neochloris oleoabundans using box–behnken experimental design. 89(4):932–939
287. Xiao Y et al (2016) Effect of small-scale turbulence on the physiology and morphology of two bloom-forming cyanobacteria. PLoS One 11(12):e0168925
288. Carr NG, Whitton BA (1982) The biology of cyanobacteria, vol 19. University of California Press
289. Salim S et al (2011) Harvesting of microalgae by bio-flocculation. J Appl Phycol 23(5): 849–855
290. Griffiths MJ, van Hille RP, Harrison ST (2012) Lipid productivity, settling potential and fatty acid profile of 11 microalgal species grown under nitrogen replete and limited conditions. J Appl Phycol 24(5):989–1001
291. Granados M et al (2012) Evaluation of flocculants for the recovery of freshwater microalgae. Bioresour Technol 118:102–110
292. Harith ZT et al (2009) Effect of different flocculants on the flocculation performance of flocculation performance of microalgae, Chaetoceros calcitrans, cells. Afr J Biotechnol 8(21)
293. Wu Z et al (2012) Evaluation of flocculation induced by pH increase for harvesting microalgae and reuse of flocculated medium. Bioresour Technol 110:496–502
294. Vandamme D et al (2010) Flocculation of microalgae using cationic starch. J Appl Phycol 22(4):525–530
295. Uduman N et al (2010) Dewatering of microalgal cultures: a major bottleneck to algae-based fuels. J Renew Sustain Energy 2(1):012701
296. Sukenik A, Shelef GJB (1984) Algal autoflocculation – verification and proposed mechanism. Biotechnol Bioeng 26(2):142–147
297. Mata TM, Martins AA, Caetano NS (2010) Microalgae for biodiesel production and other applications: a review. Renew Sustain Energy Rev 14(1):217–232

298. Chen P et al (2010) Review of biological and engineering aspects of algae to fuels approach. 2(4):1–30
299. Lee SY et al (2020) Techniques of lipid extraction from microalgae for biofuel production: a review. Environ Chem Lett:1–21
300. Landels A et al (2019) Improving electrocoagulation floatation for harvesting microalgae. Algal Res 39:101446
301. Kwon H et al (2014) Harvesting of microalgae using flocculation combined with dissolved air flotation. Biotechnol Bioprocess Eng 19(1):143–149
302. Lokare P (2021) Spirulina farming: to ingrain the entrepreneurship. Priya Lokare
303. Giménez JB et al (2018) Assessment of cross-flow filtration as microalgae harvesting technique prior to anaerobic digestion: evaluation of biomass integrity and energy demand. Bioresour Technol 269:188–194
304. Safi C et al (2017) Biorefinery of microalgal soluble proteins by sequential processing and membrane filtration. Bioresour Technol 225:151–158
305. Brennan L, Owende P (2010) Biofuels from microalgae – a review of technologies for production, processing, and extractions of biofuels and co-products. Renew Sustain Energy Rev 14(2):557–577
306. Rickman M, Pellegrino J, Davis R (2012) Fouling phenomena during membrane filtration of microalgae. J Membr Sci 423:33–42
307. Milledge JJ, Heaven S (2011) Disc stack centrifugation separation and cell disruption of microalgae: a technical note. Environ Nat Resour Res 1(1):17–24
308. Dassey AJ, Theegala CS (2013) Harvesting economics and strategies using centrifugation for cost effective separation of microalgae cells for biodiesel applications. Bioresour Technol 128: 241–245
309. Al Hattab M, Ghaly A, Hammouda A (2015) Microalgae harvesting methods for industrial production of biodiesel: critical review and comparative analysis. J Fundam Renew Energy Appl 5(2):1000154
310. de Souza Sossella F et al (2020) Effects of harvesting Spirulina platensis biomass using coagulants and electrocoagulation–flotation on enzymatic hydrolysis. Bioresour Technol 311:123526
311. Huang W-C, Kim J-D (2013) Cationic surfactant-based method for simultaneous harvesting and cell disruption of a microalgal biomass. Bioresour Technol 149:579–581
312. Różyło R (2020) Recent trends in methods used to obtain natural food colorants by freeze-drying. Trends Food Sci Technol
313. Alavijeh RS et al (2020) Combined bead milling and enzymatic hydrolysis for efficient fractionation of lipids, proteins, and carbohydrates of Chlorella vulgaris microalgae. Bioresour Technol:123321
314. Carullo D et al (2018) Effect of pulsed electric fields and high pressure homogenization on the aqueous extraction of intracellular compounds from the microalgae Chlorella vulgaris. Algal Res 31:60–69
315. Zhang R et al (2019) Effect of ultrasonication, high pressure homogenization and their combination on efficiency of extraction of bio-molecules from microalgae Parachlorella kessleri. Algal Res 40:101524
316. Leonhardt L et al (2020) Bio-refinery of Chlorella sorokiniana with pulsed electric field pre-treatment. Bioresour Technol 301:122743
317. Luengo E et al (2014) Effect of pulsed electric field treatments on permeabilization and extraction of pigments from Chlorella vulgaris. J Membr Biol 247(12):1269–1277
318. Parniakov O et al (2015) Pulsed electric field and pH assisted selective extraction of intracellular components from microalgae Nannochloropsis. Algal Res 8:128–134
319. Kapoore RV et al (2018) Microwave-assisted extraction for microalgae: from biofuels to biorefinery. Biology 7(1):18
320. Gunerken E et al (2015) Cell disruption for microalgae biorefineries. Biotechnol Adv 33(2): 243–260

321. Lee AK, Lewis DM, Ashman PJ (2012) Disruption of microalgal cells for the extraction of lipids for biofuels: processes and specific energy requirements. Biomass Bioenergy 46:89–101
322. D'hondt E et al (2017) Cell disruption technologies. In: Microalgae-based biofuels and bioproducts. Elsevier, pp 133–154
323. Kermanshahi-pour A et al (2014) Enzymatic and acid hydrolysis of Tetraselmis suecica for polysaccharide characterization. 173:415–421
324. Beale SI, Cornejo J (1984) Enzymatic heme oxygenase activity in soluble extracts of the unicellular red alga, Cyanidium caldarium. Arch Biochem Biophys 235(2):371–384
325. Gavalás-Olea A et al (2020) Enzymatic synthesis and characterization of chlorophyllide derivatives as possible internal standards for pigment chromatographic analysis. Algal Res 46:101688
326. Banerjee M et al (2021) Functional and mechanistic insights into the differential effect of the toxicant 'Se (IV)'in the cyanobacterium Anabaena PCC 7120. Aquat Toxicol 236:105839
327. Prabakaran P, Ravindran AD (2011) A comparative study on effective cell disruption methods for lipid extraction from microalgae. Lett Appl Microbiol 53(2):150–154
328. Lin J-Y, Ng I-S (2021) Production, isolation and characterization of C-phycocyanin from a new halo-tolerant Cyanobacterium aponinum using seawater. Bioresour Technol:125946
329. Strati IF, Gogou E, Oreopoulou V (2015) Enzyme and high pressure assisted extraction of carotenoids from tomato waste. Food Bioprod Process 94:668–674
330. Patrignani F, Lanciotti R (2016) Applications of high and ultra high pressure homogenization for food safety. Front Microbiol 7:1132
331. Azencott HR, Peter GF, Prausnitz MR (2007) Influence of the cell wall on intracellular delivery to algal cells by electroporation and sonication. Ultrasound Med Biol 33(11): 1805–1817
332. Keris-Sen UD et al (2014) An investigation of ultrasound effect on microalgal cell integrity and lipid extraction efficiency. Bioresour Technol 152:407–413
333. Lupatini AL et al (2017) Potential application of microalga Spirulina platensis as a protein source. J Sci Food Agric 97(3):724–732
334. Gerde JA et al (2013) Optimizing protein isolation from defatted and non-defatted Nannochloropsis microalgae biomass. Algal Res 2(2):145–153
335. de Boer K et al (2012) Extraction and conversion pathways for microalgae to biodiesel: a review focused on energy consumption. J Appl Phycol 24(6):1681–1698
336. Jaeschke DP et al (2019) Extraction of valuable compounds from Arthrospira platensis using pulsed electric field treatment. Bioresour Technol 283:207–212
337. Akaberi S et al (2020) Impact of incubation conditions on protein and C-phycocyanin recovery from Arthrospira platensis post-pulsed electric field treatment. Bioresour Technol 306:123099
338. Zheng H et al (2011) Disruption of Chlorella vulgaris cells for the release of biodiesel-producing lipids: a comparison of grinding, ultrasonication, bead milling, enzymatic lysis, and microwaves. Appl Biochem Biotechnol 164(7):1215–1224
339. Laurens L et al (2015) Acid-catalyzed algal biomass pretreatment for integrated lipid and carbohydrate-based biofuels production. Green Chem 17(2):1145–1158
340. Kulkarni S, Nikolov Z (2018) Process for selective extraction of pigments and functional proteins from Chlorella vulgaris. Algal Res 35:185–193
341. Levine SN et al (1997) Phosphorus, nitrogen, and silica as controls on phytoplankton biomass and species composition in Lake Champlain (USA-Canada). J Great Lakes Res 23(2):131–148
342. Sierra LS, Dixon CK, Wilken LR (2017) Enzymatic cell disruption of the microalgae Chlamydomonas reinhardtii for lipid and protein extraction. Algal Res 25:149–159
343. Grima EM, González MJI, Giménez AG (2013) Solvent extraction for microalgae lipids. In: Algae for biofuels and energy. Springer, pp 187–205
344. Llewellyn CA et al (2019) Deriving economic value from metabolites in cyanobacteria. In: Grand challenges in algae biotechnology. Springer, pp 535–576
345. Taher H et al (2014) Effective extraction of microalgae lipids from wet biomass for biodiesel production. Biomass Bioenerg 66:159–167

346. Michalak I, Chojnacka K (2014) Algal extracts: technology and advances. Eng Life Sci 14(6): 581–591
347. Rodrigues RDP et al (2018) Ultrasound-assisted extraction of phycobiliproteins from Spirulina (Arthrospira) platensis using protic ionic liquids as solvent. Algal Res 31:454–462
348. Furuki T et al (2003) Rapid and selective extraction of phycocyanin from Spirulina platensis with ultrasonic cell disruption. J Appl Phycol 15(4):319–324
349. Jubeau S et al (2013) High pressure disruption: a two-step treatment for selective extraction of intracellular components from the microalga Porphyridium cruentum. J Appl Phycol 25(4): 983–989
350. Denery JR et al (2004) Pressurized fluid extraction of carotenoids from Haematococcus pluvialis and Dunaliella salina and kavalactones from Piper methysticum. Anal Chim Acta 501(2):175–181
351. Machmudah S et al (2006) Extraction of astaxanthin from Haematococcus pluvialis using supercritical CO2 at high pressure
352. Marchal L et al (2013) Centrifugal partition extraction of β-carotene from Dunaliella salina for efficient and biocompatible recovery of metabolites. Bioresour Technol 134:396–400
353. Hosseini SRP, Tavakoli O, Sarrafzadeh MH (2017) Experimental optimization of SC-CO2 extraction of carotenoids from Dunaliella salina. J Supercrit Fluids 121:89–95
354. Sachindra N, Bhaskar N, Mahendrakar N (2006) Recovery of carotenoids from shrimp waste in organic solvents. Waste Manag 26(10):1092–1098
355. Kumar P, Ramakritinan C, Kumaraguru A (2010) Solvent extraction and spectrophotometric determination of pigments of some algal species from the shore of Puthumadam, southeast coast of India. Int J Oceans Oceanogr 4(1):29–34
356. Metivier R, Francis F, Clydesdale F (1980) Solvent extraction of anthocyanins from wine pomace. J Food Sci 45(4):1099–1100
357. Henriques M, Silva A, Rocha J (2007) Extraction and quantification of pigments from a marine microalga: a simple and reproducible method. In: Communicating current research and educational topics and trends in applied microbiology, vol 2, pp 586–593
358. Desai RK et al (2016) Novel astaxanthin extraction from Haematococcus pluvialis using cell permeabilising ionic liquids. Green Chem 18(5):1261–1267
359. Hernández D et al (2014) Biofuels from microalgae: lipid extraction and methane production from the residual biomass in a biorefinery approach. Bioresour Technol 170:370–378
360. Chang Y-K et al (2018) Isolation of C-phycocyanin from Spirulina platensis microalga using ionic liquid based aqueous two-phase system. Bioresour Technol 270:320–327
361. Sanchez-Laso J et al (2021) A successful method for phycocyanin extraction from Arthrospira platensis using [Emim][EtSO4] ionic liquid. Biofuels Bioprod Biorefin 15(6):1638–1649
362. Nobre B et al (2006) Supercritical carbon dioxide extraction of astaxanthin and other carotenoids from the microalga Haematococcus pluvialis. Eur Food Res Technol 223(6):787–790
363. Macıas-Sánchez M et al (2005) Supercritical fluid extraction of carotenoids and chlorophyll a from Nannochloropsis gaditana. J Food Eng 66(2):245–251
364. Sun M, Temelli F (2006) Supercritical carbon dioxide extraction of carotenoids from carrot using canola oil as a continuous co-solvent. J Supercrit Fluids 37(3):397–408
365. Nobre B et al (2006) Supercritical carbon dioxide extraction of pigments from Bixa orellana seeds (experiments and modeling). Braz J Chem Eng 23(2):251–258
366. Jeffrey S (1974) Profiles of photosynthetic pigments in the ocean using thin-layer chromatography. Mar Biol 26(2):101–110
367. Cohen Z et al (1993) Production and partial purification of γ-linolenic acid and some pigments fromSpirulina platensis. J Appl Phycol 5(1):109–115
368. Reis A et al (1998) Production, extraction and purification of phycobiliproteins from Nostoc sp. Bioresour Technol 66(3):181–187
369. Abalde J et al (1998) Purification and characterization of phycocyanin from the marine cyanobacterium Synechococcus sp. IO9201. Plant Sci 136(1):109–120
370. Ranjitha K, Kaushik B (2005) Purification of phycobiliproteins from Nostoc muscorum

371. Jiang Z (2002) Recent research on extration and purification of phycobiliproteins [J]. Food Sci 11
372. Chaiklahan R et al (2018) Stepwise extraction of high-value chemicals from Arthrospira (Spirulina) and an economic feasibility study. Biotechnol Rep 20:e00280
373. Sivakaminathan S et al (2020) Light guide systems enhance microalgae production efficiency in outdoor high rate ponds. Algal Res 47:101846
374. Wolf J et al (2016) Multifactorial comparison of photobioreactor geometries in parallel microalgae cultivations. Algal Res 15:187–201
375. Huang J, Hankamer B, Yarnold J (2019) Design scenarios of outdoor arrayed cylindrical photobioreactors for microalgae cultivation considering solar radiation and temperature. Algal Res 41:101515
376. Boussiba S, Richmond AE (1980) C-phycocyanin as a storage protein in the blue-green alga Spirulina platensis. Arch Microbiol 125(1):143–147
377. Chaiklahan R, Chirasuwan N, Bunnag B (2012) Stability of phycocyanin extracted from Spirulina sp.: influence of temperature, pH and preservatives. Process Biochem 47(4):659–664
378. Torres GF et al (2021) Microalgae strain catalogue: a strain selection guide for microalgae users: cultivation and chemical characteristics for high added-value products
379. Jacob-Lopes E et al (2019) Bioactive food compounds from microalgae: an innovative framework on industrial biorefineries. Curr Opin Food Sci 25:1–7
380. Hu I-C (2019) Production of potential coproducts from microalgae. In: Biofuels from Algae. Elsevier, pp 345–358
381. Kirnev P et al (2020) Technological mapping and trends in photobioreactors for the production of microalgae. World J Microbiol Biotechnol 36(3):1–9
382. Hoffman J et al (2017) Techno-economic assessment of open microalgae production systems. Algal Res 23:51–57
383. Thomassen G et al (2016) A techno-economic assessment of an algal-based biorefinery. Clean Techn Environ Policy 18(6):1849–1862
384. Silva SC et al (2020) Microalgae-derived pigments: a 10-year bibliometric review and industry and market trend analysis. Molecules 25(15):3406
385. Torzillo G et al (1986) Production of Spirulina biomass in closed photobioreactors. Biomass 11(1):61–74
386. Schenk PM et al (2008) Second generation biofuels: high-efficiency microalgae for biodiesel production. Bioenergy Res 1(1):20–43
387. Guidi F et al (2021) Long-term cultivation of a native Arthrospira platensis (Spirulina) Strain in Pozo Izquierdo (Gran Canaria, Spain): technical evidence for a viable production of food-grade biomass. PRO 9(8):1333
388. Mohan SV et al (2020) Algal biorefinery models with self-sustainable closed loop approach: trends and prospective for blue-bioeconomy. Bioresour Technol 295:122128
389. Kumar AK et al (2020) Techno-economic analysis of microalgae production with simultaneous dairy effluent treatment using a pilot-scale high volume V-shape pond system. Renew Energy 145:1620–1632
390. Roles J et al (2021) Techno-economic evaluation of microalgae high-density liquid fuel production at 12 international locations. Biotechnol Biofuels 14(1):1–19
391. Barbosa MJ, Hoogakker J, Wijffels RH (2003) Optimisation of cultivation parameters in photobioreactors for microalgae cultivation using the A-stat technique. Biomol Eng 20(4-6): 115–123
392. Milledge JJ (2010) The challenge of algal fuel: economic processing of the entire algal biomass. Condensed Matter Mater Eng Newslett:4–6
393. Skjånes K, Rebours C, Lindblad P (2013) Potential for green microalgae to produce hydrogen, pharmaceuticals and other high value products in a combined process. Crit Rev Biotechnol 33(2):172–215
394. Nations U (2020) World economic situation and prospects 2020. [Cited 2020 28/12/20]; Available from: https://unctad.org/system/files/official-document/wesp2020_en.pdf

395. Kim JYH et al (2016) Microfluidic high-throughput selection of microalgal strains with superior photosynthetic productivity using competitive phototaxis. Sci Rep 6(1):1–11
396. Guillard RR (2005) Purification methods for microalgae. Algal Culturing Tech:117
397. Day JG (2007) Cryopreservation of microalgae and cyanobacteria. In: Cryopreservation and freeze-drying protocols. Springer, pp 141–151
398. Bui TV et al (2013) Impact of procedural steps and cryopreservation agents in the cryopreservation of chlorophyte microalgae. PLoS One 8(11):e78668
399. Sun A et al (2011) Comparative cost analysis of algal oil production for biofuels. Energy 36(8):5169–5179
400. WHO (2013) Global action plan for the prevention and control of noncommunicable diseases 2013-2020. World Health Organization
401. Mourelle ML, Gómez CP, Legido JL (2017) The potential use of marine microalgae and cyanobacteria in cosmetics and thalassotherapy. Cosmetics 4(4):46
402. Morone J et al (2019) Revealing the potential of cyanobacteria in cosmetics and cosmeceuticals – a new bioactive approach. Algal Res 41:101541
403. BCCResearch (2018) The global market for carotenoids. [Cited 2021 21/04/2021]; Available from: https://www.bccresearch.com/market-research/food-and-beverage/the-global-market-for-carotenoids.html
404. AstaReal (2021) AstaReal – natural astaxanthin. [Cited 2021 14/01/2021]; Available from: http://www.astareal.se/?informationmodal=true
405. Markets MA (2020) Carotenoids market by type (astaxanthin, beta-carotene, lutein, lycopene, canthaxanthin, and zeaxanthin), application (feed, food & beverages, dietary supplements, cosmetics, and pharmaceuticals), source, formulation, and region – global forecast to 2026. [Cited 2021 04/01/2021]; Available from: https://www.marketsandmarkets.com/Market-Reports/carotenoid-market-158421566.html
406. D'Alessandro EB, Antoniosi Filho NR (2016) Concepts and studies on lipid and pigments of microalgae: a review. Renew Sustain Energy Rev 58:832–841
407. Priyadarshani I, Rath B (2012) Commercial and industrial applications of micro algae–a review. J Algal Biomass Util 3(4):89–100

Adv Biochem Eng Biotechnol (2023) 183: 253–280
https://doi.org/10.1007/10_2023_216

Published online: 3 April 2023

Photobiohydrogen Production and Strategies for H_2 Yield Improvements in Cyanobacteria

Wanthanee Khetkorn, Wuttinun Raksajit, Cherdsak Maneeruttanarungroj, and Peter Lindblad

Contents

W. Khetkorn
Division of Biology, Faculty of Science and Technology, Rajamangala University of Technology, Thanyaburi, Pathum Thani, Thailand
e-mail: wanthanee_k@rmutt.ac.th

W. Raksajit
Faculty of Veterinary Technology, Program of Animal Health Technology, Kasetsart University, Bangkok, Thailand
e-mail: cvtwnr@ku.ac.th

C. Maneeruttanarungroj
Department of Biology, School of Science, King Mongkut's Institute of Technology Ladkrabang, Bangkok, Thailand

Bioenergy Research Unit, School of Science, King Mongkut's Institute of Technology Ladkrabang, Bangkok, Thailand
e-mail: cherdsak.ma@kmitl.ac.th

P. Lindblad (✉)
Microbial Chemistry, Department of Chemistry-Ångström, Uppsala University, Uppsala, Sweden
e-mail: peter.lindblad@kemi.uu.se

Abstract Hydrogen gas (H_2) is one of the potential future sustainable and clean energy carriers that may substitute the use of fossil resources including fuels since it has a high energy content (heating value of 141.65 MJ/kg) when compared to traditional hydrocarbon fuels [1]. Water is a primary product of combustion being a most significant advantage of H_2 being environmentally friendly with the capacity to reduce global greenhouse gas emissions. H_2 is used in various applications. It generates electricity in fuel cells, including applications in transportation, and can be applied as fuel in rocket engines [2]. Moreover, H_2 is an important gas and raw material in many industrial applications. However, the high cost of the H_2 production processes requiring the use of other energy sources is a significant disadvantage. At present, H_2 can be prepared in many conventional ways, such as steam reforming, electrolysis, and biohydrogen production processes. Steam reforming uses high-temperature steam to produce hydrogen gas from fossil resources including natural gas. Electrolysis is an electrolytic process to decompose water molecules into O_2 and H_2. However, both these two methods are energy-intensive and producing hydrogen from natural gas, which is mostly methane (CH_4) and in steam reforming generates CO_2 and pollutants as by-products. On the other hand, biological hydrogen production is more environmentally sustainable and less energy intensive than thermo-chemical and electrochemical processes [3], but most concepts are not yet developed to production scale.

Graphical Abstract

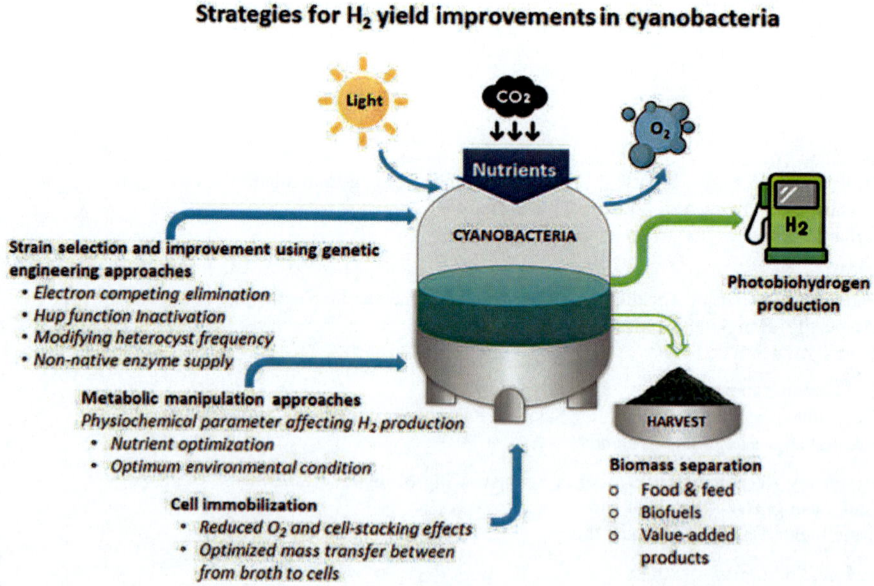

Keywords Photohydrogen, Cyanobacteria, Metabolic manipulation, Genetic engineering, Cell immobilization

1 Introduction

Hydrogen gas (H_2) is one of the potential future sustainable and clean energy carriers that may substitute the use of fossil resources including fuels since it has a high energy content (heating value of 141.65 MJ/kg) when compared to traditional hydrocarbon fuels [1]. Water is a byproduct of combustion being a most significant advantage of H_2 being environmentally friendly with the capacity to reduce global greenhouse gas emissions. H_2 is used in various applications. It generates electricity in fuel cells, including applications in transportation, and can be applied as fuel in rocket engines [2]. Moreover, H_2 is an important gas and raw material in many industrial applications. However, the high cost of the H_2 production processes requiring the use of other energy sources is a significant disadvantage. At present, H_2 can be prepared in many conventional ways, such as steam reforming, electrolysis, and biohydrogen production processes. Steam reforming uses high-temperature steam to produce hydrogen gas from fossil resources including natural gas. Electrolysis is an electrolytic process to decompose water molecules into O_2 and H_2. However, both these two methods are energy-intensive and producing hydrogen from natural gas, which is mostly methane (CH_4) and in steam reforming generates CO_2 and pollutants as by-products. On the other hand, biological hydrogen production is more environmentally sustainable and less energy intensive than thermochemical and electrochemical processes [3], but most concepts are not yet developed to production scale.

The production of H_2 using microorganisms has attracted public interest due to its potential as a renewable energy carrier that can be produced using nature's plentiful resources. There are various approaches for biological H_2 production using microorganisms such as green algae, cyanobacteria, photosynthetic anoxic bacteria, and dark fermentative bacteria. These microorganisms are physiologically very diverse, occupy different ecological niches, and use distinct metabolic pathways generating H_2. Cyanobacteria, a group of microorganisms performing an oxygenic photosynthesis, can be utilized for H_2 production via biophotolysis [4–8]. They are autotrophic organisms and thereby fix CO_2 from the atmosphere as carbon source. In addition, many strains have the capacity to reduce atmospheric N_2. This chapter addresses and discusses H_2 metabolic pathways involved in cyanobacterial H_2 production and summarizes available and future potential strategies for H_2 yield improvements. The focus is on metabolic manipulation and genetic engineering approaches and on immobilization technologies for enhancing H_2 productivity in cyanobacteria.

2 Biophotolysis and H_2 Metabolism in Cyanobacteria

Cyanobacteria are photoautotrophic organisms that use sunlight as energy source together with atmospheric CO_2 and water for growth. The thylakoid membranes in the cytoplasm of cyanobacteria contain pigment molecules such as chlorophyll *a*, phycocyanin, phycoerythrin, and allophycocyanin used to absorb light energy (i.e., photons) for oxygenic photosynthesis. The photosynthetic electron transfer reaction is divided into two parts, the light and dark reaction, respectively. The light reaction is involved in transferring electrons through an electron transport chain from PSII to plastoquinone (PQ) pool, cytochrome b_6f complex (Cyt b_6f), photosystem I (PSI), and ferredoxin (Fd), respectively, for generating ATP and reductants, NAD(P)H. For the dark reaction, CO_2 is fixed and reduced into organic compounds using chemical energy obtained from the light reaction [2].

Cyanobacteria constitute a highly diverse group of prokaryotes that have different morphologies, unicellular to heterocystous and non-heterocystous filamentous forms. They are potential microbial chassis for H_2 production by biophotolysis [9]. Biophotolysis is a process that involves the use of water as an electron donor, leading to the generation of O_2 and H_2 in the biological systems in a photosynthetic process. It can be divided into two pathways: direct and indirect biophotolysis pathways (Fig. 1). During direct biophotolysis, H_2 is derived from the electrons generated by water splitting at PSII, whereas for indirect biophotolysis, protons and electrons are mainly supplied for hydrogen generation by degradation of intracellular carbon compound(s), the so-called fermentation [3].

3 H_2-Catalyzing Enzymes in Cyanobacteria

In cyanobacteria, there may be three enzymes directly involved in H_2 metabolism [5, 6]. (1) Nitrogenase catalyzes the fixation of atmospheric N_2 to produce ammonia (NH_3) under limiting nitrogen condition and concomitantly produces H_2 as a by-product. (2) Uptake (Hup) hydrogenase catalyzes the consumption of the H_2 evolved during N_2-fixation, which reduces the energy loss during nitrogenase catalysis. (3) Bidirectional (Hox) hydrogenase catalyzes both consumption and production of H_2. Both nitrogenase and hydrogenase are highly O_2 sensitive and have been a popular target for enzyme improvement. Figure 2 shows an overview of the structural organization of the different hydrogen catalyzing enzymes in cyanobacteria.

Nitrogenase in N_2-fixing, diazotrophic, cyanobacteria is a multiprotein enzyme complex consisting of the dinitrogenase (MoFe protein) and dinitrogenase reductase (Fe protein). The MoFe protein is a heterotetramer $\alpha_2\beta_2$ with a molecular weight of about 220–240 kDa encoded by *nifD* and *nifK* for α and β subunits, respectively. However, it has also been found that several strains of *Anabaena*, including *Anabaena variabilis*, are able to synthesize an alternative nitrogenase, encoded by the *vnf* gene cluster, where molybdenum is replaced by vanadium in the active center of the enzyme [13]. The function of dinitrogenase is a reduction of N_2 bonds leading to the formation of ammonia (NH_3). The Fe protein is a homodimer with a molecular weight of about 60–70 kDa and encoded by *nifH.* It transfers electrons from the external electron donor to the dinitrogenase protein [13]. This enzyme catalyzes the reduction of atmospheric N_2 to NH_3 and is also responsible for reducing protons

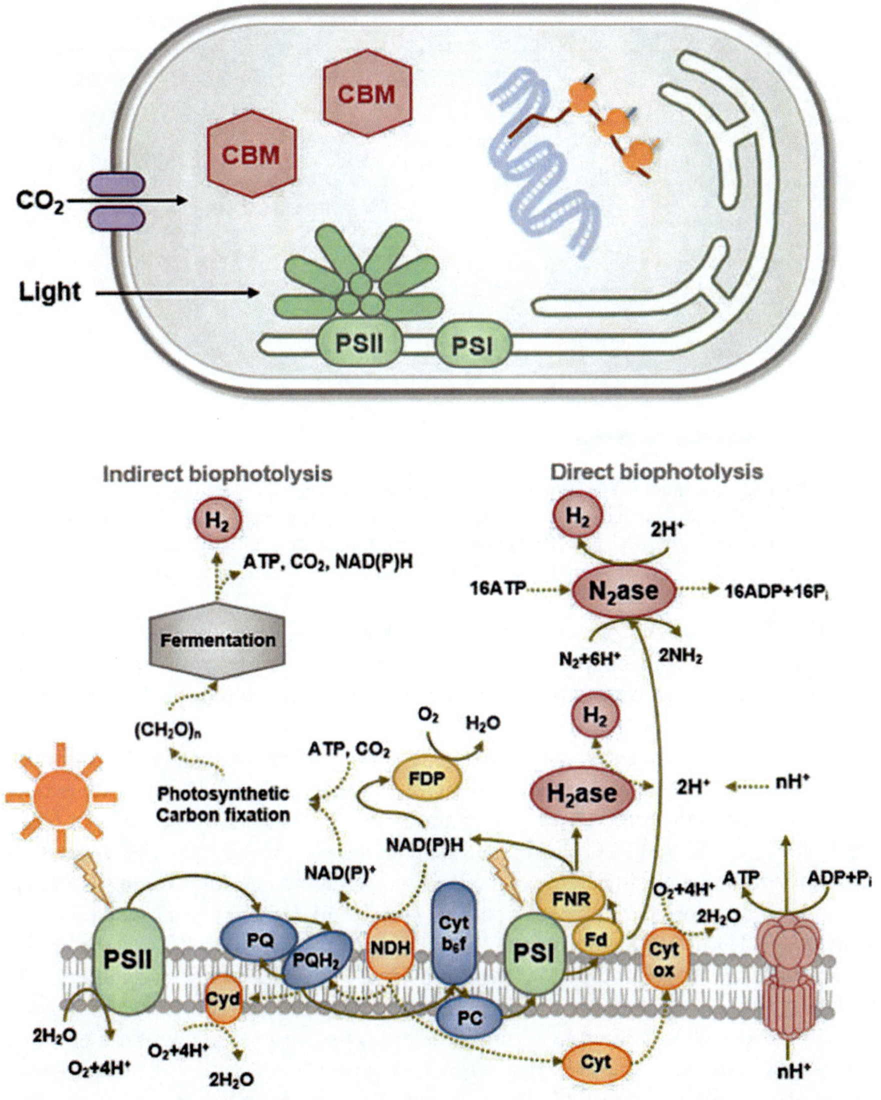

Fig. 1 Simplified view of direct and indirect biophotolysis for hydrogen metabolism involving photosynthetic system in thylakoid membrane of cyanobacterial cell. *PSII* photosystem II, *PSI* photosystem I, *Cyd* cytochrome *bd* quinol oxidase, *PQH_2/PQ* plastoquinol/plastoquinone, *Cyt b_6f* cytochrome b_6f complex, *PC* plastocyanin, *Fd* ferredoxin, *FNR* ferredoxin NAD(P) reductase, *NDH* NAD(P)H dehydrogenase, *N_2ase* nitrogenase, *H_2ase* hydrogenase, *H_2* hydrogen (This view was modified from previous articles [10, 11])

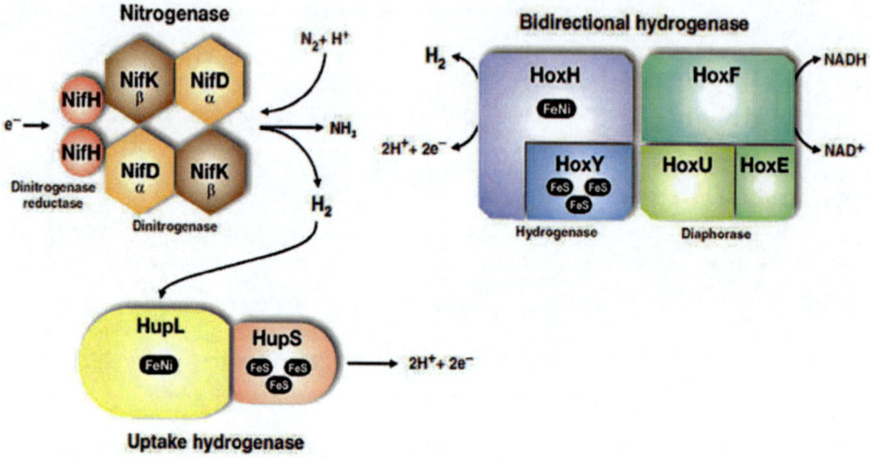

Fig. 2 The enzymes involved in H_2 metabolism in cyanobacteria. Nitrogenase catalyzes N_2-fixing from the atmosphere to produce ammonia and H_2 as a by-product. The produced H_2 is consumed by the uptake Hup-hydrogenase. The bidirectional Hox-hydrogenase can either consume or produce a molecule of H_2 depending on the redox potential (Modified from Tamagnini et al. [12])

(H^+) into H_2. However, nitrogenases have a rather low turnover rate [14] and H_2 production by nitrogenase requires a considerable number of electrons, reductants, and ATP molecules provided from photosynthesis or by carbohydrate degradation in the cell.

Moreover, nitrogenases are extremely O_2 sensitive, and diazotrophic cyanobacteria have evolved several strategies to separate the photosynthetic evolution of O_2 from the process of N_2 fixation. In filamentous cyanobacteria (e.g., *Anabaena variabilis* ATCC 29413 and *Anabaena* sp. PCC 7120), the vegetative cells can differentiate into heterocyst cells (Fig. 3). Mature heterocysts are individual cells providing a microaerobic environment suitable for the enzymes involved in N_2 fixation and H_2 metabolism. Heterocysts contain a thick cell wall and lack active photosystem II (PSII) complexes resulting in the absence of photosynthetic O_2 evolution [16]. The vegetative cells perform photosynthetic and CO_2 fixing processes, whereas CO_2 fixation is absent in heterocysts due to the lack of the primary CO_2 fixing enzyme ribulose bisphosphate carboxylase (Rubisco). Heterocysts import carbohydrates, most likely as sugars, from vegetative cells and use the oxidative pentose phosphate (OPP) pathway for carbohydrate degradation to generate energy and reduce power for nitrogen fixation. In return, the heterocysts export nitrogen in the form of glutamine to the vegetative cells through the GS-GOGAT pathway (Fig. 3) [17, 18]. In some unicellular cyanobacteria, such as *Cyanothece* sp. and *Trichodesmium* sp., N_2 fixation may be controlled by the circadian clock. They separate the production of O_2 and H_2 by performing oxygenic photosynthesis during the daytime and nitrogen fixation at night [19–21].

Uptake (Hup) hydrogenase has been reported for all known N_2-fixing cyanobacteria [6, 22]. It is a heterodimeric enzyme consisting of at least two

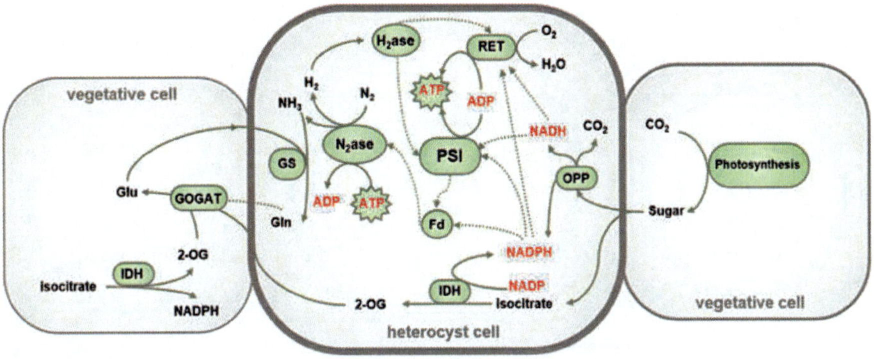

Fig. 3 Simplified view of heterocyst metabolism and exchange with vegetative cells of filamentous heterocystous cyanobacteria under nitrogen starvation. Carbohydrates are imported from vegetative cells into the heterocyst, where they supply reducing power for N_2-fixation. In turn, N_2 is bound in glutamine and exported into vegetative cells through the GOGAT pathway. Dotted lines represent a flow of reducing equivalents. *PSI* photosystem I, *OPP* oxidative pentose phosphate pathway, *RET* respiratory electron transport chain, *IDH* isocitrate dehydrogenase, *GS* glutamine synthetase, *GOGAT* glutamate synthase, *Fd* ferredoxin, *Gln* glutamine, *Glu* glutamate, *2-OG* 2-oxoglutarate, *N_2ase* nitrogenase, *H_2ase* uptake-Hup-hydrogenase (Modified from Lindberg [15])

subunits: HupS (encoded by *hupS*) and HupL (encoded by *hupL*). The HupS subunit has a molecular weight of about 35 kDa containing three FeS clusters. The HupL subunit containing the active site is about twice as large with about 60 kDa. It consists of four conserved cysteine residues involved in coordinating the metallic NiFe at the center of the active site [23, 24]. The uptake (Hup) hydrogenase is involved in the efficient recycling or consumption of the H_2 produced by the nitrogenase. Utilization of H_2 in N_2-fixing cyanobacteria is associated with (1) providing additional reducing equivalents to PSI and various cell functions, (2) generating ATP from oxyhydrogen reaction, and (3) preventing inactivation of nitrogenase by removing O_2 [25]. The structural *hupS* and *hupL* genes have been characterized in many cyanobacteria such as *Nostoc* sp. PCC 73102, *Anabaena variabilis* ATCC 29413, and *Gloeothece* sp. ATCC 27152 [26–28]. *hupS* is usually located upstream of *hupL*. The analysis of gene expression using RT-PCR technique revealed that *hupS* and *hupL* are co-transcription and an enhanced transcription level was found when cells were grown under N_2-fixing condition or the addition of external Ni^{2+} in the culture medium [12]. In some N_2-fixing cyanobacteria, e.g., *Anabaena* sp. PCC 7120, *hupL* in the vegetative cells is interrupted by a DNA element which is excised during heterocyst differentiation by a site-specific recombinase (XisC) resulting in a contiguous *hupL* (Fig. 4) [22, 29, 30].

Bidirectional (Hox) hydrogenase is commonly, though not universally, found in cyanobacteria, catalyzing both consumption and production of molecular H_2 [31, 32]. It is a heteropentamer encoded by *hoxEFUYH* and consists of two protein complexes: a hydrogenase complex (HoxY and HoxH) and a diaphorase complex (HoxE, HoxF, and HoxU). The large subunit, HoxH contains the active metal NiFe

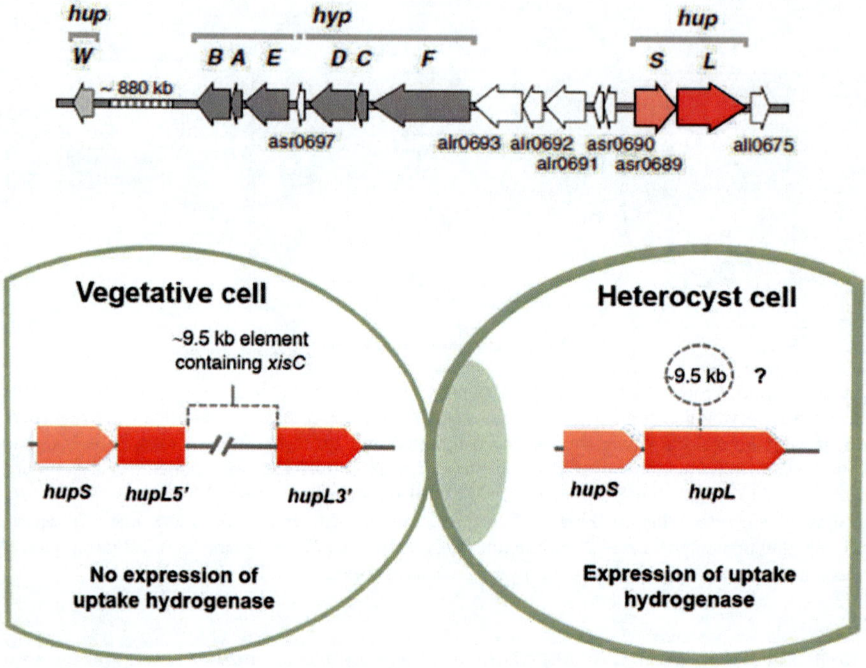

Fig. 4 Schematic representation of the *hupL* rearrangement occurring in *Anabaena* sp. PCC 7120. In the vegetative cell, *hupL* is interrupted by a 9.5-kb DNA element containing site-specific recombination (*xisC*). In contrast, the structure of the *hupL* gene is restored, allowing its expression only in the heterocyst cell. The question marks indicate unclear data explanation (Modified from Tamagnini et al. [12])

center like the uptake hydrogenase (Fig. 2). The physiological role of this enzyme is still under debate. It was found that the bidirectional hydrogenase of *Synechocystis* sp. PCC 6803 acts as an electron sink, storing excess electrons from PSI in the form of hydrogen [33]. Gutekunst et al. [34] reported that Hox-hydrogenase probably functions as an electron sink for reduced ferredoxin/flavodoxin under mixotrophic and nitrate-limiting condition. In addition, this enzyme has been proposed to be a mediator in the release of excess reducing power under anaerobic conditions [35]. Studies in *Synechocystis* PCC 6803 found that the enzyme was insensitive to light, reversibly inactivated by O_2, and quickly reactivated by NADH or NADPH [36].

4 Strategies for H_2 Yield Improvements in Cyanobacteria

Cyanobacteria are potential H_2-producers, as they can produce H_2 from water as a result of solar energy conversion. However, the main obstacle for the biotechnological process is the low yield of cyanobacteria strains producing H_2 (in the range of 0.06–31.8 μmol H_2/mg Chl *a*/h). Increasing the H_2-productivity by cell improvement has been widely studied using diverse technologies. This section summarizes recent improvements of H_2-metabolism in cyanobacteria by focusing on metabolic manipulation and genetic engineering approaches to understand the metabolic pathways further and increase their respective H_2 yields. An overview of selected cyanobacterial strains and their corresponding rates of H_2 production are summarized in Tables 1 and 2.

4.1 Metabolic Manipulation Approaches

4.1.1 Physiochemical Parameters Affecting H_2 Production

Several parameters may enhance H_2 production, such as nutrient and culture compositions, inorganic mineral supplements, the pH and temperature of culture media, and light intensity. Carbon (C), nitrogen (N), phosphorus (P), and sulfur (S) are all required nutrients for cyanobacterial growth and have been examined for optimizing cellular H_2 production by various microalgae, see Table 1. Changes in the composition of nutrients affect the H_2 production rates. Addition of a carbon source supports by providing energy for cell metabolism. Some cyanobacteria can consume organic carbon sources such as glucose, fructose, galactose, lactose, mannitol, sorbitol, sucrose [39, 44, 49], acetate, succinate, and malate [55] having an effect on hydrogenase or nitrogenase activity and thus on H_2 production. In *Synechocystis* PCC 6803 it was shown that addition of glucose increases the level of reduced NAD (P) which is beneficial for bidirectional Hox-hydrogenase activity, resulting in enhanced H_2 production [49]. Besides, in *Anabaena* sp. PCC 7120, fructose mediated an increase of H_2-production with increased nitrogenase activity and *nifD* expression, in conjunction with elevated electron flow from utilization of fructose through the oxidative pentose phosphate pathway [39].

Although nitrogen and sulfur are essential nutrients for microbial growth, an enhanced H_2 production rate was detected when cells were grown in the nitrogen- or sulfur-deprived condition. This phenomenon was observed in several cyanobacteria such as *Aphanothece halophytica* [40], *Anabaena siamensis* [38], *Arthrospira* sp. PCC8005 [43], *Gloeocapsa alpicola*, and *Synechocystis* sp. PCC 6803 [46]. Furthermore, exogenously added nitrogen sources inhibit nitrogenase activity [13]. Phosphorus (P) is an essential heteroelement in compounds such as ATP, NAD(P)H, nucleic acids, and sugar phosphates, all of which play important roles

Table 1 H_2 production in different cyanobacteria and their optimum environmental condition for enhanced H_2 production

Strains	Maximum H_2 production	Growth condition	H_2 production condition	References
Anabaena siamensis TISTR 8012	31.79 ± 0.54 μmol H_2/mg Chl*a*/h	Air, $BG11_o$, 30°C, 40 μE/m^2/s	Ar, $BG11_o$, 30°C, 40 μE/m^2/s, 0.5% fructose, 200 μE/m^2/s	[37]
Anabaena siamensis TISTR 8012	0.057 μmol H_2/mg Chl*a*/h	Air, $BG11_o$, 30°C, 30 μE/m^2/s	Ar, $BG11_o$, 30°C, 30 μE/m^2/s, 4 μMNi2$^+$	[38]
Anabaena sp. PCC 7120	21.69 μmol H_2/mg Chl*a*/h	Air, $BG11_o$, 30°C, 40 μE/m^2/s	Ar, $BG11_o$, 30°C, 40 μE/m^2/s, 60 mM fructose	[39]
Aphanothece halophytica	13.804 ± 0.373 μmol H_2/mg Chl*a*/h	Air, BG11 with Turk Island salt solution, 30°C, 30 μmol photons/m^2/s	Ar, $BG11_o$, 30°C, 30 μE/m^2/s, 0.5 M NaCl, 0.4 μMFe^{3+}	[40]
Arthrospira maxima CS-328	4.5–5.2 ml H_2/dry wt/day	Air, Zarrouk medium, 1 μM Ni^{2+}, 30°C, 12 h light/dark	Ar, Zarrouk medium, 1 μM Ni^{2+}, darkness	[41]
Arthrospira sp. PCC 8005	5.91 ± 0.14 μmol H_2/mg Chl*a*/h	Air, Zarrouk medium, 30°C, 40 μE/m^2/s	Air, ZN_o-S-deprived, 0.15 mM Fe^{2+}, β-mercaptoethanol, 30°C	[42]
Arthrospira sp. PCC 8005	7.24 ± 0.25 μmol H_2/mg Chl*a*/h	Air, Zarrouk medium, 32°C, 40 μE/m^2/s	Air, ZN_o, 0.17 μM Ni^{2+}, 30°C, darkness	[43]
Calothrix elenkinii	3.21 ± 0.19 μmol H_2/mg Chl*a*/h	Air, $BG11_o$, 30°C, 50 μmol photons/m^2/s	Ar, $BG11_o$, 30°C, 0.3% glucose, 50 μmol photons/m^2/s	[44]
Fischerella muscicola	8.73 ± 0.43 μmol H_2/mg Chl*a*/h	Air, $BG11_o$, 30°C, 50 μmol photons/m^2/s	Ar, $BG11_o$, 30°C, 0.3% glucose, 50 μmol photons/m^2/s	[44]
Fischerella muscicola TISTR 8215	0.32 ± 0.01 mmol H_2/L	Air, $BG11_o$, 30°C, 40 μmol photons/m^2/s	Ar, $BG11_o$, 30°C, 40 μmol photons/m^2/s	[45]
Gloeocapsa alpicola	140 nmol H_2/mg protein/h	Air, $BG11_o$, 24°C, 25 μE/m^2/s	CH_4, $BG11_o$ with S-deprived, darkness	[46]
Lyngbya perelegans	18.9 ± 0.28 mmol H_2/kg dry wt/h	Air, BG11, 3,000 lx, pH 8.0, 27°C	Ar, BG11, (2,000 lx), light: Dark (21:3 h), pH 8.0, 25°C	[47]
Nostoc calcicola	4.27 ± 0.17 μmol H_2/mg Chl*a*/h	Air, $BG11_o$, 30°C, 50 μmol photons/m^2/s	Ar, $BG11_o$, 30°C, 50 μmol photons/m^2/s, 0.3% glucose	[44]
Nostoc muscorum	9.3 nmol H_2/mg dry mass/h	Arnon's medium, 3,000 lx, 16 h light/8 h dark, 25°C	Arnon's medium combined N-free, 3,000 lx, 16 h light/8 h dark, 40°C	[48]

(continued)

Table 1 (continued)

Strains	Maximum H_2 production	Growth condition	H_2 production condition	References
Nostoc punctiforme ATCC 29133	20.7 ± 0.72 μmol H_2/mg Chl*a*/h	Air, $BG11_o$, 30°C, 40 μE/m^2/s	Ar, $BG11_o$, 30°C, 40 μE/m^2/s	[37]
Scytonema bohneri	7.63 ± 0.26 μmol H_2/mg Chla/h	Air, $BG11_o$, 30°C, 50 μmol photons/m^2/s	Ar, $BG11_o$, 30°C, 0.3% glucose 50 μmol photons/m^2/s	[44]
Synechocystis sp. PCC 6803	0.12 ± 0.01 μmol H_2/mg Chl*a*/h	Air, BG11, 30°C, 30 μE/m^2/s	Ar, BG11, 30°C, 0.1% glucose, darkness	[49]
Tolypothrix distorta	10.95 ± 0.22 μmol H_2/mg Chl*a*/h	Air, $BG11_o$, 30°C, 50 μmol photons/m^2/s	Ar, $BG11_o$, 30°C, 0.3% glucose, 50 μmol photons/m^2/s	[44]

in photosynthesis. NAD(P)H is the electron donor to the bidirectional Hox-hydrogenase in cyanobacteria [56].

Generally, trace elements act as essential cofactors, which play an important role in activities of both hydrogenase and nitrogenase enzymes involved in H_2 evolution. For example, a culture of *Fischerella muscicola* TISTR 8215 grown with higher levels of Mo^{6+} showed increased nitrogenase activity leading to increased H_2 production [45]. Additionally, the relevance of concentrations of Fe^{3+}, Ni^{2+}, and Mo^{2+} ions for H_2 production has been investigated and optimized for several strains of cyanobacteria [38, 42, 43, 57], with results suggesting that availability of these elements is a critical factor in controlling H_2 production and N_2 fixation, including effects on expression of hydrogenase and nitrogenase genes.

Furthermore, pH and temperature are crucial parameters influencing the H_2 production process. The pH ranges from 6 to 9 were examined for enhanced H_2 production in several cyanobacteria. In tests using *Lyngbya perelegans* the highest H_2 production was obtained at pH 8.0 [47]. Regarding the temperature, the optimum temperature for H_2 production for most cyanobacteria varies between 23 and 40°C but with differences between strains. *Nostoc muscorum* and *Lyngbya perelegans* showed optimum hydrogen production at 40°C [47, 48] whereas in *Arthrospira* sp. PCC 8005 the maximum rate of H_2 production was observed at 30°C [42]. Moreover, *Calothrix* sp., *Nodularis* sp., and *Microcystis* sp. showed optimum H_2 production at 23°C [58].

Light intensity is a most critical factor affecting the efficiency of cyanobacterial H_2 production. Under artificial illumination, microalgal cultivation under different light intensities alters the metabolic capacity of the cells. Photobiological H_2 production in microalgae and cyanobacteria results from the contribution of a direct and an indirect electron transfer pathway [59–61]. The direct biophotolysis involves a PSII-dependent pathway, which links water-splitting activity to H_2 production. In indirect biophotolysis, electrons, which are derived from the degradation of stored carbohydrates entering the electron chain at the plastoquinone pool are hereafter

Table 2 H_2 production comparison among cyanobacteria using immobilization techniques vs. cell suspension (free cell)

Strains	Matrix	Media	No. cycle	Maximum H_2 yield/rate					References
				μmol/g DW	nmol/mg DW/h	mL H_2	mmol/L	μmol/mg Chl *a*/h	
Synechocystis sp. PCC 6803	Alginate bead	BG110	2			5.8 (144 h)			[50]
Lyngbya perelegans	Alginate bead	BG11	1		21.5				[51]
	Agar cube	BG11	1		19.8				
Anabaena sp. PCC 7120	Alginate film	Z8x	3					9	[52]
	Free cell	Z8x	1					13	
Anabaena sp. PCC 7120 mutant strain Δ*hupL*	Alginate film	Z8x	3					30	[52]
	Free cell	Z8x	2					30	
Calothrix 336/3	Alginate film	Z8x	6					25	[52]
	Free cell	Z8x	3					35	
Aphanothece halophytica	Agar cube	BG11	3	3,700					[53]
	Free cell	BG11	3	1,200					
Fischerella muscicola TISTR 8215	Agar cube	BG110	3				7.5		[45]
	Free cell	BG110	–				0.3		
Microcystis aeruginosa	Agar cube	MA	3						[54]

transferred to the hydrogenase to produce H_2 [62]. Previous studies reported that the impact of light intensity varies among different species and strains. The heterocystous cyanobacteria *Nostoc muscorum* and *Anabaena* PCC 7120 produce H_2 from nitrogenase in heterocysts under light conditions [48, 63]. Enhanced light intensity resulted in increased H_2 production in *A. siamensis* TISTR 8012 with a saturation at 200 μE/m^2/s of light intensity after 12 h. The cells generated less H_2 above 200 E/m^2/s, along with decreased chlorophyll *a* and cell lysis [37].

4.1.2 Cell Immobilization for Reduced O_2 and Cell-Stacking Effects

Hydrogenase catalyzes the incorporation of two protons and two electrons to form H_2, which is the smallest molecule in the universe. H_2 is produced inside the cytoplasm of the cell, diffuses toward liquid broth through the lipid bilayer cell membrane, and finally to the headspace, driven by the partial pressure of the gas. The cytoplasm has the highest H_2 partial pressure, followed by liquid broth and headspace, accordingly. Consequently, the H_2 yield can be determined by quantifying the gas in the headspace using, e.g., gas chromatography. O_2, the strong competitive hydrogenase inhibitor generated in PSII by the water-splitting reaction, has highly similar physicochemical properties, which makes it challenging to separate both gases. Apart from being a strong inhibitor for most H_2 producing enzymes, it also forms an explosive mixture with H_2 (Knallgas reaction) and thus poses a significant safety issue if it comes to scale-up. The amount of molecular oxygen released by the photosynthetic activity depends on the cellular respiration process consuming O_2 as the final electron acceptor. Therefore, one strategy to keep O_2 levels low in H_2 producing cultures is the balancing of PSII activity and cellular respiration [41].

Apart from the parameters discussed above also the cell concentration in the culture was reported to affect the H_2 yield, which is decreasing with increasing cell densities [45, 51]. Too dense cells culture led to the so-called cell-stacking effect, in which cells shade each other and thus run into a light limitation (Fig. 5), which is also difficult to solve by vigorous shaking or mixing [64]. Especially for filamentous cyanobacteria strong mixing is not an option, as the filaments prevent a homogenous mixing, and will also be negatively impacted by high shear forces. Cell immobilization may be a promising solution to relieve the problem of cell-stacking.

Cell immobilization is an essential technique for reducing the cell-stacking effect. For immobilization, the cells are embedded into a supporter material, which is polymerizing during the process. Depending on the physicochemical properties and the concentration of the supporter material, in situ gas removal strategies (like O_2 or H_2) can be implemented into the system. To obtain the highest H_2 yield many studies have been performed with production cycles. Several common biomaterials have been used as cell support. These include carrageenan [53], agar [45, 51, 53, 54], agarose [53], and alginate [52, 65–68]. After immobilizing the cells in a selected support, the mixtures can be molded into different shapes. Thin films [52, 67], cubes [45, 53, 54], and beads [51, 65, 66, 68] are commonly used. Reported yields and corresponding production system parameters are summarized in Table 2.

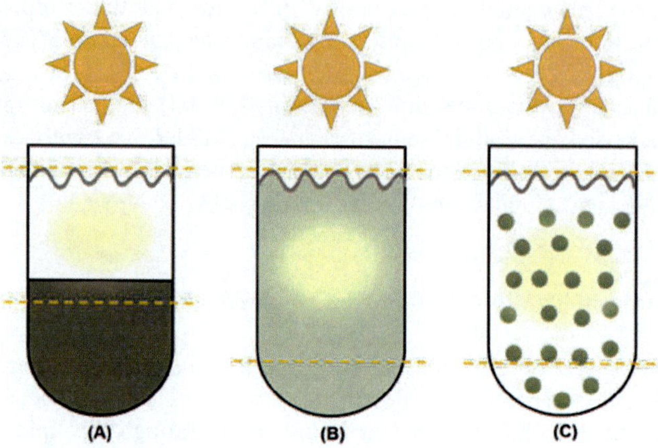

Fig. 5 Schematic shows light penetration power to stacked cells/precipitated cells (**a**), shaking cells (**b**), and immobilized cells (**c**). Dotted lines represent the light penetration path from the surface toward the center of a container

Rashid et al. [54] demonstrated that the unicellular cyanobacterium *Microcystis aeruginosa* immobilized in 1.5% agar in cubes can sustain a hydrogen production phase for up to 95 h with a yield of about 65–70 mL H_2/L culture. This can be increased by the addition of glucose to the culture, which may be degraded through glycolytic pathways, generating the reducing equivalent NADH, which support the flow of electrons to the plastoquinone pool between PSII and PSI and thereby increase the yield of H_2. Wuthithien et al. [45] also reported that immobilizing cells of the N_2-fixing filamentous cyanobacterium *Fischerella muscicola* TISTR 8215 in 1.5% agar improved the H_2 yield significantly, and increasing the Mo^{6+} ion concentration also resulted in an increase in H_2 production rate. It seems that stimulation of nitrogenase activity occurs through an addition of molybdenum into their active site [38].

The beneficial effect of immobilization on H_2 production may be explained as follows: (1) The immobilization matrix reduced the O_2 concentration in the direct environment of the cells [69]. (2) The cell-stacking effect was reduced by the immobilization resulting in improved light supply. (3) Optimized mass-transfer between nutrients from broth to cells [70]. (4) The initial cell numbers to agar concentration was appropriate for increasing agar mechanical stability [70].

Pansook et al. [53] compared different materials for immobilization and reported that immobilized unicellular cyanobacterium *Aphanothece halophytica* in 3% (*w/v*) agar showed the highest H_2 production compared to carrageenan, agarose, and free cells. Carrageenan might encounter the problem of low stability during gel formation due to the presence of NaCl, as previously reported [71], whereas agarose showed lower stability than agar in agreement with Semenchuk et al. [72]. The high production rate in cells immobilized in agar was related to: (1) improved cell survival rate and mechanical stability, (2) better nutrient diffusion rate from broth to cells, and

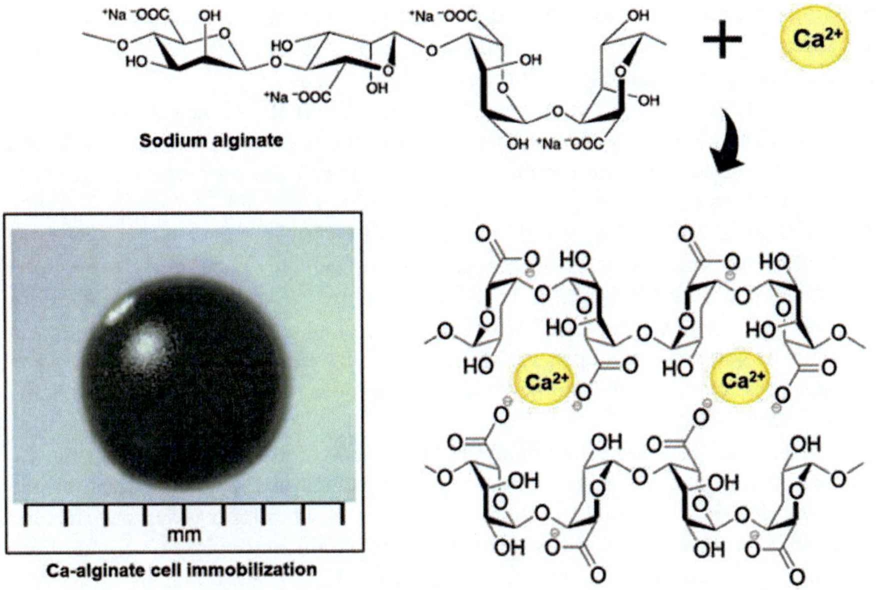

Fig. 6 Cells immobilized in calcium alginate beads prepared using sodium alginate dissolved in algal medium, and Ca-alginate formation from Na-alginate. Negatively charges of Na-alginate chains repulse each other, leading to a uniform structure. Once Ca^{2+} ions are present, positively charges attempt to combine each negative strand close to each other, forming a gel structure (Modified from Touloupakis et al. [50])

(3) small size of the immobilization particles (0.125 cm^3) facilitating H_2 and O_2 diffusion from cells toward the bulk.

Another commonly used immobilization matrix is alginate. It is a water-soluble carbohydrate polymer, which will polymerize when interacting with $CaCl_2$ in solution. Ca^{2+} will replace Na^+ ions and cross-interact with carboxylate groups ($-COO^-$) and negatively polar groups (-OH), leading to carbohydrate strand incorporations and gel formation over time (Fig. 6). Immobilized filamentous cyanobacterium *Lyngbya perelegans* in 4% agar cubes was studied and compared to alginate beads [51]. There were only slight differences between the two materials tested, but both showed about 1.8 times higher H_2 productivity than free cells. Interestingly, this immobilized culture was used to investigate the impact of various gas mixtures, and it was found that a CH_4:Ar (11:2) mix resulted in the highest productivities [51]. Furthermore, Leino et al. [52] screened for H_2 producing cyanobacteria from the University of Helsinki Culture Collection and identified the N_2-fixing heterocystous filamentous cyanobacterium *Calothrix* 336/3 as the strain with the highest H_2 production. Immobilized in a Na-Alginate thin film, it showed a maximum H_2 production rate higher than the rate from free cells. In a study using the model strain *Anabaena* PCC 7120 wildtype and a Δ*hupL* mutant of the same strain, there were only small differences between free cells and cells immobilized in alginate films. The study also found that periodically purging the system with CO_2 balanced Ar led to increased H_2 yield as CO_2 was used as a signal to enhance N_2 fixation [52], consequently prolonging H_2 production.

One of the essential parameters for H_2 production is the photosynthetic activity since PSII generates O_2 with an electron flow to the bidirectional hydrogenase and nitrogenase. Restoring photosynthetic activity between the production cycles thus plays a vital role in prolonging H_2 production as observed in *Calothrix* 336/3, *Anabaena* PCC 7120 wildtype, and its Δ*hupL* strain [67].

Finally, cell immobilization enables cell retention and the recycling of the cellular biocatalyst for multiple batches. This facilitates process optimization, as various reaction conditions can be tested with one batch of biocatalysts, like purging with inert gas or gas mixtures, applying different media, etc. Furthermore, it allows to operate the reactors in a continuous or semi-continuous mode, positively influencing process economics.

In summary, cyanobacterial immobilization is an interesting option to enhance H_2 production and process stability by facilitating gas (H_2 and O_2) removal from the cultures. However, optimal conditions in terms of immobilization material and reaction environment will differ from strain to strain and we are still far from defining general process parameters for optimized H_2 production, as up to now only case (strain) specific examples are reported and general operation protocols are missing.

4.2 *Genetic Engineering Approaches*

4.2.1 Eliminating of Electron Competing Pathways for Promoting H_2 Metabolism

The principal reason for H_2 metabolism through bidirectional Hox-hydrogenase in cyanobacteria may be a disposal of excess reducing equivalents during fermentative metabolism associated with photosynthesis or/and dark anaerobic conditions. Therefore, the bidirectional Hox-hydrogenase requires numerous electrons and reductants as substrates supporting its activities. However, electrons generated through oxygenic photosynthesis are under most conditions not primarily shuttled to H_2 metabolism. Instead, these electrons can be transferred to other competing pathways, such as the respiratory electron transport chain, nitrogen assimilation, and carbohydrate metabolism, shown in Fig. 7. Therefore, diverse genetic engineering strategies for enhanced H_2 production by re-direction of electrons flow toward H_2 metabolism have been extensively examined (Table 3). In *Synechocystis* sp. PCC 6803, interruption of all respiratory terminal oxidases (Δ*ctaI*, Δ*ctaII*, and Δ*cyd*) induce the bidirectional Hox-hydrogenase activity leading to a higher H_2-production rate than in wildtype cells under light condition [73]. Moreover, inactivation of type I NADPH-dehydrogenase complex (NDH I) by deleting the large subunit NdhB in a mutant *Synechocystis* strain M55 resulted in prolonged H_2-production and a lower level of O_2 being produced under light condition [36]. Engineering strains with

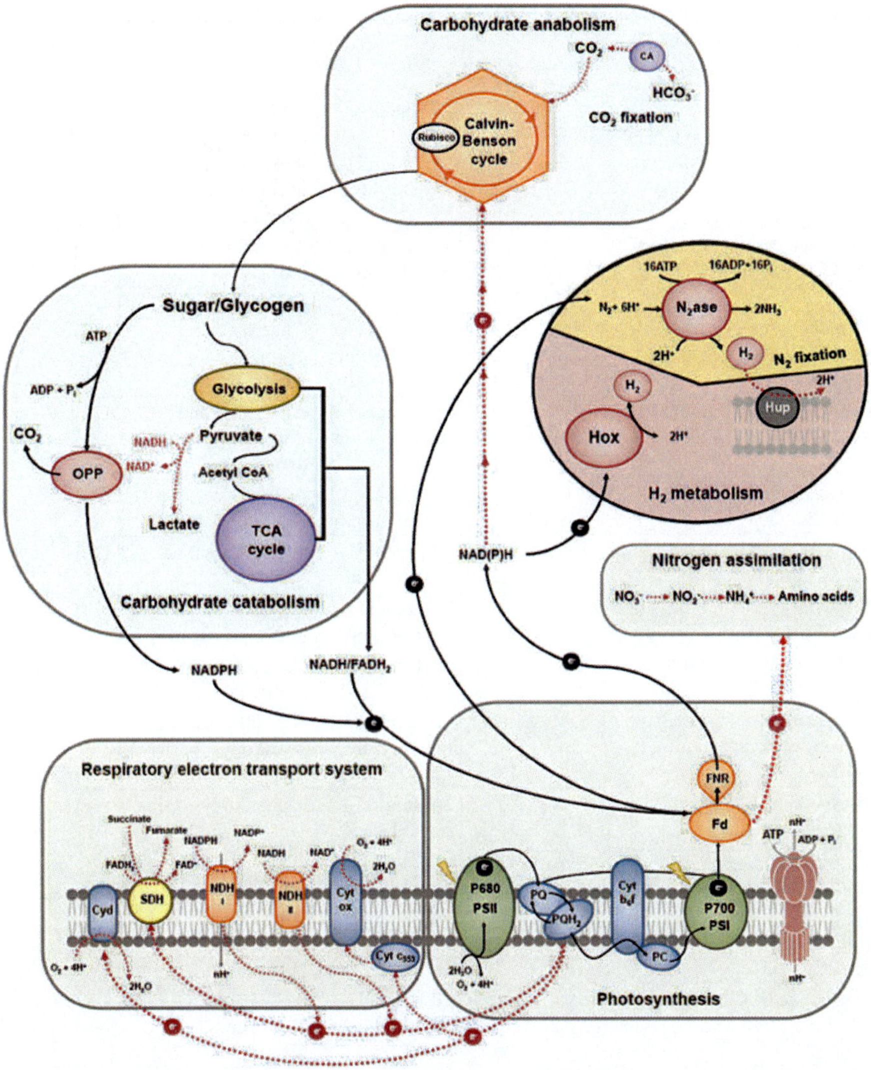

Fig. 7 Different pathways of electron flow involved in H_2 metabolism of cyanobacteria. Dotted lines represent electrons that can be transferred to other assimilatory or competing pathways. *Cyd* quinol oxidase, *Cyt b₆f* cytochrome b_6f, *Cyt c553* cytochrome c553, *Cyt ox* cytochrome c oxidase, *Fd* ferredoxin, *FNR* ferredoxin-NADP reductase, *Hox* bidirectional Hox-hydrogenase, *Hup* uptake Hup-hydrogenase, *N_2ase* nitrogenase, *NDH* NADPH dehydrogenase, *OPP* oxidative pentose phosphate pathway, *PC* plastocyanin, *PSI* photosystem I, *PSII* photosystem II, *PQ* plastoquinone pool, *Rubisco* ribulose-1,5-bisphosphate carboxylase oxygenase, *SDH* succinate dehydrogenase (Modified from Khetkorn et al. [10])

Table 3 H_2 production in engineered cyanobacterial strains using different strategies (Modified from Khetkorn et al. [10])

Strains	Engineered genes	H_2 production rate	H_2 production condition	References
Synechocystis strain M55	*ndhB*	200 nmol H_2/mg chl *a*/min	Anaerobic and nitrogen deprivation	[36]
Synechocystis sp. PCC 6803	*ctaI/cyd*	190 nmol H_2/mg chl *a*/min	Anaerobic and nitrogen deprivation	[73]
Synechocystis sp. PCC 6803	*ctaII/cyd*	115 nmol H_2/mg chl *a*/min	Anaerobic and nitrogen deprivation	[73]
Synechocystis sp. PCC 6803	*ctaI/ctaII/cyd*	100 nmol H_2/mg chl *a*/min	Anaerobic and nitrogen deprivation	[73]
Synechocystis sp. PCC 6803	*narB*	86 nmol H_2/mg chl *a*/min	Ar, darkness, nitrogen deprivation	[73]
Synechocystis sp. PCC 6803	*nirA*	174 nmol H_2/mg chl *a*/min	Ar, darkness, nitrogen deprivation	[74]
Synechocystis sp. PCC 6803	*narB/nirA*	300 nmol H_2/mg chl *a*/min	Ar, darkness, nitrogen deprivation	[74]
Synechococcus sp. PCC 7002	*ldhA*	14.1 mol H_2 day/10^{17} cell	Dark anaerobic fermentation	[75]
Anabaena variabilis strain AVM13	*hupSL*	135 μmol H_2/mg chl *a*/h	Ar, 100 μE/m^2/s, N_2-fixing	[26]
Nostoc punctiforme strain NHM5	*hupL*	14 μmol H_2/mg chl *a*/h	Light and N_2-fixing	[76]
Anabaena sp. PCC 7120	*hupL/hoxH*	53 μmol H_2/mg chl *a*/h	Ar, 10 W/m^2, N_2-fixing	[77]
Anabaena sp. PCC 7120	*hupW*	3.3 μmol H_2/mg chl *a*/h	Ar, 10 W/m^2, N_2-fixing	[78]
Nostoc sp. PCC 7422	*hupL*	100 μmol H_2/mg chl *a*/h	Ar + 5% CO_2, 70 μE/m^2/s, N_2-fixing	[79]
Anabaena siamensis TISTR 8012	*hupS*	29.7 μmol H_2/mg chl *a*/h	Ar, 200 μE/m^2/s, N_2-fixing	[37]
Anabaena sp. PCC 7120	*hupL*	101.33 μmol H_2/mg Chl*a*/h	Ar, nitrogen deprivation, 30°C, 40 μE/m^2/s, 60 mM fructose	[39]
Synechococcus elongatus	*hydA* and maturation *operon (hydEFG)* from *Clostridium acetobutylicum*	2.8 μmol H_2/mg Chl*a*/h	Light, 5 μM DCMU, bubbling with 2.5% CO_2 and 97.5% N_2	[80]

(continued)

Table 3 (continued)

Strains	Engineered genes	H_2 production rate	H_2 production condition	References
Anabaena sp. PCC 7120	Hydrogenase operon, *hydA, hydB, hydE, hydF, hydG* along with two additional genes, S03922 and S03924, from *Shewanella oneidensis* MR-1	3.4 nmol H_2/μg chl *a*/h	Light and nitrate deprivation	[81]
Synechococcus elongatus	[NiFe] hydrogenase from *Thiocapsa roseopersicina*	~0.07 nmol H_2 mg protein/h	Anaerobic, 40 μE/m^2/s	[82]
Synechococcus elongatus	[NiFe] hydrogenase (*hynSL* along with 11 adjacent proteins) from *Alteromonas macleodii*	~4.2 nmol H_2 mg protein/h	Anaerobic, 40 μE/m^2/s	[82]
Synechocystis sp. PCC 6803	O_2-tolerant, and NAD (H)-dependent hydrogenase from *Ralstonia eutropha* (ReSH)	177.6 μmol H_2/gCDW	Anaerobic and fermentative condition, 30°C, 50 μE/m^2/s, 10 mM glucose	[83]

disrupted nitrate assimilation, either nitrate reductase (Δ*narB*) or nitrite reductase (Δ*nirA*) or both genes (Δ*narB*/Δ*nirA*), in *Synechocystis* sp. PCC 6803 were found to induce significantly higher H_2 production than in wildtype cells [74]. In addition, a mutant *Synechococcus* sp. PCC 7002 (Δ*ldhA*), lacking the enzyme for the NADH-dependent reduction of pyruvate to D-lactate, showed an increased ratio of NADPH to $NADP^+$ and a five-times higher H_2-production when compared with wildtype cells [75]. This work supported that by eliminating competing fermentative carbon metabolism such as the pathway to produce lactate it may be possible to redirect the electron flux to H_2 metabolism in cyanobacteria. Accordingly, an engineering approach by eliminating competitive electron pathways is an effective and promising method to improve cyanobacteria potential for H_2 production, which should be further explored.

4.2.2 Modifying Heterocyst Frequency for Increased H_2 Production

In heterocystous filamentous cyanobacteria, nitrogenase is a key player for H_2 production. The heterocyst provides a partially microoxic environment suitable for oxygen-sensitive enzymes such as nitrogenase since it lacks the PSII activity and has an increased respiration rate [84]. Furthermore, it is surrounded by a thick envelope limiting O_2 diffusion through the cell wall (Fig. 3). Therefore, increasing the heterocyst frequency may enhance H_2 production by promoting nitrogenase activity. The heterocyst differentiation process has been primarily studied in *Anabaena* sp. PCC 7120 in which it takes approximately 24 h to develop a mature heterocyst from a vegetative cell under nitrogen limited condition [17]. One of the key genes in the regulation of heterocyst pattern formation, *hetR*, encodes a serine-type protease, which is expressed early during heterocyst differentiation. Inactivation of *hetR* inhibits early steps in the differentiation process, while overexpression of the gene increases heterocyst frequency [85]. Recently, it was demonstrated that the addition of fructose rapidly induced the development of mature heterocysts and led to upregulation of *hetR* transcription, resulting in enhanced N_2-fixation and H_2-production in *Anabaena* sp. PCC 7120 Δ*hupL* strain [39]. HetF (a protease) influences heterocyst development by inhibiting *hetR* expression during cell differentiation [86]. PatA, a response regulator, is also known to effect post-translational modification of HetR [87]. However, a practical study with strains exhibiting a genetically engineered high heterocyst frequency with enhanced H_2 production is yet to be reported.

4.2.3 Inactivation of Uptake (Hup) Hydrogenase Function for Enhanced H_2 Production

Uptake hydrogenase activity is a major obstacle for enhanced H_2-production in N_2-fixing cyanobacteria since it catalyzes the consumption of H_2 produced by nitrogenase. Therefore, the disruption of uptake hydrogenase function has been widely studied in many N_2-fixing cyanobacteria. Generally, the structural genes encoding uptake Hup-hydrogenases are clustered in a similar physical organization forming a transcript unit, *hupS* being located upstream of *hupL* (Fig. 4). Inactivation of *xisC* in *Anabaena* sp. PCC 7120 resulted in a strain incapable of forming a functional uptake hydrogenase [29]. A mutant strain AMC 414 (Δ*xisC*) showed high potential for H_2-production compared to wildtype strain under higher light intensity [63]. Moreover, target genes (*hupS*, *hupL*, and *hupW*) that affect H_2-uptake deficiency in N_2-fixing cyanobacteria have been extensively investigated, see Table 3. All generated strains produce H_2 at significantly higher rates than their respective wildtype cells. These experiments indicate that the genetic inactivation of *hup* is an effective strategy for improving cyanobacterial H_2 production.

4.2.4 Introduction of Non-native Hydrogenase for Enhanced H_2 Productivity

Cyanobacteria produce H_2 through bidirectional Hox-hydrogenase ([NiFe]-hydrogenase) with a low rate of H_2-evolution. Therefore, the expression of non-native hydrogenase has been a focus for improving H_2 productivity in cyanobacteria. These include high turnover [FeFe] hydrogenase and some O_2-tolerant [NiFe] hydrogenases from other organisms using advanced synthetic biology techniques. However, successful heterologous expression of [FeFe]-hydrogenase in cyanobacteria remains a challenge, and to date, only very few reports are available, Table 3. The first report by Ducat et al. [80] demonstrated the expression of a [FeFe] hydrogenase (HydA) and the accessory HydEFG from the anaerobic fermentative bacterium *Clostridium acetobutylicum* into *Synechococcus elongatus* PCC 7942. Interestingly, the results showed both in vitro and in vivo activity of non-endogenous hydrogenase connected to the light-dependent reactions of the electron transport chain. Gärtner et al. [81] have been successfully expressed the FeFe-hydrogenase operon (*hydA*, *hydB*, *hydE*, *hydF*, *hydG*) and two additional genes, S03922 and S03924, from *Shewanella oneidensis* MR-1 into the filamentous cyanobacterium *Anabaena* sp. PCC 7120. Avilan et al. [88] expressed a clostridial [FeFe]-hydrogenase specifically in the heterocysts together with a GlbN cyanoglobin to decrease the O_2 levels in the cell. The obtained strain showed H_2 production concomitantly with oxygenic photosynthesis in the vegetative cells of the filaments. Furthermore, Weyman et al. [82] reported expressing [NiFe] hydrogenases from *Thiocapsa roseopersicina*, as well as *hynSL* along with 11 adjacent proteins from *Alteromonas macleodii* in *Synechococcus elongatus*. The advantage of using [NiFe] homolog over the [FeFe] hydrogenases was their increased half-life and enhanced tolerance toward oxygen stress [89]. The results showed in vitro activity of the expressed protein. Expression of such oxygen-tolerant hydrogenases in photosynthetic systems may open new avenues in cyanobacterial H_2 production. Recently, another strategy that circumvents the biological maturation of [FeFe]-hydrogenase by an artificial synthetic activation of a heterologously expressed HydA protein in living cells of, e.g., *Synechocystis* PCC 6803 was developed. A functional HydA was created by the addition of a synthetic analogue of the [2Fe] subcluster mimicking the active site outside the cells [7]. The experiments showed that the non-native, semisynthetic FeFe-hydrogenase retain its H_2 production capacity for several days after synthetic activation with a regulation of activity based on availability of electrons. The artificial activation technology was expanded to a newly discovered [FeFe]-hydrogenase which when expressed in *Synechocystis* showed stable expression and significant H_2 production under different environmental conditions [8]. The developed technology opens up unique possibilities to investigate not only [FeFe]-hydrogenases but also other metalloenzymes in a photosynthetic microbial cell environment, completely bypassing the many challenges of, e.g., biological maturation and regulations.

In another recent development, Lupacchini et al. [83] introduced an O_2-tolerant hydrogenase from *Ralstonia eutropha* (ReSH) into *Synechocystis* genome. The resulting engineered strain was able to produce H_2 in the dark under fermentative conditions, as well as in the light, under conditions promoting intracellular NADH excess. This opens new possibilities for efficient cyanobacterial H_2 production also under O_2 replete conditions.

5 Conclusions and Perspectives

Due to the growing emphasis on developing renewable energy sources, cyanobacteria have been intensively studied as green cell factories for sustainable H_2 production. Researchers are concentrating their efforts on the main native processes of cyanobacterial photosynthesis, fermentative metabolism, and on the enzymes involved in H_2-metabolism, which holds great promise in terms of gaining fundamental knowledge and practical applications in biotechnology. The majority of research focuses on applying various metabolic manipulation strategies to enhance H_2 yield in cyanobacteria. Additionally, genetic engineering is used to increase the H_2 yield as well as the technology of cell immobilization for H_2 scale-up challenges. Despite the enormous theoretical potential of cyanobacterial based H_2 production, there are still significant barriers to its commercialization. The prospects of the biohydrogen energy sector will be determined by the combined efforts of scientists and engineers, state political support, and substantial R&D efforts.

References

1. Perry JH (1963) Chemical engineers' handbook. McGraw-Hill, New York
2. Allahverdiyeva Y, Aro EM, Kosourov SN (2014) Chapter 21 – Recent developments on cyanobacteria and green algae for biohydrogen photoproduction and its importance in CO_2 reduction. In: Gupta VK, Tuohy MG, Kubicek CP, Saddler J, Xu F (eds) Bioenergy research: advances and applications. Elsevier, Amsterdam, pp 367–387. https://doi.org/10.1016/B978-0-444-59561-4.00021-8
3. Mathews J, Wang G (2009) Metabolic pathway engineering for enhanced biohydrogen production. Int J Hydrog Energy 34(17):7404–7416. https://doi.org/10.1016/j.ijhydene.2009.05.078
4. Hansel A, Lindblad P (1998) Towards optimization of cyanobacteria as biotechnologically relevant producers of molecular hydrogen. Appl Microbiol Biotechnol 50:153–160. https://doi.org/10.1007/s002530051270
5. Kosourov S, Böhm M, Senger M, Berggren G, Stensjö K, Mamedov F et al (2021) Photosynthetic hydrogen production: novel protocols, promising engineering approaches and application of semi-synthetic hydrogenases. Physiol Plant 173(2):555–567. https://doi.org/10.1111/ppl.13428
6. Lindblad P (2018) Hydrogen production using novel photosynthetic cell factories. Cyanobacterial hydrogen production: design of efficient organisms. In: Microalgal hydrogen production: achievements and perspectives. G Royal Society of Chemistry, pp 323–334
7. Wegelius A, Khanna N, Esmieu C, Barone GD, Pinto F, Tamagnini P, Berggren G, Lindblad P (2018) Generation of a functional, semisynthetic [FeFe]-hydrogenase in a photosynthetic microorganism. Energy Environ Sci 11(11):3163–3167. https://doi.org/10.1039/C8EE01975D
8. Wegelius A, Land X, Berggren G, Lindblad P (2021) Semisynthetic [FeFe]-hydrogenase with stable expression and H_2 production capacity in a photosynthetic microbe. Cell Rep Phys Sci 2(3):100376. https://doi.org/10.1016/j.xcrp.2021.100376
9. Pinto FAL, Troshina O, Lindblad P (2002) A brief look at three decades of research on cyanobacterial hydrogen evolution. Int J Hydrog Energy 27(11–12):1209–1215. https://doi.org/10.1016/S0360-3199(02)00089-7
10. Khetkorn W, Khanna N, Incharoensakdi A, Lindblad P (2013) Metabolic and genetic engineering of cyanobacteria for enhanced hydrogen production. Biofuels 4(5):535–561. https://doi.org/10.4155/bfs.13.41
11. Sadvakasova AK, Kossalbayev BD, Zayadan BK, Bolatkhan K, Alwasel S, Najafpour MM, Allakhverdiev SI (2020) Bioprocesses of hydrogen production by cyanobacteria cells and possible ways to increase their productivity. Renew Sust Energ Rev 133:110054. https://doi.org/10.1016/j.rser.2020.110054

12. Tamagnini P, Leitão E, Oliveira P, Ferreira D, Pinto F, Harris DJ, Heidorn T, Lindblad P (2007) Cyanobacterial hydrogenases: diversity, regulation and applications. FEMS Microbiol Rev 31(6):692–720. https://doi.org/10.1111/j.1574-6976.2007.00085.x
13. Bothe H, Schmitz O, Yates MG, Newton WE (2010) Nitrogen fixation and hydrogen metabolism in cyanobacteria. Microbiol Mol Biol Rev 74(4):529–551. https://doi.org/10.1128/MMBR.00033-10
14. Hallenbeck PC, Benemann JR (2002) Biological hydrogen production; fundamentals and limiting processes. Int J Hydrog Energy 27(11):1185–1193. https://doi.org/10.1016/S0360-3199(02)00131-3
15. Lindberg P (2003) Cyanobacteial hydrogen metabolism-uptake hydrogenase and hydrogen production by nitrogenase in filamentous cyanobacteria. Doctor of Philosophy, Uppsala University, Uppsaly, Sweden
16. Golden JW, Yoon HS (1998) Heterocyst formation in *Anabaena*. Curr Opin Microbiol 1(6): 623–629. https://doi.org/10.1016/s1369-5274(98)80106-9
17. Curatti L, Flores E, Salerno G (2002) Sucrose is involved in the diazotrophic metabolism of the heterocyst-forming cyanobacterium *Anabaena* sp. FEBS Lett 513(2–3):175–178. https://doi.org/10.1016/s0014-5793(02)02283-4
18. Thomas J, Meeks JC, Wolk CP, Shaffer PW, Austin SM (1977) Formation of glutamine from [13N] ammonia,[13N] dinitrogen, and [14C] glutamate by heterocysts isolated from *Anabaena cylindrica*. J Bacteriol 129(3):1545–1555. https://doi.org/10.1128/jb.129.3.1545-1555.1977
19. Bandyopadhyay A, Stöckel J, Min H, Sherman LA, Pakrasi HB (2010) High rates of photobiological H2 production by a cyanobacterium under aerobic conditions. Nat Commun 1:139. https://doi.org/10.1038/ncomms1139
20. Berman-Frank I, Lundgren P, Chen Y-B, Küpper H, Kolber Z, Bergman B, Falkowski P (2001) Segregation of nitrogen fixation and oxygenic photosynthesis in the marine cyanobacterium *Trichodesmium*. Science 294(5546):1534–1537. https://doi.org/10.1126/science.1064082
21. Compaoré J, Stal LJ (2010) Oxygen and the light–dark cycle of nitrogenase activity in two unicellular cyanobacteria. Environ Microbiol 12(1):54–62. https://doi.org/10.1111/j.1462-2920.2009.02034.x
22. Tamagnini P, Costa JL, Almeida L, Oliveira MJ, Salema R, Lindblad P (2000) Diversity of cyanobacterial hydrogenases, a molecular approach. Curr Microbiol 40(6):356–361. https://doi.org/10.1007/s002840010070
23. Houchins JP, Burris RH (1981) Light and dark reactions of the uptake hydrogenase in *Anabaena* 7120. Plant Physiol 68(3):712–716. https://doi.org/10.1104/pp.68.3.712
24. Lindblad P, Sellstedt A (1990) Occurrence and localization of an uptake hydrogenase in the filamentous heterocystous cyanobacterium *Nostoc* PCC 73102. Protoplasma 159:9–15. https://doi.org/10.1007/BF01326630
25. Bothe H, Winkelmann S, Boison G (2008) Maximizing hydrogen production by cyanobacteria. Z Naturforsch C 63(3–4):226–232. https://doi.org/10.1515/znc-2008-3-412
26. Happe T, Schütz K, Böhme H (2000) Transcriptional and mutational analysis of the uptake hydrogenase of the filamentous cyanobacterium *Anabaena variabilis* ATCC 29413. J Bacteriol 182(6):1624–1631. https://doi.org/10.1128/jb.182.6.1624-1631.2000
27. Oliveira P, Leitao E, Tamagnini P, Moradas-Ferreira P, Oxelfelt F (2004) Characterization and transcriptional analysis of *hupSLW* in *Gloeothece* sp. ATCC 27152: an uptake hydrogenase from a unicellular cyanobacterium. Microbiology 150(11):3647–3655. https://doi.org/10.1099/mic.0.27248-0
28. Oxelfelt F, Tamagnini P, Lindblad P (1998) Hydrogen uptake in *Nostoc* sp. strain PCC 73102. Cloning and characterization of a *hupSL* homologue. Arch Microbiol 169(4):267–274. https://doi.org/10.1007/s002030050571
29. Carrasco CD, Buettner JA, Golden JW (1995) Programmed DNA rearrangement of a cyanobacterial hupL gene in heterocysts. Proc Natl Acad Sci 92(3):791–795. https://doi.org/10.1073/pnas.92.3.791

30. Carrasco CD, Holliday SD, Hansel A, Lindblad P, Golden JW (2005) Heterocyst-specific excision of the *Anabaena* sp. strain PCC 7120 *hupL* element requires *xisC*. J Bacteriol 187(17):6031–6038. https://doi.org/10.1128/jb.187.17.6031-6038.2005
31. Ghirardi ML, Posewitz MC, Maness PC, Dubini A, Yu J, Seibert M (2007) Hydrogenases and hydrogen photoproduction in oxygenic photosynthetic organisms. Annu Rev Plant Biol 58:71–91. https://doi.org/10.1146/annurev.arplant.58.032806.103848
32. Kentemich T, Casper M, Bothe H (1991) The reversible hydrogenase in *Anacystis nidulans* is a component of the cytoplasmic membrane. Naturwissenschaften 78:559–560. https://doi.org/10.1007/BF01134448
33. Appel J, Phunpruch S, Steinmüller K, Schulz R (2000) The bidirectional hydrogenase of *Synechocystis* sp. PCC 6803 works as an electron valve during photosynthesis. Arch Microbiol 173(5):333–338. https://doi.org/10.1007/s002030000139
34. Gutekunst K, Chen X, Schreiber K, Kaspar U, Makam S, Appel J (2014) The bidirectional NiFe-hydrogenase in *Synechocystis* sp. PCC 6803 is reduced by flavodoxin and ferredoxin and is essential under mixotrophic, nitrate-limiting conditions. J Biol Chem 289(4):1930–1937. https://doi.org/10.1074/jbc.M113.526376
35. Troshina O, Serebryakova L, Sheremetieva M, Lindblad P (2002) Production of H_2 by the unicellular cyanobacterium *Gloeocapsa alpicola* CALU 743 during fermentation. Int J Hydrog Energy 27(11):1283–1289. https://doi.org/10.1016/S0360-3199(02)00103-9
36. Cournac L, Guedeney G, Peltier G, Vignais PM (2004) Sustained photoevolution of molecular hydrogen in a mutant of *Synechocystis* sp. strain PCC 6803 deficient in the type I NADPH-dehydrogenase complex. J Bacteriol 186(6):1737–1746. https://doi.org/10.1128/jb.186.6.1737-1746.2003
37. Khetkorn W, Lindblad P, Incharoensakdi A (2010) Enhanced biohydrogen production by the N2-fixing cyanobacterium *Anabaena siamensis* strain TISTR 8012. Int J Hydrog Energy 35(23):12767–12776. https://doi.org/10.1016/j.ijhydene.2010.08.135
38. Taikhao S, Phunpruch S (2017) Effect of metal cofactors of key enzymes on biohydrogen production by nitrogen fixing cyanobacterium *Anabaena siamensis* TISIR 8012. Energy Procedia 138:360–365. https://doi.org/10.1016/j.egypro.2017.10.166
39. Khetkorn W, Lindblad P, Incharoensakdi A (2020) Enhanced H_2 production with efficient N_2-fixation by fructose mixotrophically grown *Anabaena* sp. PCC 7120 strain disrupted in uptake hydrogenase. Algal Res 47:101823. https://doi.org/10.1016/j.algal.2020.101823
40. Taikhao S, Junyapoon S, Incharoensakdi A, Phunpruch S (2013) Factors affecting biohydrogen production by unicellular halotolerant cyanobacterium *Aphanothece halophytica*. J Appl Phycol 25(2):575–585. https://doi.org/10.1007/s10811-012-9892-3
41. Ananyev G, Carrieri D, Dismukes GC (2008) Optimization of metabolic capacity and flux through environmental cues to maximize hydrogen production by the cyanobacterium "*Arthrospira (Spirulina) maxima*". Appl Environ Microbiol 74(19):6102–6113. https://doi.org/10.1128/AEM.01078-08
42. Raksajit W, Satchasataporn K, Lehto K, Mäenpää P, Incharoensakdi A (2012) Enhancement of hydrogen production by the filamentous non-heterocystous cyanobacterium *Arthrospira* sp. PCC 8005. Int J Hydrog Energy 37(24):18791–18797. https://doi.org/10.1016/j.ijhydene.2012.10.011
43. Raksajit W, Maneeruttanarungroj C, Mäenpää P, Lehto K, Incharoensakdi A (2020) Upregulation of Hox-hydrogenase gene expression by nutrient adjustment in the filamentous non-heterocystous cyanobacterium *Arthrospira* sp. PCC 8005. J Appl Phycol 32(6): 3799–3807. https://doi.org/10.1007/s10811-020-02217-x
44. Yodsang P, Raksajit W, Aro EM, Mäenpää P, Incharoensakdi A (2018) Factors affecting photobiological hydrogen production in five filamentous cyanobacteria from Thailand. Photosynthetica 56(1):334–341. https://doi.org/10.1007/s11099-018-0789-5
45. Wutthithien P, Lindblad P, Incharoensakdi A (2019) Improvement of photobiological hydrogen production by suspended and immobilized cells of the N_2-fixing cyanobacterium *Fischerella*

muscicola TISTR 8215. J Appl Phycol 31(6):3527–3536. https://doi.org/10.1007/s10811-019-01881-y

46. Antal TK, Lindblad P (2005) Production of H_2 by sulphur-deprived cells of the unicellular cyanobacteria *Gloeocapsa alpicola* and *Synechocystis* sp. PCC 6803 during dark incubation with methane or at various extracellular pH. J Appl Microbiol 98(1):114–120. https://doi.org/10.1111/j.1365-2672.2004.02431.x
47. Kaushik A, Anjana K (2011) Biohydrogen production by *Lyngbya perelegans*: influence of physico-chemical environment. Biomass Bioenergy 35(3):1041–1045. https://doi.org/10.1016/j.biombioe.2010.11.024
48. Shah V, Garg N, Madamwar D (2003) Ultrastructure of the cyanobacterium *Nostoc muscorum* and exploitation of the culture for hydrogen production. Folia Microbiol 48(1):65–70. https://doi.org/10.1007/bf02931278
49. Baebprasert W, Lindblad P, Incharoensakdi A (2010) Response of H_2 production and hox-hydrogenase activity to external factors in the unicellular cyanobacterium Synechocystis sp. strain PCC 6803. Int J Hydrog Energy 35(13):6611–6616. https://doi.org/10.1016/j.ijhydene.2010.04.047
50. Touloupakis E, Rontogiannis G, Silva Benavides AM, Cicchi B, Ghanotakis DF, Torzillo G (2016) Hydrogen production by immobilized *Synechocystis* sp. PCC 6803. Int J Hydrog Energy 41(34):15181–15186. https://doi.org/10.1016/j.ijhydene.2016.07.075
51. Anjana K, Kaushik A (2014) Enhanced hydrogen production by immobilized cyanobacterium *Lyngbya perelegans* under varying anaerobic conditions. Biomass Bioenergy 63:54–57. https://doi.org/10.1016/j.biombioe.2014.01.019
52. Leino H, Kosourov SN, Saari L, Sivonen K, Tsygankov AA, Aro E-M, Allahverdiyeva Y (2012) Extended H_2 photoproduction by N_2-fixing cyanobacteria immobilized in thin alginate films. Int J Hydrog Energy 37(1):151–161. https://doi.org/10.1016/j.ijhydene.2011.09.088
53. Pansook S, Incharoensakdi A, Phunpruch S (2019) Enhanced dark fermentative H_2 production by agar-immobilized cyanobacterium *Aphanothece halophytica*. J Appl Phycol 31(5):2869–2879. https://doi.org/10.1007/s10811-019-01822-9
54. Rashid N, Song W, Park J, Jin H-F, Lee K (2009) Characteristics of hydrogen production by immobilized cyanobacterium Microcystis aeruginosa through cycles of photosynthesis and anaerobic incubation. J Ind Eng Chem 15(4):498–503. https://doi.org/10.1016/j.jiec.2008.12.013
55. Touloupakis E, Poloniataki EG, Ghanotakis DF, Carlozzi P (2021) Production of biohydrogen and/or poly-β-hydroxybutyrate by *Rhodopseudomonas* sp. using various carbon sources as substrate. Appl Biochem Biotechnol 193(1):307–318. https://doi.org/10.1007/s12010-020-03428-1
56. Khanna N, Lindblad P (2015) Cyanobacterial hydrogenases and hydrogen metabolism revisited: recent progress and future prospects. Int J Mol Sci 16(5):10537–10561. https://doi.org/10.3390/ijms160510537
57. Tuo SH, Rodriguez IB, Ho TY (2020) H_2 accumulation and N_2 fixation variation by Ni limitation in *Cyanothece*. Limnol Oceanogr 65(2):377–386. https://doi.org/10.1002/lno.11305
58. Allahverdiyeva Y, Leino H, Saari L, Fewer DP, Shunmugam S, Sivonen K, Aro E-M (2010) Screening for biohydrogen production by cyanobacteria isolated from the Baltic Sea and Finnish lakes. Int J Hydrog Energy 35(3):1117–1127. https://doi.org/10.1016/j.ijhydene.2009.12.030
59. Benemann JR (2000) Hydrogen production by microalgae. J Appl Phycol 12(3):291–300. https://doi.org/10.1023/A:1008175112704
60. Dutta D, De D, Chaudhuri S, Bhattacharya SK (2005) Hydrogen production by cyanobacteria. Microb Cell Factories 4(1):36. https://doi.org/10.1186/1475-2859-4-36
61. Touloupakis E, Torzillo G (2019) Photobiological hydrogen production. In: Solar hydrogen production. Academic Press, pp 511–525. https://doi.org/10.1016/B978-0-12-814853-2.00014-X

62. Nagarajan D, Lee D-J, Kondo A, Chang J-S (2017) Recent insights into biohydrogen production by microalgae – from biophotolysis to dark fermentation. Bioresour Technol 227:373–387. https://doi.org/10.1016/j.biortech.2016.12.104
63. Lindblad P, Christensson K, Lindberg P, Fedorov A, Pinto F, Tsygankov A (2002) Photoproduction of H_2 by wildtype *Anabaena* PCC 7120 and a hydrogen uptake deficient mutant: from laboratory experiments to outdoor culture. Int J Hydrog Energy 27(11): 1271–1281. https://doi.org/10.1016/S0360-3199(02)00111-8
64. Krujatz F, Illing R, Krautwer T, Liao J, Helbig K, Goy K, Opitz J, Cuniberti G, Bley T, Weber J (2015) Light-field-characterization in a continuous hydrogen-producing photobioreactor by optical simulation and computational fluid dynamics. Biotechnol Bioeng 112(12):2439–2449. https://doi.org/10.1002/bit.25667
65. Castro-Ceseña AB, del Pilar Sánchez-Saavedra M (2016) Effect of glycerol and PEGMA coating on the efficiency of cell holding in alginate immobilized *Synechococcus elongatus*. J Appl Phycol 28(1):63–71. https://doi.org/10.1007/s10811-015-0552-2
66. Kaklamani G, Cheneler D, Grover LM, Adams MJ, Bowen J (2014) Mechanical properties of alginate hydrogels manufactured using external gelation. J Mech Behav Biomed Mater 36:135–142. https://doi.org/10.1016/j.jmbbm.2014.04.013
67. Kosourov S, Leino H, Murukesan G, Lynch F, Sivonen K, Tsygankov AA et al (2014) Hydrogen photoproduction by immobilized N_2-fixing cyanobacteria: understanding the role of the uptake hydrogenase in the long-term process. Appl Environ Microbiol 80(18): 5807–5817. https://doi.org/10.1128/aem.01776-14
68. Therien JB, Zadvornyy OA, Posewitz MC, Bryant DA, Peters JW (2014) Growth of Chlamydomonas reinhardtii in acetate-free medium when co-cultured with alginate-encapsulated, acetate-producing strains of *Synechococcus* sp. PCC 7002. Biotechnol Biofuels 7(1):1–8. https://doi.org/10.1186/s13068-014-0154-2
69. Kosourov SN, Seibert M (2009) Hydrogen photoproduction by nutrient-deprived *Chlamydomonas reinhardtii* cells immobilized within thin alginate films under aerobic and anaerobic conditions. Biotechnol Bioeng 102(1):50–58. https://doi.org/10.1002/bit.22050
70. Seol E, Manimaran A, Jang Y, Kim S, Oh Y-K, Park S (2011) Sustained hydrogen production from formate using immobilized recombinant *Escherichia coli* SH5. Int J Hydrog Energy 36(14):8681–8686. https://doi.org/10.1016/j.ijhydene.2010.05.118
71. Nguyen BT, Nicolai T, Benyahia L, Chassenieux C (2014) Synergistic effects of mixed salt on the gelation of κ-carrageenan. Carbohydr Polym 112:10–15. https://doi.org/10.1016/j.carbpol.2014.05.048
72. Semenchuk IN, Taranova LA, Kaleniuk AA, Il'iasov PV, Reshetilov AN (2000) Effect of various methods of immobilization on stability of a microbial biosensor based on *Pseudomonas rathonis* T during detection of surfactants. Prikl Biokhim Mikrobiol 36(1):80–84
73. Gutthann F, Egert M, Marques A, Appel J (2007) Inhibition of respiration and nitrate assimilation enhances photohydrogen evolution under low oxygen concentrations in *Synechocystis* sp. PCC 6803. Biochim Biophys Acta (BBA)-Bioenerget 1767(2):161–169. https://doi.org/10.1016/j.bbabio.2006.12.003
74. Baebprasert W, Jantaro S, Khetkorn W, Lindblad P, Incharoensakdi A (2011) Increased H_2 production in the cyanobacterium *Synechocystis* sp. strain PCC 6803 by redirecting the electron supply via genetic engineering of the nitrate assimilation pathway. Metab Eng 13(5):610–616. https://doi.org/10.1016/j.ymben.2011.07.004
75. McNeely K, Xu Y, Bennette N, Bryant DA, Dismukes GC (2010) Redirecting reductant flux into hydrogen production via metabolic engineering of fermentative carbon metabolism in a cyanobacterium. Appl Environ Microbiol 76(15):5032–5038. https://doi.org/10.1128/aem.00862-10
76. Lindberg P, Schütz K, Happe T, Lindblad P (2002) A hydrogen-producing, hydrogenase-free mutant strain of *Nostoc punctiforme* ATCC 29133. Int J Hydrog Energy 27(11):1291–1296. https://doi.org/10.1016/S0360-3199(02)00121-0

77. Masukawa H, Mochimaru M, Sakurai H (2002) Disruption of the uptake hydrogenase gene, but not of the bidirectional hydrogenase gene, leads to enhanced photobiological hydrogen production by the nitrogen-fixing cyanobacterium *Anabaena* sp. PCC 7120. Appl Microbiol Biotechnol 58(5):618–624. https://doi.org/10.1007/s00253-002-0934-7
78. Lindberg P, Devine E, Stensjö K, Lindblad P (2012) HupW protease specifically required for processing of the catalytic subunit of the uptake hydrogenase in the cyanobacterium *Nostoc* sp. strain PCC 7120. Appl Environ Microbiol 78(1):273–276. https://doi.org/10.1128/aem.05957-11
79. Yoshino F, Ikeda H, Masukawa H, Sakurai H (2007) High photobiological hydrogen production activity of a *Nostoc* sp. PCC 7422 uptake hydrogenase-deficient mutant with high nitrogenase activity. Mar Biotechnol 9(1):101–112. https://doi.org/10.1007/s10126-006-6035-3
80. Ducat DC, Sachdeva G, Silver PA (2011) Rewiring hydrogenase-dependent redox circuits in cyanobacteria. Proc Natl Acad Sci 108(10):3941–3946. https://doi.org/10.1073/pnas.1016026108
81. Gärtner K, Lechno-Yossef S, Cornish AJ, Wolk CP, Hegg EL (2012) Expression of *Shewanella oneidensis* MR-1 [FeFe]-hydrogenase genes in *Anabaena* sp. strain PCC 7120. Appl Environ Microbiol 78(24):8579–8586. https://doi.org/10.1128/aem.01959-12
82. Weyman PD, Vargas WA, Tong Y, Yu J, Maness PC, Smith HO, Xu Q (2011) Heterologous expression of *Alteromonas macleodii* and *Thiocapsa roseopersicina* [NiFe] hydrogenases in *Synechococcus elongatus*. PLoS One 6(5):e20126. https://doi.org/10.1371/journal.pone.0020126
83. Lupacchini S, Appel J, Stauder R, Bolay P, Klähn S, Lettau E, Adrian L, Lauterbach L, Bühler B, Schmid A, Toepel J (2021) Rewiring cyanobacterial photosynthesis by the implementation of an oxygen-tolerant hydrogenase. Metab Eng 68:199–209. https://doi.org/10.1016/j.ymben.2021.10.006
84. Wolk CP, Ernst A, Elhai J (1994) Heterocyst metabolism and development. In: The molecular biology of cyanobacteria. Springer, Dordrecht, pp 769–823
85. Buikema WJ, Haselkorn R (2001) Expression of the Anabaena hetR gene from a copper-regulated promoter leads to heterocyst differentiation under repressing conditions. Proc Natl Acad Sci 98(5):2729–2734. https://doi.org/10.1073/pnas.051624898
86. Wong FC, Meeks JC (2001) The *hetF* gene product is essential to heterocyst differentiation and affects HetR function in the cyanobacterium *Nostoc punctiforme*. J Bacteriol 183(8):2654–2661. https://doi.org/10.1128/jb.183.8.2654-2661.2001
87. Muñoz-García J, Ares S (2016) Formation and maintenance of nitrogen-fixing cell patterns in filamentous cyanobacteria. Proc Natl Acad Sci 113(22):6218–6223. https://doi.org/10.1073/pnas.1524383113
88. Avilan L, Roumezi B, Risoul V, Bernard CS, Kpebe A, Belhadjhassine M, Rousset M, Brugna M, Latifi A (2018) Phototrophic hydrogen production from a clostridial [FeFe] hydrogenase expressed in the heterocysts of the cyanobacterium *Nostoc* PCC 7120. Appl Microbiol Biotechnol 102(13):5775–5783. https://doi.org/10.1007/s00253-018-8989-2
89. Vargas WA, Weyman PD, Tong Y, Smith HO, Xu Q (2011) [NiFe] hydrogenase from *Alteromonas macleodii* with unusual stability in the presence of oxygen and high temperature. Appl Environ Microbiol 77(6):1990–1998. https://doi.org/10.1128/aem.01559-10

Adv Biochem Eng Biotechnol (2023) 183: 281–302
https://doi.org/10.1007/10_2022_212

Published online: 29 November 2022

Utilizing Cyanobacteria in Biophotovoltaics: An Emerging Field in Bioelectrochemistry

Hans Schneider, Bin Lai, and Jens Krömer

Contents

Abstract Anthropogenic global warming is driven by the increasing energy demand and the still dominant use of fossil energy carriers to meet these needs. New carbon-neutral energy sources are urgently needed to solve this problem. Biophotovoltaics, a member of the so-called bioelectrochemical systems family, will provide an important piece of the energy puzzle. It aims to harvest the electrons from sunlight-driven water splitting using the natural oxygenic photosystem (e.g., of cyanobacteria) and utilize them in the form of, e.g., electricity or hydrogen. Several key aspects of biophotovoltaics have been intensively studied in recent years like physicochemical properties of electrodes or efficient wiring of microorganisms to electrodes. Yet, the exact mechanisms of electron transfer between the biocatalyst and the electrode remain unresolved today. Most research is conducted on micro-scale reactors generating small currents over short time-scales, but multiple experiments have shown biophotovoltaics great potential with lab-scale reactors producing currents over weeks to months. Although biophotovoltaics is still in its

H. Schneider (✉), B. Lai, and J. Krömer
Department of Solar Materials, Helmholtz Center for Environmental Research, Leipzig, Germany
e-mail: hans.schneider@ufz.de; bin.lai@ufz.de; jens.kroemer@ufz.de

infancy with many open research questions to be addressed, new promising results from various labs around the world suggest an important opportunity for biophotovoltaics in the decades to come. In this chapter, we will introduce the concept of biophotovoltaics, summarize its recent key progress, and finally critically discuss the potentials and challenges for future rational development of biophotovoltaics.

Graphical Abstract

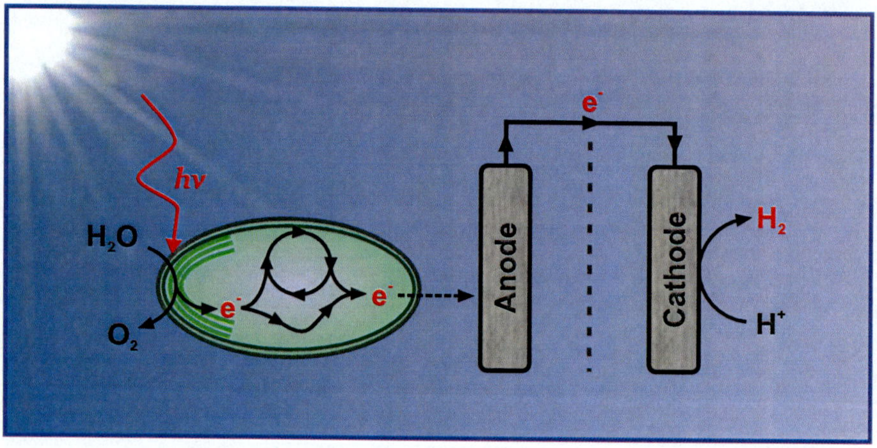

Keywords Biophotovoltaics, Extracellular electron transfer, Oxygenic photosynthesis, Photosynthetic electron transport chain, Renewable energy

1 Introduction: Biophotovoltaic and Other Light Harvesting Bioelectrochemical Systems

A fundamental driver of biological processes is the generation of chemical energy through oxidation and reduction reactions and the generation of charge gradients across membranes. Microorganisms in natural environments require for this an oxidizable substrate and a terminal electron acceptor. The process is limited by the amount of electron donors and acceptors and the ratio of both, since cells cannot accumulate free electrons. Sources and sinks can be organic compounds, metals, metal ions in solution or gases. Remarkably, electrode surfaces can also serve as electron sources or sinks in technical systems. Examples of such systems are bioelectrochemical systems (BESs), where anodes accept electrons from the microbes, or cathodes can provide electron input. Historically, the first BESs dating back to 1911 [1] were termed microbial fuel cells (MFC) due to their working principle. In MFCs chemotrophic microbes generate electrons by the oxidation of organic compounds, which are harvested at the anode and on their way to the cathode generate a current [2, 3]. Where oxygenic phototrophic microorganisms are employed, the electrons used to generate an electric current are derived from the water splitting reaction fueled by light [4–6]. By now, there is a great variety of light harvesting BESs and the corresponding research field developed rather broadly investigating various aspects of biological and/or technical nature necessary to generate currents using light energy [7–9]. Major developments are the establishment of photosynthetic MFCs (photoMFC) and more recently the whole-cell *biophotovoltaic (BPV)* systems, which will be introduced shortly in the following paragraphs.

A classical photoMFC typically utilizes non-oxygenic photosynthetic microorganism (e.g., purple bacteria) or a combination of heterotrophic and phototrophic (both oxygenic and non-oxygenic) microorganisms [6, 10–12]. Here, the phototrophic organisms use the light energy to lift the energy status of the intracellular electrons and thus drive the electron flux toward an external electrode or release reduced organic substances that can be used by chemotrophic electrogenic microbes for current production (Fig. 1). The different system configurations at early phase were reviewed by Rosenbaum et al. [11]. Depending on the format of the biocatalysts, photoMFCs can be subdivided into three categories: whole-cell, complex, and sub-cellular photoMFCs. Whole-cell photoMFCs utilize pure living chemoautotroph microbes, often purple non-sulfur bacteria for light-dependent generation of electricity. The whole-cell photoMFC is relatively resilient, capable

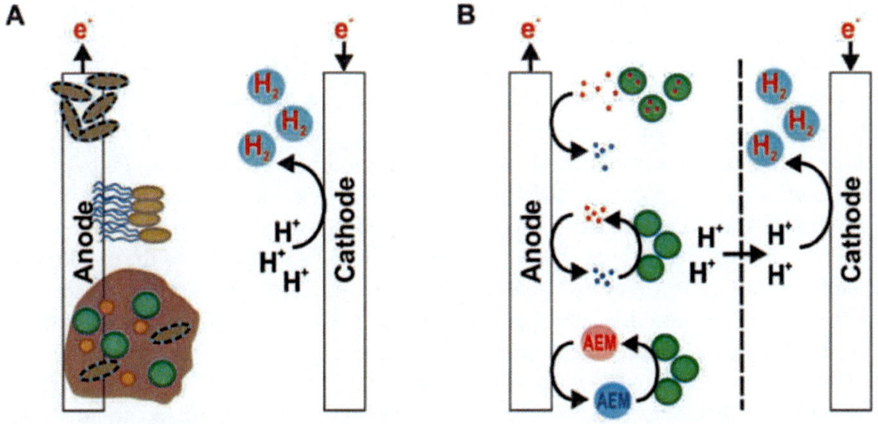

Fig. 1 Different mechanisms of microbial EET in BPV. The basic mechanisms for EET to the anode are either direct (**a**) or indirect (**b**). Direct EET includes (**a**, top to bottom) direct contact between cells and anode and electron transfer by cell surface redox proteins, direct contact by conductive cellular appendages, and conductive biofilms of mixed species on the anode that allow transfer of electrons along the nanowires or surface proteins of the exoelectrogenic chemotrophs. Indirect EET is based on (**b**, top to bottom) oxidation of secreted products or metabolites without recycling, cycling of endogenous electron mediators produced by the microbes, or artificially added AEMs. Depending on the microbial species or consortium used and the setup of the BES all mechanisms may exist in a single BES concomitantly

of reproduction and self-repair. Exogenous reducing equivalents (e.g., acetate) are typically required as electron source, while light energy, saved in the form of intracellular reducing equivalents, could be utilized for current production even in dark conditions. In contrast, complex photoMFCs contain both heterotrophic and phototrophic microorganisms or even plants. In general, the phototrophs harvest the sunlight energy and produce carbon-based compounds, which are then used by the heterotrophs as nutrient source. However, undefined microbial consortia are typically present in these systems and the interspecies interactions between different biocatalysts as well as the interactions between microbes and electrodes remain largely unknown. Finally, sub-cellular photoMFCs utilize purified, non-oxygenic photosynthetic enzyme immobilized on the electrode surface. For instance, the photosynthetic reaction center from *Rhodobacter sphaeroides* [13] and photosystem I (PSI) from *Synechococcus elongatus* [14] were attached on anode surfaces and achieved a light-dependent current output. In the past decade, advances in this field, like probing photocurrents of up to 10 pA from single PSI complexes, have extended the understanding about kinetics of photocatalytic reaction centers greatly [7, 14].

Unlike the photoMFC, BPVs use solely oxygenic photosynthetic organisms (e.g., cyanobacteria, microalgae, etc.) and target the direct use of photosynthetic electron fluxes from light-driven water splitting. In this chapter, we focus on the BPV system harboring cyanobacteria. BPV is a fairly new concept, and apart from BPV, several other nomenclatures have been also used, e.g. bio-photoelectrolysis cell, photo-bioelectrocatalytic cell, biohybrid photoelectrochemical cell, biophotovoltaic electrochemical cell, or microbial solar cell, etc. BPVs have the same system configurations as typical bioelectrochemical systems, consisting of one- or two-chambered reactors with a two- or three-electrode setup depending on the application purpose. Briefly, cyanobacterial cells split water in the anodic chamber using sunlight, and then release free electrons that can be harvested by the anode. The electrons are then transferred to the cathode via an external electric circuit, where the counter reaction (e.g., the reduction of protons into hydrogen) takes place. A reference electrode can be used to quantitatively measure or control the redox potential of the whole cell or individual electrodes. Among the proof-of-concept studies, current output has been observed for BPV systems with planktonic cyanobacterial cells or cyanobacterial "biofilm" (i.e., pseudo-biofilm where the biomass is pasted and dried on the electrode surface [15]). Reports from literature suggest that cyanobacterial cells could transfer electrons to an anode directly or via self-secreted redox compounds [7]. Nevertheless, artificial electron mediators (AEMs), like ferricyanide, phenazines, or quinones, were also used in multiple studies, typically leading to higher current densities [16–23]. An ion exchange membrane or a salt bridge is required if AEMs are introduced, to avoid the redox short-cut of the AEM between anode and cathode [4, 23, 24].

In general, the efficiency of electron transfer from bacterial cells to anode in BPV is much lower compared to those achieved with MFCs or photoMFCs. For MFCs the electron yields from oxidation of organic substrates range from 10 to 90% depending on experimental setups and condition [2, 25–27]. A Coulombic efficiency (efficiency of charge transfer facilitating an electrochemical reaction in a system) of nearly

100% was even reported for biofilms containing the exoelectrogenic bacteria *G. sulfurreducens* fueled with acetate in a two-chambered BES [28]. The highest Coulombic efficiencies reported for photoMFCs were around 70–80% [29–31]. Such high efficiencies have been achieved for electrons transferred from the cellular respiration by oxidation of organic substances either artificially added or secreted by photosynthetic microorganisms. In BPV systems where electrons could be retrieved directly from the water splitting reaction in the cyanobacterial cell, the typical light-to-energy efficiencies are often well below 2%. But BPV systems show a great potential with a theoretically light-to-energy efficiency reaching over 20% when the photosystem II (PSII) is directly coupled to an electrode. On the other hand, when electrons derived from photosynthesis are stored in biomass and subsequently released via respiration (i.e., in photoMFCs or MFCs), the theoretical maximum light-to-energy efficiency decreases to a theoretical maximum 4.5% [32–34].

Photocurrent and Darkcurrent In BPV systems, all electrons are originating from the water splitting reaction during the photosynthetic light reaction. Illumination of the BPV system activates photosynthesis and thus the photosynthetic electron transport chain (PETC) resulting in an increased current – so-called light- or photocurrent, photo response, or photo power output (hereafter termed *photocurrent*). Darkness results in a decreased power output of the BPV system but the current still remains significantly higher than the abiotic background of the respective system [9, 24]. This so-called *darkcurrent* can be considered a delayed photocurrent, where the electrons stored in metabolites, such as glycogen, are utilized [35–37]. The electrons nevertheless originate from water splitting during illumination. The level and duration of this dark current is thus dependent on the amount of electrons stored in the cell and can vary a lot between different experimental setups, especially the physiological status of the microbial cells. For example, Lai et al. showed a dark current more than three times higher than the abiotic background current for 48 h [24].

2 Cyanobacterial Electron Transfer Pathways and Exoelectrogenesis

This chapter is addressing BPV systems containing cyanobacteria, in particular the model strain *Synechocystis sp.* PCC 6803 (hereafter *Synechocystis*). Cyanobacteria are usually classified as gram-negative, due to their cell envelope consisting of a cytoplasmic membrane and an outer membrane separated by a peptidoglycan layer [38]. Some cyanobacteria like *Synechocystis* also show gram-positive characteristics, such as a considerably thicker peptidoglycan layer, compared to typical gram-negative bacteria, with a higher degree of cross-linking between the peptidoglycan chains [39, 40]. Another unique feature is the presence of an internal system of membranes which form the thylakoids. The conversion of solar radiation into chemical energy is catalyzed by two reaction centers: PSII and PSI, which are the two key components of the PETC. PETC is a chain consisting of several redox enzymes and electron carriers including, e.g., quinones, ferredoxin, plastocyanin, and cytochrome b_6f. In cyanobacteria all protein complexes and electron mediators involved in the PETC are mainly located in the thylakoids. These cellular compartments are located in the cytoplasm where they form stacks of membranes. The tips of the thylakoids merge in biogenic regions and form the so-called convergence membranes, which come in close contact, but do not merge, with the cytoplasmic membrane [41, 42]. The cytoplasmic membrane of *Synechocystis* lacks both a cytochrome-*c* oxidase and the cytochrome b_6f complex [43]. Electrons can be shuttled from the cellular electron carriers to external electron acceptors via different mechanisms. The origin of electrons and their route across the cell membrane are key questions in BPV research.

Extracellular Electron Transfer (EET) The process of microbial cells exchanging electrons with external electrodes is called EET [44]. In natural environments, similar processes are the transfer of electrons to and from metal ore, playing an important part of natural geochemical cycles [45]. This naturally occurring microbial EET is called exoelectrogenesis and is used by humankind, for instance, in biohydrometallurgy processes [46]. Generally, EET can be classified into two groups: direct EET (DET), where the EET relies on the outer membrane redox proteins in physical contact with an electrode, whereas for indirect EET (IDET) a soluble redox carrier is necessary to convey the electron flux between the microbial cells and the electrodes (Fig. 1). The majority of the reported photoMFCs were operated employing mixed cultures of photo- and chemotrophic organisms, and the performed DET was facilitated by the co-cultured exoelectrogenic chemotrophic bacteria like *Geobacter sulfurreducens* via electrically conductive extracellular appendages [47], but also some phototrophs were suspected to supply electrons indirectly by soluble redox shuttles [48].

DET by *Synechocystis* There is an ongoing debate about the ability of *Synechocystis* to perform DET. Several studies have shown current outputs in BPV without additionally added AEMs where the biocatalysts were in direct contact with the anode [5, 15, 49, 50]. However, uncertainties about the possible self-secreted mediator in such setups were not able to be ruled out [51, 52]. *Synechocystis* lacks the pili-based DET pathways known for the exoelectrogenic bacteria *Shewanella oneidensis* and *Geobacter sp.* [53, 54]. The cells were reported to produce type IV pili for, e.g., phototaxis and natural transformation competency [55, 56], and some studies suggested they are essential for iron reduction [57] and are electrochemically conductive under CO_2 limitation condition [53]. However, photocurrent production by wild-type *Synechocystis* and its pili-deficient mutant in a BPV system did not show a significant difference, suggesting that pili are likely not essential for the EET at least under the tested conditions [49]. Furthermore, conductivity measurements via atomic force microscopy of native *Synechocystis* pili showed no evidence for conductivity of these structures [54], which is in contradiction to what was previously reported by Gorby et al. [53]. Hereby, two main hypotheses remain explaining the EET transfer in a BPV system without AEMs:

- DET by exoelectrogenesis mechanisms that employ c-type cytochromes at and/or beyond the outer surface of the organism, which have been shown for different gram-negative and -positive prokaryotes in BESs [58–60];
- IDET by excretion of endogenously produced redox mediators shuttling electrons between cyanobacteria and the electrode [49, 52, 61].

IDET for *Synechocystis* Electrons are shuttled between cells and an electrode via soluble redox mediators. These redox shuttles can be categorized into two groups: endogenous electron mediators and AEMs [62]. In a BPV system using planktonic *Synechocystis* cells without the addition of an AEM, a correlation between current generation and the presence of a small unidentified molecule was shown [52]. In this work, the cells were pressurized (10–15 psi) by a microfluidizer before inoculation, and the resulting cells showed a 3 times higher photocurrent response, both in the light and in darkness, compared to untreated *Synechocystis* cells. Further analysis suggested an increasing secretion of an undefined small molecule (<3 kDa) that served as an endogenous mediator, which was mainly secreted under illuminated conditions. In a follow-up study from the group [51], they proposed the molecule to be NADPH via two-dimensional fluorescence mapping, but more evidence is needed to confirm if NADPH is cycling between the cells and anode and how. Also, uncertainties are still present for the *Synechocystis* cells without being pressurized, which also produce certain photocurrent output.

AEMs are often used in BPV research to improve the EET rate and thus the redox driving force from the anode on microbial physiology changes. Furthermore, the addition of a specific mediator (e.g., polarity, redox potential, etc.) in principle allows targeting specific points along the cellular (either photosynthetic or respiratory) electron transport chains (ETCs) for harvesting electrons. The selection of a suitable AEM is therefore not only dependent on the molecule's toxicity and effects on cell physiology but also on the mid-point potential and the point of access to the electron transport chains. One widely used AEM is ferricyanide which is lipid insoluble, therefore it should be unable to penetrate at least the cytoplasmic membrane, which is supported by its low toxicity under the applied conditions [63–65]. Other AEMs used in BPV research include 1,4-Benzoquinone [66–68], 2-hydroxy-1,4-naphthoquinone [6, 20, 66, 69], 2,6-dimethyl-1,4-benzoquinone [66, 67, 70], diaminodurene [70, 71], among others. A recent study examined the rerouting of electron flow from the PETC of the algae *Chlamydomonas reinhardtii* to the anode via different quinones as AEMs [22] and found 2,3-dimethyl-5-phenyl-1,4-benzoquinone as the best AEM candidate for future optimization, in terms of the stability and density of photocurrents. However, the authors acknowledged the difficulty to comparatively evaluate different mediators due to, e.g., the balance between mediator toxicity and current density along with the increasing amount of mediator [22]. A general limitation when using AEMs is scalability. When thinking about future large-scale applications of BPV systems the use of AEMs poses general restrictions in terms of, e.g., operation costs and downstream processing. So far not much work has been done on the recycling and reuse of AEMs, which has to be tackled for future application [72]. One possible solution could be to engineer or enhance the production and secretion of biodegradable endogenous electron mediators from cyanobacteria. For example, Clifford et al. demonstrated that cyanobacteria can be genetically engineered to produce and secrete phycocyanin, which acted as an AEM in the BPV system increasing the power output fourfold [21]. Another example of this approach is the aforementioned pressure treatment of cells for increased NADPH secretion [51]. However, this approach is at the cost of adding additional metabolic burden to the cell's metabolism, and the loss of electrons while producing these redox compounds from CO_2 would also lower the overall light-to-energy efficiency of the cyanobacterial cells in a BPV system.

3 State of the Art of BPV

The typical BPV refers to the system using living oxygenic photosynthetic microbes to produce current with electrons originating from the splitting of water. Besides whole cells, also isolated fractions of the oxygenic photosystem apparatus (PSI, PSII, or even the whole thylakoid membrane) were used in some BPV studies and good current outputs have been achieved thereby [73–77]. For example, Calkins et al. immobilized thylakoids on carbon nanotubes and reported the highest power outputs of 53 mW m^{-2} (at 250 mA m^{-2}) yet achieved with thylakoids. But these systems are intrinsically restricted by the low stability of the respective enzymes and the current outputs only lasted a short time (up to tens of minutes) [7]. Photocurrents of BPV systems applying whole cyanobacterial cells have been shown to be stable over multiple days up to several weeks [10, 24], making the overall system much more robust and suitable for optimization and ultimately field application. Increased photocurrents are generated during illumination, while comparably lower darkcurrents are generated by respiratory breakdown of carbon storage molecules during darkness [5, 65]. In BPV systems photosynthetic prokaryotic microbes are typically favored, due to simplicity of cell structure and accessibility of the photosystems, compared to eukaryotes. Living cyanobacteria are regarded as the best compromise in this regard as they are more robust compared to sub-cellular systems while having a simpler physiology, lower basal energy demand, and less compartmented cell structure compared to eukaryotic microorganisms. In BPV research various cyanobacterial species have been used including *Synechocystis* [10, 15, 24, 52, 61, 64, 78, 79], *Nostoc sp.* [10, 61, 62, 67], *Phormidium* spp. [19], *Anabaena variabilis M-2* [6, 10, 23], and *Synechococcus* [10, 23, 66, 69–71]. However, several research groups utilize the unicellular algae *Chlamydomonas reinhardtii* as it is, like *Synechocystis*, a model organism for photosynthesis which has been extensively studied, i.e. in regard to interactions with exogenous quinones [22, 37, 68].

Current outputs of BPV systems vary significantly depending on the applied biocatalyst, electrode material, light intensity and wavelength, temperature, and system configurations [7, 9, 79]. There are indications that power outputs of BPV systems might be influenced by the presence of O_2 in the media [80]. Oxygen, generated during photosynthesis, is postulated to be a limiting factor for power outputs in BPV systems by reacting with electrons and thus competing with anodic electron transfer with comparably higher current outputs observed for anaerobic-anode MFCs [80]. The question, whether oxygen as an electron sink (in terms of, e.g., Mehler reaction, respiration, oxidation of reduced mediator, etc.) affects photo- and dark current outputs significantly, is not clear yet and remains for further investigation under different physiological conditions.

Variety of BPV Setups BPV is a fairly new concept and there is little standardization regarding experimental reactors leading to a diversity of BPV setups. Microfluidic or chip-based systems enjoy great popularity in BPV research, due to the advantages of miniaturized systems like fast fabrication, low energy and resource demand, and parallelization. Many miniaturized BPV systems rely on sedimenting cells on the anode bringing them in close physical proximity to the electrode [52, 61, 78, 81]. In large volume reactors gas availability becomes a bigger concern, compared to micro scale (or continuous) reactors, and therefore require aeration or stirring of the media thus complicating the defined sedimentation of cells [24]. To ensure cells are in close proximity of the anode in lab-scale reactors, anodes can be coated with a biofilm of cyanobacteria: either naturally grown or by drying cells on the electrode before assembling the BPV reactors [5, 61, 67, 82]. Polymer or redox hydrogel coating was used in some studies to improve the biofilm formation on the anode surface [83]. When cells are kept in suspension the working volume, and thus the amount of biomass, can be scaled up easily from milliliter to liter scale systems allowing to sample appropriate cell amounts or media volume for detailed cell physiological investigations like dry mass analysis, metabolomics, or proteomics [24].

Electrolytes Most early BPV systems used simple buffers for electrochemical measurements to ensure a low bias in the electrochemical measurements. Buffers included, for example, tricine-KOH [23], phosphate [6], or $H_3BO_3 - Na_2CO_3 -$ KCl buffer [19]. With developments in the field of BPV research and cell physiology moving to the center of attention, more and more researchers used common growth media like BG11 to keep BPV conditions more similar to growth conditions. One problem with BG11 is that it differs in composition for many labs and the first recipe had some undefined components [84–88]. Furthermore, several components in BG11 media were recently found to significantly influence bioelectrochemical measurements, particularly due to organic pH buffers and unchelated Mn^{2+} ions in the trace solution [24]. Thus, the classical BG11 media was modified and optimized and an nBG11 medium recipe was generated which is more suited for electrochemical working conditions while allowing for similar growth behavior of cyanobacteria as in BG11 media.

Electrodes A wide range of anode materials has been used in BPV research so far, including PET [71], SnO_2 [19], or Au-mesh [75] (Fig. 2). Several studies make use of indium–tin oxide due to its electrical conductivity and optical transparency [21, 73, 89–91]. Because electrochemical charge transfer rates at the electrode are significantly influenced by physicochemical properties of the electrode surfaces, several research groups focused on surface modulation, i.e. increasing electrode porosity [61, 92]. Yet, more research is needed regarding the impact of electrical conductivity, chemical stability, biocompatibility, surface area/porosity of electrodes on the kinetics of metabolic reactions in the biotic compartment [93]. A more in-depth analysis of photoanode materials used in biophotovoltaic systems can be found in the review published by Anam et al. [93].

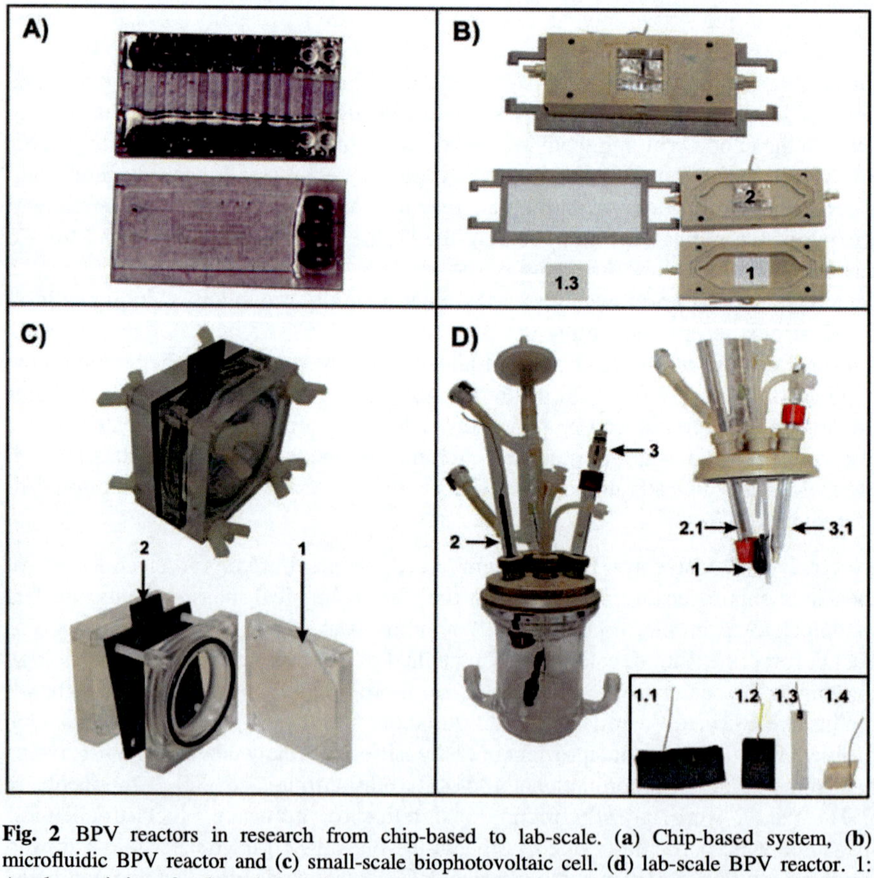

Fig. 2 BPV reactors in research from chip-based to lab-scale. (**a**) Chip-based system, (**b**) microfluidic BPV reactor and (**c**) small-scale biophotovoltaic cell. (**d**) lab-scale BPV reactor. 1: Anode consisting of carbon cloth (1.1), graphite (1.2), indium–tin oxide (1.3), or Au-mesh (1.4), 2: Cathode, 2.1 Cathode chamber, 3: Reference electrode and 3.1 Reference electrode chamber. See Table 1 for experimental parameters of comparable systems in research

Some of the most efficient BPV systems were achieved using carbon-based electrodes [78]. Besides the suitable electrochemical properties, carbon-based electrodes offer environmental advantages regarding sustainability, especially in regard to upscaling and industrial application.

Table 1 lists properties and achieved power densities representative for different BPV setups depicted in Fig. 2. A comprehensive comparison of power outputs and efficiencies from various BPV systems and photoMFCs can also be found in the publications by Tschörtner et al. [9] and McCormick et al. [7].

4 Unraveling the EET Pathways

One of the main bottlenecks for the rational improvement and application of BPV systems is the poor understanding of the proteins and mechanisms involved in EET. One approach to study the EET pathways is to use synthetic biology tools to target specific respiratory electron transport chain (RETC) or PETC components. For instance, mutants of *Synechocystis* lacking some or all three terminal oxidase (TO) enzyme complexes, in all combinations, were compared in a BPV system regarding their potential to reduce the electron acceptor ferricyanide [64]. *Synechocystis* mutants lacking either TO, i.e. bd-quinol, cytochrome-c, or alternative respiratory TO, showed a significantly higher ferricyanide reduction rate in the dark compared to wild-type cells. The highest reduction rate and thus highest peak current were achieved for the triple TO mutant. The authors concluded that for the triple TO mutant a significant amount of electrons were redirected from the respiration to the electrode, accounting to 10% of electrons from the RETC compared to less than 0.2% in wild-type *Synechocystis* [64]. In this study, no significant increase in photocurrent output was observed for any of the TO mutants. With most parts of cellular ETCs being essential for cell viability, the downregulation of respective ETC functions is likely a more feasible approach to study EET compared to a complete gene deletion. For both approaches the (untargeted) physiological changes induced by genetic engineering have to be determined for their potential impacts on the EET, using systems biology tools. Furthermore, genetic engineering of model cyanobacteria is still far more challenging compared to model heterotrophic organisms [99]. Thus, strain engineering of cyanobacteria to improve their performance in BPV systems is yet limited to the removal of competing electron sinks like TOs [4, 64, 100].

Another, more straight-forward, approach to study electron transfer mechanisms is to use specific ETC inhibitors during BPV experiments. To investigate whether the water splitting reaction is the source of photocurrent, DCMU has been used in multiple BPV studies. Technically, DCMU inhibits electron transfer at PSII and thus obstructs electron flow from the excited PSII into the PQ pool (Fig. 3). DCMU can either incompletely or completely inhibit the activity of PSII (e.g., via binding to the D1 subunit which is one of the core subunits of PSII binding all the redox-active cofactors involved in the energy conversion process [101, 102]), leading to a

Table 1 Performance of representative BPV systems with different configurations. Peak current, anode area, working volume, anode material, cathode material, mediator, organism, application of the biocatalyst, and type of BPV system are compared. IO-ITO = Inverse opal indium-tin oxide

Anode material	Cathode material	Applied AEM	Organism	Application of biocatalyst	Working volume [mL]	Peak current [μA]	Anode surface [mm^2]	Type of BPV system	Reference
InSnBi alloy	Platinum	No AEM	*Synechocystis*	Sedimented on anode	0.0004	0.0882	0.03	Chip-based (Fig. 2a)	[78]
Carbon (screen printed)	Carbon (screen printed)	No AEM	*Synechocystis*	Cells in suspension	0.05	0.0237	0.79	Droplet on electrode	[51]
ITO or carbon paper	Platinum wire	No AEM	*Synechocystis* slr0688i	Sedimented on anode	4	94.2	314	Small-scale (Fig. 2b or c)	[94]
IO-ITO	Platinum wire	DCBQ	*Synechocystis*	Biofilm on anode	4.5	3.375	75	Small-scale (Fig. 2c)	[21, 95, 96]
ITO	Platinum wire	No AEM	*Synechococcus sp.*	Cells in suspension	5	0.0465	310	Small-scale (Fig. 2b or c)	[90]
ITO	Pt-coated glass	No AEM	*Synechococcus sp.*	Biofilm on anode	12.56	120.9	1,300	Small-scale (Fig. 2c)	[5]
Carbon cloth	Carbon cloth	Duro-quinol	*Synechocystis*	Cells in suspension	50	122.5	1,225	Lab-scale (Fig. 2d)	[97]
Carbon cloth	Steel-mesh	Ferricyanide	*Synechocystis*	Planktonic cultures	220	20	2,500	Lab-scale (Fig. 2d)	[24]
Aluminum (recycled can)	Carbon paper	No AEM	*C. sorokiniana*	Cells in suspension	2000	50	Approx. 4,300	Conventional PET-bottle	[98]

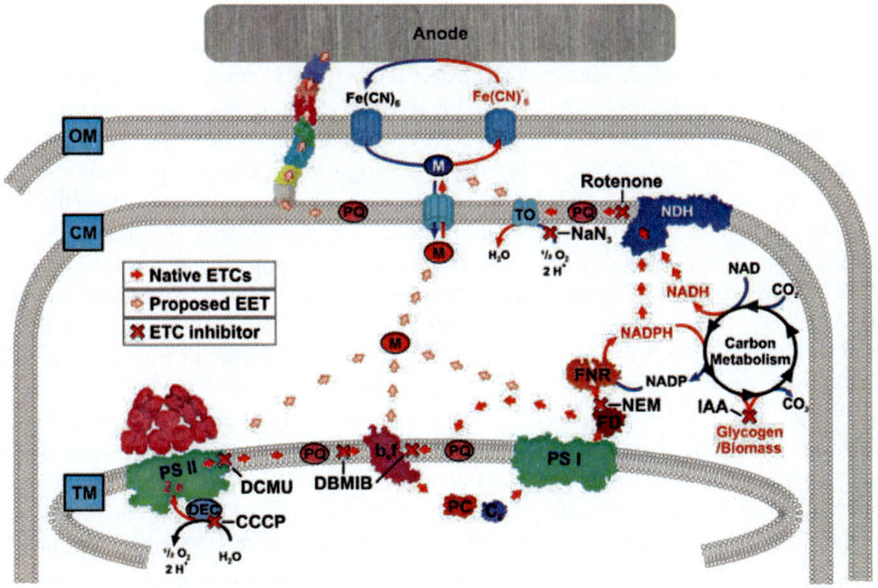

Fig. 3 Schematic representation of electron transfer key elements between respiratory and photosynthetic electron transport complexes. Depicted are the PETC, including: PSII, plastoquinone (PQ), cytochrome b6f (b_6f), plastocyanin (PC) and cytochrome c_6 (C_6), PSI, ferredoxin (FD) and ferredoxin-NADP+ reductase (FNR), and RETC, including: NADH dehydrogenase (NDH) and TOs, with native electron flow (red arrows) and suggested EET pathways (light red arrows) and their location in outside of the three membranes: outer, cytoplasmic, and thylakoid membrane (OM, CM, and TM respectively). Red X indicates the point of action for the inhibitors DCMU, DBMIB, NEM, CCCP, NaN3, Rotenone, and IAA [9, 105–108]. A suggested endogenous electron mediator is depicted (M) which might be able to move through the CM via a hypothesized channel protein. A detailed report about cyanobacterial electron transfer pathway was presented by Lea-Smith and colleagues [109]

decreased or completely extinguished electron flux from PSII [103]. However, conflicted results were obtained while applying DCMU in different BPV studies. A complete inhibition of photocurrents by biofilms of *Nostoc sp.* and *Lyngbya* was achieved using low concentrations (25 μM) of DCMU in a BPV system [62]. In another study, using *Synechocystis* cells in a BPV system, oxygen evolution was completely inhibited but photocurrent outputs were only reduced by 63 ± 17% with 15 μM DCMU [65]. The residual current was suggested to be a result of increased electron flow from the RETC due to the PETC inhibition by DCMU. In contrast to this, a mediator-less BPV system with *Synechocystis* showed no reduction in photocurrent after adding DCMU in the system [52]. On the contrary, the system reached the peak current much faster in presence of DCMU than without DCMU. The study also showed prolonged photocurrent lifetime by addition of glucose regardless of the addition of DCMU, suggesting DCMU to affect electron flow but not photocurrent stability. Furthermore, DCMU was found to be essential for photocurrent production by *Synechocystis* mutants lacking the binding site for DCMU in PSII with photocurrents dropping to 14% in absence of DCMU [52]. Increasing photocurrents in the presence of DCMU were attributed to higher release of the endogenous redox mediator by *Synechocystis*, suggesting either inhibition of forward electron flow in RETC or enhancement of mediator release. Although DCMU is widely used to verify the origin of a proposed photocurrent in several studies, results and their interpretation were diverse which could possibly be due to the application of different conditions (i.e., Organisms, AEMs, or BPV working conditions) and dynamic effects depending on cellular physiology.

The second PETC inhibitor, DBMIB, binds to the quinol oxidation site of cytochrome b_6f and subsequently prevents electron transfer between the PQ pool and the cytochrome b_6f complex [104]. Due to a light-dependent translocation of DBMIB across the b_6f complex, a resulting inhibition of the PETC is suggested. In a BPV system using *Synechococcus sp.* as biocatalyst, the addition of DBMIB, after initially decreasing dark currents, increased currents back to the control level with successive addition of higher DBMIB concentrations [69]. Another study found a three-fold increase of photocurrent outputs after exposure of *Nostoc* and *Lyngbya* cells to high concentrations of DBMIB [62]. The authors (of both aforementioned studies) thus suggest that DBMIB could likely act as a redox mediator, take up electrons from the PQ pool, and result in increased power outputs in both dark and light phases. Therefore, DBMIB might not be a good candidate for EET studies [9], unless being combined with other approaches (e.g., tracing the electron flux, etc.) to distinguish these two contradictory effects on the current output.

Another PETC inhibitor, CCCP, acts as a proton uncoupler inhibiting ATP synthesis, as well as inhibiting electron transfer at the reducing site of PSII [69, 110]. A significant decrease of power outputs from *Lyngbya* biofilms was shown by gradually adding CCCP to the electrolyte [10]. Other experiments showed a temporary inhibition of photocurrents produced by *Lyngbya* and *Nostoc* biofilms with lower concentrations of CCCP, and complete inhibition with increased concentrations [62]. The authors accounted the decrease in power output with CCCP to the inhibition of electron flow from PSII and thus concluded the origin of the current to be from the PETC. Nevertheless, electrons derived from RETC could also be impacted by these substances as the decreased ATP synthesis could impact both PETC and RETEC.

Azides selectively inhibit two respiratory TOs [108]. A significant reduction of BPV photocurrents generated by *Nostoc* but only minor influence on currents generated by *Lyngbya* were observed with NaN_3 [62]. The authors contributed the different results for both microorganisms to changing presence of one TO depending on growth conditions or a non-substantial role of this TO for electrogenic activity in cyanobacteria.

Although the inhibitory mechanism of DCMU is comparably well understood and it is widely used in BPV research to confirm PSII and the splitting of water as source of the observed photocurrent, multiple studies had varying results (using different cyanobacteria in different BPV system with or without the addition of AEMs) leading to conflicting hypotheses regarding the origin of photocurrents. The varying results of BPV inhibitor studies using DBMIB, CCCP, or azides illustrate the difficulty of interpretation of such experiments. The effects of ETC inhibitor seem to be more dynamic and closely linked with the physiological status of the organisms, this has to be considered in systematic ETC inhibitor studies. A key limitation here, as with ETC mutant studies, is the inhibitors' influence on cell physiology and viability and thus attribution of decreasing current outputs to specific ETC locations is not trivial. A standardization of the reactor setups and working conditions should be pursued. Also, the untargeted effects of these chemicals on other biological functionalities should also be taken into account.

5 Outlook

Lab-scale applications of BPV system have been reported in a few cases by connecting several BPV systems in series to power small devices like watches or calculators [5, 111]. Bombelli et al. showed that they can power a widely used microprocessor continuously with a simple BPV system using metal electrode for over 6 months [112]. Yet, major challenges for BPV systems have to be addressed and overcome in order to allow large-scale long-term applications in the future.

Recent Advances and Remaining Challenges for BPV Systems One of the key challenges for BPV research and potential application is overcoming the currently low EET kinetics from cells to the anode. A lot of advances from BESs research regarding electrode materials or reactor design can be transferred to BPV. Yet, challenges essential for BPV systems, like targeting specific positions in the PETC for EET, have to be elucidated further to improve BPVs potential for industrial application. One of the key factors hindering progress in BPV research is the poor understanding of effects from artificial electron sinks on cellular metabolisms and the redirection of the complex intracellular electron flows. Several research groups addressed these questions by experimentation and computing. For example, a quantitative analysis of the effects of morphological changes on EET rates in cyanobacteria was recently published [16]. A BPV system with cyanobacterial mutants has shown that a TO-deletion mutant has the highest known current outputs and further deletion or restriction of electron sinks could steer even more electrons to the anode. A study found major increase in current generation by cyanobacteria in a BPV system by in situ biosynthesis of gold nanoparticles inside and around the cells [113]. Here the gold nanoparticles increased light harvesting and promoted EET rates.

Improvement of BPV systems by strain engineering is limited by the remaining challenges of manipulating ETCs mainly located on the thylakoid membrane and the still-needed efficient gene editing tools for cyanobacteria [99]. This could change in the future with new tools for genetic manipulation like CRISPRi, which allows for tunable inactivation of specific genes in vivo [114, 115]. With this approach, genes coding for essential proteins for cell growth, like core components of PETC or

RETC, could be (partially) inactivated while observing the effects on BPV performance. Generally, there is little knowledge about the influence of BPV conditions (i.e., mediators, applied potential, additional electron acceptors) on cyanobacterial physiology. This is largely missing in current BPV studies and should be a main focus for the future of BPV research. On the other hand, the common approach to investigate EETs with ETC inhibitors could be combined with methods like metabolomics, proteomics, or fluxomics to draw more meaningful information about electron flows in the cell. New tools like the DUAL-KLAS-NIR, a further development of the PAM chlorophyll fluorometer, allow the observation of redox changes in plastocyanin, the PSI reaction center, and in ferredoxin in vivo [116, 117]. A combination of genetic modification, inhibition, and analysis of redox states of cyanobacterial ETCs could facilitate a systematic understanding and rational improvement of the current EET process in BPV systems.

The realization of BPV systems in an industrial scale also poses some unique challenges. Compared to the well-established biotechnological processes based on heterotrophs, light would be a major limitation for the scaling of BPV processes, which is the same for other photobiotechnology processes. The limited total irradiation per unit surface area poses an upper boundary for the system and consequently maximizing the light-use efficiency in time and space becomes an extremely critical parameter for the application of this technique in future. In addition, the use of anodes as external electron sinks for the microbes further complicates the multi-interfacial interactions present in photobiotechnology and raise, more challenges to balance interfacial electron and mass transfer. Revolutions in reactor and process engineering are necessary to realize large-scale phototrophic processes, e.g. via the concepts of flat-panel or capillary reactors [118–121].

Potential for BPV Systems In general, BPV systems show a great potential for carbon-neutral energy production with the most commonly used materials being considered not limited, namely: water and salts (possibly seawater), steel and carbon for electrodes, CO_2, and sunlight. To date, use of AEMs like ferricyanide stays relevant for research of EET, and its applicability in large-scale applications remains to be determined by comparing the rising operating cost and the increased EET efficiency (i.e., productivity of the system). The operation cost might be tackled with possible endogenous production of redox mediators by cyanobacteria [122, 123]. In general, increasing conductivity of cells/cellular membranes is one of the key obstacles in BPV research. Several approaches can be implemented to deal with this obstacle, for example: increasing membrane conductivity by employing membrane soluble mediators or genetic engineering to produce cytosolic/membrane bound redox proteins [50].

The possibility for carbon-neutral (or even negative) hydrogen production is probably the most promising application for BPV systems in the future. Hydrogen is anticipated to be a key component for sustainable energy supply in the near future [124], and BPV provides a unique carbon-sink approach for its production using water, sunlight, and CO_2. Furthermore, BPV systems could be used to additionally produce high value carbon compounds or biomass while producing electricity or hydrogen.

References

1. Potter MC (1997) Electrical effects accompanying the decomposition of organic compounds. Proc R Soc Lond Ser B Contain Pap Biol Character 84(571):260–276
2. Obileke K et al (2021) Microbial fuel cells, a renewable energy technology for bio-electricity generation: a mini-review. Electrochem Commun 125
3. Santoro C et al (2017) Microbial fuel cells: from fundamentals to applications. A review. J Power Sources 356:225–244
4. McCormick AJ et al (2013) Hydrogen production through oxygenic photosynthesis using the cyanobacterium Synechocystis sp. PCC 6803 in a bio-photoelectrolysis cell (BPE) system. Energ Environ Sci 6(9)
5. McCormick AJ et al (2011) Photosynthetic biofilms in pure culture harness solar energy in a mediatorless bio-photovoltaic cell (BPV) system. Energ Environ Sci 4(11):4699–4709
6. Tanaka K, Kashiwagi N, Ogawa T (1988) Effects of light on the electrical output of bioelectrochemical fuel-cells containing Anabaena variabilis M-2: mechanism of the post-illumination burst. J Chem Technol Biotechnol 42(3):235–240
7. McCormick AJ et al (2015) Biophotovoltaics: oxygenic photosynthetic organisms in the world of bioelectrochemical systems. Energ Environ Sci 8(4):1092–1109
8. Wey LT et al (2019) The development of biophotovoltaic Systems for Power Generation and Biological Analysis. ChemElectroChem 6(21):5375–5386
9. Tschortner J, Lai B, Kromer JO (2019) Biophotovoltaics: green power generation from sunlight and water. Front Microbiol 10:866
10. Pisciotta JM, Zou Y, Baskakov IV (2010) Light-dependent electrogenic activity of cyanobacteria. PLoS One 5(5):e10821
11. Rosenbaum M, He Z, Angenent LT (2010) Light energy to bioelectricity: photosynthetic microbial fuel cells. Curr Opin Biotechnol 21(3):259–264
12. Kaushik S, Goswami P (2018) Bacterial membrane depolarization-linked fuel cell potential burst as signal for selective detection of alcohol. ACS Appl Mater Interfaces 10(22): 18630–18640
13. Janzen AF, Seibert M (1980) Photoelectrochemical conversion using reaction-Centre electrodes. Nature 286(5773):584–585
14. Gerster D et al (2012) Photocurrent of a single photosynthetic protein. Nat Nanotechnol 7(10): 673–676
15. Cereda A et al (2014) A bioelectrochemical approach to characterize extracellular electron transfer by Synechocystis sp. PCC6803. PLoS One 9(3):e91484
16. Okedi TI, Fisher AC, Yunus K (2020) Quantitative analysis of the effects of morphological changes on extracellular electron transfer rates in cyanobacteria. Biotechnol Biofuels 13:150
17. Pinhassi RI et al (2016) Hybrid bio-photo-electro-chemical cells for solar water splitting. Nat Commun 7:12552
18. Sekar N et al (2016) Enhanced photo-bioelectrochemical energy conversion by genetically engineered cyanobacteria. Biotechnol Bioeng 113(3):675–679
19. Ochiai H et al (1983) Properties of semiconductor electrodes coated with living films of cyanobacteria. Appl Biochem Biotechnol 8(4):289–303
20. Zou Y et al (2009) Photosynthetic microbial fuel cells with positive light response. Biotechnol Bioeng 104(5):939–946
21. Clifford ER et al (2021) Phenazines as model low-midpoint potential electron shuttles for photosynthetic bioelectrochemical systems. Chem Sci 12(9):3328–3338
22. Sayegh A et al (2021) Finding adapted quinones for harvesting electrons from photosynthetic algae suspensions. ChemElectroChem 8(15):2968–2978
23. Yagishita T et al (1998) Performance of photosynthetic electrochemical cells using immobilized Anabaena variabilis M-3 in discharge/culture cycles. J Ferment Bioeng 85(5): 546–549

24. Lai B et al (2021) Technical-scale biophotovoltaics for long-term photo-current generation from Synechocystis sp. PCC6803. Biotechnol Bioeng 118(7):2637–2648
25. Chen W et al (2021) Improved electricity generation, coulombic efficiency and microbial community structure of microbial fuel cells using sodium citrate as an effective additive. J Power Sources 482:228947
26. Cao B et al (2021) Silver nanoparticles boost charge-extraction efficiency in Shewanella microbial fuel cells. Science 373(6561):1336–1340
27. Xin S et al (2020) Electricity generation and microbial community of single-chamber microbial fuel cells in response to Cu2O nanoparticles/reduced graphene oxide as cathode catalyst. Chem Eng J 380
28. Nevin KP et al (2008) Power output and columbic efficiencies from biofilms of Geobacter sulfurreducens comparable to mixed community microbial fuel cells. Environ Microbiol 10(10):2505–2514
29. Xing D et al (2008) Electricity generation by Rhodopseudomonas palustris DX-1. Environ Sci Technol 42(11):4146–4151
30. Bazdar E et al (2018) The effect of different light intensities and light/dark regimes on the performance of photosynthetic microalgae microbial fuel cell. Bioresour Technol 261:350–360
31. Gunaseelan K et al (2021) Blending of microbial inocula: an effective strategy for performance enhancement of clayware biophotovoltaics microbial fuel cells. Bioresour Technol 323:124564
32. Dau H, Zaharieva I (2009) Principles, efficiency, and blueprint character of solar-energy conversion in photosynthetic water oxidation. Acc Chem Res 42(12):1861–1870
33. Barber J (2009) Photosynthetic energy conversion: natural and artificial. Chem Soc Rev 38(1):185–196
34. Michel H (2012) Editorial: the nonsense of biofuels. Angew Chem Int Ed Engl 51(11):2516–2518
35. Tanaka K, Tamamushi R, Ogawa T (1985) Bioelectrochemical fuel-cells operated by the cyanobacterium, Anabaena variabilis. J Chem Technol Biotechnol Biotechnol 35(3):191–197
36. Yagishita T et al (1997) Behavior of glucose degradation in Synechocystis sp. M-203 in bioelectrochemical fuel cells. Bioelectrochem Bioenerg 43(1):177–180
37. Beauzamy L et al (2020) Mediator-microorganism interaction in microbial solar cell: a fluo-electrochemical insight. Anal Chem 92(11):7532–7539
38. Dijkstra AJ, Keck W (1996) Peptidoglycan as a barrier to transenvelope transport. J Bacteriol 178(19):5555–5562
39. Glauner B, Höltje J, Schwarz U (1988) The composition of the murein of Escherichia coli. J Biol Chem 263(21):10088–10095
40. Al-Amoudi A et al (2004) Cryo-electron microscopy of vitreous sections. EMBO J 23(18):3583–3588
41. Rast A et al (2019) Biogenic regions of cyanobacterial thylakoids form contact sites with the plasma membrane. Nat Plants 5(4):436–446
42. Pisareva T et al (2011) Model for membrane organization and protein sorting in the cyanobacterium Synechocystis sp. PCC 6803 inferred from proteomics and multivariate sequence analyses. J Proteome Res 10(8):3617–3631
43. Schultze M et al (2009) Localization of cytochrome b6f complexes implies an incomplete respiratory chain in cytoplasmic membranes of the cyanobacterium Synechocystis sp. PCC 6803. Biochim Biophys Acta 1787(12):1479–1485
44. Schröder U, Harnisch F, Angenent LT (2015) Microbial electrochemistry and technology: terminology and classification. Energ Environ Sci 8(2):513–519
45. Kato S (2016) Microbial extracellular electron transfer and its relevance to iron corrosion. J Microbial Biotechnol 9(2):141–148
46. Bosecker K (1997) Bioleaching: metal solubilization by microorganisms. FEMS Microbiol Rev 20(3–4):591–604

47. Nishio K, Hashimoto K, Watanabe K (2013) Digestion of algal biomass for electricity generation in microbial fuel cells. Biosci Biotechnol Biochem 77(3):670–672
48. Cao X et al (2008) Electricity generation by an enriched phototrophic consortium in a microbial fuel cell. Electrochem Commun 10(9):1392–1395
49. Thirumurthy MA et al (2020) Type IV pili-independent photocurrent production by the cyanobacterium synechocystis sp. PCC 6803. Front Microbiol 11:1344
50. Gonzalez-Aravena AC et al (2018) Tapping into cyanobacteria electron transfer for higher exoelectrogenic activity by imposing iron limited growth. RSC Adv 8(36):20263–20274
51. Shlosberg Y et al (2021) NADPH performs mediated electron transfer in cyanobacterial-driven bio-photoelectrochemical cells. iScience 24(1):101892
52. Saper G et al (2018) Live cyanobacteria produce photocurrent and hydrogen using both the respiratory and photosynthetic systems. Nat Commun 9(1):2168
53. Gorby YA et al (2006) Electrically conductive bacterial nanowires produced by Shewanella oneidensis strain MR-1 and other microorganisms. Proc Natl Acad Sci U S A 103(30): 11358–11363
54. El-Naggar MY et al (2010) Electrical transport along bacterial nanowires from Shewanella oneidensis MR-1. Proc Natl Acad Sci U S A 107(42):18127–18131
55. Bhaya D et al (2000) Type IV pilus biogenesis and motility in the cyanobacterium Synechocystis sp. PCC6803. Mol Microbiol 37(4):941–951
56. Yoshihara S et al (2001) Mutational analysis of genes involved in pilus structure, motility and transformation competency in the unicellular motile cyanobacterium Synechocystis sp. PCC 6803. Plant Cell Physiol 42(1):63–73
57. Lamb JJ et al (2014) Functional role of PilA in iron acquisition in the cyanobacterium Synechocystis sp. PCC 6803. PLoS One 9(8):e105761
58. Richter K, Schicklberger M, Gescher J (2012) Dissimilatory reduction of extracellular electron acceptors in anaerobic respiration. Appl Environ Microbiol 78(4):913–921
59. Wrighton KC et al (2011) Evidence for direct electron transfer by a gram-positive bacterium isolated from a microbial fuel cell. Appl Environ Microbiol 77(21):7633–7639
60. Inoue K et al (2011) Specific localization of the c-type cytochrome OmcZ at the anode surface in current-producing biofilms of Geobacter sulfurreducens. Environ Microbiol Rep 3(2): 211–217
61. Wenzel T et al (2018) Porous translucent electrodes enhance current generation from photosynthetic biofilms. Nat Commun 9(1):1299
62. Pisciotta JM, Zou Y, Baskakov IV (2011) Role of the photosynthetic electron transfer chain in electrogenic activity of cyanobacteria. Appl Microbiol Biotechnol 91(2):377–385
63. Jones RW, Garland PB (1977) Sites and specificity of the reaction of bipyridylium compounds with anaerobic respiratory enzymes of Escherichia coli. Effects of permeability barriers imposed by the cytoplasmic membrane. Biochem J 164(1):199–211
64. Bradley RW et al (2013) Terminal oxidase mutants of the cyanobacterium Synechocystis sp. PCC 6803 show increased electrogenic activity in biological photo-voltaic systems. Phys Chem Chem Phys 15(32):13611–13618
65. Bombelli P et al (2011) Quantitative analysis of the factors limiting solar power transduction by Synechocystis sp. PCC 6803 in biological photovoltaic devices. Energ Environ Sci 4(11)
66. Torimura M et al (2001) Electrochemical investigation of cyanobacteria Synechococcus sp. PCC7942-catalyzed photoreduction of exogenous quinones and photoelectrochemical oxidation of water. J Electroanal Chem 496(1–2):21–28
67. Sekar N, Umasankar Y, Ramasamy RP (2014) Photocurrent generation by immobilized cyanobacteria via direct electron transport in photo-bioelectrochemical cells. Phys Chem Chem Phys 16(17):7862–7871
68. Longatte G et al (2018) Investigation of photocurrents resulting from a living unicellular algae suspension with quinones over time. Chem Sci 9(43):8271–8281

69. Yagishita T, Horigome T, Tanaka K (2007) Effects of light, CO2 and inhibitors on the current output of biofuel cells containing the photosynthetic organism Synechococcus sp. J Chem Technol Biotechnol 56(4):393–399
70. Tsujimura S et al (2001) Photosynthetic bioelectrochemical cell utilizing cyanobacteria and water-generating oxidase. Enzyme Microb Technol 29(4–5):225–231
71. Martens N, Hall EAH (2008) Diaminodurene as a mediator of a photocurrent using intact cells of cyanobacteria. Photochem Photobiol 59(1):91–98
72. Gemünde A et al (2022) Redox mediators in microbial electrochemical systems. ChemElectroChem 9(13)
73. Kato M et al (2012) Photoelectrochemical water oxidation with photosystem II integrated in a mesoporous indium-tin oxide electrode. J Am Chem Soc 134(20):8332–8335
74. Carpentier R et al (1989) A photoelectrochemical cell using immobilized photosynthetic membranes. Bioelectrochem Bioenerg 22(3):391–401
75. Lam KB et al (2006) A MEMS photosynthetic electrochemical cell powered by subcellular plant photosystems. J Microelectromech Syst 15(5):1243–1250
76. Calkins JO et al (2013) High photo-electrochemical activity of thylakoid–carbon nanotube composites for photosynthetic energy conversion. Energ Environ Sci 6(6)
77. Ruff A, Conzuelo F, Schuhmann W (2019) Bioelectrocatalysis as the basis for the design of enzyme-based biofuel cells and semi-artificial biophotoelectrodes. Nat Catal 3(3):214–224
78. Bombelli P et al (2015) A high power-density, mediator-free, microfluidic biophotovoltaic device for cyanobacterial cells. Adv Energy Mater 5(2):1–6
79. Madiraju KS et al (2012) Carbon neutral electricity production by Synechocystis sp. PCC6803 in a microbial fuel cell. Bioresour Technol 110:214–218
80. Chen CY, Chen TY, Chung YC (2014) A comparison of bioelectricity in microbial fuel cells with aerobic and anaerobic anodes. Environ Technol 35(1–4):286–293
81. Thorne RJ et al (2015) Iron reduction by the cyanobacterium Synechocystis sp. PCC 6803. Bioelectrochemistry 105:103–109
82. Kaushik S, Sarma MK, Goswami P (2017) FRET-guided surging of cyanobacterial photosystems improves and stabilizes current in photosynthetic microbial fuel cell. J Mater Chem A 5(17):7885–7895
83. Hasan K et al (2014) Photo-electrochemical communication between cyanobacteria (Leptolyngbia sp.) and osmium redox polymer modified electrodes. Phys Chem Chem Phys 16(45):24676–24680
84. Allen MM, Stanier RY (1968) Growth and division of some unicellular blue-green algae. J Gen Microbiol 51(2):199–202
85. Hughes EO, Gorham PR, Zehnder A (1958) Toxicity of a unialgal culture of Microcystis aeruginosa. Can J Microbiol 4(3):225–236
86. Shcolnick S, Shaked Y, Keren N (2007) A role for mrgA, a DPS family protein, in the internal transport of Fe in the cyanobacterium Synechocystis sp. PCC6803. Biochim Biophys Acta 1767(6):814–819
87. Rippka R et al (1979) Generic assignments, strain histories and properties of pure cultures of cyanobacteria. Microbiology 111(1):1–61
88. Kratz WA, Myers J (1955) Nutrition and growth of several blue-green algae. Am J Bot 42(3): 282
89. Thong CH et al (2022) Metal organic frameworks (MOFs) as potential anode materials for improving power generation from algal biophotovoltaic (BPV) platforms. Catal Today 397: 419–427
90. Tokunou Y et al (2021) Synechococcus and other bloom-forming cyanobacteria exhibit unique redox signatures. ChemElectroChem 8(2):360–364
91. Tanaka K et al (2021) Ferrihydrite reduction by photosynthetic Synechocystis sp. PCC 6803 and its correlation with electricity generation. Front Microbiol 12:650832
92. Zou Y, Pisciotta J, Baskakov IV (2010) Nanostructured polypyrrole-coated anode for sun-powered microbial fuel cells. Bioelectrochemistry 79(1):50–56

93. Anam M et al (2021) Evaluation of photoanode materials used in biophotovoltaic systems for renewable energy generation. Sustain Energy Fuels 5(17):4209–4232
94. Kusama S et al (2022) Order-of-magnitude enhancement in photocurrent generation of Synechocystis sp. PCC 6803 by outer membrane deprivation. Nat Commun 13(1):3067
95. Zhang JZ et al (2018) Photoelectrochemistry of photosystem II in vitro vs in vivo. J Am Chem Soc 140(1):6–9
96. Wey LT et al (2021) A biophotoelectrochemical approach to unravelling the role of cyanobacterial cell structures in exoelectrogenesis. Electrochim Acta 395
97. Lewis CM et al (2022) Electrochemically driven photosynthetic electron transport in cyanobacteria lacking photosystem II. J Am Chem Soc 144(7):2933–2942
98. Bateson P et al (2018) Electrochemical characterisation of bio-bottle-voltaic (BBV) systems operated with algae and built with recycled materials. Biology (Basel) 7(2)
99. Berla BM et al (2013) Synthetic biology of cyanobacteria: unique challenges and opportunities. Front Microbiol 4:246
100. Saar KL et al (2018) Enhancing power density of biophotovoltaics by decoupling storage and power delivery. Nat Energy 3(1):75–81
101. Inoué H et al (1990) Reversible inhibition of DCMU on the oxidizing side of photosystem II. Curr Res Photosynth:917–920
102. Ferreira KN et al (2004) Architecture of the photosynthetic oxygen-evolving center. Science 303(5665):1831–1838
103. Hsu BD, Lee JY, Pan RL (1986) The two binding sites for DCMU in photosystem II. Biochem Biophys Res Commun 141(2):682–688
104. Yan J, Kurisu G, Cramer WA (2006) Intraprotein transfer of the quinone analogue inhibitor 2,5-dibromo-3-methyl-6-isopropyl-p-benzoquinone in the cytochrome b6f complex. Proc Natl Acad Sci U S A 103(1):69–74
105. Dzelzkalns VA et al (1994) Deletion of the structural gene for the NADH-dehydrogenase subunit 4 of Synechocystis 6803 alters respiratory properties. Plant Physiol 106(4):1435–1442
106. Mi H et al (1994) NAD(P)H dehydrogenase-dependent cyclic electron flow around photosystem I in the cyanobacterium Synechocystis PCC 6803: a study of dark-starved cells and Spheroplasts. Plant Cell Physiol 35(2):163–173
107. Schmidt MM, Dringen R (2009) Differential effects of iodoacetamide and iodoacetate on glycolysis and glutathione metabolism of cultured astrocytes. Front Neuroenerg 1:1
108. Pils D, Schmetterer G (2001) Characterization of three bioenergetically active respiratory terminal oxidases in the cyanobacterium Synechocystis sp. strain PCC 6803. FEMS Microbiol Lett 203(2):217–222
109. Lea-Smith DJ et al (2016) Photosynthetic, respiratory and extracellular electron transport pathways in cyanobacteria. Biochim Biophys Acta 1857(3):247–255
110. Kimimura M et al (1971) Inhibitory site of carbonyl cyanide m-chlorophenylhydrazone in the electron transfer system of the chloroplasts. Biochim Biophys Acta (BBA) – Bioenerg 234(1): 92–102
111. Ieropoulos I, Greenman J, Melhuish C (2008) Microbial fuel cells based on carbon veil electrodes: stack configuration and scalability. Int J Energy Res 32(13):1228–1240
112. Bombelli P et al (2022) Powering a microprocessor by photosynthesis. Energ Environ Sci 15(6):2529–2536
113. Liu L, Choi S (2021) Enhanced biophotoelectricity generation in cyanobacterial biophotovoltaics with intracellularly biosynthesized gold nanoparticles. J Power Sources 506
114. Yao L et al (2016) Multiple gene repression in cyanobacteria using CRISPRi. ACS Synth Biol 5(3):207–212
115. Yao L et al (2020) Pooled CRISPRi screening of the cyanobacterium Synechocystis sp PCC 6803 for enhanced industrial phenotypes. Nat Commun 11(1):1666
116. Theune ML et al (2021) In-vivo quantification of electron flow through photosystem I – cyclic electron transport makes up about 35% in a cyanobacterium. Biochim Biophys Acta (BBA) – Bioenerg 1862(3):148353

117. Sétif P et al (2020) Identification of the electron donor to flavodiiron proteins in Synechocystis sp. PCC 6803 by in vivo spectroscopy. Biochim Biophys Acta (BBA) – Bioenerg 1861(10): 148256
118. Lippi L et al (2018) Exploring the potential of high-density cultivation of cyanobacteria for the production of cyanophycin. Algal Res 31:363–366
119. Karande R, Schmid A, Buehler K (2016) Applications of multiphasic microreactors for biocatalytic reactions. Org Process Res Dev 20(2):361–370
120. David CC, Karande R, Bühler K (2021) Cyanobacterial biofilms. In: Natural and synthetic environments
121. Bozan M, Schmid A, Buhler K (2022) Evaluation of self-sustaining cyanobacterial biofilms for technical applications. Biofilms 4:100073
122. Rabaey K et al (2005) Microbial phenazine production enhances electron transfer in biofuel cells. Environ Sci Technol 39(9):3401–3408
123. von Canstein H et al (2008) Secretion of flavins by Shewanella species and their role in extracellular electron transfer. Appl Environ Microbiol 74(3):615–623
124. Balat M (2008) Potential importance of hydrogen as a future solution to environmental and transportation problems. Int J Hydrogen Energy 33(15):4013–4029

Adv Biochem Eng Biotechnol (2023) 183: 303–352
https://doi.org/10.1007/10_2022_214

Published online: 27 December 2022

Process Technologies of Cyanobacteria

Marco Witthohn, Dorina Strieth, Jonas Kollmen, Anna Schwarz, Roland Ulber, and Kai Muffler

Contents

Abstract Although the handling and exploitation of cyanobacteria is associated with some challenges, these phototrophic bacteria offer great opportunities for innovative biotechnological processes. This chapter covers versatile aspects of working with cyanobacteria, starting with up-to-date *in silico* and in vitro screening methods for bioactive substances. Subsequently, common conservation techniques and vitality/viability estimation methods are compared and supplemented by own

M. Witthohn, A. Schwarz, and K. Muffler
Department of Life Sciences and Engineering, University of Applied Sciences Bingen, Bingen, Germany

D. Strieth, J. Kollmen, and R. Ulber (✉)
Chair of Bioprocess Engineering, Technical University of Kaiserslautern, Kaiserslautern, Germany
e-mail: ulber@mv.uni-kl.de

data regarding the non-invasive vitality evaluation via pulse amplitude modulated fluorometry. Moreover, novel findings about the influence the state of the pre-cultures have on main cultures are presented. The following sub-chapters deal with different photobioreactor-designs, with special regard to biofilm photobioreactors, as well as with heterotrophic and mixotrophic cultivation modes. The latter topic provides information from literature on successfully enhanced cyanobacterial production processes, augmented by own data.

Graphical Abstract

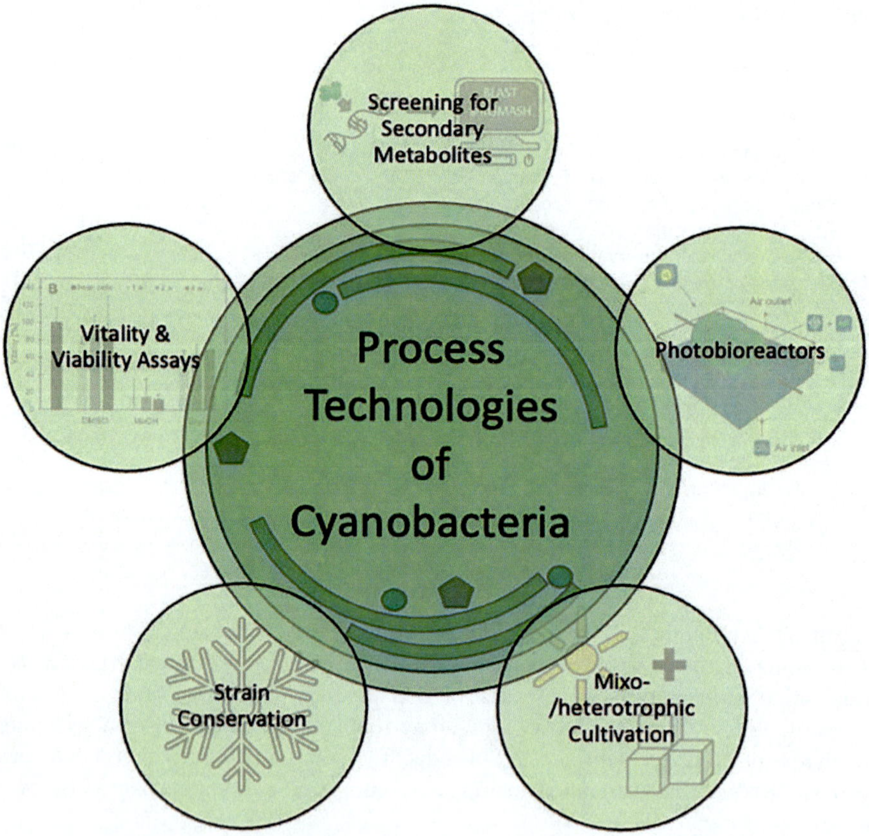

Keywords Cell characterization, Cyanobacteria, Mixo-/Heterotrophic cultivation, Photobioreactors, Strain conservation

1 Introduction

Bioeconomy describes the transformation from a petroleum-based market economy to a sustainable, post-fossil market economy. The focus is on securing nutrition of a growing population (Sustainable Development goal "SGD" 2: Zero Hunger), the sustainable use of resources (SGD 12: Responsible Consumption and Production) and climate change (SGD 13: Climate Action). The vision of a post-fossil market economy replaces fossil resources with renewable raw materials, which are used both energetically and materially. However, the production of renewable resources requires intensive cultivation of agricultural land, which leads to a loss of biodiversity. In addition, the availability of agricultural land is limited (see Fig. 1), which creates a conflict between tank, trough and plate.

However, renewable resources are most often coupled to the intensive use of fertilizer. This adds substantially to the climate issue, as in most cases the Haber Bosch process is used for ammonia production. It comes along with a significant climate footprint and it is estimated that 1–2% of the global energy consumption and 1.44% of global CO_2 emissions are attributed to this process [2]. Further, the use of artificial fertilizer releases high amounts of nitrous oxide (greenhouse gas) [3]. Innovative and new concepts are needed here to meet the challenges in agricultural research such as yield security, climate friendliness, environmental protection and biodiversity conservation. Microalgae can be part of the solution. Photo-autotrophy is a form of life in which, simplified, organisms use light as an energy source to produce oxygen and carbohydrates (sugars) from water and CO_2 and thereby contribute to SGDs 13 and 2. According to current projections, they bind around 30% of atmospheric CO_2 via the Calvin Benson Bassham cycle (CBB cycle). Cyanobacteria, on which we will focus in this chapter, are prokaryotes, performing an oxygenic photosynthesis, although they are commonly referred to as microalgae. Some cyanobacterial strains form biofilms, meaning they live embedded in a self-produced matrix of extracellular polymeric substances (EPS) that protectively

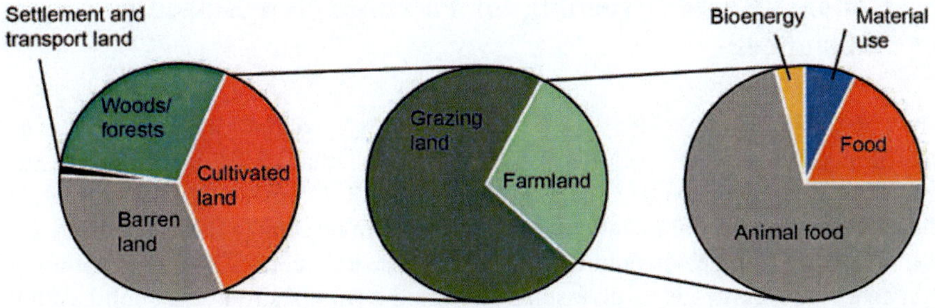

Fig. 1 Current global land use. 37% are used for agricultural cultivation, 29.1% are woods and forests, 32.1% are barren (mountains, desert, etc.) and 1.5% are used for settlement and transportation [1]. The share of cultivated agriculture land is further divided into 71% used for grazing and 29% farmland, which in turn is divided into land used for the production of raw materials (6.9%), food (17.9%), feed (71%) and bioenergy (3.8%). Figure created after [1]

surround them [4]. Depending on their origin, aquatic and terrestrial cyanobacteria can be distinguished. In addition to CO_2, some cyanobacteria also bind atmospheric nitrogen and convert it into bioavailable nitrogen [5], making them interesting to be used as a natural living fertilizer in agriculture as a co-culture with plants [6, 7]. Furthermore, they are also able to bind particulate matter and can be cultivated under flue gas [8]. They live in the most extreme habitats (deserts, hot springs, Antarctica) and are tolerant to fluctuating biotic and abiotic conditions. This makes them easy cultivation organisms, since the common monitoring and controlling of critical cultivation parameters, as temperature and pH, is often not necessary. On the other hand, light is one of the most important substrates which is not dispersible. The intensity and quality of the light can change during cultivation due to, e.g., mutual shading and thus influences cell physiology and in a consequence growth rate as well as physiology [9]. Each cyanobacterium has organism-specific requirements based on its origin and – depending on the desired target product – in terms of light intensity and light composition as well as temperature [10]. In addition, cyanobacteria can be used to produce a variety of interesting products such as natural dyes, biopolymers, polysaccharides and fatty acids. The biomass itself can be used as a food (supplement), animal feed or fertilizer. Moreover, many cyanobacterial strains are known as producers of various bioactive substances of different chemical classes with, amongst others, antibacterial [11–13], antifungal [14], antiviral [15] and anti-tumoural [16] activity. Several reviews summarize the potential of cyanobacteria as well as potential biotechnological products [17–19].

In this chapter, an overview of modern genome-based screening methods is given. In addition, different ways to preserve and cultivate cyanobacteria are discussed as well as the application of state-of-the-art photobioreactors. Furthermore, alternative ways to increase productivity by heterotrophic and mixotrophic cultivation, and methods for determining cell vitality and biofilm growth are presented.

2 Genome-Based Screening for Potential New Bioactive Substances

The potential of cyanobacteria, especially regarding secondary metabolites, has not yet been exhausted due to the fact that only a small fraction of all existing cyanobacteria have been described up to now [20, 21]. Especially the discovery of bioactive cyanobacterial metabolites is a hot topic in research, which is indicated by the increasing number of publications in this research field [22]. The structure of secondary metabolites is as diverse as their field of application. They fulfil various roles including defence against other organisms (antimicrobial substances), metal transporting agents, facilitators for symbiosis, antioxidants, photoprotectors and others [23, 24]. Over the last decades the excessive use of antibiotics has led to the rise of multi-drug resistant bacteria. Thus, the identification of new antibacterial compounds, as well as antifungal and anticancer substances is of high interest [25, 26].

Standard in vitro screening is based on the phenotypic properties of cyanobacteria: i.e., cyanobacterial biomass is collected or grown and usually an extract is prepared, which is then tested in bioactivity assays for its anticancer, antifungal or antibacterial properties. Commonly used bioactivity tests in lab-scale are summarized by Strieth et al. [27]. However, this type of screening can be very laborious and time-consuming, especially in view of the preceding slow growth of the cyanobacteria. This disadvantage is only partially overcome by High-Throughput-Screening (HTS) methods. HTS allows to screen automatically and robot supported of up to 100,000 or more samples per day. Such methods are mainly used to identify compounds (so-called hits) with pharmacological or biological activity. Using HTS only a single concentration (commonly 10 μM) of the compound is tested, which may or may not be sufficient to achieve a response. In quantitative HTS (qHTS), multiple concentrations of the respective compound are tested, which allows the determination of the "minimal inhibitory concentration" (MIC) [28].

It is known that the synthesis pathways leading to such active molecules are mostly encoded by large gene clusters [29]. Due to the ever-increasing number of sequenced genomes and the fact that availability and accessibility of genome data has highly improved, there has been an expansion of genomic screening in the last decade [30–32]. Two examples of genome data collections are the GEO DataSets from National Center for Biotechnology Information (NCBI; https://www.ncbi.nlm.nih.gov) that collects and provides, for example, biomedical and genomic information and the CyanoMetDB. The latter was developed by Jones et al. [22] and is an openly-accessible database of cyanobacterial secondary metabolites summarized from 850 peer-reviewed articles published between 1967 and 2020.

Genomic-based screening methods can be divided into in vivo and *in silico* methods. While in vivo screening methods are mainly based on the detection of the gene(s) of interest using polymerase chain reaction (PCR) *in silico* methods rely on genome mining approaches. Thereby, a lot of information about an organism can be obtained [33]. Many secondary metabolites have the advantage of being synthesized from the so-called bioactive gene clusters (BGC) whose structure is similar within their molecular class. It is advantageous that the corresponding nucleotide sequences are very similar across different species [34]. Thus, within a substance family, highly conserved sequences are often found within these gene clusters. For example, LanC is involved in the modification of lantibiotics [35]. Different tools are available for identifying respective BGCs and potential cyanobacterial producers in the genome data space, e.g. antiSMASH (Antibiotics and Secondary Metabolite Analysis Shell) [36].

In some cases these analysis tools even can predict the possible structural formula of a molecule without an existing extract and/or structure [33]. If a strain has the LanC, it has the potential to produce lantibiotics. Weber & UK give an overview of existing tools like BAGEL (BActeriocin GEnome Mining tool), Evo Mining and

RODEO that can be used to investigate secondary (bioactive) compounds and their corresponding gene clusters (http://secondarymetabolites.org/) [37]. Additional, Basic Local Alignment Search Tool (BLAST) by NCBI can be used to compare unknown DNA or protein sequences to already annotated sequences in various databases. Methods like blastn (for nucleotide sequences) or blastx (for amino acid sequences) are readily available. Although the number of known cyanobacterial genomes is slowly increasing, only a small part of the existing genome space is covered up to date. To gain more information about potential strains, it is important that as many cyanobacteria as possible get sequenced and included in the existing genome databases. It is known that a majority of bioactive compounds reported from cyanobacteria are cyclic or linear peptides [38]. Peptide derived compounds can be synthesized in different ways, either through non-ribosomal peptide synthetases (NRPS) or via ribosomally synthesized and post-translationally modified peptides (RiPPs). Additional structure diversity is afforded by mixing NRPS and polyketide synthases (PKS) [39, 40]. Vestola et al. characterized the biosynthetic pathway of numerous novel glycosylated lipopeptides in *Anabaena sp.* SYKE748A [41]. The glycolipopeptides were similar to antimycotic hassalidins A and B found in *Hassallia* sp. Further analysis showed that 20 strains of the genus *Anabaena* carry the hassalidin (*has*) synthetase genes, whereby the structure of antimycotic compound differs, but all inhibit the growth of *Candida albicans*. Finally, they could find the hassalidin synthetase genes in four other cyanobacterial genii [41].

In most cases the genome of a cyanobacterium is not sequenced. In this case analysis can be performed in vivo using PCR. For example, the hassalidin NRPS genes indicate the presence of the *has* gene cluster. Thus, this sequence could be used to screen for potential hassalidin producers [41]. Often a mixture of similar primers is used, so-called degenerated primers, to amplify homologous (in different species) or paralogous (within a species) genes. Degenerated primers have a high but not complete sequences similarity to their target, which allows them binding to similar sequences and enables in vivo detection of conserved regions of biosynthetic gene clusters [34]. Thereby, Vestola et al. detected the *has* NRPS gene in 31 cyanobacteria. It could be shown that although all investigated strains carried the *has* NRPS gene, five strains were not synthesizing the antimycotic compound [41]. In a further study Neilan et al. reported that 70% of the investigated cyanobacteria contain NRPS gene clusters, which does not necessarily mean that an antimicrobial compound is synthesized by the organisms [42]. Genomic screening for synthesis pathways of bioactive compounds facilitates the identification of cyanobacterial workhorses, but verification of the actual product is crucial. For this reason, genome mining should be coupled to a subsequent bioactivity assay as well as isolation, purification and characterization of the product [43, 44]. Up to now, mainly phenotypic screening methods are used due to the small number of available genomes of cyanobacteria.

3 Strain Conservation of Cyanobacteria

The phylum "cyanobacteria" includes 323 genera with approximately 2,698 known species [45] and represents one of the largest sub-groups of Gram-negative prokaryotes [46]. They occur on almost every place on earth, from oceans over freshwaters to deserts, rocks and hot springs. Due to the many different habitats cyanobacteria live in, they possess a great variety in form, metabolism and growth demands. Therefore, it is a challenging task to implement a strain conservation technique that fits every single strain. Techniques suiting a wide range of different cyanobacteria were already described.

3.1 Cryopreservation of Cyanobacteria

In the area of biosciences, cryopreservation is the predominantly used method for strain conservation. Since it was (accidently) discovered in 1948 that glycerol can prevent fowl sperm cells from freezing injuries, this discovery significantly promoted the field of cell conservation [47, 48]. For a long time, it was a common opinion that cryopreservation is not applicable for cyanobacterial strains, or at least only for a very limited number. After thawing of the cells, vitality often decreases dramatically and cells may suffer from freezing damages and die rapidly. A great improvement in cell vitality can be achieved by the two-step cooling method, using a controlled rate temperature decrease and, in most cases, a cell penetrating cryoprotective agent (CPA) [49].

In the process of cryopreservation, there are mainly two critical effects affecting cell viability. The first one being the formation of intracellular ice and the second one being osmotic stress, eventually followed by plasmolysis, caused by cell dehydration. Both effects are directly connected to the cooling rate in the course of cryopreservation. During slow cooling rates, the concentration of extracellular solutes increases, as pure water freezes more rapidly and gets eliminated from the solute-solvent system [50]. Consequently, intracellular water is osmotically driven out of the cells, which have a lower freezing point caused by a higher concentration of dissolved solutes. These physical mechanisms lead to a reduction in cell size and osmotic stress of the cells. However, a too rapid or uncontrolled cooling rate prevents proper dehydration of the cells, which dramatically increases the danger of intracellular ice formation. Cell penetrating CPAs like dimethyl sulfoxide (DMSO), glycerol or methanol (MeOH) are mostly used for the cryopreservation of cyanobacteria. They passively enter the cell over the cytoplasmic membrane, replace the intracellular water, and thus prevent the formation of cell-damaging ice crystals. Moreover, cell permeating CPAs prevent cells from excessive dehydration by equilibrating the osmotic pressure between the intra- and extracellular room. A certain incubation time of the cells after addition of the respective CPA at room temperature and under low light intensity assures the osmotic equilibration between solute concentrations outside and inside of the cells [51]. However, a prolonged incubation time can also cause cell damages, especially if potentially toxic compounds as DMSO and MeOH are used as cryo-preservative [52]. In order to overcome the risks of cell damages during cryopreservation, most recent protocols use a two-step freezing method. As a first step, the prepared cryo-vials are frozen to a sub-zero temperature (usually between −30 and −40°C), using a defined cooling rate. The cooling rate depends on the used strains, usually a decrease of −1°C min^{-1} is used. In the course of cooling the cells get slowly dehydrated, which prevents the formation of intracellular ice crystals. The second step implies a rapid cooling to −196°C with liquid nitrogen. After the freezing process, samples should be stored at ≤ −139°C, as recrystallization processes can occur at higher temperatures [53]. However, not every laboratory has access to liquid nitrogen, so also higher storage temperatures (mostly −80°C) are frequently used for cyanobacteria. The success of cryopreservation highly depends on the used strains. Cell viabilities can vary a lot by application of different CPAs and cooling protocols. In Table 1, several cryopreservation approaches are listed, sorted by the respective cyanobacterial order. It shall provide a quick overview about successful cryopreservation methods (with and without usage of liquid nitrogen) and can give hints on appropriate approaches for not yet cryopreserved strains.

Usually, 3–5% DMSO (v/v) is the most appropriate CPA for the cryopreservation of cyanobacteria. Only few strains maintain a high viability with MeOH as CPA.

During the thawing process, it is essential that a quick warming rate is applied to the cells (e.g. water bath, 35–45°C). This is proposed by almost all cryopreservation protocols, as rapid thawing minimizes the risk of intracellular water recrystallization. After thawing, the CPA should be quickly removed, particularly if cytotoxic substances as MeOH and DMSO were used. Glycerol generally won't harm cyanobacterial cells, but it can be used as carbon source by heterotrophic contaminants, if non-axenic cyanobacteria were cryopreserved. During the washing steps, light intensity should be maintained low and centrifugation steps should be avoided or reduced to minimum speed/duration.

3.2 *Alternative Conservation Approaches for Cyanobacteria*

There are also interesting alternatives to cryopreservation, two of which are shortly summarized in the following chapter.

3.2.1 Lyophilization

In the process of lyophilization, frozen samples are dehydrated under vacuum and at low temperatures. Through sublimation, ice crystals are directly converted to a gaseous state, without getting liquid. Lyophilization is a rather rarely used method for the conservation of cyanobacteria. It has been shown only for a very limited number of strains/genera. Most filamentous, heterocysts-forming strains, like *Nostoc* or *Anabaena*, are relatively robust and survive freeze-drying.

Table 1 Overview of different (successful) cryopreservation approaches sorted by cyanobacterial orders. 1S = one-step freezing with controlled rate; 2S = two-step freezing with controlled rate, 2S LN = two-step freezing with controlled rate, involves Liquid Nitrogen; SF = one-step freezing without controlled rate. +, generally viable

Cyanobacterial order	No. of strains	Vitality (vt), viability (vb) [%]	(Best) cryoprotectant	Cooling protocol	Reference
Chroococcales	9	>90 (8/9) (vt)	DMSO 3–5% (v/v)	RT → −40°C (−4°C min^{-1}) → −80°C (−4°C min^{-1}) 2S	[54]
	1	>75 (vb)	DMSO 5% (v/v)	RT → −80°C SF	[55]
	74	>60 (62/74) (vb)	DMSO 3% (v/v)	RT → −30°C (−1°C min^{-1}) → −196°C 2S LN	[56]
	6	100 (vb)	None	RT → −60°C SF	[57]
	9	+ (8/9) (vb)	DMSO 15% (v/v)	RT → −40°C (−3°C min^{-1}) → −196°C 2S LN	[57]
Chroococcidiopsidales	3	>90 (vt)	DMSO/ MeOH 3–5% (v/v)	RT → −40°C (−4°C min^{-1}) → −80°C (−4°C min^{-1}) 2S	[54]
Nostocales	34	>90 (32/34) (vt)	DMSO 3–5% (v/v)	RT → −40°C (−4°C min^{-1}) → −80°C (−4°C min^{-1}) 2S	[54]
	2	78–97 (vb)	Glycerol 10% (v/v)	RT → −80°C SF	[55]
	28	>60 (11/28) (vb)	DMSO 3% (v/v)	RT → −30°C (−1°C min^{-1}) → −196°C 2S LN	[56]
	2	0–80 (vb)	None	RT → −60°C SF	[57]
	8	+ (7/8) (vb)	DMSO 15% (v/v)	RT → −40°C (−3°C min^{-1}) → −196°C 2S LN	[58]
	10	50–80 (9/10) (vt)	DMSO 5% (v/v)	RT → −80°C (−1°C min^{-1}) 1S	This work
Oscillatoriales	13	>90 (12/13) (vt)	DMSO 3–5% (v/v)	RT → −40°C (−4°C min^{-1}) → −80°C (−4°C min^{-1}) 2S	[54]
	1	>90 (vb)	Glycerol 10% (v/v)	RT → −80°C SF	[55]
	39	>60 (24/39) (vb)	DMSO 3% (v/v)	RT → −30°C (−1°C min^{-1}) → −196°C 2S LN	[56]

(continued)

Table 1 (continued)

Cyanobacterial order	No. of strains	Vitality (vt), viability (vb) [%]	(Best) cryoprotectant	Cooling protocol	Reference
	1	100 (vb)	None	RT → −60°C SF	[57]
	3	+ (7/8) (vb)	DMSO 15% (v/v)	RT → −40°C (−3°C min^{-1}) → −196°C 2S LN	[58]
	2	>50 (vt)	DMSO 5% (v/v)	RT → −80°C (−1°C min^{-1}) 1S	This work
Pleurocapsales	2	>90 (vt)	DMSO 3–5% (v/v)	RT → −40°C (−4°C min^{-1}) → −80°C (−4°C min^{-1}) 2S	[54]
	2	>60 (1/2) (vb)	DMSO 3% (v/v)	RT → −30°C (−1°C min^{-1}) → −196°C 2S LN	[56]
Synechococcales	29	>90 (27/29) (vt)	DMSO 3–5% (v/v)	RT → −40°C (−4°C min^{-1}) → −80°C (−4°C min^{-1}) 2S	[54]
	1	>80 (vb)	DMSO 1% (v/v)	RT → −80°C (− 1°C min^{-1}) 1S	[59]
	1	80 (vt)	DMSO 5% (v/v)	RT → −80°C (−1°C min^{-1}) 1S	This work
Stigonematales	1	87 (vb)	Glycerol 10% (v/v)	RT → −80°C SF	[55]
	1	>60 (vb)	DMSO 3% (v/v)	RT → −30°C (−1°C min^{-1}) → −196°C 2S LN	[56]

Cyanobacteria that produce akinetes, thick-walled, cold and desiccation resistant spores are also described to be insensitive towards lyophilization [60]. The production of a thick layer of extracellular polymeric substances (EPS) can also lead to a higher viability rate after re-culturing [55]; Table 2).

When a suitable strain conservation technique needs to be chosen, it could be worth considering lyophilization. If the concerning strains are amenable to this technique and show high viability rates afterwards, lyophilization is a useful approach. Freeze-dried cells do not need to be stored at sub-zero temperatures and can be revived more quickly than cryopreserved cells. Moreover, lyophilization reduces the risk of contaminations and does not promote growth of heterotrophic contaminants [55]. This can be the case after cryopreservation, with CPAs like glucose.

Table 2 Overview of different (successful) lyophilization approaches with different cyanobacterial strains. +, generally vital

Strain/s	Viability [%]	Lyoprotectant/suspending medium	Reference
Nostoc sp.	<4	BG-11	[55]
Stigonema sp.	>60	BG-11	[55]
Synechococcus cedrorum S1C1/ J3C1	+	Lamb serum	[60]
Synechococcus elongatus S2C1	+	Human ascites fluid/beef serum	[60]
Anacystis nidulans Ac1C1	+	Human ascites fluid/lamb serum	[60]
Microcystis aeruginosa J1/M2/M4	+	Lamb serum/Foetal bovine serum	[60]
Merismopedia elegans Me1	+	Human ascites fluid/lamb serum	[60]
Gloeocapsa calcarea G1	+	Lamb serum	[60]
Oscillatoria subbrevis Os1	+	Lamb serum	[60]
Anabaena flos-aque Ab1C1	+	Human ascites fluid/lamb serum	[60]
Anabaena viariabilis Ab2C1	+	Lamb serum	[60]
Nostoc muscorum N1C1	+	Human ascites fluid /lamb serum	[60]
Tolypothrix tenuis	58–92%	Human serum albumin	[61]
Calothrix brevissima	42–96%	Human serum albumin	[61]
Lyngbya sp. 487/488	+	Horse serum	[62]
Lyngbya versicolor	+	Horse serum	[62]
Nostoc sp. 387/389	+	Horse serum	[62]
Nostoc ellipsosporum	+	Horse serum	[62]
Phormidium luridum	+	Horse serum	[62]

3.2.2 Immobilization

The immobilization of cyanobacterial cells has already been used for a wide range of applications. In photobioreactors, immobilized cyanobacteria can be used for the continuous production of valuable bioproducts [63–65]. Further advantages are the simplification of downstream processes; higher cell densities, combined with improved production effectivity, and the applicability of higher dilution rates [63]. Immobilized cyanobacteria can as well be used for bioremediation processes [66, 67], or the detection of pollutants [68]. However, immobilization can also be applied for the conservation of cyanobacteria. In 1988, Lukavský immobilized six cyanobacterial strains (among several eukaryotic algae) in 2% agar and stored them under low light intensity at 10°C. After 32 months, the cells were transferred to fresh cell medium and showed good growth at standard culturing conditions. Overlying the agar tubes with paraffin oil clearly decreased the recovery rate [69]. Also alginate beads (hardened with $CaCl_2$) are suited as shown for the filamentous cyanobacterium *Pseudoanabaena galeata*, which could be preserved for 14–18 months at 4°C in the dark, without a decrease in growth rate or alteration of physiological characteristics [70]. Thereby, cyanobacterial cell metabolism was drastically decreased, or completely stopped. This was indicated by a constant number of cells and no significant alterations in C, H, N and P content [70]. Thus, it is highly likely that genetic changes through mutations or selection of subspecies do not occur. Conservation of cyanobacteria by immobilization with agar or alginate is an interesting alternative to established cryopreservation. It has the advantages of being cheap and accessible for any laboratory. Furthermore, cells can be re-cultured more rapidly than after cryopreservation. Even an encapsulation device for the automated and continuous production of alginate beads has been proposed [71]. However, there is not much data about which cyanobacterial strains are suited for this kind of conservation method nor is it known, how long cells can be stored this way.

3.2.3 Commonly Used Pre-Culture Technologies in Algae Biotechnology

In general, cryopreserved cultures are used to inoculate pre-cultures in lab-scale. To subsequently inoculate the main culture, the cells are harvested in the exponential phase. This procedure minimizes differences in performance, ensures identical starting conditions and minimizes DNA mutations. Different protocols for cryopreservation of microalgae are established and described in the previous chapters, but since microalgae are growing very slow this standard procedure would be very time-consuming. Traditionally, cyanobacteria are preserved as metabolically active serial subcultures, which must constantly be transferred to fresh culture medium in intervals depending on the growth rates of the respective strains. Often, cyanobacteria are immobilized on agar plates and stored at 4°C and low light intensities resulting in low biomass formation by simultaneously high viability of the cells (up to 80%) [72]. In regular intervals, new medium is added to provide the necessary nutrients. This biomass is then used to start a pre-culture. This kind of strain conservation provides the constant availability of vital cyanobacterial cell mass, which is especially helpful for slow-growing strains. However, for medium to large strain collections, this method has the clear disadvantage of being extremely time-consuming and labour-intensive. Additionally, it was reported that continuous serial subcultures can lead to alterations in the morpho- and genotype due to selection towards subspecies and DNA mutation [73, 74]. Thus, serial subcultures can be useful for a limited period, if fresh cell material is constantly needed. But for long time storage of cyanobacterial strains, cryopreservation or immobilization should be the method of choice.

However, since serial sub-culturing is often used the influence on the main culture needs to be investigated. It is unknown if different growth phases and the fluctuating nutrient concentrations of the pre-culture lead to a different performance in the main culture. To investigate the impact of the age of pre-culture on main cultures, *Nostoc* sp. was cultivated for 56 days in shaking flasks without baffles in a shaking incubator at 24°C without any medium exchange. After 21, 28, 35, 42, 49 and 56 days of cultivation biomass was harvested and used to start the main cultures that were then cultivated for 14 days in shaking flasks at different light intensities. The typical growth curve of cyanobacteria could be detected including a lag-, an exponential as well as the beginning of the stationary phase (see Fig. 2a).

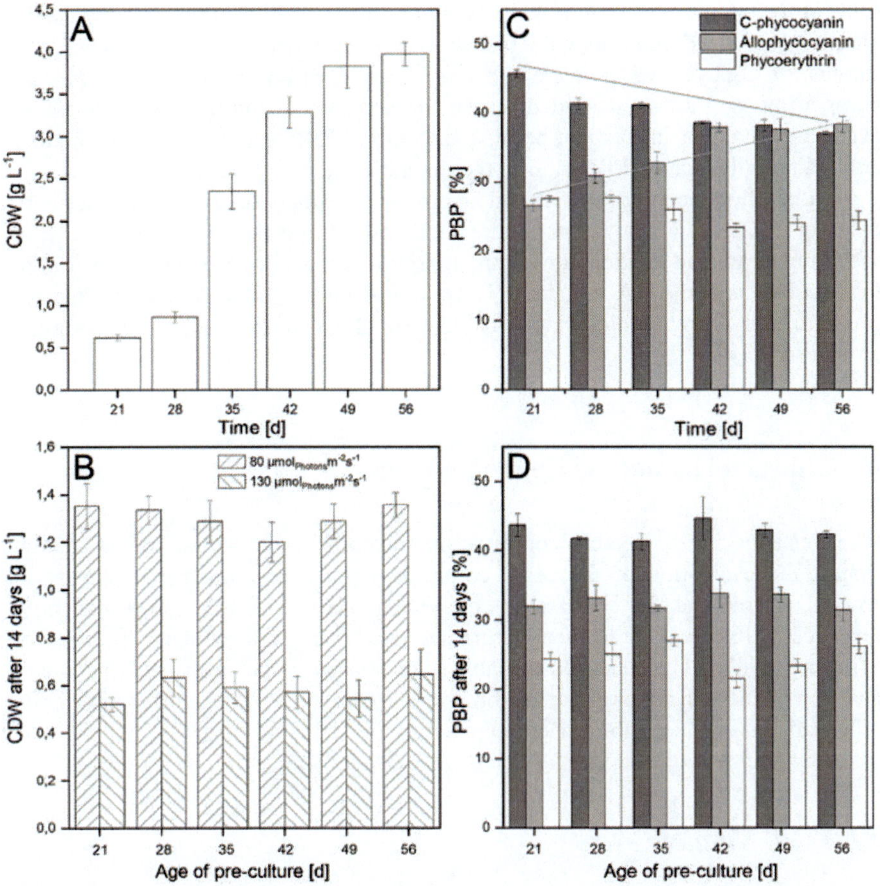

Fig. 2 Influence of different growth phases on the main culture of *Nostoc* sp. (**a**) Cell dry weight (CDW) of the pre-culture over cultivation time. (**b**) CDW of the respective main cultures inoculated with pre-culture of different age. CDW was determined after 14 days of cultivation at 80 and 130 $\mu mol_{photons}\ m^{-2}\ s^{-1}$, respectively. (**c**) Influence of the pre-culture cultivation conditions on phycobiliprotein ratio in per cent. (**d**) Ratio of phycobiliproteins in the main culture after 14 days of cultivation. Cultivation parameters: 300 mL shaking flasks without baffles, inoculation with 0.1 g cell wet weight (CWW) of *Nostoc* sp. per 50 mL BG-11 medium, 24°C, 120 rpm (eccentricity 2.5 cm), 130 $\mu mol_{photons}\ m^{-2}\ s^{-1}$, except the main cultures, n_b (biological replicates) = 3

In this study, the age of pre-culture had no influence on biomass formation under two different conditions (see Fig. 2b). For vitality determination the resazurin assay was used [75]. No influence on vitality could be detected over cultivation time (data not shown). EPS, Pigments and PBP were determined using the method described by Strieth and Stiefelmaier et al. [76]. Again, the age of pre-culture had no influence on

EPS formation and phycobiliproteins (see Fig. 2d) as well as pigment composition (data not shown). Furthermore, the total amount of PBP decreased, whereby also the amount of C-phycocyanin decreased over time. C-phycocyanin is also used as nitrogen storage is thus degraded due to the decrease of nitrogen in the medium over time. That was interesting, because the ratio of PBP changed in the pre-culture but had no influence on PBP ratio in the main culture.

In conclusion the age of culture had no influence on the productivities of the slow-growing organism of the main culture in terms of biomass formation, but serial subcultures can lead to alterations in the morpho- and genotype due to formation of subpopulations and DNA mutation. Thus strains have to be regularly checked for genomic and morphological integrity and should also be available as original cryopreserves .

4 Characterization of Cyanobacteria

When establishing a suitable conservation method for individual cyanobacterial strains, the evaluation of the cells' condition before and after the conservation process is essential. In contrast to cell vitality tests, viability assays can only superficially differentiate between living and dead cells. Intracellularly impaired or dying cells will still be identified as viable cells. However, such tests are fast and will give first quick indications about cell fitness. For more reliable data, time-consuming cell vitality assays need to be accomplished.

4.1 *Cell Vitality*

As cyanobacteria tend to have slow to very slow growth rates, a vitality check through this variable can be a time-consuming matter. In addition, the calculation of growth rates by absorption measurement is often not applicable for cyanobacteria. Especially cell aggregates interfere with spectrophotometric methods. In this subchapter, alternative methods for determining cell vitality are introduced.

4.1.1 In Vivo Growth Fluorometry

A widespread method for the vitality determination of cyanobacteria is the quantification of cell growth by measuring the in vivo fluorescence of chlorophyll-a. This method is used for diatoms [77], cyanobacteria [78] and green algae [79]. Fluorometric measurements can be performed with a spectrophotometer. A modified variant has been implemented by Karsten et al. [77], in which microalgal cells can be measured in petri dishes [78]. The method's principle remains the same, chlorophyll-a gets excited by a light source at 435–470 nm [78, 80] and emits light of slightly lower energy. The light intensity is measured by a detector and used for the calculation of growth rates, without need for calibration. An excitation wavelength of 630 nm was proposed for the measurement of phycocyanin and allophycocyanin in cyanobacteria [81]. An advantage of the method is the low amount of needed cell mass. Karsten et al. [77] only used 0.5 μg L^{-1} per run that minimizes the risk of self-shading and scattering effects [80]. Another benefit is the specificity of the method, as only living cells are measured and heterotrophic contaminants can be distinguished from cyanobacterial cells [82]. However, although single measurements are only a matter of seconds, the vitality test is still based on the determination of culture growth. Thus, for slowly growing strains several days of culturing are required. Another difficulty of using fluorometry for cyanobacterial growth monitoring is that many strains form cell clusters and aggregates with EPS and are therefore not evenly dispersed in the culture medium. These clusters tend to sediment quickly, which prevents the collection of samples and the quantitative measurement from being reproducible.

4.1.2 Resazurin Assay

The resazurin assay, or Alamar Blue Assay, is based on the reduction of the barely fluorescent, dark purple dye resazurin (redox and pH indicator) to resorufin, which is pink and highly fluorescent. This reaction occurs in a vital cell. If the metabolic activity of the cell is reduced (less vital), less resazurin is converted. By measuring the fluorescence, conclusions can be drawn about the cells vitality. This method is mainly used as bioactivity assay where different amounts of extracts are tested against, e.g., *Escherichia coli*. The test strain is resuspended in a buffer solution and placed together with the respective extract and resazurin into a microtitre plate for cultivation. After incubation a colour change indicates no inhibition of the test strain by the used extract [83]. This method cannot only be used as bioactivity assay. Mehring et al. showed that the test is also usable to detect vitality of cyanobacteria and is transferable to heterotrophic bacteria and callus cells [75]. In this case, a certain amount of biomass is resuspended in a buffer solution and resazurin is added. The reaction to resorufin only takes place in vital cells and the intensity of the resulting fluorescence can be correlated to metabolic activity. This is a medium to high throughput method for fast and easy determination of cell vitality, which can also be well automated.

4.1.3 Vitality Determination by pO_2 Measurements

The evaluation of cyanobacterial cell vitality should include more classifications than "alive" or "dead", as important metabolic functions or cell compartments can be damaged without leading to cell death, but hampering growth. Therefore, growth experiments are a reliable basis to judge cell vitality. However, they have the disadvantage of being very time-consuming in case of cyanobacteria.

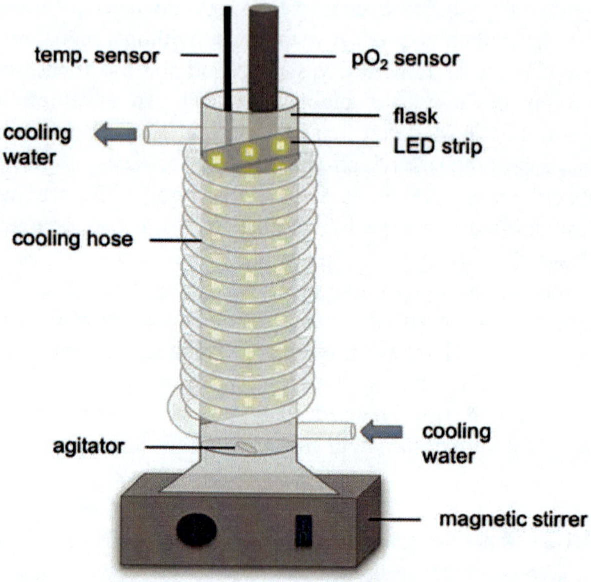

Fig. 3 pO_2 measurement set-up for the in vivo vitality determination of cyanobacteria. The working volume is 20–50 mL. Modified according to Witthohn et al. [84]. Reproduction of this figure is granted by a Creative Commons Attribution License

Since these bacteria perform an oxygenic photosynthesis oxygen is produced during photosynthetic activity. Consequently, oxygen is a good indicator for vital, growing cells. This fact was used to establish an easy vitality test, based on pO_2 measurements [84] (Fig. 3).

A cell wet mass pellet of 0.5 g cryopreserved and thawed *Nostoc* sp. was resuspended in BG-11 medium and applied to the measuring flask. The culture was stirred at about 500 rpm and heated to 27°C. The LED strip provides light for photosynthesis; the pO_2 increase was measured by a sensor inside the culture medium. The slope of the resulting graph can be compared to the one obtained by fresh, not cryopreserved cells (Fig. 4a) [84]. In this way, different CPAs can be tested in a relatively short time. For example, the results shown in Fig. 4b can theoretically be obtained in 5–6 h, as one run takes about 30 min.

Thereby it could be shown that DMSO is the most appropriate CPA for cryopreservation of *Nostoc* sp. Moreover, the comparison of these results with data from "classic" growth experiments demonstrates the reliability of the pO_2 measurement technique (Fig. 5).

The vitality determination by means of pO_2 increase constitutes an easy and functional approach for the quick evaluation of cyanobacterial cell states. The data shown for *Nostoc* sp. could reflect the growth behaviour in shaking flasks. Although this method can give no hints on specific cell damages as vital staining with different dyes does, it nevertheless can be used to predict the anticipatory cell growth. As the proof of concept was only presented for one strain, more data with different cyanobacteria would be interesting.

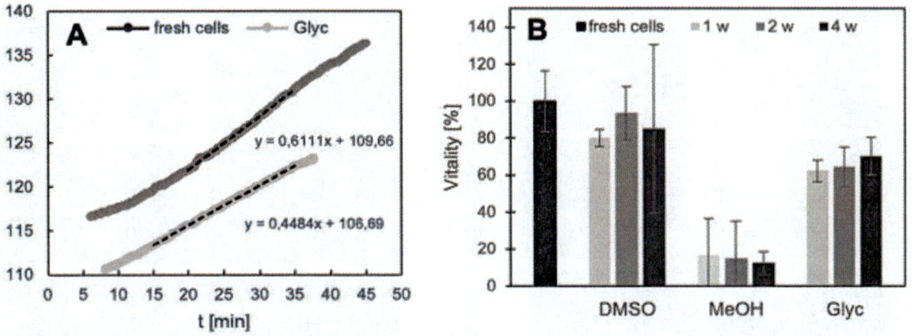

Fig. 4 (**a**) pO_2 increase for vitality determination of *Nostoc* sp. (formerly referred to as *Trichocoleus sociatus*) cells cryopreserved with glycerin as CPA. Measurements have been performed in the device shown in Fig. 3. (**b**) obtained vitality data of *Nostoc* sp. cells cryopreserved with different CPAs. Modified according to Witthohn et al. [84]. Reproduction of this figure is granted by a Creative Commons Attribution License

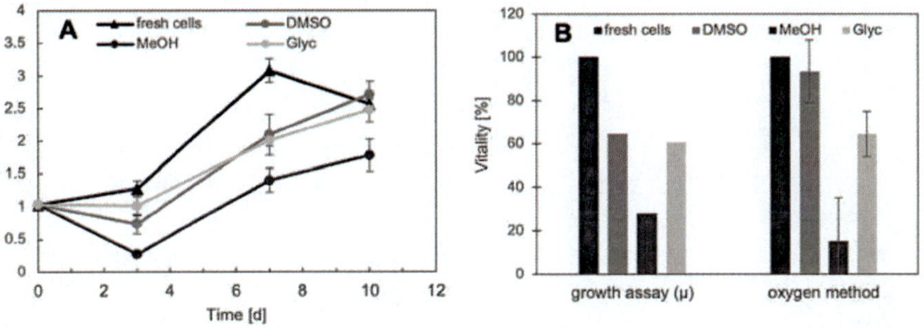

Fig. 5 (**a**) Growth assay with *Nostoc* sp. (formerly referred to as *Trichocoleus sociatus*) by determination of CDM. The cells were cryopreserved for 2 weeks with different CPAs (DMSO/MeOH 5%, Glyc 15% v/v). (**b**) comparison of growth assay and pO2 measurement as vitality determination methods. Modified according to Witthohn et al. [84]. Reproduction of this figure is granted by a Creative Commons Attribution License

4.1.4 Spectral Domain Optical Coherence Tomography (sdOCT) and Pulse Amplitude Modulated (PAM)-Fluorometry

In this chapter, two further non-invasive methods to characterize cyanobacterial growth are presented: the spectral domain optical coherence tomography (sdOCT) and pulse amplitude modulated (PAM)-fluorometry (Fig. 6). The growth behaviour under identical cultivation parameters can be seen as indicator for cell vitality. Cyanobacterial growth on surfaces can be measured by means of chlorophyll activity using a PAM fluorometer (Imaging-PAM). A saturation pulse in the form of red light (620 nm) is given via the PAM fluorometer, which excites the chlorophyll molecules. This raises the electrons in the electron transport chain to a higher energy level.

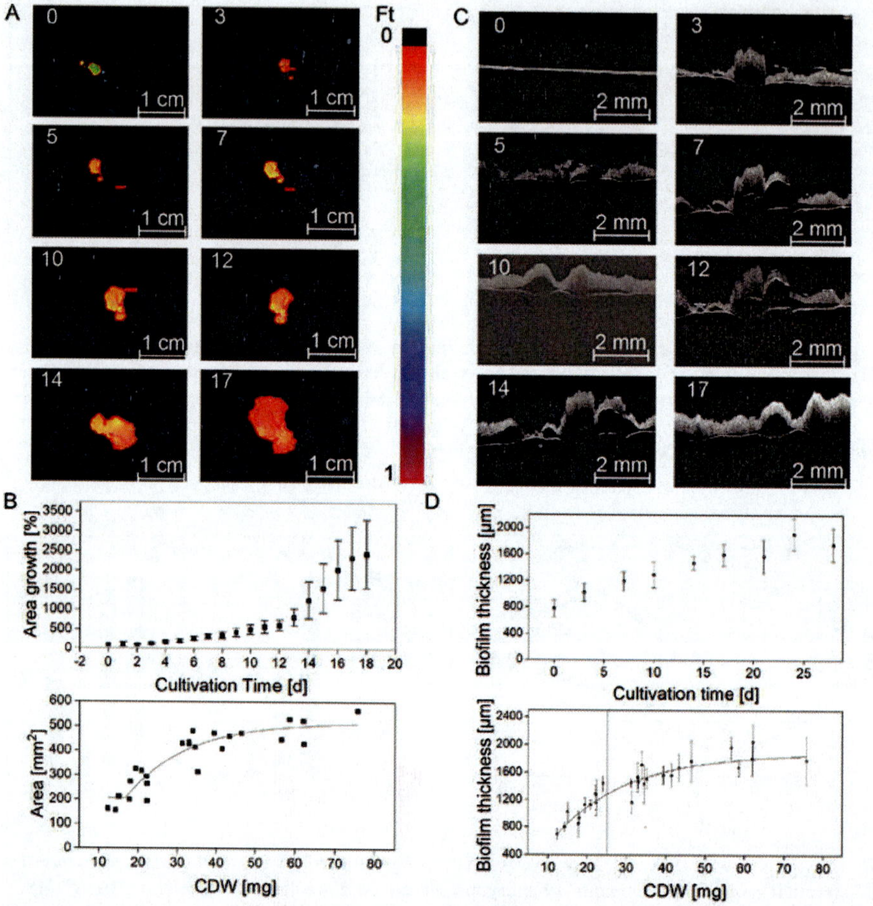

Fig. 6 Non-invasive methods for cell growth characterization as indicator for cell vitality by the example of *Nostoc* sp. **(a)** Area growth determined by chlorophyll-a fluorescence using PAM fluorometry over a cultivation period of 17 days. n_b (biological replicates) = 3 **(b)** Area growth in per cent over cultivation time and correlation of area with cell dry weight (CDW). **(c)** Biofilm thickness measured using OCT over a cultivation period of 17 days. $n_b = 3$, n_t (technical replicates) = 30 **(d)** Biofilm thickness over cultivation time and correlation of biofilm thickness with cell dry weight (CDW). Cultivation parameters: Solid BG-11 medium, 24°C, 100 $\mu mol_{photons}\ m^{-2}\ s^{-1}$, 400 ppm CO_2, 30 days, $n_b = 3$

When the electrons fall back to their ground state, energy is released in the form of heat or radiation. This energy can also be used for photochemical processes. Accordingly, radiation is at its maximum, when the energy used for photochemical processes is at its minimum. The reaction centre of photosystem II can be inactivated, or the electron acceptors reduced by a short, strong flash of light, allowing the radiated heat to be measured [85]. The Imaging-Win software (Heinz Walz GmbH, Effeltrich, Germany) can be used to record the emitted radiation and thus determine the spreading of the biofilm in two-dimensional space via the activity of the reaction centre (photosystem II). By this, it is possible to describe growth curves of surface-associated growing cyanobacteria in a short time (about 1 min per plate of 5 colonies). The technique is thus applicable for the evaluation of cyanobacterial cell viability and cell vitality after cryopreservation, for example.

In principle, the method relies on measuring the surface covered by cyanobacteria when growing attached to a surface as a biofilm, like on an agar plate. This is accomplished by measuring chlorophyll-a emission when excited at 620 nm (Fig. 6a). Both the fluorescence intensity and the area from which fluorescence occurs can be measured. Only the absolute area is used to calculate growth. The areas can be reproducibly measured by defining fixed thresholds for fluorescence. The cyanobacterial growth is described as the increase in the surface covered by the bacteria (Fig. 6b). Typical growth phases of the biofilm area over the cultivation time could be reported [86]. The correlation between area and biomass was linear up to 10 days, afterwards deviations between replicates increased (Fig. 6b). This can be explained by the increasing biofilm thickness that can be determined using non-invasive sdOCT (Fig. 6c). Here, the contrast is achieved by the different light scattering properties of the biofilm and thus provides information about its microstructure and thickness without any use of contrast agents [87]. A problem when using sdOCT to determine the layer thickness is that above a certain biofilm thickness (also depending on the pigment content), mutual shadowing occurs, and the biofilm cannot be completely imaged. In this case, it is no longer possible to distinguish between cavity and shadowing, which makes the evaluation of the data more difficult. Furthermore, it is not possible to distinguish between water deposits and biofilm as well as between cells and EPS [86]. Similar to the spreading of the biofilm, biofilm thickness could also not be linearly correlated with CDW meaning (see Fig. 6d), real growth rates cannot be determined. This is probably because the ratio of biomass to EPS changes over the cultivation period (data not shown), which allows the biofilm to become thicker without forming cell mass. Furthermore, the thickness of the biofilm depends on the stored water in the EPS. Therefore, this method can be used to determine re- and dehydration of biofilms (Fig. 7). Biofilm thickness changes and the associated water loss of *Nostoc* sp. biofilms growing on borosilicate glass, PMMA and silicone at 24°C and a relative humidity of 30% (typical cultivation conditions in aerosol-based photobioreactors (for description of the PBR, see Sect. 5.4) without aerosol supply) were documented by sdOCT (Fig. 7). To better compare the influence of different substrates on the dehydration of the biofilm, the respective half-lives were calculated at which the biofilm showed 50% of its initial thickness. Respective values determined were 30.27 ± 7.26 min (borosilicate glass), 27.60 ± 5.13 min (PMMA) and 18.90 ± 6.70 min (silicone). In comparison, a water film reduced its thickness by 50% after only 8 min (data not shown).

This shows that EPS protects the biofilm from dehydration and that the change in layer thickness can be determined reliably and with low deviations using sdOCT. Based on the results obtained, sdOCT is suitable for visualizing the surface morphology and for dehydration and rehydration experiments of biofilms, assuming that

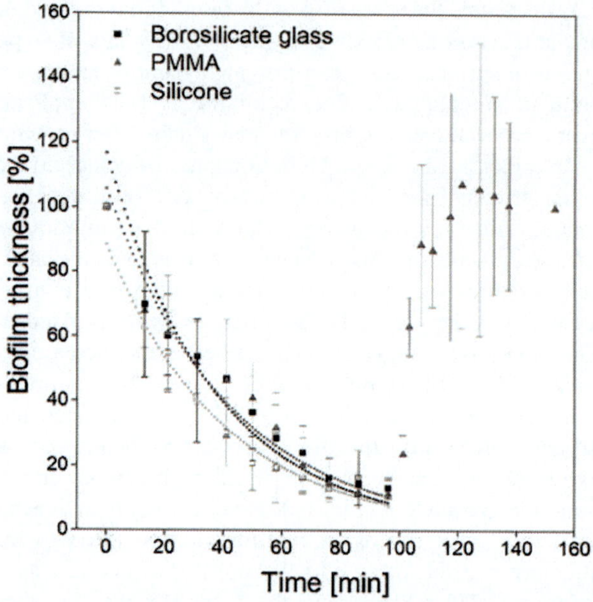

Fig. 7 De- and rehydration of *Nostoc* sp. biofilms. Dehydration was performed on different materials (borosilicate glass, PMMA, and silicone) and rehydration was only performed on silicone in an aerosol-based photobioreactor. The biofilm thickness was recorded by OCT at 24°C and 30% relative humidity and plotted as a percentage ($n_b = 5$)

no cavities are formed. Additional, growth curves can be used as additional information source to gain more information about the state of the cells. Furthermore, characterization of growth behaviour using PAM fluorometry or sdOCT under the same cultivation conditions can be applied to determine cell vitality and viability, since growth only occurs when the cells are viable and biomass formation depends on metabolism activity (cell vitality). Both methods are suitable for a fast and simple characterization of the surface-associated growth of cyanobacteria [86]. It should be mentioned again that for the calculation of growth rates a linear correlation between horizontal spreading and CDW is essential, which is not possible with this method. However, sdOCT is suited for characterization of surface attached cyanobacteria and allows to obtain qualitative data on biofilm development and cell vitality.

4.2 *Cell Viability*

In contrast to cell vitality tests, viability assays can only superficially differentiate between living and dead cells. Intracellularly impaired or dying cells will still be identified as viable cells. However, such tests are fast and will give first quick indications about cell fitness. For more reliable data, time-consuming cell vitality assays as discussed above need to be accomplished.

4.2.1 Staining Methods

In many cases, the number of living cells is used as an indicator for culture viability after cryopreservation. For this purpose, the cells are stained and evaluated by microscopy. In this process either only dead cells are stained, as the colourants can cross their damaged cell membrane, or only living cells are stained because of enzymatic activation of the dye. A staining approach addressing cell vitality by combining different stains was developed by Tashyreva et al. [100]. They used a series of different dyes to determine the physiological state of *Phormidium* cells. With SYTOX green, damaged cell membranes could be revealed; by staining with 4′,6-diamidino-2-phenylindole (DAPI), degraded DNA was shown and with 2-(4-Iodo-phenyl)-3-(4-nitrophenyl)-5-phenyltetrazolium chloride (INT) and 5-Cyano-2,3-Ditolyl Tetrazolium Chloride (CTC) the respiratory function of the cells could be verified.

There are several staining methods for cell viability testing established for bacteria, however, not all work for cyanobacteria. For example, it was found that propidium iodide, which is meant to stain only non-viable cells, also stained viable cells of filamentous cyanobacteria [88]. It was postulated that this is due to intracellular channels between the cells [89, 90]. In the following, compounds frequently used for viability staining of cyanobacteria are presented. Depending on the used substance, stained cells can be observed either via light or fluorescence microscopy. The bisazo dye trypan blue [54, 91] and the triphenylmethane compound erythrosine b [92] are staining agents detectable via light microscopy. Trypan blue binds to proteins of cells with an impaired cell wall; the staining procedure takes 5–10 min [91]. Intact cell walls of viable cells are not permeative for the dye. A prolonged incubation should be avoided since trypan is cytotoxic and can thus lead to false positive results. Moreover, this substance should be handled with care, as it is teratogenic [93] and carcinogenic [94]. Erythrosine b, on the other hand, is used as a food colouring and therefore a non-toxic compound. The staining process is as quick as with trypan blue and it also acts on proteins of cells with damaged cell walls [95].

Fluorescein diacetate (FDA) [96–99] and SYTOX green are fluorescent stains frequently used for cyanobacteria [100–102]. FDA is a non-fluorescent molecule which can cross the bacterial cell membrane of living cells. Upon entering the cell the compound is hydrolysed to the yellow-green fluorescent compound fluorescein, which can be detected under ultraviolet light [103]. It is often used in combination with propidium iodide, an analog of ethidium bromide, which stains DNA of dead cells [97]. SYTOX green also acts as nucleic acid stain of cells with damaged plasma membranes [104]. It can be excited at 488 nm and emits light of 523–530 nm. However, just as propidium iodide, it was described as non-applicable for filamentous cyanobacteria [89].

5 Photobioreactors

Due to their diversity, cyanobacteria can be cultivated in many different ways. Heterotrophic cultivation in stainless steel fermenters, for example, is possible, although it has some disadvantages compared to phototrophic cultivation [105]. For example, there is a risk of contamination by adherent heterotrophic bacteria, which cannot be completely removed from the cultures even with complex isolation methods [106–108]. Another major advantage of phototrophic and mixotrophic cultivation is that CO_2 can be used as a carbon source. This not only allows the use of cheaper cultivation media, but is also of great interest, especially in the current times when global climate goals depend on a reduction of carbon dioxide emissions. Section 6 gives a more detailed view on the advantages of the different cultivation modes of cyanobacteria, with a focus on mixotrophic cultivation. Furthermore, the type of cultivation has an influence on the profile of the synthesized secondary metabolites, making some processes depended on photo-autotrophic growth. The systems for phototrophic cultivation can be divided into open and closed systems based on their design. Open systems in the form of open ponds are still the most commonly used form for large-scale cultivation. This is mainly due to the fact that they are cheap and easy to set up and operate. However, due to the open design, there is a high evaporation rate of water and additionally a risk of contamination. Furthermore, gas exchange is poor, leading to low biomass productivities, which in turn adds to the bad biomass to land ratio of such facilities. Therefore, open pond systems are not covered further here, instead this chapter is focussing on closed photobioreactors (PBRs).

5.1 *Closed Photobioreactors (PBRs)*

Closed systems have several advantages over open systems. The risk of contamination is significantly reduced, and the closed design enables more diverse construction options. This allows the surface area to volume ratio to be optimized, which is particularly important with regard to an optimal light supply for phototrophic organisms. As light is not dispersible, systems with a low surface area to volume ratio lead to inhomogeneous light distribution. A good compromise must be found here, as a large surface area to volume ratio automatically results in a large footprint of the reactor. The light can be supplied either as artificial or natural light and also the material of the PBR will heavily impact light intensities and quality due to different refractive indices. Optimized light supply significantly increases biomass productivity compared to open ponds. Furthermore, the use of resources such as water is also improved, as the problem of evaporation is eliminated. In this chapter, only a brief description of the different submerged PBR designs is given, as the focus here will be on biofilm reactors. The interested reader is referred to existing reviews in the literature [109–111].

The simplest form of a submerged PBR is the stirred tank, which due to its geometry offers poor light distribution and is therefore less suitable for phototrophic cultivation. Flat panel PBRs are characterized by small depth, improving light supply. Beside these, there are multiple PBR designs based on tubular geometry. Vertical tubular PBRs are operated as a bubble column or airlift reactor and the mixing in these reactors can be controlled by adjusting the gassing rate. Horizontal tubular PBRs consist of tubes that can be arranged in different shapes and lengths. The horizontal arrangement simplifies the illumination by natural light [109]. However, submerged systems are not the optimal cultivation method, especially for biofilm-forming cyanobacteria, as they do not grow submerged by nature but surface-associated.

5.2 *Attached (Biofilm) Cultivation of Cyanobacteria*

Biofilm formation usually follows the following principle [110] (1) initial adhesion of cyanobacteria to the surface by adsorption (reversible), (2) adhesion by the formation of EPS (irreversible), (3) formation of EPS, (4) biofilm growth by attachment of additional cyanobacteria and organisms, and (5) partial detachment of the biofilm due to loss of integrity. The formation of a biofilm and the resulting growth in this form can be utilized with the help of specially designed reactors. Cultivation in the form of biofilms can thus bring a multitude of advantages, like the potential of reduced water to biomass ratio [112] and a reduction in costs compared to submerged PBRs. The reduced costs are the result of several factors. These include the already mentioned reduced water consumption as well as increased biomass production. Furthermore, the harvesting of cyanobacteria as biofilm is significantly simplified, as the biomass can, for example, simply be scraped off the surface and the separation of the biomass from the process water is much easier than for planktonic cultures. Depending on biofilm thickness, cells benefit from better light availability especially in the outer regions of the biofilm. The lower layers may become light limited if the biofilm is too thick, which can reduce the productivity of the cells in the biofilm [113]. This problem can be avoided by regular harvesting, which in turn is not a major problem if considered in the design of the reactor [114]. Another advantage of regular partial harvesting is a faster re-growth of the biofilm, which in turn can increase productivity [115]. A disadvantage of biofilm cultivation is the unwanted, spontaneous detachment of biofilm, which then continues to grow in the medium, or biofouling and clogging of the complete PBR system.

In the cultivation of biofilms, a distinction can be made between submerged systems, in which cultivation takes place in a liquid medium, and surface-associated systems, in which the biofilms grow air-exposed. These systems will be discussed separately in the following. Table 3 gives an overview on existing systems for biofilm cultivation attached to surfaces. The table shows that most systems are aimed at either optimizing biomass productivity or maximizing the production of lipids, which can be used, e.g., for the production of biofuels. In addition, the most important application is the treatment of wastewater, to remove high concentrations of pollutants, like nitrogen and phosphorus.

Table 3 Comparison of different photobioreactors for the cultivation of cyanobacteria or microalgae attached to surfaces. (PBR – photobioreactor, NA – not applicable)

Reactor	Biofilm placement	Cyanobacteria/microalgae	Product	Liquid reservoir (l)	Growth surface (m^2)	Biomass productivity ($g\ m^{-2}\ day^{-1}$)	Reference
Vertical PBR	Air-exposed	*Phormidium, Pseudanabaena, Nitzschia, Scenedesmus*	Removal of nitrogen and phosphorus	–	0.125	6.70–7.20	[114]
Multiple plates PBR	Air-exposed	*Botryococcus braunii*	Lipid and hydrocarbon production	–	0.54	49.10	[116]
Multiple layer vertical plate attached PBR	Air-exposed	*Scenedesmus obliquus*	Lipids	–	0.54	50–80	[117]
Multi-skin sheet emerse PBR (MSSePBR)	Air-exposed	*Coleofasciculus chtonoplastes, Nostoc* sp.	Biomass	–	0.046	1.7	[118]
Capillary-driven PBR (CPBR)	Air-exposed	*Scenedesmus* sp.	Lipid production, removal of nitrogen and phosphorus	–	0.077–0.538	10	[119]
Twin-layer PBR	Air-exposed	*Isochrysis* sp., *Tetraselmis suecica, Phaeodactylum tricornutum, Nannochloropsis* sp.	Biomass	–	10.72	0.6–1.8	[120]
Biofilm cultivation system	Air-exposed	*Haematococcus pluvialis*	Optimization of water footprint	–	0.001	6.0	[121]
Attached biofilm reactor	Air-exposed	*Pseudochlorococcum*	Biomass	–	0.00025	6–8	[122]
Emerse PBR (ePBR)	Air-exposed	*Nostoc* sp.	Biomass and EPS	–	0.0025	2.4	[123]
Twin-layer system	Air-exposed	*Halochlorella rubescens*	Removal of nitrogen and phosphorus	–	NA	6.3	[124]

Attached PBR	Air-exposed	*Scenedesmus* sp.	Removal of NH_4Cl, $CuSO_4$, tetracycline, norfloxacin and sulfadimidine	–	NA	6.2	[125]
Fixed-bed biofilm reactor (FBR)	Air-exposed	*Chlamydomonas* sp.	Biomass, removal of nitrogen, phosphorus and Cu(II)	–	0.8	49.70	[126]
Horizontal flow lanes	Intermittently submerged	*Phormidium*, *Pseudanabaena*, *Nitzschia*, *Scenedesmus*	Removal of nitrogen and phosphorus	–	0.140	4.50–9.90	[114]
Revolving algal biofilm (RAB)	Intermittently submerged	*Chlorella vulgaris*	Biomass	–	2.94 or 1.85	18.90	[127]
Rotating algal biofilm reactor (RABR)	Intermittently submerged	Mixed algal culture	Removal of nitrogen and phosphorus	8, 535, 8,000	0.24, 2.72, 4.26	5.50, 20.00, 31.00	[128]
Rotating biological contactor (RBC)	Intermittently submerged	*Chlorella sorokiniana*	Biomass	11	0.362	20.1	[129]
Photorotating biological contactor (PRBC)	Intermittently submerged	*Klebsormidium* sp.	Biomass	15	1.57	0.45	[130]
Rotating biological contactor (RBC)	Intermittently submerged	Mixed algal culture	Treatment of aqueous effluent containing diesel oil	4	0.83	NA	[131]
Rotating flat plate (RFP) PBR	Intermittently submerged	*Chlorella vulgaris*	Biomass, lipids	8	0.286	2.99	[132]
Rotating algal biofilm reactor (RABR)	Intermittently submerged	Mixed algal culture	Removal of wastewater nutrients, lipid production	8	NA	0.96	[133]
Parallel plate air lift (PPAL)	Submerged	*S. obliquus*, *C. vulgaris*, *Coccomyxa* sp., *Nannochloris* sp., *Nitzschia palea*, *Oocystis* sp., *Oocystis polymorpha*	Lipids	15	0.064	1.10–2.08	[134]

(continued)

Table 3 (continued)

Reactor	Biofilm placement	Cyanobacteria/microalgae	Product	Liquid reservoir (l)	Growth surface (m^2)	Biomass productivity (g m^{-2} day^{-1})	Reference
Biofilm membrane PBR (BMPBR)	Submerged	*Chlorella vulgaris*	Secondary effluent treatment	280	–	NA	[135]
Attached algal culture system	Submerged	Mixed algal culture	Fatty acids production, removal of nitrogen and phosphorus	0.2	0.0136	0.58–2.57	[115]
Attached algal cultivation	Submerged	*Scenedesmus*, *Chlorella*, *Pediastrum*, *Nitzschia*, *Cosmarium*, filamentous microalgae and others	Lipids	8,000	33.1	9.10	[136]
Semi-continuous flat plate parallel horizontal PBR	Submerged	*Scenedesmus obliquus*, *Nitzschia palea*	Lipids	0.288	0.072	2.1–2.8	[137]
Algae biofilm PBR	Submerged	*Botryococcus braunii*	Lipids	NA	0.275	0.71	[112]
Moving bed PBR	Submerged	*Nostoc* sp.	Biomass	65	11.26	NA	[138]
Roof-installed parallel plate microalgae biofilm reactor	Submerged	*Scenedesmus obliquus*	Wastewater treatment	5	0.5	2.5	[139]
Algal turf scrubber	Submerged	Mixed algal culture	Removal of nitrogen, phosphorus and chemical oxygen demand	200	1	5.0	[140]
Horizontal flat panel (HFP) PBR	Submerged	Mixed algal culture	Removal of phosphorus	NA	2	12.21	[141]
Algal biofilm reactor (ABR)	Submerged	*Chlorella, Phormidium*	Biomass	3	0.063	4.0	[142]

Algal turf scrubber (ATS)	Submerged	Mixed algal culture	Phosphorus removal	–	2.67	33–39	[143]
Biofilm capillary reactor	Submerged	*Synechocystis* sp.	Biomass	NA	NA	NA	[144]
Attached PBR	Submerged	Mixed algal culture	Biomass, nutrient removal	15	0.171	1.57–1.91	[145]
Tubular biofilm PBR	Submerged	*Chlorella sorokiniana*	Removal of carbon, nitrogen and phosphorus	7,5	1.1	NA	[146]
Algal-based immobilization reactor	Submerged	*Scenedesmus*	Removal of nitrogen and phosphorus	96	NA	NA	[147]

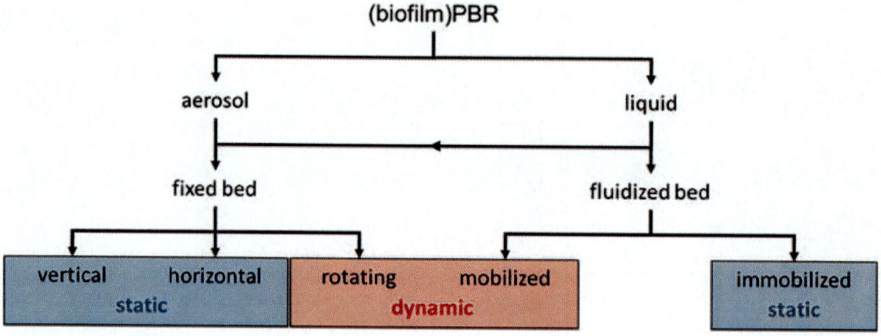

Fig. 8 Classification of (biofilm) photobioreactors

5.3 (Partly-)Submerged Biofilm PBRs

The reactor systems for submerged or partly-submerged cultivation of biofilms can be further divided (see Fig. 8). One possible classification is into fixed bed, in which the biofilm grows on a fixed surface, and fluidized bed reactors, in which the support material behaves like a fluid. The fixed bed can be further subdivided into vertical, horizontal and rotating reactors, depending on the orientation of the support material. For fluidized bed reactors, a further distinction can be made between mobilized and immobilized systems [148]. A further possible division of (partly-)submerged biofilm PBRs is the distinction between dynamic and stationary systems, whereby dynamic systems include all reactors in which the substratum is moved [149]. This categorization of reactors is also applied here.

5.3.1 Dynamic Systems

Dynamic biofilm PBRs are defined by the surface on which the biofilm is cultivated and thus also the biofilm itself is moved in the medium. The movement serves to simulate natural growth conditions, for example by imitating the tide. Furthermore, in systems where the biofilm is not constantly submerged, the gas exchange can be improved. In the attached algal culture system, the supporting material made from polystyrene foam is located on the bottom of a growth chamber, which is fixed on a rocking mechanism (see Fig. 9a) [115]. The growth chamber is shaken gently by 15° via the horizontal axis and illumination is continuous with 110–120 $\mu mol_{photons}\ m^{-2}\ s^{-1}$ from the top. Manure wastewater is used as medium. Harvesting is accomplished by scraping the biofilm off the surface. Another dynamic biofilm PBR is the rotating biological contactor (see Fig. 9b) [131].

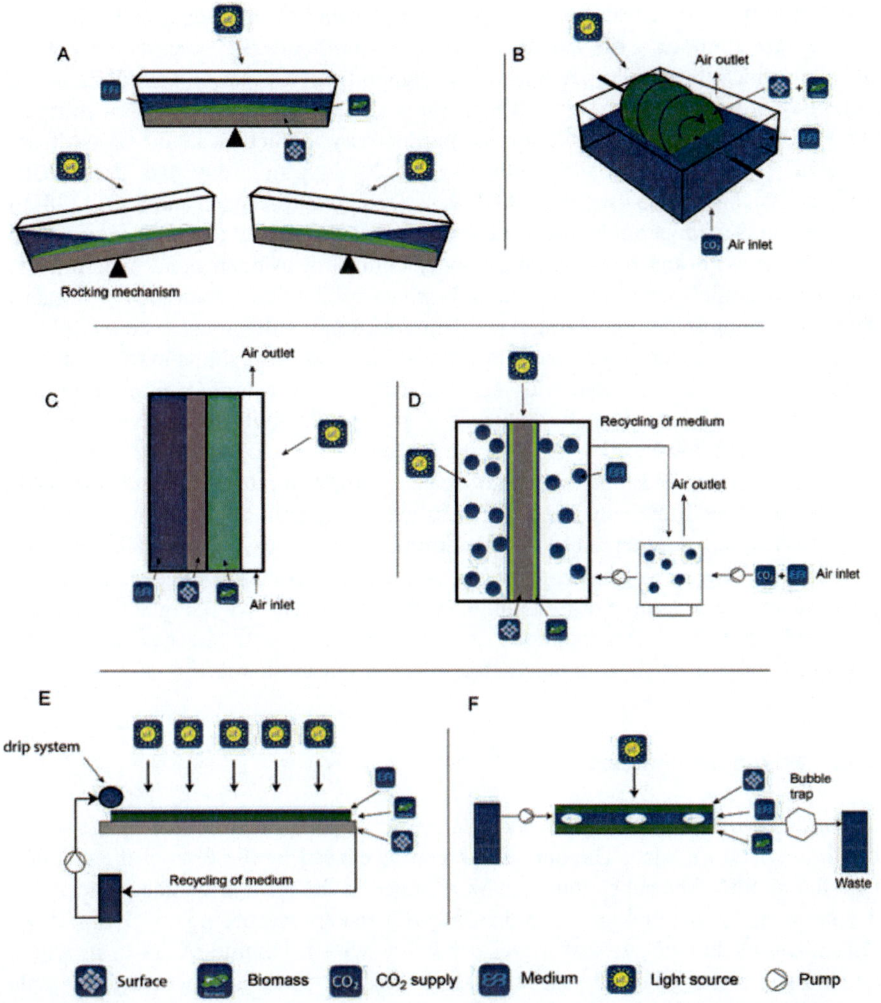

Fig. 9 Overview of Biofilm Photobioreactors. (**a**) Attached algal culture system with a rocking mechanism (modified from Johnson and Wen [115]), (**b**) Rotating biological contactor (RBC) (modified according to Mukherji and Chavan [131]), (**c**) Porous substrate bioreactor (PSBR) (modified according to Podola et al. [152]), (**d**) Emerse Photobioreactor (ePBR) (modified according to Strieth et al. [7]), (**e**) Algal biofilm photobioreactor system (modified according to Ozkan et al. [112]), F: Biofilm capillary reactor (modified according to Heuschkel et al. [144])

The RBC consists of 27 acrylic discs mounted on a PVC shaft, which serve as growth substratum. 35% of the respective disc surface is submerged in the cultivation medium. Due to the rotation of the discs with the biofilm, it is only temporarily submerged and otherwise exposed to the air. The rotating algal biofilm reactor

(RABR) [128] is based on the same principle. Instead of individual disks, it has a rotating tube supporting the growth substratum. Another design consists of a paddle wheel, in which the paddles are made of the supporting material. In both designs, the respective growth surfaces were 40% submerged. Compared to suspended cultures, the RABR achieved higher biomass productivity, which was increased from 7.4 g m^{-2} day^{-1} in suspended cultivation to 20–31 g m^{-2} day^{-1} in the RABR. Hodges et al. [150] also used the RABR described by Christenson and Sims [128] to remove solids from petrochemical wastewater. They observed a significant increase of solids removal and biomass productivity compared to open pond experiments. Another dynamic system for increasing biomass productivity combined with facilitated harvesting is the Revolving Algal Biofilm (RAB) cultivation system [127]. In this reactor, the supporting material is stretched around drive shafts in the form of a flexible mat. Different geometries can be achieved by a triangular or a vertical arrangement of the drive shafts, whereby higher productivity is achieved with the latter. Only the lower drive shaft and thus only a small part of the surface is submerged in the medium. On a pilot-scale, biomass productivity with the RAB was increased by 302% compared to a classic raceway pond (8.5 m^2). Walther et al. [138] developed a submerged biofilm reactor based on a moving bed bioreactor. The carriers were made of high-density polyethylene (HDPE) with a size of 1–5 cm. The glass reactor has a volume of 65 L and mixing is achieved by gassing at the bottom. To avoid dead zones, an inclined plate is installed next to the gassing unit. A cultivation of *Nostoc* sp. was successfully carried out in the reactor.

5.3.2 Stationary Systems

In stationary biofilm cultivation systems, the supporting material and thus also the biofilm is fixed in place. The only movement is caused by the flow of the medium over the biofilm. The substratum can be arranged in the form of vertical plates, as is the case in the parallel plate air lift (PPAL) reactor according to Genin et al. [134]. The reactor consists of a glass chamber with a volume of 15 L, in which two vertical plates made of acrylic glass are located. Various supporting materials can be attached to these plates. The gas supply is located at the bottom of the reactor and between the two plates. The lighting is provided from the side. The multi-layered photobioreactor (MLPR) consists of several alternating layers of cell suspension layers and transparent medium layers, separated only by membranes [151]. The incident light is diffused by the medium layers and thus evenly distributed in the MLPR, providing illumination over a larger area. A simpler vertical system for attached cultivation was used by Lee et al. [136]. In this system, the attaching material was suspended in the form of several rectangular nylon meshes in a raceway pond, so that the flow runs across the length of the mesh. Lee et al. compared growth directly with a suspended culture and achieved a 2.8-fold increase in biomass and total lipid productivity in the attached system. In addition to a vertical arrangement, a horizontal one is of course also conceivable, such as in the flow lane biofilm reactor

[114]. This consists of horizontal channels of different depths. The medium flows over the biofilm by itself due to a slight tilt of the reactor. The lighting is also provided from above. Boelee et al. simulated a wave effect by pouring the medium at regular intervals from a reservoir. The algae biofilm photobioreactor (see Fig. 9e) by Ozkan et al. [112] also relies on the independent flow of the medium through a slight negative slope. The reactor consists of a concrete plate that serves as a growth surface. The medium is fed at the highest point and collected and recirculated at the lower end. Schnurr et al. [137] use a flow forced by pumps in their semi-continuous flat plate parallel horizontal PBR. This allows them to precisely adjust the flow velocity and thus also the shear stress for the biofilm. The reactor consists of 18 small parallel PBRs, which are all operated with the same parameters. A system based neither on vertical nor on horizontal flat growth surfaces is described by Gao et al. [135] in the form of a biofilm membrane PBR. As substratum, flexible fibre bundles were used. They were completely submerged in the medium. The fibres are submerged in a 0.5 m deep reactor made of plexiglass and the lighting is provided from the outside. In the biofilm capillary reactor (see Fig. 9f) [144], the medium is transported by capillary forces through a thin reaction chamber with the biofilm on its inner surface. A segmented flow can be used to alternately supply the biofilm with medium and air.

5.4 *Air-Exposed Biofilm PBRs*

Air-exposed cultivation of biofilms in reactors is not as well studied as submerged cultivation, but in recent years it has become increasingly popular. Especially for terrestrial cyanobacteria, this type of cultivation is advantageous, as their natural habitat is imitated. The supply of media can be carried out primarily in two ways: i) supply via a liquid medium, which is available to the biofilm on one side, while the other side of the biofilm is exposed to air and ii) supply via a nutrient mist (aerosol). The first type includes a multiple layer vertical plate PBR described by Liu et al. [117], for example. The supporting material consists of filter paper fixed on glass. The medium is passed through the filter paper, so that the biofilm grows exposed to the air on the outside, which optimizes gas exchange and light absorption. The light is diluted between the individual surfaces, which are arranged in an array fashion. The same reactor was also used by Cheng et al. [116] and in addition in a horizontal arrangement. The biomass productivity of the vertical multi-layer PBR was about 10 times higher than in the horizontal reactor. A simpler construction of an air-exposed biofilm reactor was shown by Boelee et al. [114]. A vertical plate made of different layers of geotextiles serves as substratum. Nutrients were supplied by continuously adding the liquid medium to the biofilm at the upper edge. Cultivation was carried out with continuous illumination. Harvesting can be done in this reactor by simply scraping the biofilm from the surface. Xu et al. [119] used a capillary-driven PBR (CPBR) consisting of polyester microfibres that were vertically attached in bundles. The lower end was placed in the medium, which was distributed over the fibres by capillary forces. The illumination was from above and the biofilm grows completely exposed to air. Porous substrate bioreactors (PSBRs) (see Fig. 9c) are another alternative for the emerse cultivation of biofilms [152]. The biofilm grows on a porous substrate, which, on the one hand, serves as a barrier to the liquid medium and as a growth surface, but, on the other hand, also allows the transport of water and nutrients. Scherer et al. [118] developed a multi-skin sheet PBR for the emerse cultivation of terrestrial cyanobacteria as biofilm. The design was optimized for later application in facades. In this case, nutrients are supplied via an aerosol. The biomass productivity could be increased in comparison with suspended cultures. The emerse photobioreactor (ePBR) (see Fig. 9d) developed by Kuhne et al. [153] and further improved by Strieth et al. [123, 154] is an aerosol-based PBR specifically designed for the cultivation of terrestrial cyanobacteria. The ePBR was fully characterized in terms of aerosol distribution to ensure optimal nutrient supply. Through the optimization, the biomass formation of *Nostoc* sp. could be almost tripled. An influence of the surface on the growth of the biofilms with regard to biomass productivity could not be observed. Another version of the ePBR is the hexagonal ePBR developed by Stiefelmaier et al. [155] which differs in its geometry.

6 Cultivation Modes of Cyanobacteria

Among the prokaryotes, cyanobacteria are the only organisms that are capable of oxygenic photosynthesis. Just as algae and higher plants, cyanobacteria also possess photosystems I and II. However, in difference to plants, photosynthesis and cell respiration can be performed simultaneously at the thylakoids [156]. Moreover, the CO_2 fixation efficiency is 10- to 50-fold higher than in plants [157]. Besides using CO_2 as sole source of carbon and energy, many cyanobacteria can also metabolize organic carbon sources, like glucose. In cyanobacteria, all known glycolytic pathways could be identified [158]. Carbohydrates can be metabolized via the oxidative pentose phosphate pathway (OPP), the Entner-Doudoroff (ED) pathway [159], as well as via the phosphoketolase [160] and the Embden-Meyerhof-Parnas (EMP) pathway. All pathways eventually result in acetyl-CoA, which enters the tricarboxylic acid (TCA) cycle. In the past, it was assumed that cyanobacteria possess an incomplete TCA cycle, missing the α-ketoglutarate-dehydrogenase [161]. However, through synthesis of a 2-oxoglutarate decarboxylase and a succinic semialdehyde dehydrogenase, which were first identified in *Synechococcus* sp. PCC 7002, the cycle is closed. These two enzymes catalyse the conversion of 2-oxoglutarate to succinate, with succinate semialdehyde as intermediate product [162] (Fig. 10).

Moreover, the γ-aminobutyrate (GABA) shunt, which catalyses the conversion of glutamate to succinate, also contributes to a fully functional TCA cycle [163].

The mixotrophic cultivation of cyanobacteria confers great advantages over phototrophic cultivation. Several studies report that many cyanobacteria show a clearly enhanced growth if an organic carbon source is concomitantly applied with light. However, many cyanobacterial strains are contaminated with heterotrophic

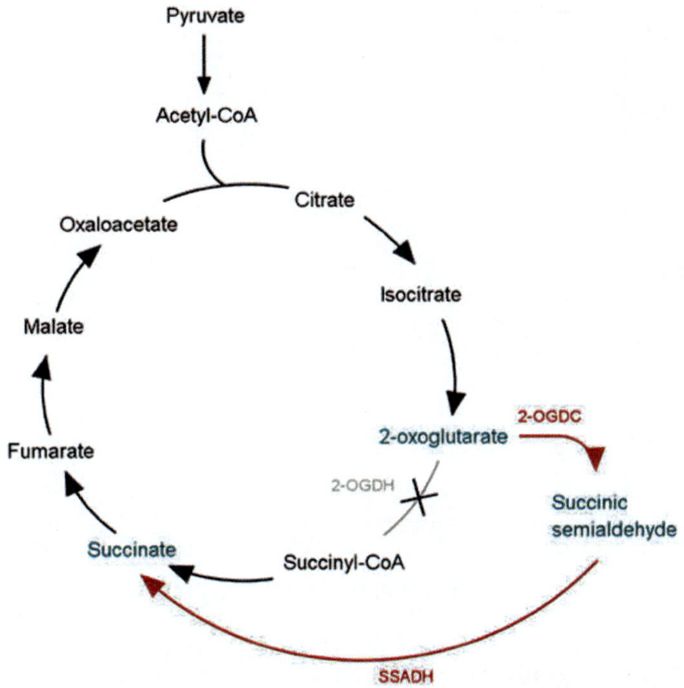

Fig. 10 Schematic diagram of the citric acid cycle in cyanobacteria. SSADH, succinic semialdehyde dehydrogenase; 2-OGDC, 2-oxoglutarate decarboxylase; 2-OGDH, 2-oxoglutarate dehydrogenase. Modified according to Zhang and Bryant [161]

bacteria or fungal species. These organisms start growing when an organic carbon source is provided. For an estimation of contaminant share in cyanobacterial cultures, Walther et al. developed a qPCR method, which enables a differentiation between cyanobacterial cells and those of heterotrophic bacterial contaminants by means of specific DNA primers [164]. They could show that heterotrophic cultivation of the terrestrial cyanobacterium *Nostoc* sp. (formerly referred to as *Trichocoleus sociatus*) does not lead to a high concentration of contaminant cells. The cdw partition of *Nostoc* sp. shortly dropped to 90% after 2 days of cultivation and quickly rose again to about 100% of total cdw [164]. Similar results were shown for heterotrophic batch, mixotrophic batch and mixotrophic fed-batch cultivations of *Nostoc* sp. and *Desmonostoc muscorum* (formerly referred to as *Nostoc muscorum*) with different carbon sources [165].

By addition of 0.5 g L^{-1} glucose, *Spirulina* sp. reached growth rates of >0.05 h^{-1}, compared to >0.02 h^{-1} at photo-autotrophic conditions [166]. It was simultaneously observed that photoinhibition, which occurred from approximately 30 W m^{-2} (approx. 138 $\mu mol_{photons}$ m^{-2} s^{-1}) in autotrophic cultures, was completely unascertainable at mixotrophic conditions. By addition of glucose to the cultivation medium, light intensities of 50 W m^{-2} (approx. 230 $\mu mol_{photons}$ m^{-2} s^{-1}) could be applied.

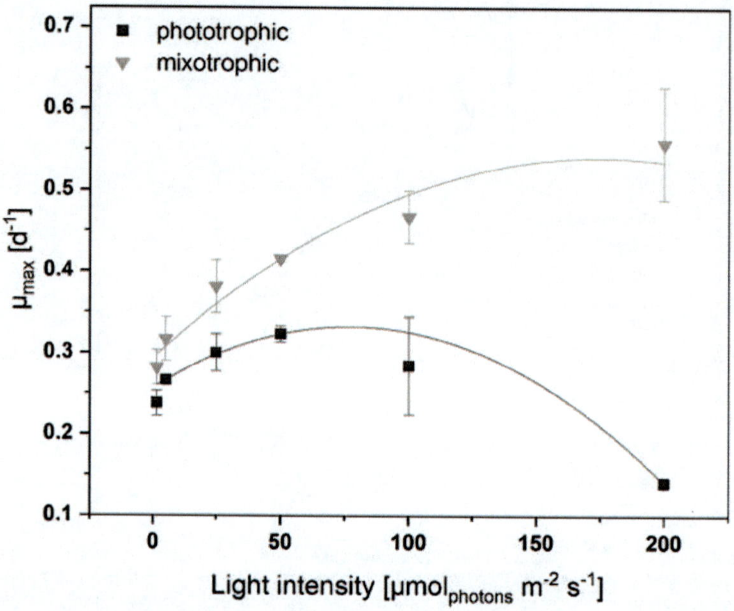

Fig. 11 Growth rates of *Nostoc* sp., phototrophic/mixotrophic cultivation (2.5 g L^{-1} raffinose), as function of light intensity. BG-11 medium, pH 7, $t = 2$ days, $T = 27°C$, $n = 120$ rpm, $N = 5$

Above this value, growth could not be further increased, but also no drop in growth rates was noticed [166]. Similar results could be obtained with the terrestrial cyanobacterium *Nostoc* sp. (Fig. 11, this work).

At mixotrophic cultivation, a linear growth rate increase between 1.5 and 100 $\mu mol_{photons}\ m^{-2}\ s^{-1}$ was measured. At higher light intensities, a plateau was reached. In solely phototrophic cultures, photoinhibition at light intensities over 100 $\mu mol_{photons}\ m^{-2}\ s^{-1}$ was noticed. This effect is caused by an excessive photon flux in the cell, which cannot be consumed by the Calvin-Benson-Basham (CBB) cycle. These electrons react with water molecules and form cell-damaging hydrogen peroxide [166]. It has been suggested that dissolved carbohydrates have a protective impact against photoinhibition [166, 167]. Moreover, addition of carbon sources protects the cells from photoinhibition by significantly diminishing the chlorophyll content [168].

Schwarz et al. tested growth of two terrestrial cyanobacteria, *Nostoc* sp. and *Desmonostoc muscorum*, under phototrophic, heterotrophic and mixotrophic conditions [165]. The latter cultivation mode was also tested in combination with fed-batch cultivation. Especially *Nostoc* sp. showed significantly increased growth with fructose, glucose, galactose and raffinose (0.25% w/v, respectively), when the cells were once again supplied with the respective organic carbon source after 5 days of cultivation.

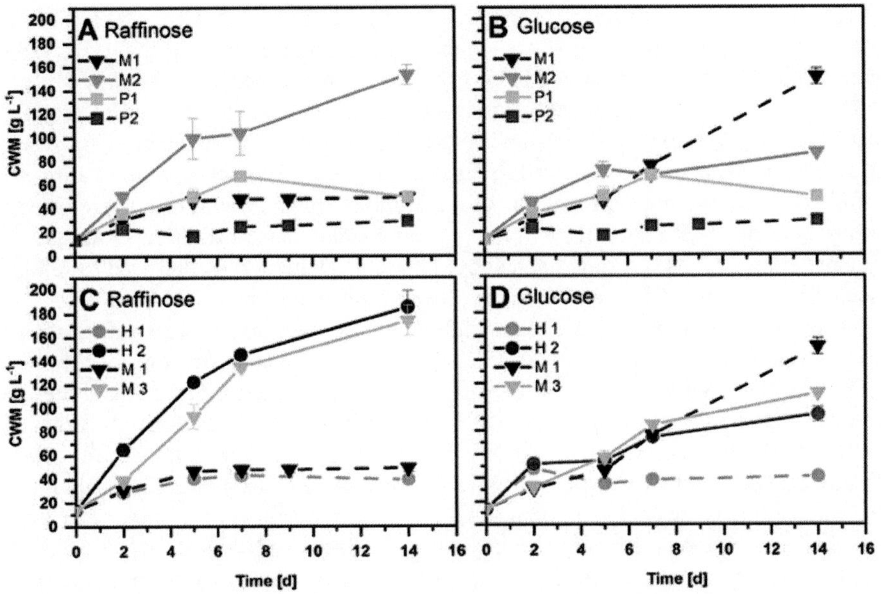

Fig. 12 Growth behaviour of *Nostoc* sp. in BG-11 medium at different light intensities and raffinose (**a**, **c**), or glucose concentrations (**b**, **d**). Initial pH 7, M1 = mixotrophic $c_{C\text{-source}} = 2.5\ \text{g L}^{-1}$, $E_v = 5\ \mu\text{mol}_{\text{photons}}\ \text{m}^{-2}\ \text{s}^{-1}$, M2 = mixotrophic $c_{C\text{-source}} = 2.5\ \text{g L}^{-1}$, $E_v = 200\ \mu\text{mol}_{\text{photons}}\ \text{m}^{-2}\ \text{s}^{-1}$, M3 = mixotrophic $c_{C\text{-source}} = 50\ \text{g L}^{-1}$, $E_v = 5\ \mu\text{mol}_{\text{photons}}\ \text{m}^{-2}\ \text{s}^{-1}$, P1 = phototrophic, $E_v = 5\ \mu\text{mol}_{\text{photons}}\ \text{m}^{-2}\ \text{s}^{-1}$, P2 = phototrophic, $E_v = 200\ \mu\text{mol}_{\text{photons}}\ \text{m}^{-2}\ \text{s}^{-1}$, H1 = heterotrophic $c_{C\text{-source}} = 2.5\ \text{g L}^{-1}$, H2 = heterotroph $c_{C\text{-source}} = 50\ \text{g L}^{-1}$. CWM = cell wet mass. Cultivation parameters: t = 14 days, T = 27°C, n = 120 rpm, N = 3, phototrophic/mixotrophic L:D 24:0

By addition of raffinose, a cell dry weight (cdw) of 1.32 g L^{-1} at heterotrophic cultivation and 1.49 g L^{-1} at mixotrophic cultivation could be reached after already 2 days of cultivation. This is 1.9/2.1 times more cdw than by phototrophic cultivation. *Nostoc* sp. showed very promising growth under hetero−/mixotrophic conditions with glucose or raffinose, further experiments with different concentrations of organic C-source and de-/increased light intensity were done for this subchapter (Fig. 12).

Not surprisingly, phototrophic growth increased along with an increase of light intensity from 5 (Fig. 12a, b; P1) to 200 $\mu\text{mol}_{\text{photons}}\ \text{m}^{-2}\ \text{s}^{-1}$ (P2). In case of raffinose, this effect can also be observed at mixotrophic conditions, although the effect is much more significant (Fig. 12a; M1, M2). However, an interesting difference to mixotrophic cultivation with glucose can be noticed in Fig. 12b. Here, the increase of light intensity shows the opposite effect. While at 5 $\mu\text{mol}_{\text{photons}}\ \text{m}^{-2}\ \text{s}^{-1}$, a cell wet mass (cwm) of about 150 g L^{-1} was reached, at 200 $\mu\text{mol}_{\text{photons}}\ \text{m}^{-2}\ \text{s}^{-1}$ only about 80 g L^{-1} was obtained. In Fig. 12c it can be seen that *Nostoc* sp. possesses a high affinity for the metabolization of the trisaccharide raffinose. Moreover, the added concentration seems to be a crucial factor for growth.

While at 2.5 g L^{-1} raffinose (Fig. 12c, H1) a growth rate of 0.217 day^{-1} and a maximum cwm of about 40 g L^{-1} were reached, an addition of 50 g L^{-1} (Fig. 12c, H2) resulted in a growth rate of 0.34 day^{-1} and a preliminary maximum cwm of >180 g L^{-1} after 14 days of cultivation. *Nostoc* sp. showed slightly poorer growth with 5 $\mu mol_{photons}$ m^{-2} s^{-1} light intensity and constant raffinose supply (Fig. 12c, M3). The opposite can be observed in Fig. 12d. As *Nostoc* sp. seems to have a lower affinity for the metabolization of glucose, and mixotrophic growth is thus preferred, a supply of light with 5 $\mu mol_{photons}$ m^{-2} s^{-1} and simultaneous addition of 50 g L^{-1} glucose (Fig. 12d, M3) results in an about 20 g L^{-1} higher cwm compared to solely heterotrophic cultivation with the same glucose concentration.

The noticed differences between cultivation of *Nostoc* sp. with raffinose or glucose could be explained by consideration of the molecular structure of the two carbohydrates. While glucose is a simple monosaccharide, raffinose constitutes a trisaccharide composed of glucose, galactose and fructose. As such, it possesses a relatively high molecular weight of 594.5 g/mL and an entrance into the cyanobacterial cell by diffusion is highly unlikely. As a consequence, it must either be extracellularly degraded or imported by an active transport system. The first possibility includes the energy consuming synthesis and export of specialized enzymes, without a previously transmitted signal for transcription of the corresponding genes. The second option implies an active transport over the cell membrane. Although no specific raffinose transporter has been described in cyanobacteria so far, a number of ATP-binding cassette (ABC) type transporter systems have been identified. In *Anabaena* sp. ATCC 29413, the uptake of fructose is conferred by such a system (frtABC) [169] just like in *Nostoc* sp. ATCC 29133 [170]. The genome of *Synechocystis* sp. PCC 6803 contains genes coding for an ABC transporter that is responsible for the export of polysaccharides and thus for the development of exopolysaccharide layers [171]. Consequently, the import of raffinose could indeed be granted by an ATP-dependent ABC transporter in *Nostoc* sp. In phototrophically grown cells, ATP is synthesized by the electron transport chain which powers the ATP-synthase. At low light intensity, while only small amounts of ATP are produced, less of these nucleotides can be spent on the transport of raffinose. This could explain the effects seen in Fig. 12a, **M1**. By application of higher light intensities, more energy can be delivered and raffinose gets imported and degraded in higher amounts (Fig. 12a, **M2**). However, these explanations alone cannot explain the data shown in Fig. 12c. By supply of higher raffinose concentrations (50 g L^{-1}, M3), *Nostoc* sp. cells show a significantly better growth compared to Fig. 12a, **M1**, despite only 5 $\mu mol_{photons}$ m^{-2} s^{-1} light intensity. This can be explained by a signal cascade, triggered by the increased carbohydrate availability in the culture medium and a consequent release of extracellularly enzymes with α-galactosidase activity. A thermostable glycosidase that also shows galactosidase activity was found in the extracellular matrix of *Nostoc commune* [172]. The resulting degradation products galactose and sucrose could enter the cell through permeases, or TonB-dependent transporters [173, 174] and promote the significantly improved growth seen in Fig. 12c.

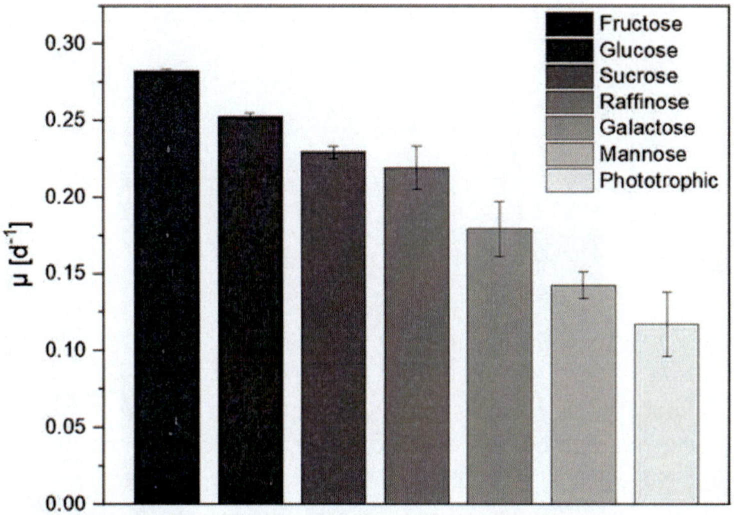

Fig. 13 Growth rates of *C. cubana* in BG-11 medium enriched with different carbon sources. Initial pH 7. $E_v = 160\ \mu mol_{photons}\ m^{-2}\ s^{-1}$, $c_{C\text{-}source} = 5\ g\ L^{-1}$, $t = 13$ days, $T = 30°C$, $n = 120$ rpm, $N = 3$, phototrophic/mixotrophic L:D 24:0

In further studies regarding mixotrophic growth of cyanobacteria, cells of *Chroococcidiopsis cubana* were cultivated with addition of different carbohydrates (5 g L^{-1}, respectively; Fig. 13).

The presented results show that *C. cubana* is capable of metabolizing a wide range of carbohydrates. As seen in the growth assays with *Nostoc* sp., this terrestrial cyanobacterium also shows a significantly enhanced growth at mixotrophic conditions, compared to solely phototrophic cultivation. By supplementation of almost each organic carbon source, growth rates could be at least doubled (Fig. 13). As the highest amount of biomass could be gained by cultivation with fructose, further experiments for growth optimization were performed with this monosaccharide.

The reduction of light supply, or the application of light-dark-periods, can significantly diminish cultivation costs. In Fig. 12 it was shown that in some cases lower light intensities can even improve cyanobacterial cell growth when combined with mixotrophic growth. This phenomenon was also observed in cultivations with *C. cubana* (Fig. 14).

While *C. cubana* only shows slight growth differences at different light intensities under solely phototrophic conditions, significant growth rate alterations were determined in mixotrophic cultivations. By decreasing the light intensity from 160 to 60 $\mu mol_{photons}\ m^{-2}\ s^{-1}$, a growth rate increase of 15.8% was measured for fructose; for glucose, an increase of even 25% was observed. How can these observations be explained? Under phototrophic conditions, NADPH is generated through the photosynthetic electron transport between photosystems II and I at the thylakoid membrane. The NADPH is needed for carbon fixation in the CBB cycle. Under mixotrophic conditions, NAD(P)H can also be gained through several glycolytic pathways.

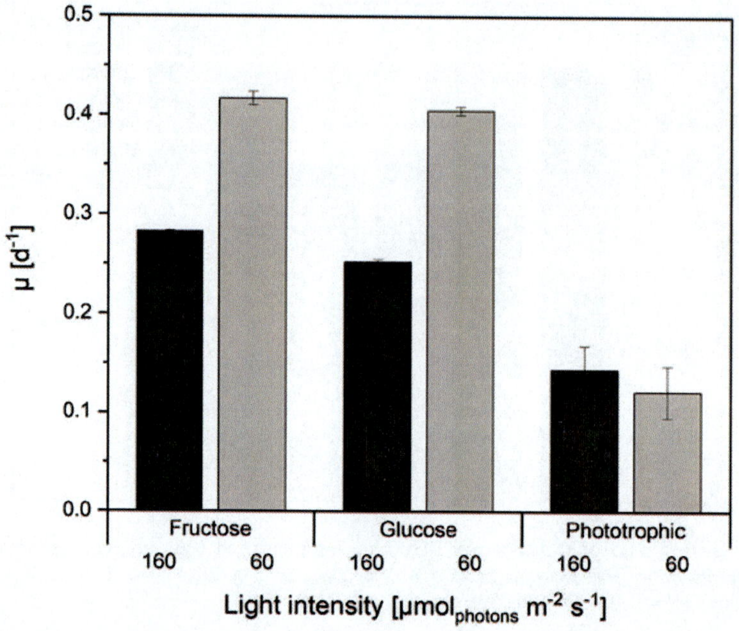

Fig. 14 Growth rates of *C. cubana* in BG-11 medium with different carbon sources and under alternating light intensities. Initial pH 7. E_v = 60/160 $\mu mol_{photons}\ m^{-2}\ s^{-1}$, $c_{\text{-source}} = 5\ g\ L^{-1}$, $t = 13$ days, $T = 30°C$, $n = 120$ rpm, $N = 3$, phototrophic/mixotrophic L:D 24:0

The most abundantly used is the oxidative pentose phosphate pathway (OPP), which can generate 5.33 NAD(P)H per molecule of glucose [158]. This pathway can be upregulated under light limiting conditions [175]. As a result, organic carbohydrates are metabolized much more effectively, which causes an improved growth at lower light intensities. Consequently, mixotrophic cultivation of cyanobacteria does lead not only to enhanced cell growth, but also to upregulation of glycolytic pathways under certain cultivation conditions.

In recent literature, there are several promising examples of cyanobacterial bioprocesses, where a heterotrophic or mixotrophic cultivation mode greatly increased product yields (Table 4). Apparently, not only a general biomass productivity increase could be demonstrated for different strains, but also a significant enhancement of process productivity concerning different target products. This was, e.g., found for biopolymers as poly-β-hydroxybutyrate (PHB) and the co-polymer Poly(3-hydroxybutyrate-co-3-hydroxyvalerate) [P(3HB-co-3HV)], produced by *Synechocystis* sp. PCC 6803 [176] and *Nostoc muscorum* Agardh [176], respectively. With *Synechocystis*, the addition of 0.4% acetate to a culture pre-grown in BG-11 supplemented with 0.1% glucose led to a 29% higher PHB accumulation (w/w dry cell mass), compared to phototrophically grown cells. In case of *Nostoc muscorum*, a supplementation with 0.4% (w/v) fructose, glucose or acetate led to a respective share of 19.2%, 26% and 28% of total dry cell weight, compared to only 8.4% in the phototrophic control.

Table 4 Examples for cyanobacterial bioprocesses enhanced by mixotrophic/heterotrophic cultivation

Strain	Product	Cultivation mode	C-source (conc.)	Max product yield [g L^{-1}]	Yield increase [%]	Literature
Arthrospira platensis	Biomass	Mixotrophic/fed-batch	Acetate (387 mg L^{-1} day^{-1})	1.769	39	[180]
Spirulina platensis	Biomass	Mixotrophic/batch	Molasses (0.75 g L^{-1})	2.94	n.a.	[181]
Synechocystis sp. PCC 6803	PHB	Mixotrophic/batch	Acetate (0.4%)	29 (% w/w)	29	[176]
Spirulina platensis	Phycocyanin	Mixotrophic/fed-batch	Glucose (2 g L^{-1})	0.795	284	[177]
Nostoc flagelliforme	Biomass	Mixotrophic/heterotrophic/batch	Glucose (2.53 g L^{-1})	1.67/0.731	499/218	[182]
Anabaena sp. PCC 7120	Biomass	Mixotrophic	Glucose (18 g L^{-1})	3.1	451	[183]
Nostoc muscorum Agardh	PHB/P(3HB-*co*-3 HV)	Mixotrophic/batch	Acetate/fructose/glucose (0.2–0.6%)	0.13/0.145/0.165	283/315/359	[184]
Leptolyngbya subtilis	Lipids	Mixotrophic/batch	Glycerol	56 (% w/w)	448	[178]
Synechococcus elongatus PCC 7942	2,3-butanediol	Mixotrophic/batch	Glucose (15 g L^{-1})	12.6	95	[179]

Mixotrophic cultivation also showed a positive impact on the production of the phycobiliprotein phycocyanin with *Spirulina platensis* [177]. Through implementation of a fed-batch process with glucose (2 g L^{-1}), the maximum biomass concentration could be increased from 2 to 10.2 g L^{-1} and the maximum phycocyanin production rose from 280 to 795 mg L^{-1}.

A more recent study dealt with the maximization of *Leptolyngbya subtilis* JUCHE1 cell lipid concentration and lipid productivity for the production of biofuels [178]. Under photoheterotrophic (mixotrophic) conditions of 2.5 kLux (approx. 88 $\mu mol_{photons}$ m^{-2} s^{-1}) light illumination and a glycerol concentration equivalent to 5% (v/v) CO_2, a maximum lipid productivity of 0.0702 g L^{-1} day^{-1} could be obtained – a 4.66-fold higher value than by solely phototrophic cultivation. This study could also show that not only biomass formation was enhanced by mixotrophic cultivation (1.47-fold), but also particularly the lipid productivity.

Through genetic engineering of *Synechococcus elongatus* PCC 7942, Kanno et al. managed to greatly improve glucose utilization under concomitant light supply [179]. The modifications in glycolytic pathways and the CBB cycle led to a 2,3-butanediol production rate of 1.1 g L^{-1} day^{-1}. The theoretical maximum yield from solely glucose was significantly exceeded by 36%, suggesting actual mixotrophic growth with concurrent metabolization of an organic and an inorganic carbon source. Under diurnal conditions, a theoretical maximum yield increase of even 95% was reached. These results impressively show what mixotrophic cultivation can achieve when combined with metabolic engineering strategies. Such attempts could eventually lead to an industrially relevant use of cyanobacteria in diverse biotechnological production processes.

The examples show that the great metabolic versatility of cyanobacteria allows a variety of possible cultivation modes. For many cyanobacterial strains, mixotrophic cultivation is described to yield the highest densities in cell mass. But solely heterotrophic processes as well show very promising results. These also have the significant advantage that cheap carbohydrates can be used in combination with regular, non-illuminated bioreactors. Cyanobacteria are potentially able to metabolize a wide range of carbohydrates. Consequently, it might be worth to test several organic carbon sources prior to coping with low cell densities in phototrophic cultivation.

7 Conclusion

Cyanobacteria offer great chances for biotechnical processes and gain more and more attention. Novel in silico screening possibilities and an increasing availability of sequenced genomes open new doors for genome mining attempts and lead to the detection of valuable bioactive metabolites. For a save conservation of specific strain characteristics, cryoconservation should be considered for long time storage of cyanobacteria. Contrary to common opinions, many strains are capable to survive the process, if the right cryo-protectant and -protocol is used. For a cell condition evaluation, a variety of vitality and viability tests can be chosen. As a cell viability confirmation does not necessarily result into growing cells, a vitality assay should be considered to prevent cultivation failures. These tests do not mandatorily require high-tech hardware – the determination of cell vitality by resazurin-assay, or by means of pO2 increase, can be conducted in a quick, easy and economic way. Because of their unique chacteristics, cyanobacteria often require specialized methods and cultivation conditions. This can be challenging, but also led and leads to the development of intriguing photobioreactor systems. In this chapter, a special focus was placed on the relatively new and heterogenous group of biofilm-based cultivation systems. By an improved light supply through lower self-shading of the cells and optimal conditions for biofilm producing terrestrial cyanobacteria, the productivity could be strongly improved, and expensive cell harvest steps can be avoided. Together with the promising mixotrophic cultivation attempts, these systems could be a way to overcome the commonly low productivity rates of cyanobacteria and to prepare the ground for industrial applications.

Acknowledgements This study was supported by funding from the State of Rhineland-Palatinate (project "iProcess" and the "Forschungsinitiative Rheinland-Pfalz"), the DFG (project STR1650/1-1), the Carl-Zeiss-Foundation, the TU Nachwuchsring and Forschungsinitiative Rheinland-Pfalz: "NanoKat – Nanostrukturierte Katalysatoren – Systeme für den Rohstoffwandel".

References

1. Jering A, Klatt A, Seven J et al (2013) Globale Landflächen und Biomasse nachhaltig und ressourcenschonend nutzen. Umweltbundesamt, Dessau-Roßlau
2. Kyriakou V, Garagounis I, Vourros A et al (2020) An electrochemical Haber-Bosch process. Joule 4:142–158. https://doi.org/10.1016/j.joule.2019.10.006
3. Robertson GP, Dale VH, Doering OC et al (2008) Sustainable biofuels redux. Science 322:49–50. https://doi.org/10.1126/science.1161525
4. Tamaru Y, Takani Y, Yoshida T, Sakamoto T (2005) Crucial role of extracellular polysaccharides in desiccation and freezing tolerance in the terrestrial cyanobacterium *Nostoc commune*. Appl Environ Microbiol 71:7327–7333. https://doi.org/10.1128/AEM.71.11.7327
5. Morris JJ, Schniter E (2018) Black queen markets: commensalism, dependency, and the evolution of cooperative specialization in human society. J Bioecon 20:69–105. https://doi.org/10.1007/s10818-017-9263-x
6. Kollmen J, Strieth D (2022) The beneficial effects of cyanobacterial co-culture on plant growth. Life 12:1–21. https://doi.org/10.3390/life12020223
7. Strieth D, Di Nonno S, Stiefelmaier J et al (2021) Co-cultivation of diazotrophic terrestrial cyanobacteria and *Arabidopsis thaliana*. Eng Life Sci 21:126–136. https://doi.org/10.1002/elsc.202000068
8. Choi SY, Sim SJ, Ko SC et al (2020) Scalable cultivation of engineered cyanobacteria for squalene production from industrial flue gas in a closed photobioreactor. J Agric Food Chem 68:10050–10055. https://doi.org/10.1021/acs.jafc.0c03133
9. Wang B, Lan CQ, Horsman M (2012) Closed photobioreactors for production of microalgal biomasses. Biotechnol Adv 30:904–912. https://doi.org/10.1016/j.biotechadv.2012.01.019

10. Lakatos M, Strieth D (2017) Terrestrial microalgae: novel concepts for biotechnology and applications. Progr Bot:269–312. https://doi.org/10.1007/124_2017_10
11. Ghasemi Y, Tabatabaei Yazdi M, Shafiee A et al (2004) Parsiguine, a novel antimicrobial substance from *Fischerella ambigua*. Pharm Biol 42:318–322. https://doi.org/10.1080/13880200490511918
12. Volk RB, Furkert FH (2006) Antialgal, antibacterial and antifungal activity of two metabolites produced and excreted by cyanobacteria during growth. Microbiol Res 161:180–186. https://doi.org/10.1016/j.micres.2005.08.005
13. Witthohn M, Strieth D, Eggert S et al (2021) Heterologous production of a cyanobacterial bacteriocin with potent antibacterial activity. CRBIOT 3:281–287. https://doi.org/10.1016/j.crbiot.2021.10.002
14. Shishido TK, Humisto A, Jokela J et al (2015) Antifungal compounds from cyanobacteria. Mar Drugs 13:2124–2140
15. Carpine R, Sieber S (2021) Antibacterial and antiviral metabolites from cyanobacteria: their application and their impact on human health. Curr Res Biotechnol 3:65–81. https://doi.org/10.1016/j.crbiot.2021.03.001
16. Silva-Stenico ME, Souza C, Silva P et al (2011) Non-ribosomal peptides produced by Brazilian cyanobacterial isolates with antimicrobial activity. Microbiol Res 166:161–175. https://doi.org/10.1016/j.micres.2010.04.002
17. Abed RMM, Dobretsov S, Sudesh K (2009) Applications of cyanobacteria in biotechnology. J Appl Microbiol 106:1–12. https://doi.org/10.1111/j.1365-2672.2008.03918.x
18. Demay J, Bernard C, Reinhardt A, Marie B (2019) Natural products from cyanobacteria: focus on beneficial activities. Mar Drugs 17:1–49. https://doi.org/10.3390/md17060320
19. Singh R, Parihar P, Singh M et al (2017) Uncovering potential applications of cyanobacteria and algal metabolites in biology, agriculture and medicine: current status and future prospects. Front Microbiol 8:1–37. https://doi.org/10.3389/fmicb.2017.00515
20. Norton TA, Melkonian M, Andersen RA (1996) Algal biodiversity. Phycologia 35:308–326. https://doi.org/10.2216/i0031-8884-35-4-308.1
21. Shalaby EA (2011) Algal biomass and biodiesel production. In: Biodiesel – feedstocks and processing technologies. InTech Croatia, pp 111–132
22. Jones MR, Pinto E, Torres MA et al (2021) CyanoMetDB, a comprehensive public database of secondary metabolites from cyanobacteria. Water Res 196:117017. https://doi.org/10.1016/j.watres.2021.117017
23. Pereira DA, Giani A (2014) Cell density-dependent oligopeptide production in cyanobacterial strains. FEMS Microbiol Ecol 88:175–183. https://doi.org/10.1111/1574-6941.12281
24. Demain AL, Fang A (2000) The natural functions of secondary metabolites. In: Fiechter A (ed) History of modern biotechnology I. Springer, Berlin, pp 1–39
25. Laxminarayan R, Duse A, Wattal C et al (2013) Antibiotic resistance – the need for global solutions. Lancet Infect Dis 13:1057–1098
26. de With K (2015) Antibiotic Stewardship – Maßnahmen zur Optimierung der antibakteriellen Therapie. Internist:1264–1270. https://doi.org/10.1007/s00108-015-3706-z
27. Strieth D, Lenz S, Ulber R (2022) In vivo and in silico screening for antimicrobial compounds from cyanobacteria. Authorea. https://doi.org/10.22541/au.164436121.14229912/v1
28. Attene-Ramos MS, Austin CP, Xia M (2014) High throughput screening. In: Wexler P (ed) Encyclopedia of toxicology3rd edn. Academic Press, Oxford, pp 916–917
29. Naughton LM, Romano S, Gara FO, Dobson ADW (2017) Identification of secondary metabolite gene clusters in the Pseudovibrio genus reveals encouraging biosynthetic potential toward the production of novel bioactive compounds. Front Microbiol 8:1–15. https://doi.org/10.3389/fmicb.2017.01494
30. Micallef ML, Agostino PMD, Sharma D et al (2015) Genome mining for natural product biosynthetic gene clusters in the subsection V cyanobacteria. BMC Genomics 1–20. https://doi.org/10.1186/s12864-015-1855-z

31. Micallef ML, D'Agostino PM, Al-Sinawi B et al (2015) Exploring cyanobacterial genomes for natural product biosynthesis pathways. Mar Genomics 21:1–12
32. Singh SP, Klisch M, Sinha RP, Häder D (2010) Genome mining of mycosporine-like amino acid (MAA) synthesizing and non-synthesizing cyanobacteria: a bioinformatics study. Genomics 95:120–128. https://doi.org/10.1016/j.ygeno.2009.10.002
33. Ziemert N, Alanjary M, Weber T (2016) Natural product reports the evolution of genome mining in microbes – a review. Nat Prod Rep 33:988–1005. https://doi.org/10.1039/c6np00025h
34. Sarkar A, Soueidan H, Nikolski M (2011) Identification of conserved gene clusters in multiple genomes based on synteny and homology. BMC Bioinformatics 12:1–10
35. Mayer H, Bauer H, Prohaska R (2001) Organization and chromosomal localization of the human and mouse genes coding for LanC-like protein 1 (LANCL1). Cytogenet Cell Genet 104:100–104
36. Weber T, Blin K, Duddela S et al (2015) antiSMASH 3.0 – a comprehensive resource for the genome mining of biosynthetic gene clusters. Nucleic Acids Res 43:237–243. https://doi.org/10.1093/nar/gkv437
37. Weber T, Kim HU (2016) The secondary metabolite bioinformatics portal: computational tools to facilitate synthetic biology of secondary metabolite production. Synth Syst Biotechnol 1:69–79. https://doi.org/10.1016/j.synbio.2015.12.002
38. Welker M, Von Döhren H (2006) Cyanobacterial peptides – nature's own combinatorial biosynthesis. FEMS Microbiol Rev 30:530–563. https://doi.org/10.22541/au.164436121.14229912/v1
39. Agrawal S, Acharya D, Adholeya A et al (2017) Nonribosomal peptides from marine microbes and their antimicrobial and anticancer potential. Front Pharmacol 8:1–26. https://doi.org/10.3389/fphar.2017.00828
40. Swain SS, Paidesetty SK, Padhy RN (2017) Antibacterial, antifungal and antimycobacterial compounds from cyanobacteria. Biomed Pharmacother 90:760–776
41. Vestola J, Shishido TK, Jokela J et al (2014) Widespread among cyanobacteria and are the end-product of a nonribosomal pathway. Proc Natl Acad Sci U S A:E1909–E1917. https://doi.org/10.1073/pnas.1320913111
42. Neilan BA, Dittmann E, Rouhiainen LEO et al (1999) Nonribosomal peptide synthesis and toxigenicity of cyanobacteria. J Bacteriol 181:4089–4097
43. Mohimani H, Liu W, Kersten RD et al (2014) NRPquest: coupling mass spectrometry and genome mining for nonribosomal peptide discovery. J Nat Prod 77:1902–1909. https://doi.org/10.1021/np500370c
44. Sigrist R, Paulo BS, Angolini CFF, De Oliveira LG (2020) Mass spectrometry-guided genome mining as a tool to uncover novel natural products. JoVE. https://doi.org/10.3791/60825
45. Nabout JC, da Silva RB, Carneiro FM, Sant'Anna CL (2013) How many species of cyanobacteria are there? Using a discovery curve to predict the species number. Biodivers Conserv 22:2907–2918. https://doi.org/10.1007/s10531-013-0561-x
46. Rippka R, Deruelles J, Waterbury JB et al (1979) Generic assignments, strain histories and properties of pure cultures of cyanobacteria. Microbiology 111:1–61. https://doi.org/10.1099/00221287-111-1-1
47. Parkes AS (1956) Preservation of living cells and tissues at low temperatures. In: III international congress on animal reproduction, Cambridge, 25th–30th June, 1956. Cambridge, pp 69–75
48. Pegg DE (2002) The history and principles of cryopreservation. Semin Reprod Med 20:5–13. https://doi.org/10.1055/s-2002-23515
49. Morris GJ (1978) Cryopreservation of 250 strains of chlorococcales by the method of two-step cooling. B Phycol J 13:15–24. https://doi.org/10.1080/00071617800650031
50. Day JG, Brand JJ (2005) Cryopreservation methods for maintaining microalgal cultures. In: Algal culturing techniques. Academic Press, New York, pp 165–187

51. Brand JJ, Diller KR (2004) Application and theory of algal cryopreservation:175–189. https://doi.org/10.1127/0029-5035/2004/0079-0175
52. Day JG, Glyn NS (2007) Day JG, Stacey G (eds) Cryopreservation and freeze-drying protocols. Humana Press., 347 p. https://doi.org/10.1007/978-1-59745-362-2
53. Morris GJ (1981) Cryopreservation: an introduction to cryopreservation in culture collections. Institute of Terrestrial Ecology, Cambridge
54. Rastoll MJ, Ouahid Y, Martín-Gordillo F et al (2013) The development of a cryopreservation method suitable for a large cyanobacteria collection. J Appl Phycol 25:1483–1493. https://doi.org/10.1007/s10811-013-0001-z
55. Esteves-Ferreira AA, Corrêa DM, Carneiro APS et al (2013) Comparative evaluation of different preservation methods for cyanobacterial strains. J Appl Phycol 25:919–929. https://doi.org/10.1007/s10811-012-9927-9
56. Mori F, Erata M, Watanabe MM (2002) Cryopreservation of cyanobacteria and green algae in the NIES-collection. Microbiol Cult Coll 18:45–55Mi
57. Park H (2006) Long-term preservation of bloom-forming cyanobacteria by cryopreservation. Algae 21:125–131. https://doi.org/10.4490/ALGAE.2006.21.1.125
58. Wood SA, Rhodes LL, Adams SL et al (2010) Maintenance of cyanotoxin production by cryopreserved cyanobacteria in the New Zealand culture collection. N Z J Mar Freshw Res 42: 277–283. https://doi.org/10.1080/00288330809509955
59. Racharaks R, Peccia J (2019) Cryopreservation of *Synechococcus elongatus* UTEX 2973. J Appl Phycol 31:2267–2276. https://doi.org/10.1007/s10811-018-1714-9
60. Corbett LL, Parker DL (1976) Viability of lyophilized cyanobacteria (blue green algae). Appl Environ Microbiol 32:777–780. https://doi.org/10.1128/aem.32.6.777-780.1976
61. Watanabe A (1959) Some devices for preserving blue-green algae in viable state. J Gen Appl Microbiol 5:153–157
62. Daily WA, McGuire JM (1954) Preservation of some algal cultures by lyophilization. Butl Univ Bot Stud 11:139–143
63. Brouers M, Hall DO (1986) Ammonia and hydrogen production by immobilized cyanobacteria. J Biotechnol 3:307–321. https://doi.org/10.1016/0168-1656(86)90012-X
64. Chetsumon A, Maeda I, Umeda F et al (1995) Continuous antibiotic production by an immobilized cyanobacterium in a seaweed-type bioreactor. J Appl Phycol 7:135–139. https://doi.org/10.1007/BF00693059
65. Pansook S, Incharoensakdi A, Phunpruch S (2019) Enhanced dark fermentative H_2 production by agar-immobilized cyanobacterium *Aphanothece halophytica*. J Appl Phycol 31:2869–2879. https://doi.org/10.1007/s10811-019-01822-9
66. Gardea-Torresdey JL, Arenas JL, Francisco NMC et al (1998) Ability of immobilized cyanobacteria to remove metal ions from solution and demonstration of the presence of Metallothionein genes in various strains. J Hazard Subst Res 1:1–18. https://doi.org/10.4148/1090-7025.1001
67. Cui J, Xie Y, Sun T et al (2021) Deciphering and engineering photosynthetic cyanobacteria for heavy metal bioremediation. Sci Total Environ 761:144111. https://doi.org/10.1016/j.scitotenv.2020.144111
68. Avramescu A, Rouillon R, Carpentier R (1999) Potential for use of a cyanobacterium *Synechocystis* sp. immobilized in poly(vinylalcohol): application to the detection of pollutants. Biotechnol Tech 13:559–562. https://doi.org/10.1023/A:1008991531206
69. Lukavský J (1988) Long-term preservation of algal strains by immobilization. Arch Protistenkd 135:65–68. https://doi.org/10.1016/S0003-9365(88)80054-X
70. Romo S, Pérez-Martínez C (1997) The use of immobilization in alginate beads for long-term storage of algae. J Phycol 1076:1073–1076. https://doi.org/10.1111/j.0022-3646.1997.01073.x
71. Serp D, Cantana E, Heinzen C et al (2000) Characterizetion of an encapsulation device for the production of monodisperse alginate beads for cell immobilization. Biotechnol Bioeng 70:41–53. https://doi.org/10.1002/1097-0290(20001005)70:1<41::aid-bit6>3.0.co;2-u

72. Prasad RN, Sanghamitra K, Antonia G-M et al (2013) Isolation, identification and germplasm preservation of different native *spirulina* species from Western Mexico. Am J Plant Sci 04:65–71. https://doi.org/10.4236/ajps.2013.412a2009
73. Lorenz M, Friedl T, Day JG (2005) Perpetual maintenance of actively metabolizing microalgal cultures. In: Anderson RA (ed) Algal culturing techniques, pp 145–156
74. Day JG, Ller JM, Comte K et al (2006) Phenotypic and genotypic stability ofcryopreserved algal and cyanobacterial cultures: a prerequisite for taxonomic and systematic studies. Cryobiology 53:430. https://doi.org/10.1016/j.cryobiol.2006.10.152
75. Mehring A, Erdmann N, Walther J et al (2021) A simple and low-cost resazurin assay for vitality assessment across species. J Biotechnol 333:63–66. https://doi.org/10.1016/j.jbiotec.2021.04.010
76. Strieth D, Stiefelmaier J, Wrabl B et al (2020) A new strategy for a combined isolation of EPS and pigments from cyanobacteria. J Appl Phycol 32:1729–1740. https://doi.org/10.1007/s10811-020-02063-x
77. Karsten U, Schumann R, Rothe S et al (2006) Temperature and light requirements for growth of two diatom species (Bacillariophyceae) isolated from an Arctic macroalga. Polar Biol 29: 476–486. https://doi.org/10.1007/s00300-005-0078-1
78. Karsten U, Klimant I, Holst G (1996) A new in vivo fluorimetric technique to measure growth of adhering phototrophic microorganisms. Appl Environ Microbiol 62:237–243. https://doi.org/10.1128/aem.62.1.237-243.1996
79. Skjelbred B, Edvardsen B, Andersen T (2012) A high-throughput method for measuring growth and loss rates in microalgal cultures. J Appl Phycol 24:1589–1599
80. Gustavs L, Schumann R, Eggert A, Karsten U (2009) In vivo growth fluorometry: accuracy and limits of microalgal growth rate measurements in ecophysiological investigations. Aquat Microb Ecol 55:95–104. https://doi.org/10.3354/ame01291
81. Jakob T, Schreiber U, Kirchesch V et al (2005) Estimation of chlorophyll content and daily primary production of the major algal groups by means of multiwavelength-excitation PAM chlorophyll fluorometry: performance and methodological limits. Photosynth Res 83:343–361. https://doi.org/10.1007/s11120-005-1329-2
82. Bramucci AR, Labeeuw L, Mayers TJ et al (2015) A small volume bioassay to assess bacterial/phytoplankton co-culture using WATER-pulse-amplitude-modulated (WATER-PAM) fluorometry. J Vis Exp:e52455. https://doi.org/10.3791/52455
83. Hamid R, Rotshteyn Y, Rabadi L et al (2004) Comparison of alamar blue and MTT assays for high through-put screening. Toxicol In Vitro 18:703–710. https://doi.org/10.1016/j.tiv.2004.03.012
84. Witthohn M, Schwarz A, Walther J et al (2020) Novel method enabling a rapid vitality determination of cyanobacteria. 20:580–584. https://doi.org/10.1002/elsc.201900164
85. Schulze ED, Beck E, Müller-Hohenstein K (2005) Plant ecology. Springer, Berlin
86. Stiefelmaier J, Strieth D, Di Nonno S et al (2020) Characterization of terrestrial phototrophic biofilms of cyanobacterial species. Algal Res 50:101996. https://doi.org/10.1016/j.algal.2020.101996
87. Chen Y, Liang C-P, Liu Y et al (2012) Review of advanced imaging techniques. J Pathol Inf 3: 22. https://doi.org/10.4103/2153-3539.96751
88. Johnson TJ, Hildreth MB, Gu L et al (2015) Testing a dual-fluorescence assay to monitor the viability of filamentous cyanobacteria. J Microbiol Methods 113:57–64. https://doi.org/10.1016/j.mimet.2015.04.003
89. Johnson TJ et al (2016) Evaluating viable cell indicators for filamentous cyanobacteria and their application. J Microbiol Biotechnol Food Sci 6:886–893. https://doi.org/10.15414/jmbfs.2016/17.6.3.886-893
90. Mullineaux CW, Mariscal V, Nenninger A et al (2008) Mechanism of intercellular molecular exchange in heterocyst-forming cyanobacteria. EMBO J 27:1299–1308. https://doi.org/10.1038/emboj.2008.66

91. Strober W (2001) Trypan blue exclusion test of cell viability. Curr Protoc Immunol 3:2–3. https://doi.org/10.1002/0471142735.ima03bs21
92. Tessarolli LP, Day JG, Vieira AAH (2017) Establishment of a cryopreserved biobank for the culture collection of freshwater microalgae (CCMA-UFSCar), São Paulo, Brazil. Biota Neotrop 17. https://doi.org/10.1590/1676-0611-bn-2016-0299
93. Turbow MM (1966) Trypan blue induced teratogenesis of rat embryos cultivated in vitro. JEEM 15:387–395
94. Gillman T, Hallowes RC (1972) Ultrastructural changes in rat livers induced by repeated injections of trypan blue. Cancer Res 32:2393–2399
95. Ganesan L, Buchwald P (2013) The promiscuous protein binding ability of erythrosine B studied by metachromasy (metachromasia). J Mol Recognit 26:181–189. https://doi.org/10.1002/jmr.2263
96. Gaget V, Chiu YT, Lau M, Humpage AR (2017) From an environmental sample to a long-lasting culture: the steps to better isolate and preserve cyanobacterial strains. J Appl Phycol 29:309–321. https://doi.org/10.1007/s10811-016-0945-x
97. Jones KH, Senft JA (1985) An improved method to determine cell viability by simultaneous staining with fluorescein diacetate-propidium iodide. J Histochem Cytochem 33:77–79. https://doi.org/10.1177/33.1.2578146
98. Xiao X, Chen YX, Liang XQ et al (2010) Effects of Tibetan hulless barley on bloom-forming cyanobacterium (*Microcystis aeruginosa*) measured by different physiological and morphologic parameters. Chemosphere 81:1118–1123. https://doi.org/10.1016/j.chemosphere.2010.09.001
99. Azevedo R, Rodriguez E, Figueiredo D et al (2012) Methodologies for the study of filamentous cyanobacteria by flow cytometry. Fresen Environ Bull 21:679–684
100. Tashyreva D, Elster J, Billi D (2013) A novel staining protocol for multiparameter assessment of cell heterogeneity in phormidium populations (cyanobacteria) employing fluorescent dyes. PLoS One 8. https://doi.org/10.1371/journal.pone.0055283
101. Hughes C, Franklin DJ, Malin G (2011) Iodomethane production by two important marine cyanobacteria: *Prochlorococcus marinus* (CCMP 2389) and *Synechococcus* sp. (CCMP 2370). Mar Chem 125:19–25. https://doi.org/10.1016/j.marchem.2011.01.007
102. Mikula P, Zezulka S, Jancula D, Marsalek B (2012) Metabolic activity and membrane integrity changes in *Microcystis aeruginosa* – new findings on hydrogen peroxide toxicity in cyanobacteria. Eur J Phycol 47:195–206. https://doi.org/10.1080/09670262.2012.687144
103. Swisher R, Carroll GC (1980) Fluorescein diacetate hydrolysis as an estimator of microbial biomass on coniferous needle surfaces. Microb Ecol 226:217–226
104. Lebaron P, Catala P, Parthuisot N (1998) Effectiveness of SYTOX green stain for bacterial viability assessment. Appl Environ Microbiol 64:2697–2700. https://doi.org/10.1128/aem.64.7.2697-2700.1998
105. Stal LJ, Moezelaar R (1997) Fermentation in cyanobacteria. FEMS Microbiol Rev 21:179–211. https://doi.org/10.1016/S0168-6445(97)00056-9
106. Ferris MJ, Hirsch CF (1991) Method for isolation and purification of cyanobacteria. Appl Environ Microbiol 57:1448–1452. https://doi.org/10.1128/aem.57.5.1448-1452.1991
107. Temraleeva AD, Dronova SA, Moskalenko SV, Didovich SV (2016) Modern methods for isolation, purification, and cultivation of soil cyanobacteria. Microbiology 85:389–399. https://doi.org/10.1134/S0026261716040159
108. Palinska KA, Krumbein WE (1995) Electrophoretic separation of two unicyanobacterial strains leading to purification. J Microbiol Methods 24:41–48. https://doi.org/10.1016/0167-7012(95)00052-6
109. Singh RN, Sharma S (2012) Development of suitable photobioreactor for algae production – a review. Renew Sustain Energy Rev 16:2347–2353. https://doi.org/10.1016/j.rser.2012.01.026
110. Gupta PL, Lee S-M, Choi H-J (2015) A mini review: photobioreactors for large scale algal cultivation. World J Microbiol Biotechnol 31:1409–1417. https://doi.org/10.1007/s11274-015-1892-4

111. Chang J-S, Show P-L, Ling T-C et al (2017) Photobioreactors. In: Current developments in biotechnology and bioengineering. Elsevier, pp 313–352
112. Ozkan A, Kinney K, Katz L, Berberoglu H (2012) Reduction of water and energy requirement of algae cultivation using an algae biofilm photobioreactor. Bioresour Technol 114:542–548. https://doi.org/10.1016/j.biortech.2012.03.055
113. Gross M, Jarboe D, Wen Z (2015) Biofilm-based algal cultivation systems. Appl Microbiol Biotechnol 99:5781–5789. https://doi.org/10.1007/s00253-015-6736-5
114. Boelee NC, Janssen M, Temmink H (2014) The effect of harvesting on biomass production and nutrient removal in phototrophic biofilm reactors for effluent polishing. J Appl Phycol 26: 1439–1452. https://doi.org/10.1007/s10811-013-0178-1
115. Johnson MB, Wen Z (2010) Development of an attached microalgal growth system for biofuel production. Appl Microbiol Biotechnol 85:525–534. https://doi.org/10.1007/s00253-009-2133-2
116. Cheng P, Ji B, Gao L et al (2013) The growth, lipid and hydrocarbon production of Botryococcus braunii with attached cultivation. Bioresour Technol 138:95–100. https://doi.org/10.1016/j.biortech.2013.03.150
117. Liu T, Wang J, Hu Q et al (2013) Attached cultivation technology of microalgae for efficient biomass feedstock production. Bioresour Technol 127:216–222. https://doi.org/10.1016/j.biortech.2012.09.100
118. Scherer K, Stiefelmaier J, Strieth D et al (2020) Development of a lightweight multi-skin sheet photobioreactor for future cultivation of phototrophic biofilms on facades. J Biotechnol 320: 28–35. https://doi.org/10.1016/j.jbiotec.2020.06.004
119. Xu X-Q, Wang J-H, Zhang T-Y et al (2017) Attached microalgae cultivation and nutrients removal in a novel capillary-driven photo-biofilm reactor. Algal Res 27:198–205. https://doi.org/10.1016/j.algal.2017.08.028
120. Naumann T, Çebi Z, Podola B, Melkonian M (2013) Growing microalgae as aquaculture feeds on twin-layers: a novel solid-state photobioreactor. J Appl Phycol 25:1413–1420. https://doi.org/10.1007/s10811-012-9962-6
121. Yin S, Wang J, Chen L, Liu T (2015) The water footprint of biofilm cultivation of *Haematococcus pluvialis* is greatly decreased by using sealed narrow chambers combined with slow aeration rate. Biotechnol Lett 37:1819–1827. https://doi.org/10.1007/s10529-015-1864-7
122. Ji B, Zhang W, Zhang N et al (2014) Biofilm cultivation of the oleaginous microalgae *Pseudochlorococcum* sp. Bioprocess Biosyst Eng 37:1369–1375. https://doi.org/10.1007/s00449-013-1109-x
123. Strieth D, Weber A, Robert J et al (2021) Characterization of an aerosol-based photobioreactor for cultivation of phototrophic biofilms. Life 11:1–13. https://doi.org/10.3390/life11101046
124. Shi J, Podola B, Melkonian M (2007) Removal of nitrogen and phosphorus from wastewater using microalgae immobilized on twin layers: an experimental study. J Appl Phycol:417–423. https://doi.org/10.1007/s10811-006-9148-1
125. Cheng P, Osei-Wusu D, Zhou C et al (2020) The effects of refractory pollutants in swine wastewater on the growth of Scenedesmus sp. with biofilm attached culture. Int J Phytoremediation 22:241–250. https://doi.org/10.1080/15226514.2019.1658706
126. Shen Y, Yu T, Xie Y et al (2019) Attached culture of *Chlamydomonas* sp. JSC4 for biofilm production and TN/TP/Cu(II) removal. Biochem Eng J 141:1–9. https://doi.org/10.1016/j.bej.2018.09.017
127. Gross M, Wen Z (2014) Yearlong evaluation of performance and durability of a pilot-scale revolving algal biofilm (RAB) cultivation system. Bioresour Technol 171:50–58. https://doi.org/10.1016/j.biortech.2014.08.052
128. Christenson LB, Sims RC (2012) Rotating algal biofilm reactor and spool harvester for wastewater treatment with biofuels by-products. Biotechnol Bioeng 109:1674–1684
129. Blanken W, Janssen M, Cuaresma M et al (2014) Biofilm growth of *Chlorella sorokiniana* in a rotating biological contactor based Photobioreactor. Biotechnol Bioeng 111:2436–2445. https://doi.org/10.1002/bit.25301

130. Orandi S, Lewis DM, Moheimani NR (2012) Biofilm establishment and heavy metal removal capacity of an indigenous mining algal-microbial consortium in a photo-rotating biological contactor. J Ind Microbiol Biotechnol 39:1321–1331. https://doi.org/10.1007/s10295-012-1142-9
131. Mukherji S, Chavan A (2012) Treatment of aqueous effluents containing non-aqueous phase liquids in rotating biological contactor with algal bacterial biofilm. Chem Eng J 200–202:459–470. https://doi.org/10.1016/j.cej.2012.06.076
132. Melo M, Fernandes S, Caetano N, Borges MT (2018) *Chlorella vulgaris* (SAG 211-12) biofilm formation capacity and proposal of a rotating flat plate photobioreactor for more sustainable biomass production. J Appl Phycol 30:887–899. https://doi.org/10.1007/s10811-017-1290-4
133. Iman Shayan S, Agblevor FA, Bertin L, Sims RC (2016) Hydraulic retention time effects on wastewater nutrient removal and bioproduct production via rotating algal biofilm reactor. Bioresour Technol 211:527–533. https://doi.org/10.1016/j.biortech.2016.03.104
134. Genin SN, Stewart Aitchison J, Grant Allen D (2014) Design of algal film photobioreactors: material surface energy effects on algal film productivity, colonization and lipid content. Bioresour Technol 155:136–143. https://doi.org/10.1016/j.biortech.2013.12.060
135. Gao F, Yang Z, Li C et al (2015) Bioresource technology a novel algal biofilm membrane photobioreactor for attached microalgae growth and nutrients removal from secondary effluent. Bioresour Technol 179:8–12. https://doi.org/10.1016/j.biortech.2014.11.108
136. Lee S-H, Oh H-M, Jo B-H et al (2014) Higher biomass productivity of microalgae in an attached growth system, using wastewater. J Microbiol Biotechnol 24:1566–1573
137. Schnurr PJ, Espie GS, Allen DG (2013) Bioresource technology algae biofilm growth and the potential to stimulate lipid accumulation through nutrient starvation. Bioresour Technol 136: 337–344. https://doi.org/10.1016/j.biortech.2013.03.036
138. Walther J, Erdmann N, Wastian K et al (2020) Novel photobioreactor for moving bed biofilm cultivation of terrestrial cyanobacteria. Copernicus meetings
139. Zamalloa C, Boon N, Verstraete W (2013) Decentralized two-stage sewage treatment by chemical–biological flocculation combined with microalgae biofilm for nutrient immobilization in a roof installed parallel plate reactor. Bioresour Technol 130:152–160. https://doi.org/10.1016/j.biortech.2012.11.128
140. Mulbry WW, Wilkie AC (2001) Growth of benthic freshwater algae on dairy manures. J Appl Phycol 13:301–306
141. Sukačová K, Trtílek M, Rataj T (2015) Phosphorus removal using a microalgal biofilm in a new biofilm photobioreactor for tertiary wastewater treatment. Water Res 71:55–63. https://doi.org/10.1016/j.watres.2014.12.049
142. Choudhary P, Prajapati SK, Kumar P et al (2017) Development and performance evaluation of an algal biofilm reactor for treatment of multiple wastewaters and characterization of biomass for diverse applications. Bioresour Technol 224:276–284. https://doi.org/10.1016/j.biortech.2016.10.078
143. Adey W, Luckett C, Jensen K (1993) Phosphorus removal from natural waters using controlled algal production. Restor Ecol 1:29–39
144. Heuschkel I, Hoschek A, Schmid A et al (2019) Mixed-trophies biofilm cultivation in capillary reactors. MethodsX 6:1822–1831. https://doi.org/10.1016/j.mex.2019.07.021
145. de Assis LR, Calijuri ML, Assemany PP et al (2019) Evaluation of the performance of different materials to support the attached growth of algal biomass. Algal Res 39:101440
146. De Godos I, González C, Becares E (2009) Simultaneous nutrients and carbon removal during pretreated swine slurry degradation in a tubular biofilm photobioreactor. Appl Microbiol Biotechnol 82:187–194. https://doi.org/10.1007/s00253-008-1825-3
147. He S, Xue G (2010) Algal-based immobilization process to treat the effluent from a secondary wastewater treatment plant (WWTP). J Hazard Mater 178:895–899. https://doi.org/10.1016/j.jhazmat.2010.02.022
148. Rosli SS, Amalina Kadir WN, Wong CY et al (2020) Insight review of attached microalgae growth focusing on support material packed in photobioreactor for sustainable biodiesel

production and wastewater bioremediation. Renew Sustain Energy Rev 134:110306. https://doi.org/10.1016/j.rser.2020.110306
149. Strieth D, Ulber R, Muffler K (2018) Application of phototrophic biofilms: from fundamentals to processes. Bioprocess Biosyst Eng 41:295–312. https://doi.org/10.1007/s00449-017-1870-3
150. Hodges A, Fica Z, Wanlass J et al (2017) Nutrient and suspended solids removal from petrochemical wastewater via microalgal biofilm cultivation. Chemosphere 174:46–48. https://doi.org/10.1016/j.chemosphere.2017.01.107
151. Kondo T, Wakayama T, Miyake J (2006) Efficient hydrogen production using a multi-layered photobioreactor and a photosynthetic bacterium mutant with reduced pigment. Int J Hydrog 31:1522–1526. https://doi.org/10.1016/j.ijhydene.2006.06.019
152. Podola B, Li T, Melkonian M (2017) Porous substrate bioreactors: a paradigm shift in microalgal biotechnology? Trends Biotechnol 35:121–132. https://doi.org/10.1016/j.tibtech.2016.06.004
153. Kuhne S, Strieth D, Lakatos M et al (2014) A new photobioreactor concept enabling the production of desiccation induced biotechnological products using terrestrial cyanobacteria. J Biotechnol 192:28–33. https://doi.org/10.1016/j.jbiotec.2014.10.002
154. Strieth D, Schwing J, Kuhne S et al (2017) A semi-continuous process based on an ePBR for the production of EPS using *Trichocoleus sociatus*. J Biotechnol 256:6–12. https://doi.org/10.1016/j.jbiotec.2017.06.1205
155. Stiefelmaier J, Strieth D, Scherer K et al (2018) Kultivierung terrestrischer Cyanobakterien in emersen Photobioreaktoren. Chem Ing Tech 90:1248–1292. https://doi.org/10.1002/cite.201855272
156. Vermaas WF (2001) Photosynthesis and respiration II. Effect of 3-(3,4-dichlorophenyl)-1,1-dimethylurea and of partial pressure of oxygen on the rates of carbon dioxide exchange in light and in darkness of detached wheat leaves. Nature 24:344–345. https://doi.org/10.1007/BF02140811
157. Wang B, Li Y, Wu N, Lan CQ (2008) CO_2 bio-mitigation using microalgae. Appl Microbiol Biotechnol 79:707–718. https://doi.org/10.1007/s00253-008-1518-y
158. Xiong W, Cano M, Wang B et al (2017) The plasticity of cyanobacterial carbon metabolism. Curr Opin Chem Biol 41:12–19. https://doi.org/10.1016/j.cbpa.2017.09.004
159. Chen X, Schreiber K, Appel J et al (2016) The Entner-Doudoroff pathway is an overlooked glycolytic route in cyanobacteria and plants. Proc Natl Acad Sci U S A 113:5441–5446. https://doi.org/10.1073/pnas.1521916113
160. Xiong W, Lee TC, Rommelfanger S et al (2015) Phosphoketolase pathway contributes to carbon metabolism in cyanobacteria. Nat Plants 1:1–8. https://doi.org/10.1038/nplants.2015.187
161. Stanier RY, Cohen-Bazire G (1977) Phototrophic prokaryotes: the cyanobacteria. Annu Rev Microbiol 31:225–274
162. Zhang S, Bryant DA (2011) The tricarboxylic acid cycle in cyanobacteria. Science 334:1551–1553. https://doi.org/10.1126/science.1210858
163. Xiong W, Brune D, Vermaas WFJ (2014) The γ-aminobutyric acid shunt contributes to closing the tricarboxylic acid cycle in *Synechocystis* sp. PCC 6803. Mol Microbiol 93:786–796. https://doi.org/10.1111/mmi.12699
164. Walther J, Schwarz A, Witthohn M et al (2020) A qPCR method for distinguishing biomass from non-axenic terrestrial cyanobacteria cultures in hetero- or mixotrophic cultivations. J Appl Phycol 32:3767–3774. https://doi.org/10.1007/s10811-020-02282-2
165. Schwarz A, Walther J, Geib D et al (2020) Influence of heterotrophic and mixotrophic cultivation on growth behaviour of terrestrial cyanobacteria. Algal Res 52:102125. https://doi.org/10.1016/j.algal.2020.102125
166. Chojnacka K, Noworyta A (2004) Evaluation of *Spirulina* sp. growth in photoautotrophic, heterotrophic and mixotrophic cultures. Enzyme Microb Technol 34:461–465. https://doi.org/10.1016/j.enzmictec.2003.12.002

167. Vonshak A, Cheung SM, Chen F (2000) Mixotrophic growth modifies the response of *Spirulina* (*Arthrospira*) *platensis* (Cyanobacteria) cells to light. J Phycol 36:675–679. https://doi.org/10.1046/j.1529-8817.2000.99198.x
168. Nakajima Y, Ueda R (1999) Improvement of microalgal photosynthetic productivity by reducing the content of light harvesting pigment. J Appl Phycol 11:195–201. https://doi.org/10.1023/A:1008015224029
169. Ungerer JL, Pratte BS, Thiel T (2008) Regulation of fructose transport and its effect on fructose toxicity in *anabaena* spp. J Bacteriol 190:8115–8125. https://doi.org/10.1128/JB.00886-08
170. Ekman M, Picossi S, Campbell EL et al (2013) A *Nostoc punctiforme* sugar transporter necessary to establish a cyanobacterium-plant symbiosis. Plant Physiol 161:1984–1992. https://doi.org/10.1104/pp.112.213116
171. Fisher ML, Allen R, Luo Y, Curtiss R (2013) Export of extracellular polysaccharides modulates adherence of the cyanobacterium *Synechocystis*. PLoS One 8:e74514. https://doi.org/10.1371/journal.pone.0074514
172. Morsy FM, Kuzuha S, Takani Y, Sakamoto T (2008) Novel thermostable glycosidases in the extracellular matrix of the terrestrail cyanobacterium *Nostoc commune*. J Gen Appl Microbiol 54:243–252. https://doi.org/10.2323/jgam.54.243
173. Mirus O, Strauss S, Nicolaisen K et al (2009) TonB-dependent transporters and their occurrence in cyanobacteria. BMC Biol 7:68. https://doi.org/10.1186/1741-7007-7-68
174. Christman HD, Campbell EL, Meeks JC (2011) Global transcription profiles of the nitrogen stress response resulting in heterocyst or hormogonium development in *Nostoc punctiforme*. J Bacteriol 193:6874–6886. https://doi.org/10.1128/JB.05999-11
175. You L, He L, Tang YJ (2015) Photoheterotrophic fluxome in *Synechocystis* sp. strain PCC 6803 and its implications for cyanobacterial bioenergetics. J Bacteriol 197:943–950. https://doi.org/10.1128/JB.02149-14
176. Panda B, Jain P, Sharma L, Mallick N (2006) Optimization of cultural and nutritional conditions for accumulation of poly-β-hydroxybutyrate in *Synechocystis* sp. PCC 6803. Bioresour Technol 97:1296–1301. https://doi.org/10.1016/j.biortech.2005.05.013
177. Chen F, Zhang Y (1997) High cell density mixotrophic culture of *Spirulina platensis* on glucose for phycocyanin production using a fed-batch system. Enzyme Microb Technol 0229: 221–224
178. Das S, Nath K, Chowdhury R (2021) Comparative studies on biomass productivity and lipid content of a novel blue-green algae during autotrophic and heterotrophic growth. Environ Sci Pollut Res 28:12107–12118. https://doi.org/10.1007/s11356-020-09577-4
179. Kanno M, Carroll AL, Atsumi S (2017) Global metabolic rewiring for improved CO_2 fixation and chemical production in cyanobacteria. Nat Commun 8:1–11. https://doi.org/10.1038/ncomms14724
180. Matsudo MC, Moraes FA, Bezerra RP et al (2015) Use of acetate in fed-batch mixotrophic cultivation of *Arthrospira platensis*. Ann Microbiol 65:1721–1728. https://doi.org/10.1007/s13213-014-1011-z
181. Andrade MR, Costa JAV (2007) Mixotrophic cultivation of microalga *Spirulina platensis* using molasses as organic substrate. Aquaculture 264:130–134. https://doi.org/10.1016/j.aquaculture.2006.11.021
182. Yu H, Jia S, Dai Y (2009) Growth characteristics of the cyanobacterium *Nostoc flagelliforme* in photoautotrophic, mixotrophic and heterotrophic cultivation. J Appl Phycol 21:127–133. https://doi.org/10.1007/s10811-008-9341-5
183. Yu G, Shi D, Cai Z et al (2011) Growth and physiological features of cyanobacterium *Anabaena* sp. strain PCC 7120 in a glucose-mixotrophic culture. Chin J Chem Eng 19:108–115. https://doi.org/10.1016/S1004-9541(09)60185-3
184. Bhati R, Mallick N (2012) Production and characterization of poly(3-hydroxybutyrate-co-3-hydroxyvalerate) co-polymer by a N_2-fixing cyanobacterium, *Nostoc muscorum* Agardh. J Chem Technol Biotechnol 87:505–512. https://doi.org/10.1002/jctb.2737